Selected Titles in This Series

(*Continued in the back of this publication*)

Recent Advances in
Partial Differential Equations,
Venice 1996

Proceedings of Symposia in
APPLIED MATHEMATICS

Volume 54

Recent Advances in Partial Differential Equations, Venice 1996

Proceedings of a Conference in Honor of the 70th Birthdays of Peter D. Lax and Louis Nirenberg

June 10–14, 1996
Venice, Italy

Renato Spigler
Stephanos Venakides
Editors

American Mathematical Society
Providence, Rhode Island

1991 *Mathematics Subject Classification.* Primary 34–XX, 35–XX, 42–XX, 60–XX, 65–XX, 70–XX, 76–XX.

Library of Congress Cataloging-in-Publication Data
Recent advances in partial differential equations, Venice 1996 : proceedings of a conference in honor of the 70th birthdays of Peter D. Lax and Louis Nirenberg : June 10–14, 1996, Venice, Italy / Renato G. C. Spigler, Stephanos Venakides, editors.
    p. cm. — (Proceedings of symposia in applied mathematics ; v. 54)
    Includes bibliographical references.
    ISBN 0-8218-0657-2 (alk. paper)
    1. Differential equations, partial—Congresses. I. Lax, Peter D. II. Nirenberg, L. III. Spigler, Renato, 1947– . IV. Venakides, S. (Stephanos) V. Series.
QA374.R398   1997
515′.353—dc21

                                                              97-29150
                                                                 CIP

Contents

Preface

Peter D. Lax and Louis Nirenberg are two of the most distinguished mathematicians of our times. Their work on partial differential equations over the last half-century has dramatically advanced the subject and has profoundly influenced the course of mathematics. A huge part of the development in partial differential equations during this period has either been through their work, or motivated by their work, or achieved by their postdocs and students.

A large number of mathematicians honored these two exceptional scientists in a week-long Conference held in Venice, Italy, between June 10 and 14, 1996. The opportunity was offered by the 70th birthdays of both Lax and Nirenberg, who, over their entire careers, have been colleagues at the celebrated *Courant Institute of Mathematical Sciences* of New York University.

The general objective of the conference was to present the state of the art in some areas of the modern theory of nonlinear partial differential equations and their applications. Subjects including turbulence, kinetic models of a rarefied gas, vortex filaments, dispersive waves, singular limits and blow-up of solutions, conservation laws, Hamiltonian systems, and others were presented by leaders in these areas, in hour-long invited lectures. The purpose of the present Proceedings volume is to record these lectures. Unfortunately, Professor E. De Giorgi passed away a few months after the Conference, and his lecture is not included. Professor J.M. Sanz-Serna's lecture could also not be included. The large number of contributed papers which were also presented in the Conference have been recorded in a book of abstracts.

We wish to acknowledge the financial and "moral" support given by a number of Institutions. They are listed on the following pages. The event that they helped organize served as a forum for the dissemination of new scientific ideas and discoveries, enhanced scientific communication by bringing together a large number of scientists working on related fields, and allowed the international mathematics community to honor two of its outstanding members. Beyond the scientific aspect, the beauty of the historic city of Venice and the approachable personalities of Lax and Nirenberg made this a unique occasion.

We wish to express our thanks to the American Mathematical Society for their eagerness to publish this volume in admiration and affection for Peter D. Lax and Louis Nirenberg.

Renato Spigler, Università di "Roma Tre"

Stephanos Venakides, Duke University

Funding Institutions

- Duke University, Office of the Vice-Provost for Academic and International Affairs
- University of Padua
- ONR (US Office of Naval Research)
- NSF (National Science Foundation)
- DOE (US Department of Energy)
- Consiglio Nazionale delle Ricerche-Comitato per la Matematica
- Consiglio Nazionale delle Ricerche-GNAFA
- Consiglio Nazionale delle Ricerche-GNFM
- UNESCO-ROSTE, Venice Office
- Fondazione Caripuglia
- Banca Popolare Veneta
- Camera di Commercio di Padova

Other Sponsors

- University of Venice
- Courant Institute of Mathematical Sciences, New York University
- University of Lecce
- SIMAI (Società Italiana per la Matematica Applicata e Industriale)
- AMS (American Mathematical Society)
- SIAM (Society for Industrial and Applied Mathematics)
- Ateneo Veneto
- Assessorato alla Cultura, Venezia
- Azienda di promozione Turistica (APT), Venezia
- Regione del Veneto

Proceedings of Symposia in Applied Mathematics
Volume **54**, 1998

Scaling laws and vanishing viscosity limits in turbulence theory[1]

G. I. Barenblatt[2]
Alexandre J. Chorin

1. Introduction

Turbulence at very large Reynolds numbers is generally considered to be one of the happier provinces of the turbulence realm, as it is widely thought that two of its results are well-established and have a chance to enter, basically untouched, into a future complete theory; these results are the von Kármán-Prandtl logarithmic law in the wall region of wall-bounded turbulent shear flow [24],[37] and the Kolmogorov-Obukhov scaling laws for local structure [28],[36]. In addition, an elaborate statistical theory of turbulence has arisen in the last four decades, employing the powerful machinery of statistical field theory, and it is widely believed that, though this statistical theory has not been particularly fruitful in other contexts, it provides a good and sufficient explanation of the Kolmogorov-Obukhov scaling law and it is only a matter of time before it is equally successful elsewhere.

Our goal in the present paper is to explain why this well-known and widely quoted set of beliefs is in error. In particular, the appropriate scaling law for the wall region must be sought in a different direction, and the statistical theory must be built on different principles. It is satisfying to be able to say in the present venue that the correct direction and principles are intimately connected with work by Peter Lax on waves in integrable systems and on zero viscosity and zero dispersion limits.

We proceed as follows: We first present a brief discussion of the different kinds of scaling. We then discuss the vanishing-viscosity limit in hydrodynamics and its statistical version. The theory is then applied to the scaling of the wall region and of the inertial range in turbulence, with an excursion through the overlap principle in the analysis of the wall region.

[1]Supported in part by the Applied Mathematical Sciences subprogram of the Office of Energy Research, U.S. Department of Energy, under contract DE–AC03–76–SF00098, and in part by the National Science Foundation under grants DMS94–14631 and DMS89–19074.
[2]Permanent address: DAMTP,University of Cambridge, Silver St., Cambridge CB3 9EW, UK.
1991 Mathematics Subject Classification: Primary 76, Secondary 82.

2. Scaling laws and similarity

Consider first the heat equation

$$(2.1) \qquad \partial_t u = \kappa \partial_{xx} u$$

where $u = u(x,t)$ is the unknown function, ∂ denotes a derivative with respect to its subscript, $t \geq 0$ is the time, x is a space variable, $-\infty < x < +\infty$, and κ is a constant conductivity, with initial condition $u(x,0)$ of compact support. Asymptotically, as $t \to \infty$, the solution can be represented as

$$(2.2) \qquad u(x,t) = \frac{Q}{\sqrt{\kappa t}} f_1\left(\frac{x}{\sqrt{\kappa t}}\right) + \frac{M}{\kappa t} f_2\left(\frac{x}{\sqrt{\kappa t}}\right) + \dots$$

where f_1, f_2 are exponentially decaying at infinity and the coefficients are integrals of the solution, in particular, the quantities M, Q satisfy $Q(t) = \int_{-\infty}^{\infty} u(x,t)dx$, $M(t) = \int_{-\infty}^{\infty} xu(x,t)dx$ with $M(t) \equiv M(0)$, $Q(t) \equiv Q(0)$; thus these integrals are simple linear functionals of the initial data. The solution as a whole is not self-similar but each term separately is: at different times it is similar to itself provided the right variables are used. The form of the various terms in the solution can deduced from simple dimensional arguments.

In contrast, if one considers instead of (2.1) the Korteweg-de Vries equation

$$(2.3) \qquad \partial_t u + u\partial_x u + \mu\partial_{xxx} u = 0,$$

again with data of compact support, then as $t \to \infty$, the solution looks like a finite collection of solitons,

$$(2.4) \qquad u = \Sigma 3\lambda_i g_i(x - \lambda_i t + c_i),$$

where the $3\lambda_i g_i$ are solitary waves [18]. If one performs the change of variables

$$(2.5) \qquad x = ln\xi, \quad t = \ln\tau, \quad c_i = -\ln A_i,$$

then the solution (2.4) is reduced to a sum of terms of self-similar form

$$(2.6) \qquad u(\xi,t) = \Sigma 3\lambda_i G_i\left(\frac{\xi}{A_i \tau^{\lambda_i}}\right).$$

Here, however, the self-similarity is very different. The exponents λ_i cannot be obtained from dimensional analysis or from simple group-theoretical considerations; more important, though the parameters A_i, and thus c_i, are, as was shown by Lax [29], functionals of the initial data and thus integrals of the motion; they cannot be derived as simply as Q and M in the previous problem.

Another example will serve to show that such differences are ubiquitous. Consider the pair of problems:

$$(2.7) \qquad \partial_t u + u\partial_x u = \nu\partial_{xx} u,$$

with initial data $u(x,0) = u_2$ for $x \leq 0$, $u(x,0) = u_1$ for $x \geq a > 0$, and $u_2 > u_1$, $u(x,0)$ monotonically decreasing between 0 and a. In contrast to (2.7), consider the equation

$$(2.8) \qquad \partial_t u = \kappa \partial_{xx} u + f(u),$$

where $f(u) \equiv 0$ in the interval $[u_1, u_1 + \Delta]$, $f(u) > 0$ in $(u_1 + \Delta, u_2)$, $f(u_2) = 0$. Both problems have asymptotically, as $t \to \infty$, a solution of the form

$$(2.9) \qquad u(x,t) = F(x - \lambda t + c)$$

but while in the case of equation (2.7) $\lambda = \frac{1}{2}(u_1 + u_2)$, as can be easily deduced from global conservation of u, in the case of equation (2.8) the determination of λ requires the solution of a nonlinear eigenvalue problem of substantial complexity and cannot be guessed from simple considerations. It is obvious that a change of variables of the form (2.5) will map the solutions (2.9) on a form usually associated with scale invariance.

This distinction between waves whose properties can be deduced from simple properties of integrals of the motion and those for which this is not possible corresponds exactly [1] to the distinction between complete and incomplete similarity. Consider a physically meaningful relation between physical variables:

$$(2.9) \qquad y = f(x_1, x_2, \ldots, x_k, c)$$

where the arguments $x_1, x_2, \ldots$ have independent dimensions while the dimensions of y and c are monomials in the powers of the dimensions of the x_i:

$$(2.10) \qquad \begin{aligned} [y] &= [x_1]^p \ldots [x_k]^r, \\ [c] &= [x_1]^q \ldots [x_k]^s, \end{aligned}$$

where $[x]$ denotes the dimensions of the variable x and for simplicity we restrict ourselves to the case of a single independent variable c with dependent dimensions.

A physical relationship similar to (2.9) must be hold for all observers even if they use a different system of physically equivalent units. It should therefore be invariant under the transformation group:

$$(2.11) \qquad x_1' = A_1 x_1, \ldots, x_k' = A_k x_k, \quad y' = A_1^p \ldots A_k^r y, \quad c' = A_1^q \ldots A_k^s c.$$

The invariants of the group (2.11) are obviously

$$\Pi = \frac{y}{x_1^p \ldots x_k^r}, \quad \Pi_1 = \frac{c}{x_1^q \ldots x_k^s}$$

thus the invariant form of equation (2.9) is

$$(2.12) \qquad\qquad \Pi = \Phi(\Pi_1),$$

where Φ is a dimensionless function; a comparison of equations (2.9) and (2.12) shows that the function $f(x_1, \ldots x_k, c)$ has the generalized homogeneity property

$$(2.13) \qquad\qquad f(x_1, \ldots, x_k, c) = x_1^p \ldots x_k^r \Phi \left(\frac{c}{x_1^q \ldots x_k^s} \right).$$

Consider now what happens when the variable Π_1 is small, $\Pi_1 \ll 1$. In such cases one is accustomed to tell undergraduates that the function Φ can be replaced by the constant $C = \Phi(0)$ and the problem is greatly simplified. The strong implicit assumption here is that as $\Pi_1 \to 0$, Φ tends to a constant non-zero limit C. If this is indeed true, then for small enough Π_1 one can replace equation (2.9) by the simpler relation

$$(2.14) \qquad\qquad y = C x_1^p \ldots x_k^r,$$

The parameter c completely disappears from the equation for small Π_1. The powers $p, \ldots, r$ can be found by simple dimensional analysis. When this situation holds, one says that one has *complete similarity* in the parameter Π_1. However, it is obvious that in general complete similarity does not hold; in general, there is no reason to believe that Φ has a finite non-zero limit when $\Pi_1 \to 0$, and the parameter Π_1, far from disappearing, may well become essential, even when, or particularly when, it is small.

Here there is however an important special case. Assume that Φ has no non-zero finite limit when Π_1 tends to zero, but that in the neighborhood of zero one has the power asymptotics

$$(2.15) \qquad\qquad \Phi(\Pi_1) = C\Pi_1^\alpha + \ldots,$$

for some C and α. Substituting (2.15) into (2.9) for Π_1 small we find

$$(2.16) \qquad\qquad \Pi = C\Pi_1^\alpha,$$

or, returning to dimensional variables,

$$(2.17) \qquad\qquad y = C x_1^{p-\alpha q} \ldots x_k^{r-\alpha s} c^\alpha,$$

i.e., the power relations are of the same general form as in (2.14), but with two essential differences: The powers of the variables $x_i, i = 1, \ldots, k$ cannot be obtained by dimensional analysis and must be derived by an additional, separate analysis, and the argument c has not disappeared from the resulting relation. We refer to such cases as cases of *incomplete similarity in the parameter* Π_1: A scaling law (2.17) is obtained, Π_1 does not disappear but enters that law only in a certain well-defined power combination with the parameter Π. In this case, the problem has an asymptotic invariance with respect to an additional group, more complicated than (2.11):

$$(2.18) \qquad\qquad \Pi_1' = \lambda \Pi_1, \qquad \Pi' = \lambda^\alpha \Pi,$$

where λ is the parameter of the new group. This is special case of the "renormalization group" [1],[20]. Contrasting the two kinds of group invariance we have examined, we see that while in (2.11) we have

$$(2.19) \qquad\qquad c' = \lambda c, \quad y' = y, \quad x_1' = x_1, \ldots, x_k' = x_k,$$

in (2.18) we have

$$c' = \lambda c, \quad y' = \lambda^\alpha y, \quad x_1' = x_1, \ldots, x_k' = x_k.$$

Although the the determination of the parameter α requires an effort beyond dimensional analysis, the relation (2.17) has a "scaling" (power) form. Such scaling relations have a long history in engineering, where a widely-shared opinion held, until recently, that since they cannot be obtained from dimensional considerations, they were nothing more than empirical correlations. In fact they are merely a more complicated case of similarity.

3. The near-equilibrium statistical theory of turbulence

Before applying the foregoing similarity theory to turbulence we have to examine the behavior of turbulence, i.e., of ensembles of solutions of the Navier-Stokes equations, as the viscosity ν tends to zero. As is well-known from the work of Lax [30], if one adds to a hyperbolic system a viscous term with a small viscosity coefficient, one may expect, under wide conditions, that in the limit of vanishing viscosity one would recover appropriate entropy solutions of the hyperbolic system. On the other hand, if one adds to the hyperbolic equation a dispersive term, the limit of vanishing dispersion is certainly not well-behaved when the hyperbolic equation has non-smooth solutions [32]. Due to the interaction and self-interaction of vortices, the three-dimensional Navier-Stokes equations partake of both hyperbolic and dispersive properties: For example, the motion of vortex lines can be described, in a certain approximation, by an equation of Schroedinger type [21]. (For a extended analogy between turbulence and dispersive systems, see [23]). One cannot expect individual solutions of the Navier-Stokes equations to be well-behaved in the limit of vanishing viscosity, but one can expect the vanishing-viscosity limit of certain well-defined features of ensembles of solutions of the Navier-Stokes equations to be well-defined. To explain why this is so one has to make a short detour through the statistical theory of turbulence.

The goal of the statistical theory of turbulence is to understand and quantify the behavior of ensembles of solutions of the Navier-Stokes equations; experience in other parts of physics suggests that this goal may be achievable while the goal of computing individual solutions is not. The possibility of considering solutions of the Navier-Stokes equations as random is anchored in their chaotic nature [12, page 36].

It is natural to focus first on stationary random solutions of the Euler or Navier-Stokes equations, just as it is natural in the kinetic theory of gases to focus first on stationary distributions of the momenta and positions of particles. A stationary random solution in turbulence is the obvious generalization of a statistically steady state in a system of N particles; it is a collection ("ensemble") of functions, in which one has identified subsets, each with an attached probability that a function in the space belong to the subset, that probability being invariant in time (see e.g [34],[44]).

Stationary solutions are important because they may attract others — i.e., one may be able to replace long-time averages of non-stationary solutions by averages over a stationary statistical solution (i.e., over the appropriate ensemble of solutions with its time-independent probabilities), and also because non-stationary solutions depend on initial conditions and few general conclusions can be reached about them. When time averages can replaced by averages over a stationary statistical solution, the latter is called ergodic. It is understood that stationary solutions may provide only a partial description of real solutions; in turbulence, this partial description often applies to the small scales, but not exclusively so; for example, the large-scale flow outside the viscous sublayer in the center section of a long pipe can be viewed as stationary.

Statistically stationary flows come in two flavors: equilibrium and non-equilibrium. An equilibrium is what one finds after a long time in an isolated system or a portion of an isolated system. In an isolated system with only energy as an invariant, equilibrium can be characterized by a "micro-canonical" distribution, i.e., equipartition over the set of appropriate equal-energy systems; it can also be characterized by the Gibbs probability distribution, in which the probability of a collection of states is the integral of $Z^{-1}e^{-\beta H}$, where β the inverse "temperature", H the Hamiltonian (or a suitable generalization), and Z, the "partition function", is a normalizing factor. One often thinks of the Gibbs distribution as describing a system in contact with a "heat bath", but it is also possible to identify the heat bath with the remainder of the system when one considers only a portion of it. Note that the temperature, energy, entropy, etc., to be discussed, refer to the properties of the macroscopic solutions of the Navier-Stokes or Euler equation and are not necessarily related to the temperature or other thermodynamic properties of the ensemble of molecules that makes up the underlying fluid. In particular, the macroscopic "temperature" is related to the kinetic energy of the turbulence. A full discussion of such "temperatures" and entropies is given in [11].

Non-equilibrium steady states are the analogs of what one obtains in kinetic theory when one considers, for example, the distribution after a long time of velocities and momenta of gas particles between two walls at different temperatures. That distribution of momenta and locations is stationary but not Gibbsian. Unlike a Gibbsian equilibrium, it allows for the irreversible transport of mass, momentum and energy across the system.

The great discovery of Onsager, Callen, and Welton (see [10]) is that in a system not too far from a Gibbsian equilibrium, non-equilibrium properties (e.g., transport coefficients) can be evaluated on the basis of equilibrium properties. An example is heat capacity, which is perfectly well defined at equilibrium, but measures the response of the system to outside (i.e., non-equilibrium) perturbations. Most of the theory of non-equilibrium processes deals with systems not far from equilibrium; its machinery, for example the Langevin formalism, is not applicable except near equilibrium. Clearly, turbulence is not in Gibbsian equilibrium, in particular because it features an irreversible energy transfer from large to small scales or of momentum from the interior to the walls. The interesting question is: Can turbulence be viewed as a small perturbation of a suitable Gibbsian equilibrium? The key word here is "suitable". and, contrary to common assumptions, the answer is "yes".

There are two reasons why one usually thinks of turbulence as being far from equilibrium: the identification of equilibrium with the Hopf equilibrium and the historical interpretation of the Kolmogorov spectrum.

In 1952 Hopf [22] and Lee [33] constructed an "equilibrium" for incompressible fluid flow based on properties of Fourier expansions. To save writing, we present a one-dimensional version of their development. Consider the model equation $\partial_t u + \partial_x(u^2) = 0$, $\left(\partial_t = \frac{\partial}{\partial t}, \text{ etc.}\right)$, where u is periodic with period 1. Expand u in Fourier series: $u = \sum \hat{u}_k e^{ikx}$; $\hat{u}_k = \hat{u}_k(t)$ satisfies

$$(3.1) \qquad \frac{d}{dt}\hat{u}_k + ik \sum_{k' \neq k} \hat{u}_{k'} \hat{u}_{k-k'} = 0.$$

Assume $\hat{u}_0 = \int_0^1 u dx = 0$, and $\hat{u}_k = 0$ for $|k| \geq K$, where K is a cut-off. One can readily check that $E = \frac{1}{2} \sum |\hat{u}_k|^2$ is invariant under (3.1). One can further check that the uniform distribution of the set of $\hat{u}_k$'s on the sphere $E = $ constant is also invariant under (3.1) and can be viewed as a microcanonical distribution. The formal limit $K \to \infty$ produces a measure on a function space. Completely analogous constructions can be carried out for the two- and three-dimensional incompressible Euler and even Navier-Stokes equations.

The result is a measure on a space of functions, which is formally invariant under Euler flow ("formally" means that all questions of existence and convergence are disregarded). A typical "flow" in this limiting collection of flows is almost nowhere differentiable, and the energy density is infinite. Indeed, suppose that one keeps the energy E of the system constant as K increases; then a_K, the average energy per "degree of freedom", is a non-negative monotonically decreasing function of K; it has a limit, which is either zero or positive; if the limit is zero the limiting ensemble has zero energy, and if the limit of the a_K is positive the limiting ensemble has an infinite energy density. This is indeed a special case of a general theorem on the non-existence of sufficiently well-behaved measures on infinite dimensional spaces [19, page 359]. The natural reaction is: If this is equilibrium, real flow must be far from it. However, one should note that this cannot be the appropriate equilibrium. In particular, if one considers ensembles of solutions of the Navier-Stokes or Euler equations for which the energy density is initially finite, the "equipartion ensemble" cannot be ergodic. At least equally disturbing from the point of view of statistical mechanics, the truncated systems do not have the same constants of motion as the original differential equations; for our one dimensional model, the energy is a constant of motion of the truncated system but not of the original equation, while the original equation has two constants of motion that do not survive the truncation.

Another source of the belief that turbulence is far from equilibrium is the usual interpretation of the Kolmogorov-Obukhov law; that law states that in the inertial range of scales, across which energy "cascades" from the stirring scales to the dissipation scales, the energy spectrum $E(k)$, i.e., the energy E per wave number k, has the form

$$(3.2) \qquad\qquad E(k) = C\epsilon^{2/3}k^{-5/3}$$

where C is an absolute constant and ϵ is the rate of energy transfer across the spectrum. Neither the idea of a cascade nor the dimensional analysis that leads to this law prejudges the issue of distance from equilibrium. However, the presence of ϵ in the spectral law creates the impression that it is the energy transfer that creates the law. An alternate interpretation can be produced, according to which the amount of energy dissipated depends on the amount of energy present, i.e., $\epsilon = (E(k))^{3/2} k^{5/2}$, when $E(k)$ may be determined by equilibrium considerations. For a further discussion, see [14]; there are many systems which dissipate energy in an appropriately defined neighborhood of equilibrium. Be that as it may, the idea that irreversibility dominates the small scales of turbulence leads naturally to a particular formalism. The dominant effect is assumed to be the provision of energy at large scales and its removal by viscosity at small scales. Both can be represented by a linear Stokes equation with forcing, which can readily be solved. The nonlinear terms in the Navier-Stokes equations can then be formally represented as a perturbation expansion ordered by the Reynolds number R. The various terms in this expansion can be represented by Feynman diagrams, and the panoply of perturbative field theory can be used in the attempt to extract useful information. This is an awesome and uncompleted task, as the jump from $R = 0$ to $R = \infty$ is large, and it would be desirable to avoid it by constructing perturbation expansions on other premises.

More important from our present point of view, the identification of equilibrium with the Hopf-Lee equilibrium has implications for what happens to turbulence as the viscosity tends to zero. The assumed ergodicity of the Hopf-Lee "probability" (the quotes are here because this is not in fact a probability measure) introduces the need for a finite viscosity to generate the Kolmogorov spectrum. The intuitive picture is that the Euler equations strive to fill all the states available to the Hopf-Lee measure, while a finite viscosity drains the states with large wave numbers, setting up a cascade and a Kolmogorov spectrum. In this picture, the small-viscosity limit of turbulence must be very singular.

We now set out to look for other, more reasonable equilibria for the Euler equations, in the hope that turbulence can be found in their vicinity ([12],[13]). First, we should discuss why it makes any sense at all to look for equilibria. After all, turbulence is a fairly evanescent phenomenon; unless stirred, for example by shear, it soon dissipates. The relevant question is however: How fast do the small scales adjust compared to the time available to them, and do they have time enough to settle down to a statistical equilibrium. We shall give an affirmative answer in section 6 below, after we exhibit the appropriate scaling relation.

Here too, as above, we start with a discretization of the equations of motion and plan to take an appropriate limit at the end.

A general procedure for doing so would be as follows: Assume the turbulence lives in a finite volume V; divide V into small pieces of side h and volume h^3; construct a finite number of variables by integrating appropriate continuum variables, for example, the components of the vorticity vector ξ, over the small volumes. (The condition div $\xi = 0$ must be enforced, and gauge invariance provides a machinery for doing that.) The hydrodynamic energy E discretizes into a sum E_h over the boxes, and for each h, one can construct an equilibrium Gibbs distribution $Z_h^{-1}\exp(-\beta E_h)$, where β is an inverse "temperature". The question is: what happens to these equilibria as $h \to 0$? This question does not touch upon the dynamics generated by the Euler equations, to which we shall return below.

First note that the question can be asked and answered for the discretized spectral equilibria of the previous section. Given a cut-off K, standard methods show that the temperature $T = \beta^{-1}$ is proportional, for constant energy E, to $E/(K^d)$, where d is the dimension of the space. Thus $T \to 0$ as one approaches the continuum limit. This is consistent with the remarks above about the limit $K \to \infty$: If the energy is not infinite, than in the limit of vanishing T the probability of any state with a positive energy is zero.

For the systems discretized by chopping up the vorticity in physical space several things can happen. In two dimensions generally T increases as h decreases, and then T goes beyond into the "negative" (trans-infinite) region [12]. In three space dimensions a more interesting behavior may appear.

A reminder of some properties of phase transitions is needed here. At a phase transition point the correlation length of a physical system is infinite (or else the thermodynamic properties of the system are analytic in parameters such as T). Furthermore, at a critical phase transition a system is "scale invariant", which roughly means the following: If the system is discretized, or is already discrete to start with and its variables are collected into groups in a way that preserves energy, then the properties of the system are invariant under changes in the scale of the discretization or of the grouping. The relation between scale invariance and phase transition comes about because scale invariance can occur only when the correlation length is infinite.

The question whether alternate equilibria in turbulence can be found now becomes: Does the family of equilibria with parameters β and h have multiple phases with a curve separating them in the (β, h) plane? If yes, the intersection β^* of this phase transition line with the $h = 0$ axis is our candidate for a "reasonable" value of β, and the corresponding equilibrium is our statistical equilibrium.

Note that this is where the fact that we are looking for probability distributions over continuous flows impinges on the analysis. For a discrete collection of particles, invariance under a change of discretization is usually not a relevant consideration; here, however, we have to make sure that our systems have a meaningful continuum limit, and this forces us to consider systems invariant under a refinement of the discretization, and thus forces us towards phase transition points.

Statistical equilibria in vortex systems and their limiting behavior have been studied by a variety of numerical and analytical methods, some of which exploit an analogy with the vortex-dominated phase transitions that occur in superfluid and superconducting systems [12]. We will not survey this literature here; it is still lacks a completely rigorous version consistent with knowledge about the Navier-Stokes equations and their vanishing-viscosity limits. A major conclusion of the available analyses is that the postulated "other" equilibria exist, and exhibit velocity correlation and structure functions up to order 3 that are consistent with the Kolmogorov scaling (see section 6 below). Note that the construction of these measures makes no use of the Navier-Stokes equations. To construct the time evolution of measures under the Navier-Stokes equations or their vanishing viscosity limits (which of course are likely to be different from the Euler equations with zero viscosity, in particular near walls), one can proceed by perturbation methods, for example in powers of vortex fugacity. One expects that the perturbation does not affect spectral exponents nor the asymptotic small-scale behavior of structure functions. In the perturbation process, energy cascades appear naturally as the small-scale equilibria are subjected to an energy input from the larger scales and act as energy sinks. A consequence of this construction is that one can expect correlation and structure functions of low order for Navier-Stokes flows to have a well-behaved limit as the viscosity tends to zero. Note that it is not claimed that individual solutions of the Navier-Stokes equations converge to an Euler limit (and indeed, in the case of wall-bounded flow, this is clearly false [26]). Further, the nature of the convergence of the probabilities as the viscosity tends to zero remains open; all that is asserted is that certain second and third order moments (and thus also first order moments) have a limit as the viscosity tends to zero.

4. The scaling of wall-bounded shear flow

We start with the fundamental scaling relation for the intermediate region of flow in a pipe (Fig 1):

$$(4.1) \qquad \partial_y u = \frac{u_*}{y}\, \Phi\left(\eta, Re\right)$$

where Re is a properly defined Reynolds number, for instance, for pipe flow $Re = \bar{u}d/\nu$, where $\bar{u}$ is the mean velocity (discharge rate divided by the cross-section area), d is the diameter of the pipe, ν is the kinematic viscosity of the fluid, $\eta = u_*y/\nu$ is the non-dimensional distance to the wall, u_* is the "friction velocity" $u_* = \sqrt{\tau/\rho}$, τ is the stress at the wall, ρ is the density, and Φ is an unknown dimensionless function of two dimensionless arguments. This is the exact counterpart, in the case of two independent variables with dependent dimensions, of equation (2.13) above.

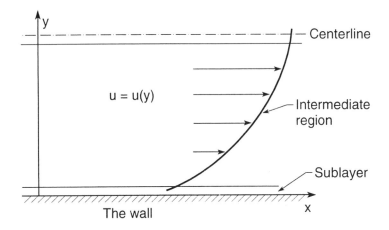

Figure 1. The intermediate region of wall-bounded shear flow (e.g., flow in a cylindrical pipe).

If one assumes that the limit of the function Φ in (4.1), when both arguments tend to infinity, exists and is finite and different from zero, (in the language of section 2, this is an assumption of complete similarity), then one can define $\kappa = 1/\Phi(\infty, \infty)$

and and an integration leads to the well-known "universal" von Kármán-Prandtl law of the wall:

$$(4.2) \qquad \frac{u}{u_*} = \frac{1}{\kappa} \log \eta + B$$

where, by the very logic of the derivation, κ and B are universal constants, independent of Reynolds number, which cannot be adjusted as the Reynolds number changes. In particular, the assumption of complete similarity causes the parameters d and ν to drop from the expression for the velocity gradient at high enough Re. The opposite self-similarity assumption, ("incomplete similarity"), leads to general scaling laws; the form suggested by the absence of a characteristic length scale is:

$$(4.3) \qquad \frac{u}{u_*} = B(Re) \left(\frac{u_* y}{\nu} \right)^{\beta(Re)}$$

An expansion of (4.3) around the state that corresponds to $Re = \infty$ in powers of $\frac{1}{\ln Re}$ (see [2],[3],[6]) then leads to:

$$(4.4) \qquad \partial_y u = \frac{u_*}{y} \left(B_0 + \frac{B_1}{\ln Re} + o\left(\frac{1}{\ln Re} \right) \right) \left(\frac{u_* y}{\nu} \right)^{\frac{\beta_1}{\ln Re} + o\left(\frac{1}{\ln Re} \right)}.$$

Both the law (4.3) and the universal logarithmic law (4.2) have precise and equally justified theoretical foundations; both are based on the assumption of self-similarity, but the logarithmic law is based on the assumption of complete similarity whereas the scaling law is based on the assumption of incomplete similarity. The question is, which of these assumptions, if any, is correct. This question can be answered in full only by further advances in the theory of Navier-Stokes equations and/or by further experimental studies, as is also the case for the local structure discussed below.

The difference between the cases of complete and incomplete similarity is significant. In the first case the experimental data should cluster, in the $\phi = u/u_*$, $\ln \eta$ plane ($\eta = u_* y/\nu$), on the universal straight line of the logarithmic law. In the second case the experimental points may occupy an area in the ϕ, $\ln \eta$ plane bounded by the envelope of the family of scaling law curves having the Reynolds number as parameter. The envelope in turn can be approximated piecewise by various straight lines that depend on the range of Reynolds numbers under consideration. Scaling laws similar to (4.3) were popular among engineers, especially before the papers of von Kármán [24] and Prandtl [37] (For recent analyses, see [3],[17],[25],[39],[41],[42]). Recently ([2],[3],[5],[7]) arguments in favor of the incomplete similarity (4.3)–(4.4) were proposed and the coefficients B_0, B_1 and β_1 were determined from the Nikuradze data [35]:

$$(4.5) \qquad B_0 = \frac{\sqrt{3}}{2}, \qquad B_1 = \frac{15}{4}, \qquad \beta_1 = \frac{3}{2}.$$

With the choice of parameters given in (4.5) the power law has the form

$$(4.6) \qquad \phi = \left(\frac{1}{\sqrt{3}} \ln Re + \frac{5}{2} \right) \eta^{\frac{3}{2 \ln Re}} , \qquad \phi = \frac{u}{u_*} , \qquad \eta = \frac{u_* y}{\nu}$$

or, equivalently,

$$(4.7) \qquad \phi = \left(\frac{1}{\sqrt{3}} \ln Re + \frac{5}{2} \right) \exp \left(\frac{3 \ln \eta}{2 \ln Re} \right) .$$

It is important to note that the analysis below does not depend on the specific values of the constants (4.5) that have been obtained by comparison with experiment; it is the form of the scaling law that matters, in particular, the fact the α is inversely proportional to $\ln Re$. The envelope of the family of curves (4.7) is a smooth curve which can be approximated by piecewise linear functions of $\ln \eta$. Asymptotically, at very large Re, the envelope is the straight line [3]

$$(4.8) \qquad \phi = \frac{\sqrt{3}}{2} \, e \ln \eta + 6.79 .$$

We now set out to extract from the scaling law (4.4) predictions for what happens at very large Reynolds number, beyond the range to which the constants were fitted. The success of this extrapolation is a validation of the scaling law (4.4). The main tool we use is *vanishing-viscosity asymptotics*, based on the idea explained above that, as long as one deals with not-too high moments of the velocity field, their limits as the viscosity tends to zero are well-defined. We have already explained the importance of assumptions about the asymptotic behavior of Φ in (4.1). Consider again equation (4.1), and its special case, equation (4.4). If one stands at a fixed distance from the wall, in a specific pipe with a given pressure gradient, one is not free to vary Re and η independently; the viscosity ν appears in both, and if ν is decreased, both arguments of Φ will vary. The appropriate limit is the limit of vanishing viscosity, whose existence we have asserted; note that here its existence can be checked independently of the statistical argument. When one takes the limit of vanishing viscosity, one effectively considers flows at ever larger η at ever larger Re; the ratio $\frac{3 \ln \eta}{2 \ln Re}$ tends to 3/2 because ν appears in the same way in both numerator and denominator. To show this in more detail, using only physically meaningful quantities, proceed as follows: Note that in experimental measurements using any probe, including the Pitot tube used by Nikuradze [35], it is impossible to approach the wall closer than a certain distance δ, say the diameter of the Pitot tube. Consider the experimental possibilities for a certain member of the family (4.7). It was shown in [7] that the experimental points presented by Nikuradze are close to the envelope. So we assume that up to some distance $\Delta > \delta$ the experimental points are close to the envelope. What happens farther? Consider the combination $3 \ln \eta / 2 \ln Re$. It can be represented in the following form

$$(4.9) \qquad \frac{3 \ln \eta}{2 \ln Re} = \frac{3 \left[\ln \frac{u_* \Delta}{\nu} + \ln \frac{y}{\Delta} \right]}{2 \left[\ln \frac{u_* \Delta}{\nu} + \ln \frac{\bar{u}}{u_*} + \ln \frac{d}{\Delta} \right]} .$$

According to [3], at small ν, i.e. large Re, $\bar{u}/u_* \sim ((1/\sqrt{3})\ln Re + 5/2)$, so that the term $\ln \bar{u}/u_*$ in the denominator of the right-hand side of (4.9) is asymptotically small, of the order of $\ln\ln Re$, and can be neglected at large Re. The crucial point is that due to the small value of the viscosity ν the first term $\ln(u_* \Delta/\nu)$ in both the numerator and denominator of (4.9) should be dominant, so that $3\ln\eta/2\ln Re$ is close to $3/2$ (y is obviously less than $d/2$). Therefore the quantity

$$1 - \ln\eta/\ln Re$$

can be considered in the intermediate region $\Delta < y < d/2$ as a small parameter, so that the factor $\exp(3\ln\eta/2\ln Re)$ is approximately equal to

$$
\begin{aligned}
\exp\left[\frac{3}{2} - \frac{3}{2}\left(1 - \frac{\ln\eta}{\ln Re}\right)\right] &\approx e^{3/2}\left[1 - \frac{3}{2}\left(1 - \frac{\ln\eta}{\ln Re}\right)\right] \\
&= e^{3/2}\left[\frac{3}{2}\frac{\ln\eta}{\ln Re} - \frac{1}{2}\right].
\end{aligned}
$$

(4.10)

According to (4.7) we have also

$$
(4.11) \qquad \eta\partial_\eta\phi = \partial_{\ln\eta}\phi = \left(\frac{\sqrt{3}}{2} + \frac{15}{4\ln Re}\right)\exp\left(\frac{3\ln\eta}{2\ln Re}\right),
$$

and the approximation (4.10) can also be used in (4.11). Thus in the intermediate asymptotic range of distances y: $y > \Delta$, but at the same time y slightly less than $d/2$, the following asymptotic relations should hold as $Re \to \infty$:

$$
(4.12) \qquad \phi = e^{3/2}\left(\frac{\sqrt{3}}{2} + \frac{15}{4\ln Re}\right)\ln\eta - \frac{e^{3/2}}{2\sqrt{3}}\ln Re - \frac{5}{4}e^{3/2},
$$

and

$$
(4.13) \qquad \partial_{\ln\eta}\phi = \left(\frac{\sqrt{3}}{2} + \frac{15}{4\ln Re}\right)e^{3/2}.
$$

Note that this law has a finite limit independent of Re. At the same time it can be easily shown that for the envelope of the power-law curves the asymptotic relation is

$$
(4.14) \qquad \partial_{\ln\eta}\phi = \left(\frac{\sqrt{3}}{2} + O\left(\frac{1}{\ln^2\eta}\right)\right)e.
$$

The difference in slopes between (4.13) and (4.14) is significant. It means that the individual members of the family (4.4) should have at large Re an intermediate part, represented in the plane $\phi, \ln\eta$ by straight lines, with a slope different from the slope of the envelope by a factor $\sqrt{e} \sim 1.65$. Therefore the graph of the individual members of the family (4.4) should have the form presented schematically in Figure 2.

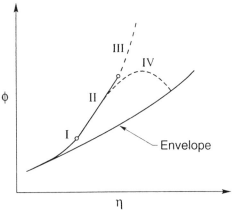

Figure 2. The individual members of the family of scaling laws (2.7)–(2.8) near the envelope in the plane ϕ, $\ln \eta$ have a straight intermediate interval with a slope essentially larger than that of the envelope. The horizontal scale is logarithmic.

I: A part close to the envelope;

II: straight intermediate part;

III: the fast growing ultimate part having no physical meaning;

IV: the region near the maximum where the scaling law is not valid.

Recently an experimental paper by Zagarola et al. [45] presented new data obtained in a high-pressure pipe flow. High pressure creates a large density and therefore a low kinematic viscosity. The experimentalists were thus able to enlarge the range of Reynolds numbers in comparison with Nikuradze's [35]. The Reynolds number is varied in these experiments by changing the pressure and thus the kinematic viscosity, exactly as is done mathematically in our our vanishing-viscosity asymptotics.

The experimental data are presented in Fig. 3, which is reproduced with permission from Fig. 4 of [45]; these data agree well with our vanishing viscosity results: At small y the deviation from the envelope is too small to be noticed (part I of the curve in Figure 2); for larger y, up to a very close vicinity of the maximum of ϕ achieved in the experiments, the data are split. To each Reynolds number corresponds its own curve with a pronounced linear part having a slope clearly larger than the slope of the envelope; the ratio of the slope of the curves of the family to that of the envelope is always larger than 1.5. Contrary to the opinion of Zagarola et al., we consider this graph to be a clear confirmation of the scaling law, and a strong argument against the universal logarithmic law according to which all the points up to a close vicinity of maxima should lie on the universal logarithmic straight line. The prediction of a difference of $\sqrt{e}$ between the slopes of the individual velocity profiles and the slope of their envelope provides an easily

verified criterion for assessing the agreement between the experimental data and
the scaling law. Note that at high Re the difference between the proposed law and
the universal logarithmic law is large enough to have a substantial impact on the
outcome of engineering calculations.

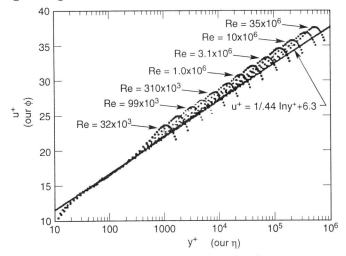

Figure 3. A graph of the velocity profiles normalized using inner scaling
variables for 13 different Reynolds numbers between 32×10^3 and 35×10^6.
Also shown is a log-law with κ and B equal to 0.44 and 6.3, respectively.
(Reproduced with permission from Zagarola et al. [45])

We have proceeded so far by considering self-similarity assumptions and their
consequences. It is also necessary to inquire about the mechanical processes that
may lead to one or the other of the two kind of self-similarity. We shall now
briefly show that the scaling law (4.4) arises because the transverse vorticity in
the wall region is intermittent; this intermittency, associated with the vorticity
bursting process, is well-documented in the experimental and numerical literature
[9],[27],[41].

Indeed, a natural measure of the length scale of the cross-section of the transverse
vortical structures near the wall, which are responsible for the vertical variation in
the velocity u, is $\ell = (\partial_y u / u_*)^{-1}$; The scaling law (4.7) gives

$$(4.15) \qquad \ell = \frac{2}{\sqrt{3} + 5\alpha}\, y^{1-\alpha}(\nu/u_*)^\alpha\,, \qquad \alpha = \frac{3}{2 \ln Re}\ .$$

Note that ℓ is proportional to $y^{1-\alpha}$ rather than to y, showing that the transverse
vortical structures are not space filling if (4.2) holds; the universal logarithmic law
produces an ℓ proportional to y. In a viscous flow the vorticity can presumable
vanish only on smooth surfaces, but one can define an essential support of the
vorticity, (see [12]), as the region where the absolute value of the vorticity exceeds
some predetermined threshold; according to (4.15), the intersection of that essential

support with a vertical line has fractal dimension $1 - \alpha$. If the essential support is statistically invariant under translations parallel to the wall, the essential support itself has dimension $3 - \alpha$. This conclusion agrees well with the data reported in [9],[27], where the more powerful streamwise vortices are indeed not space filling. One could even hypothesize that, as the streamwise vortices meander, the transverse vortices that produce u can be identified at least in part with transverse components of vortices that are mostly streamwise.

An interpretation of these observations is suggested by the discussion in [41]. The process that occurs in a wall layer is a transfer of momentum or impulse from the outer regions to the wall, or, equivalently, a transfer of impulse of opposite polarity from the wall to the interior. This transfer is intermittent, concentrated in localized bursts which create the fractalization, and thus create a vorticity scale different from y and thus plausibly explains the power law (4.4). The analogy with the discussion of the inertial range in the section 6 will be obvious.

5. Overlap arguments

We now examine in detail a well-known and beautiful argument for analyzing the structure of the flow in the region intermediate between the immediate vicinity of the wall and the region far from the wall. This argument is due to Izakson, Millikan and von Mises (IMM) (see e.g [16],[39]). In this argument, it is assumed that from the wall outward, for some distance,(see Figure 1) one has a generalized law of the wall,

$$(5.1) \qquad \phi = u/u_* = f(u_* y/\nu),$$

where f is a dimensionless function; the influence of the Reynolds number Re, which contains the external length scale (for pipe flow, the diameter d of the pipe) is neglected; heuristically, it is assumed that if the far wall is far enough, the wall one is near does not know about it. Adjacent to the axis of the pipe in pipe flow, extending to the sides, one assumes a "defect law",

$$(5.2) \qquad u_{CL} - u = u_* g(2y/d),$$

where u_{CL} is the average velocity at the centerline and g is another dimensionless function. Here the neglect of the effect of Re means that the effect of viscosity is neglected; heuristically, one assumes that near the axis, where the velocity gradients are small, the effect of a small enough viscosity is unimportant. Both assumptions taken together constitute an assumption of "separation of scales", according to which at large enough yet finite values of Re viscous scales and inviscid scales can be studied in partial isolation. Self-consistency then demands that for some interval in y the laws (5.1) and (5.2) overlap, so that

$$(5.3) \qquad u_{CL} - u = u_{CL} - u_* f(u_* y/\nu) = u_* g(2y/d).$$

After differentiation of (5.3) with respect to y followed by multiplication by y one obtains

(5.4) $$\eta f'(\eta) = \xi g'(\xi) = \frac{1}{\kappa},$$

where $\eta = u_* y/\nu$, $\xi = 2y/d$, and κ is a constant; integration then yields the law of the wall

(5.5) $$f(\eta) = \frac{1}{\kappa} \ln \eta + B,$$

as well as the defect law

(5.6) $$g(\xi) = -\frac{1}{\kappa} \ln \xi + B_*,$$

with

$$B_* = \frac{u_{CL}}{u_*} - \frac{1}{\kappa} \ln \frac{u_* d}{2\nu} - B.$$

However, the experimental data (see e.g. figure 7 in [45]) do not support the assumption of separation of scales, nor do recent numerical studies of similar problems where it has been invoked [15]. We now examine what happens if one repeats the elegant but simplistic argument we have just described without dropping the effects of Re near the wall or near the center of the pipe. We shall see that, when properly improved, much of it survives, and indeed supports our conclusions.

We begin by noting that in the nearly linear portion II of the graph of Figure 2 the flow can be described by a local logarithmic law with a Reynolds number dependent effective von Kármán constant $\kappa_{eff} = \kappa(Re)$:

(5.7) $$\kappa_{eff} = \frac{2}{\sqrt{3}e^{3/2} + \frac{15}{2}(\ln Re)e^{3/2}};$$

as $Re \to \infty$, $\kappa(Re)$ tends to the limit $\kappa_\infty = \frac{2}{\sqrt{3}e^{3/2}} \sim 0.2776\ldots$, smaller than the usual von Kármán constant $\kappa = \frac{2}{\sqrt{3}e} \sim .425\ldots$ by a factor $\sqrt{e} \sim 1.65\ldots$. With this in mind, the IMM procedure can be modified as follows: The law of the wall, equation (5.1), becomes

(5.8) $$\phi = u/u_* = f(u_* y/\nu, Re),$$

so that the influence of Re, which contains the external scale, is included. The previous defect law (5.2) is also replaced by the Reynolds number dependent defect law

(5.9) $$u_{CL} - u = u_* g(2y/d, Re),$$

so that the influence of the molecular viscosity ν is preserved. Now assume that the laws (5.8) and (5.9) overlap on some y interval:

$$u_{CL} - u = u_{CL} - u_* f(u_* y/\nu, Re) = u_* g(2y/d, Re).$$

Replacing f by its expression (4.12) yields:
(5.10)

$$g(2y/d, Re) = \phi_{CL} - \left(\frac{1}{\sqrt{3}} \ln Re + \frac{5}{2} \right) e^{3/2} - \left(\frac{\sqrt{3}}{2} + \frac{15}{4 \ln Re} \right) e^{3/2} \ln(2y/d)$$

$$+ e^{3/2} \left(\frac{\sqrt{3}}{2} + \frac{15}{4 \ln Re} \right) \ln \frac{u_*}{2\bar{u}},$$

where $\phi_{CL} = u_{CL}/u_*$. This calculation is self-consistent, and differs from the original IMM procedure by matching a Reynolds number dependent defect law to the actual curves of the scaling law (2.7) rather than to their envelope.

Another way of looking as the calculation we have just performed is to note that if one requires an overlap between a law of the wall that does not depend on d and a defect law that does not depend on ν, one obtains an overlap that depends on neither d nor ν; this enforces complete similarity and results in the von-Kármán-Prandtl law, which can be obtained by simply removing the quantities d and ν from the list of arguments in equation (2.3). On the other hand, more realistic requirements on the laws being matched leave room for incomplete similarity and are consistent with the scaling law (2.1) and with the experimental data; for example, Figure 7 in [45] exhibits clearly the dependence of the profile in the neighborhood of the centerline on ν. The matching was successfully carried out because the scaling law has an intermediate range that is approximately linear in $\ln \eta$; the success of the matching does not depend on the specific values of the constants B_0, B_1 and β_1 in (2.6).

Note several important consequences of this analysis: (i) Both linear parts of the piecewise linear chevron structure belong to the same scaling law; the constants obtained from data for the inner (i.e. closer to the wall) segment also describe the outer segment. (ii) The overlap region is the outer segment of the chevron; thus the outer segment belongs both to the wall region and to the defect region. (iii) There is no other possible locus for the overlap; as the slopes of the outer and the inner segments tend to two different constants as $\nu \to 0$, one can never get them to overlap on the inner segment. (iv) The whole chevron constituting a single law, the possibility that the inner segment is described by the universal (i.e. Re independent) logarithmic law is excluded. (v) More generally, since the defect law must be a concave-downward function of y/d, the only way there can ever be a portion of the velocity profile that is concave downward in the $(\phi, \ln \eta)$ plane, given that there must be an overlap region, is is to make the overlap region be concave upward, as we are proposing, rather than straight.

Note also that the inner and outer portions of the flow "feel" each other for all finite values of Re; the coupling disappears only in the limit of vanishing viscosity. We shall draw a similar conclusion also in the case of local structure.

6. The scaling laws in the inertial range and their consequences

The analogy between the inertial range in the local structure of developed turbulence and the intermediate range in turbulent shear flow near a wall has been noted long ago (see e.g. [11],[43, pp.147 and 263]), and we appeal to it to motivate the extension of the scaling analysis above to the case of local structure, where the experimental data are much poorer. In the inertial range of local structure the general scaling law that corresponds to (2.1) is:

$$(6.1) \qquad D_{LL} = (\langle \varepsilon \rangle r)^{\frac{2}{3}} \Phi\left(\frac{r}{\Lambda}, Re\right),$$

where $D_{LL} = \langle [u_L(\mathbf{x}+\mathbf{r}) - u_L(\mathbf{x})]^2 \rangle$ is the basic component of the second order structure function tensor which determines all the other components for incompressible flow, u_L is the velocity component along the vector $\mathbf{r}$ joining two observation points $\mathbf{x}$ and $\mathbf{x} + \mathbf{r}$, ε is the total rate of energy transfer across the spectrum, $r = |\mathbf{r}|$ is the length of the vector $\mathbf{r}$, Λ is an external length scale, e.g. the Taylor scale, and Re is a properly defined Reynolds number, for example one based on the Taylor scale. The brackets $\langle \ldots \rangle$ denote an ensemble average. By the logic of the derivation of (6.1) the function Φ should be a universal function of its arguments, identical for all flows. Formula (6.1) is assumed to hold only at very high Reynolds numbers Re and very small r/Λ. The classical "K–41" Kolmogorov theory [28] results from the assumption of complete similarity, in which, for r/Λ small enough and Re large enough, Φ can be taken as a constant different from zero, and one obtains D_{LL} proportional to $(\langle \varepsilon \rangle r)^{\frac{2}{3}}$. The spectral form of the Kolmogorov law cited in section 3 can be then obtained by a Fourier transform.

Various corrections to that law have been proposed; many of them involve the addition of an extra length scale to the problem, an addition that is hard to justify. We now explore what can be deduced from the much more plausible assumption of incomplete similarity. An expansion of Φ for flows in which Re is large in powers of $1/\ln Re$ gives, in analogy with (4.4) (see [3],[6]):

$$(6.2) \qquad D_{LL} = (\langle \varepsilon \rangle r)^{\frac{2}{3}} \left[A_0 + \frac{A_1}{\ln Re} + o\left(\frac{1}{\ln Re}\right) \right] \left(\frac{r}{\Lambda}\right)^{\frac{\alpha_1}{\ln Re} + o\left(\frac{1}{\ln Re}\right)},$$

In the present problem, the molecular viscosity ν appears only in the variable Re, so that the limit of vanishing viscosity and the limit of infinite Re coincide. Note the asymptotic covariance ([3],[6],[20]) of (6.2) in the external length scale Λ: Indeed, replacing Λ by a different length scale Λ_1, we find

$$\left(\frac{r}{\Lambda}\right)^{\alpha_1/\ln Re} = \left[\left(\frac{r}{\Lambda_1}\right)\left(\frac{\Lambda_1}{\Lambda}\right)\right]^{\alpha_1/\ln Re}$$

$$= \left(\frac{r}{\Lambda_1}\right)^{\alpha_1/\ln Re} \exp\left[\alpha_1 \ln\left(\frac{\Lambda_1}{\Lambda}\right)/\ln Re\right] \sim \left(\frac{r}{\Lambda_1}\right)^{\alpha_1/\ln Re}$$

for large Re.

The classical "K–41" Kolmogorov theory now corresponds to $A_0 \neq 0$ in (6.2); then, for large Re, the famous Kolmogorov 2/3 law is obtained

$$(6.3) \qquad D_{LL} = A_0(\langle \varepsilon \rangle r)^{\frac{2}{3}} .$$

In real measurements for finite but accessibly large Re, $\alpha_1 / \ln Re$ is small in comparison with 2/3, and the deviation in the power of r in (4.2) should be unnoticeable. On the other hand, the variations in the "Kolmogorov constant" have been repeatedly noticed (see [38],[42]). Complete similarity is possible only if $A_0 \neq 0$. If $A_0 \neq 0$ one has a well-defined turbulent state with a 2/3 law in the limit of vanishing viscosity, and finite Re effects can presumably be obtained by expansion about that limiting state. In the limit of vanishing viscosity, there are no corrections to the "K–41" scaling, as was also deduced in [12] by the statistical mechanics argument summarized in section 3 above.

Kolmogorov [28] proposed similarity relations also for the higher order structure functions:

$$D_{LL...L}(r) = \langle [u_L(\mathbf{x} + \mathbf{r}) - u_L(\mathbf{x})]^p \rangle,$$

where $LL...L$ denotes L repeated p times; the scaling gives $D_{LL...L} = C_p(\langle \varepsilon \rangle r)^{p/3}$.

Experiments, mainly by Benzi et al, see [8], apparently show some self-similarity, obviously incomplete, so that $D_{LL...L}$ is proportional to r^{ζ_p}, with exponents ζ_p always smaller then $p/3$ for $p \geq 3$, so that $\zeta_4 = 1.28$ instead of 1.33, $\zeta_5 = 1.53$ instead of 1.67, $\zeta_6 = 1.77$ instead of 2.00, $\zeta_7 = 2.01$ instead of 2.33, and $\zeta_8 = 2.23$ instead of 2.67. It is tempting to try for an explanation of the same kind as for $p = 2$:

$$(6.4) \qquad D_{LL...L} = \left(C_p^0 + \frac{C_p^1}{\ln Re} + o(\frac{1}{\ln Re}) \right) (\langle \varepsilon \rangle r)^{p/3} (r/\Lambda)^{\gamma_p / \ln Re + ...}.$$

In other words, at $Re = \infty$ the classic "K41" theory would be valid, but the experiments were performed at Reynolds numbers too small to reveal the approach to complete similarity. If this explanation were correct, the coefficients γ_p are negative starting with $p = 4$, and therefore the influence of the external scale could be very strong.

As is well-known, for $p = 3$ the Kolmogorov scaling is valid, with no corrections. For $p > 3$, however, one must proceed with caution. A numerical study of the scaling laws for $p > 3$ will be presented elsewhere, as soon as it is completed. We would like however to present a simple argument that casts a doubt on the good behavior of the structure functions for integers $p > 3$ in the vanishing-viscosity limit. Indeed, as $Re \to \infty$ the "active" regions of the flow shrink while energy is conserved. If V_0 is the portion of the fluid where the kinetic energy $\approx u^2$ is large, then $u \approx \frac{1}{\sqrt{V_0}}$; one can easily see that fourth moments such as $\langle u^4 \rangle$ diverge as $V_0 \to \infty$. This makes it likely that fourth-order structure functions also blow up (no conclusion can be drawn for $p = 3$ because for odd powers p cancellations can occur and the integrals can remain finite). Note that $p = 3$ is the power where, if both the experimental data and the general scaling law are to be believed, the sign of the power of $\frac{r}{\Lambda}$ in an expansion in powers of $\frac{1}{\ln Re}$ would change.

The obvious conclusion is that, since the structure functions for $p \geq 3$ are finite for finite viscosity ν, and their limit as the viscosity tends to zero does not exist. The simplest possibility is that the for $p \geq 4$ the constants C_0^p are zero, but equation (6.4) is still valid. However, there may well be other possibilities that we have not yet explored; for example, the inclusion of the viscosity as an additional independent variable in the basic scaling law for $p \geq 4$. A further comment relates what we have just discussed to the statistical theory of section 3: It may well be that the higher moments of u require a longer time to relax to their equilibrium values than is available in a turbulent system; in this case the near-equilibrium theory does not apply to them and one cannot expect a valid small-viscosity limit. Indeed, according the Benzi's data [8] the time scale characteristic of moments of order p increases with p for $p > 3$.

Finally, the discussion in the present section allows us to explain why, in the statistical analysis of section 3, one can assume that the small scales of turbulence have enough time to settle to an equilibrium, at least as far as an analysis of the lower order structure functions is concerned. At large Re, one can conclude from eq. (4.1) that the characteristic velocity of an "eddy" of size r is proportional to $r^{1/3}$; the characteristic time (length/velocity) is thus proportional to $r^{2/3}$ and tends to zero for small enough scales.

7. Conclusions

The following conclusions have been reached above:

(i) The customary universal logarithmic law of the wall must be jettisoned and replaced by a power law;

(ii) it is very likely that the corrections to the classical "K-41" scaling of the inertial range of local structure in fully developed turbulence are Reynolds-number dependent and disappear in the limit of infinite Reynolds number, at least for structure functions of order ≤ 3;

(iii) small-viscosity asymptotics, based on a statistical description of fully-developed turbulence in which the zero viscosity limit is well-behaved, constitutes a powerful tool for the analysis of turbulence at high Reynolds numbers.

We also wish to point out that our combination of similarity theory and of asymptotics based on a statistical theory represents a step forward in the effort to derive the properties of turbulent flow from first principles.

References

[1] G.I. Barenblatt, *Similarity, Self-Similarity and Intermediate Asymptotics*, Consultants Bureau, NY, (1979).

[2] G.I. Barenblatt, On the scaling laws (incomplete self-similarity with respect to Reynolds number) in the developed flows in turbulence, *C.R. Acad. Sc. Paris*, series II, **313**, 307–312 (1991).

[3] G.I. Barenblatt, Scaling laws for fully developed turbulent shear flows. Part 1: Basic hypotheses and analysis, *J. Fluid Mech.* **248**, 513–520 (1993).

[4] G. I. Barenblatt and A.J. Chorin, Small viscosity asymptotics for the inertial range of local structure and for the wall region of wall-bounded turbulence, *Proc. Nat. Acad. Sciences USA* **93**, 6749–6752 (1996).

[5] G.I. Barenblatt and A.J. Chorin, Scaling laws and vanishing viscosity limits for wall-bounded shear flows and for local structure in developed turbulence, *Comm. Pure Appl. Math*, bf 50, 381–398 (1997).

[6] G. I. Barenblatt and N. Goldenfeld, Does fully developed turbulence exist? Reynolds number dependence vs. asymptotic covariance, *Phys. Fluids A*, 3078–3082 (1995).

[7] G.I. Barenblatt and V.M. Prostokishin, Scaling laws for fully developed shear flows. Part 2. Processing of experimental data, *J. Fluid Mech.* **248**, 521–529 (1993).

[8] R. Benzi, C. Ciliberto, C. Baudet and G. Ruiz Chavarria, On the scaling of three dimensional homogeneous and isotropic turbulence, *Physica D* **80**, 385-398 (1995).

[9] P. Bernard, J. Thomas and R. Handler, Vortex dynamics and the production of Reynolds stress, *J. Fluid Mech.* **253**, 385–419 (1993).

[10] H.B. Callen and T.A. Welton, Irreversibility and generalized noise, *Phys. Rev.* **83**, 34-40, (1951).

[11] A.J. Chorin, Theories of turbulence, in *Berkeley Turbulence Seminar*, edited by P. Bernard and T. Ratiu, Springer, NY, (1977).

[12] A. J. Chorin, *Vorticity and Turbulence*, Springer, 1994.

[13] A.J. Chorin, Turbulence as a near-equilibrium process, *Lectures in Appl. Math.* **31**, 235–248 (1996).

[14] A.J. Chorin, Turbulence cascades across equilibrium spectra, *Physical Review E*, (in press), (1996).

[15] A.J. Chorin, Partition functions and invariant measures for two-dimensional and quasi-three-dimensional turbulence, *Physics Fluids A*, (in press) (1996).

[16] D. Coles, The law of the wall in the turbulent boundary layer, *J. Fluid Mech.* **1**, 191–226 (1956).

[17] M. Gad-el-Hak and P.R. Bandyopadhyay, Reynolds number effects in wall-bounded turbulent flows, *Appl. Mech. Rev.* **47**, 307–366 (1994).

[18] C.S. Gardner, J.M. Greene, M.D. Kruskal, and R.M. Miura, Method for

solving the Korteweg-de Vries equation, *Phys. Rev. Lett.* **19**, 1095–1097 (1967).

[19] I.M. Gelfand and N. Ya. Vilenkin, *Generalized functions*, Vol. IV, Academic, NY, 1964.

[20] Goldenfeld, N., *Lectures on Phase Transitions and the Renormalization Group*, Addison-Wesley, Reading, Mass. 1992.

[21] H. Hasimoto, A soliton on a vortex filament, *J. Fluid Mech.* **51**, 477–485 (1972).

[22] E. Hopf, Statistical hydrodynamics and functional calculus, *J. Rat. Mech. Anal.* **1**, 87-141 (1952).

[23] T.Y. Hou and P. D. Lax, Dispersive approximations in fluid dynamics, *Comm. Pure Appl. Math.* **44**, 1–14 (1991).

[24] Th. von Kármán, Mechanische Aehnlichkeit und Turbulenz, *Nach. Ges. Wiss. Goettingen Math-Phys. Klasse*, 58–76 (1932).

[25] P. Kailasnath, Reynolds number effects and the momentum flux in turbulent boundary layers, Ph.D. thesis, Yale University, 1993.

[26] T. Kato, Remarks on the zero viscosity limit for nonstationary Navier-Stokes flows with boundary, in *Seminar on Partial Differential Equations*, Edited by S.S. Chern, Springer, NY, 1984.

[27] S.J. Kline, W.C. Reynolds, F.A. Schraub, and P.W. Rundstadler, The structure of turbulent boundary layers, *J. Fluid Mech.* **30**, 741–774 (1967).

[28] A.N. Kolmogorov, Local structure of turbulence in an incompressible fluid at a very high Reynolds number, *Dokl. Akad. Nauk SSSR* **30**, 299–302 (1941).

[29] P.D. Lax, Integrals of nonlinear equations of evolution and solitary waves, *Comm. Pure Appl. Math.* **21**, 467–490 (1968).

[30] P. D. Lax, *Hyperbolic systems of conservation laws and the mathematical theory of shock waves*, SIAM Publications, Philadelphia, 1972.

[31] P.D. Lax, The zero dispersion limit, a deterministic analog of turbulence, *Comm. Pure Appl. Math.* **44**, 1047–1056 (1991).

[32] P.D. Lax and C.D. Levermore, The small dispersion limit of the Korteweg-de Vries equation III, *Comm. Pure Appl. Math*, **36**, 809–830 (1983).

[33] T.D. Lee, On some statistical properties of hydrodynamic and hydromagnetic fields, *Quarterly Appl. Math.* **10**, 69–72 (1952).

[34] A. S. Monin and A. M. Yaglom, *Statistical Fluid Mechanics*, Vol. 1, MIT Press, Boston, 1971.

[35] J. Nikuradze, Gesetzmaessigkeiten der turbulenten Stroemung in glatten Rohren, *VDI Forschungheft*, No. **356**, (1932).

[36] A.M. Obukhov, Spectral energy distribution in turbulent flow, *Dokl. Akad. Nauk SSR* **32**, 22–24 (1941).

[37] L. Prandtl, Zur turbulenten Stroemung in Rohren und laengs Platten,

Ergeb. Aerodyn. Versuch., Series 4, Goettingen, (1932).

[38] A. Praskovsky and S. Oncley, Measurements of the Kolmogorov constant and intermittency exponents at very high Reynolds number, *Phys. Fluids* **A 7**, 2778–2784 (1994).

[39] H. Schlichting, *Boundary Layer Theory*, McGraw-Hill, NY, second edition, 1968.

[40] K.R.Sreenivasan, A unified view of the origin and morphology of turbulent boundary layer structure, *IUTAM Symp.*, Bangalore, H.W.Liepmann, R.Narasimha (eds.), (1987).

[41] K.R. Sreenivasan, The turbulent boundary layer, *Frontiers in experimental fluid mechanics*, edited by M. Gad-el-Hak, Springer, 159–209, (1989).

[42] K.R. Sreenivasan, On the universality of the Kolmogorov constant, *Phys. Fluids* **A 7**, 2778–2784, (1995).

[43] H. Tennekes, and J.L. Lumley, *A First Course in Turbulence*, MIT Press, Cambridge, (1990), pp.147 and 263.

[44] M.J. Vishik and A.V. Fursikov, *Mathematical Problems of Statistical Hydrodynamics*, Kluwer, Dordrecht, 1988.

[45] M. V. Zagarola, A.J. Smits, S.A.Orszag, and V.Yakhot, Experiments in high Reynolds number turbulent pipe flow, AIAA paper 96-0654, Reno, Nevada (1996).

Department of Mathematics and
Lawrence Berkeley Laboratory
University of California
Berkeley, California 94720–3840

Proceedings of Symposia in Applied Mathematics
Volume **54**, 1998

POTENTIAL THEORY IN HILBERT SPACES

PIERMARCO CANNARSA & GIUSEPPE DA PRATO

1. Introduction

This paper is devoted to the discussion of Potential Theory in an infinite dimensional separable Hilbert space H. Let us explain the content of the paper. Let $\{e_k\}$ be a complete orthonormal system in H, and consider the following generalization of Heat equation

(1.1)
$$\begin{cases} u_t(t,x) = \dfrac{1}{2}\sum_{k=1}^{\infty}\lambda_k D_k^2 u(t,x), \ t > 0, \ x \in H, \\[2mm] u(0,x) = \varphi(x), \ x \in H, \end{cases}$$

where $\{\lambda_k\}$ is a sequence of positive numbers and D_k represents the derivative in the direction e_k. Moreover the initial datum φ belongs to $C_b(H)$, the space of all uniformly continuous and bounded mappings from H into $\mathbb{R}$. As is discussed in §3.1, a necessary condition for the existence of a solution to (1.1) for any initial datum is that

(1.2)
$$\sum_{k=1}^{\infty}\lambda_k < +\infty.$$

Equivalently, we could say that the bounded linear operator Q, defined by

$$Qx = \sum_{k=1}^{\infty}\lambda_k <x,e_k> e_k,$$

is trace–class. Using this notation, problem (1.1) can be rewritten in the following form, independent of the basis $\{e_k\}$ ([1])

(1.3)
$$\begin{cases} u_t(t,x) = \dfrac{1}{2}\ \mathrm{Tr}\ [QD^2 u(t,x)] \\[2mm] u(0,x) = \varphi(x). \end{cases}$$

1991 *Mathematics Subject Classification*. Primary 35R15; Secondary 46G12.

[1]For any trace–class operator L on a Hilbet space H we denote by $\mathrm{Tr}\ _H L$ its trace. If no confusion may arise we shall write $\mathrm{Tr}\ L$ instead of $\mathrm{Tr}\ _H L$.

Potential Theory in infinite dimensions started with the pioneering papers by L. Gross [7] and Yu. L. Daleckij [5]. Let us recall the Gross setting. Let H_G be a separable Hilbert space and let B be a separable Banach space, the inclusion $H_G \subset E$ being continuous and dense. Gross was interested, among other things, in constructing a semigroup P_t, $t \geq 0$, on the space $B_b(E)$ of all Borel bounded functions from E into $\mathbb{R}$, describing the solution to the problem

(1.4)
$$\begin{cases} u_t(t,x) = \dfrac{1}{2}\, \mathrm{Tr}\,[D_G^2 u(t,x)], \ t > 0, \ x \in H, \\[2mm] u(0,x) = \varphi(x), \end{cases}$$

where D_G^2 is the second derivative in the direction H_G. The construction of the semigroup was done with the help of Gaussian measures. As pointed out by Gross, given a function $\varphi \in C_b(H)$, $P_t\varphi$ is not Fréchet differentiable in general, but it is infinitely differentiable in all directions of H [7, Prop. 9]. Moreover he showed [7, Th. 2], that $D^2 P_t\varphi$ is not trace class, unless φ is Lipschitz continuous. Concerning the elliptic equation

(1.5)
$$\lambda\varphi(x) - \frac{1}{2}\mathrm{Tr}\,[D_G^2\varphi(x)] = g(x), \ x \in H, \ \lambda > 0,$$

Gross [7, Th. 3], proved that if g is Lipschitz continuous then equation (1.5) has a unique solution and $D^2\varphi(x)$ is of trace class for all $x \in H$.

Let now remark that our setting is included in the one by Gross just setting

$$H_G = Q^{1/2}(H), \ \ E = H,$$

the scalar product in H_G being defined by

$$< x, y >_{H_G} = < Q^{-1/2}x, Q^{-1/2}y >, \ x, y \in H_G.$$

Note that a complete orthonormal basis on H_G is given by $\{ \sqrt{\lambda_k}\, e_k \}$. Moreover the relation between the second derivative D_G^2 in the direction of H_G and our second derivative D^2 in H is given by

$$D_G^2 = Q^{1/2} D^2 Q^{1/2}$$

so that

$$\mathrm{Tr}\,_{H_G} D_G^2 u(x) = \sum_{k=1}^{\infty} < D_G^2 u(x) g_k, g_k >_{H_G} = \sum_{k=1}^{\infty} < D_G^2 u(x) e_k, e_k >_H$$

$$= \sum_{k=1}^{\infty} < Q^{1/2} D^2 u(x) Q^{1/2} e_k, e_k >_H = \sum_{k=1}^{\infty} \lambda_k < D^2 u(x) e_k, e_k >_H$$

$$= \mathrm{Tr}\,_H [Q D^2 u(x)].$$

Therefore problems (1.3) and (1.4) are equivalent.

In this paper we give a review of the functional analytic approach, introduced in P. Cannarsa and G. Da Prato in [1] and developed in [2], [3] and [4], to the costruction of a strongly continuous semigroup P_t, $t \geq 0$ on $C_b(H)$ related to problem (1.4). Moreover we present simpler proofs and some new results. In particular
(i) we give a purely analitic derivation of an integral representation formula for the semigroup P_t, $t \geq 0$,

$$(1.6) \qquad P_t\varphi(x) = \int_H \varphi(y)\mathcal{N}(x, tQ)(dy), \ \varphi \in C_b(H),$$

where $\mathcal{N}(x, tQ)$ is the Gaussian measure in H with mean x and covariance operator tQ,
(ii) we prove a maximal regularity result for the potential associated to problem (1.3).

Concerning the point (i) we remark that formula (1.6) is essentially the starting point of the quoted paper by Gross, whereas it is here derived by using functional analytic tools and the a Daniell representation theorem.

Concerning the point (ii), we define, following Gross, the potential of a function $g \in C_b(H)$ having a bounded support, by setting

$$(1.7) \qquad \varphi(x) = \int_0^{+\infty} P_t g(x) dt.$$

If φ is sufficiently regular one has

$$(1.8) \qquad -\frac{1}{2} \ \mathrm{Tr} \ [QD^2\varphi(x)] = g(x), \ x \in H.$$

If H is finite dimensional and $g \in C_b(H)$ is θ–Hölder continuous, $\theta \in]0,1[$, and with bounded support, then the classical Schauder estimates imply that φ has the second derivatives θ–Hölder continuous. In Theorem 5.1 we prove an infinite dimensional generalization of this result. We note that A. Piech, under the same assumptions, proved in [4] a weaker result, namely that φ has the second derivatives (in all directions e_k) α–Hölder continuous for any $\alpha < \theta$.

2. Preliminaries

2.1 Notation.
Throughout all the paper H will denote a separable Hilbert space (norm $|\cdot|$, inner product $< \cdot, \cdot >$), $\mathcal{L}(H)$ the Banach algebra of all bounded linear operators from H into H endowed with the norm

$$\|T\| = \sup_{x \in H, x \neq 0} \|Tx\|, \ T \in \mathcal{L}(H),$$

and $\mathcal{L}_1(H)$, the set of all trace–class operators. $\mathcal{L}_1(H)$, endowed with the norm

$$\|T\|_{\mathcal{L}_1(H)} = \ \mathrm{Tr} \ \sqrt{T^*T},$$

is a Banach space.

We will denote by $C_b(H)$ the Banach space of all mappings $\varphi : H \mapsto \mathbb{R}$ that are uniformly continuous and bounded, with the norm

$$\|\varphi\|_0 = \sup_{x \in H} |\varphi(x)|.$$

We will deal with the following subspaces of $C_b(H)$.

(i) $C_b^1(H)$ is the set of all bounded uniformly continuous functions $\varphi : H \mapsto \mathbb{R}$ which are Fréchet differentiable on H, with a bounded uniformly continuous derivative $D\varphi$. We set

$$[\varphi]_1 = \sup_{x \in H} |D\varphi(x)|, \quad \|\varphi\|_1 = \|\varphi\|_0 + [\varphi]_1.$$

If $\varphi \in C_b^1(H)$ and $x \in H$ we shall identify $D\varphi(x)$ with the element h of H such that

$$D\varphi(x)y = < h, y >, \ \forall \, y \in H.$$

(ii) $C_b^2(H)$ is the subset of all functions $\varphi : H \mapsto \mathbb{R}$ of $C_b^1(H)$ which are twice Fréchet differentiable on H, with a bounded uniformly continuous second derivative $D^2\varphi$. We set

$$[\varphi]_2 = \sup_{x \in H} |D^2\varphi(x)|$$

$$\|\varphi\|_2 = \|\varphi\|_1 + [\varphi]_2.$$

If $\varphi \in C_b^2(H)$ and $x \in H$ we shall identify $D^2\varphi(x)$ with the linear bounded operator $T \in \mathcal{L}(H)$ such that

$$D\varphi(x)(y, z) = < Ty, z >, \ \forall \, y, z \in H.$$

(iii) $C_b^{0,1}(H)$ is the subspace of $C_b(H)$ of all functions $\varphi : H \mapsto \mathbb{R}$ such that

$$[\varphi]_1 = \sup_{x,y \in H, x \neq y} \frac{|\varphi(x) - \varphi(y)|}{|x - y|} < +\infty.$$

$C_b^{0,1}(H)$ is a Banach space with the norm

$$\|\varphi\|_1 = \|\varphi\|_0 + [\varphi]_1.$$

(iv) $C_b^{1,1}(H)$ is the space of all functions $\varphi \in C_b^1(H)$ such that $D\varphi$ is Lipschitz continuous. We set

$$[\varphi]_{1,1} = \sup_{x \neq y} \frac{|D\varphi(x) - D\varphi(y)|}{|x - y|}.$$

$\varphi \in C_b^{1,1}(H)$ is a Banach space with the norm

$$\|\varphi\|_{1,1} = \|\varphi\|_1 + [\varphi]_{1,1}.$$

(v) $C_b^\alpha(H)$, $\alpha \in]0, 1[$, is the set of all functions $\varphi : H \mapsto \mathbb{R}$ such that

$$[\varphi]_\alpha = \sup_{x,y \in H, x \neq y} \frac{|\varphi(x) - \varphi(y)|}{|x - y|^\alpha} < +\infty.$$

We set

$$C_b^\alpha(H) = C_b(H) \cap C^\alpha(H).$$

$C_b^\alpha(H)$ is a Banach space with the norm

$$\|\varphi\|_\alpha = \|\varphi\|_0 + [\varphi]_\alpha.$$

(vi) $C_b^{1,\alpha}(H)$, $\alpha \in]0, 1[$, is the space of all functions $\varphi \in C_b^1(H)$ such that $D\varphi$ is α–Hölder continuous. We set

$$[\varphi]_{1,\alpha} = \sup_{x \neq y} \frac{|D\varphi(x) - D\varphi(y)|}{|x - y|^\alpha}.$$

$C_b^{1,\alpha}(H)$ is a Banach space with the norm

$$\|\varphi\|_{1,\alpha} = \|f\|_1 + [\varphi]_{1,\alpha}.$$

2.2 Approximation of continuous functions.

We are here concerned with the problem to approximate a given function $\varphi \in C_b(H)$ by more regular functions. It is well known that if dim $H < \infty$, then the space $C_b^\infty(H)$ of all functions bounded together with all their derivatives of any order is dense in $C_b(H)$.

If dim $H = \infty$, this is not true. In fact A. S. Nemirowski and S. M. Semenov [12], proved that $C_b^2(H)$ is not dense in $C_b(H)$, whereas $C_b^{1,1}(H)$ is. This last result was obtained later by J. M. Lasri and P. L. Lions [10], with a different costructive proof. They proved the following result.

Theorem 2.1. *Let* $\varphi \in C_b(H)$ *and let* φ_t, $t \in [0, 1]$ *be defined by*

$$(2.1) \qquad \varphi_t(x) = \sup_{z \in H} \left\{ \inf_{y \in H} \left[\varphi(y) + \frac{|z - y|^2}{2t} \right] - \frac{|z - x|^2}{t} \right\}.$$

Then the following statements hold.

(i) If $\varphi, \psi \in C_b(H)$ *and* $\varphi(x) \leq \psi(x)$, $\forall\, x \in H$, *then* $\varphi_t(x) \leq \psi_t(x)$, $\forall\, x \in H$.

(ii) $\varphi_t \in C_b^{1,1}(H)$ *and the following estimates hold*

$$(2.2) \qquad \|\varphi_t\|_0 \leq \|\varphi\|_0, \ \forall\, t \in [0, 1].$$

$$(2.3) \qquad [\varphi_t]_{1,1} \leq \frac{1}{t}, \ \forall\, t \in [0, 1].$$

From the theorem it follows in particular that $C_b^{1,1}(H)$ is dense $=$ in $C_b(H)$.

2.3 Gaussian measures.

Let $m \in \mathbb{R}$ and $\lambda \geq 0$. The Gaussian measure $\mathcal{N}(m, \lambda)$ is a Borel measure on $(\mathbb{R}, \mathcal{B}(\mathbb{R}))$ defined as

$$(2.4) \qquad \mathcal{N}(m, q)(d\xi) = \begin{cases} \frac{1}{\sqrt{2\pi\lambda}} \, e^{-\frac{(\xi - m)^2}{2\lambda}} d\xi & \text{if } \lambda > 0, \\[2ex] \delta_m(d\xi) & \text{if } \lambda = 0, \end{cases}$$

where δ_m is the Dirac measure concentrated in m. The following identities are easily checked.

$$(2.5) \qquad \int_{-\infty}^{+\infty} \xi \mathcal{N}(m,q)(d\xi) = m,$$

$$(2.6) \qquad \int_{-\infty}^{+\infty} (\xi - m)^2 \mathcal{N}(m,q)(d\xi) = \lambda,$$

$$(2.7) \qquad \int_{-\infty}^{+\infty} e^{i\alpha\xi} \mathcal{N}(m,q)(d\xi) = e^{i\alpha m - \frac{\lambda}{2}\alpha^2}.$$

Let now H be a separable Hilbert space, $m \in H$ and $S \in \mathcal{L}_1(H)$ with $S \geq 0$ and $S = S^*$. Let μ be a measure on $(H, \mathcal{B}(H))$. We say that μ is the Gaussian measure $\mathcal{N}(m,S)$ with *mean* m and *covariance* S if its characteristic function is given by

$$(2.8) \qquad \int_H e^{i<x,h>} \mu(dx) = e^{i<m,h> - \frac{1}{2}<Sh,h>}, \; h \in H.$$

We list now some properties of the Gaussian measure $\mathcal{N}(m,S)$ that will be needed later. We denote by $\{e_k\}$ a complete orthonormal system on H that diagonalizes S and by $\{\lambda_k\}$ the corresponding set of eigenvalues:

$$Se_k = \lambda_k e_k, \; k \in \mathbb{N}.$$

The following proposition is easy to check by computing the characteristic functions involved.

Proposition 2.1. *Let H and K be two separable Hilbert spaces, and let $F : H \mapsto K$ be defined as*

$$F(x) = Tx + a,$$

where $T \in \mathcal{L}(H,K)$ and $a \in K$. Then we have

$$(2.9) \qquad F \circ \mathcal{N}(m,S) = \mathcal{N}(Tm + a, TQT^*)$$

For any $n \in \mathbb{N}$ denote by Σ_n the mapping

$$(2.10) \qquad \Sigma_n : H \mapsto \mathbb{R}^n, \; x \mapsto \zeta = (<x,e_1>, ..., <x,e_n>).$$

Then applying Proposition 2.1 with $K = \mathbb{R}^n$ and $F = \Sigma_n$ we find

$$(2.11) \qquad \Sigma_n \circ \mathcal{N}(m,S) = \mathcal{N}((<x,e_1>, ..., <x,e_n>), \, \text{diag}\,(\lambda_1, ..., \lambda_n)).$$

Theorem 2.2. *Let $\alpha_0 = \inf_{k \in \mathbb{N}} \lambda_k^{-1}$. Then for any $\alpha < \alpha_0$ we have*

$$(2.12) \qquad \int_H e^{\frac{\alpha}{2}|x|^2}\,\mathcal{N}(0,S)(dx) = \left[\prod_{k=1}^{\infty}(1 - \alpha\lambda_k)\right]^{-1/2} = [\det(1 - \alpha\lambda_k)]^{-1/2}.$$

Proof.   We have

$$\int_H e^{\frac{\alpha}{2}|<x,e_1>|^2}\mathcal{N}(0,S)(dx) = \int_{\mathbb{R}} e^{\frac{\alpha}{2}\xi^2}\mathcal{N}(0,\lambda_1)(d\xi)$$

$$= \frac{1}{\sqrt{2\pi\lambda_1}}\int_{-\infty}^{+\infty} e^{\frac{\alpha}{2}\xi^2}e^{-\frac{\xi^2}{2\lambda_1}}\,d\xi = [1 - \alpha\lambda_1]^{-1/2}.$$

In a similar way, using (2.11) we have

$$\int_H e^{\frac{\alpha}{2}\sum_{k=1}^{n}|<x,e_k>|^2}\mathcal{N}(0,S)(dx) = \left[\prod_{k=1}^{n}(1 - \alpha\lambda_k)\right]^{-1/2}.$$

Now the conclusion follows from the Monotone Convergence Theorem.   ■
     We can now compute easily integrals of the form

$$\int_H |x|^{2m}\,\mathcal{N}(0,S)(dx),\ m \in \mathbb{N}.$$

For this we set

$$G(\alpha) = \int_H e^{\frac{\alpha}{2}|x|^2}\mathcal{N}(0,S)(dx) = [\det(1 - \alpha\lambda_k)]^{-1/2},\ \alpha < \alpha_0.$$

It is easy to check, using the elementary formula,

$$\frac{d}{d\alpha}\det(1 - \alpha\lambda_k) = -\ \text{Tr}\ \frac{S}{1 - \alpha S}\det(1 - \alpha\lambda_k),$$

that G is of C^∞ class. It follows

$$(2.13) \qquad \int_H |x|^{2m}\,\mathcal{N}(0,S)(dx) = 2^m\,\frac{d^m}{d\alpha^m}[\,\det(1 - \alpha S)]^{-1/2}|_{\alpha=0}.$$

We have in particular

$$(2.14) \qquad \int_H |x|^2\,\mathcal{N}(0,S)(dx) = \ \text{Tr}\ S,$$

$$(2.15) \qquad \int_H |x|^4\,\mathcal{N}(0,S)(dx) = 2\ \text{Tr}\ [S^2] + (\text{Tr}\ S)^2.$$

Let us give a last definition. For any $\alpha \in H$ and any $n \in \mathbb{N}$ we denote by $\xi_{\alpha,n}$ the real function in H defined as

$$\xi_{\alpha,n}(x) = <\alpha, S_n^{-1/2}x> = \sum_{k=1}^{n}\lambda_k^{-1/2}x_k\alpha_k.$$

It is a remarkable fact that the sequence $\{\xi_{\alpha,n}\}$ is convergent in $L^2(H, \mathcal{N}(0,S))$ to a function ξ_α that we denote by

$$(2.16) \qquad \xi_\alpha(x) = <\alpha, S^{-1/2}x>.$$

Note that definition (2.16) is formal since it requires that x belongs to the set $S^{1/2}(H)$ that has $\mathcal{N}(0,S)$ measure equal to 0.
     The following result holds.

Proposition 2.2. *There exists the limit*

$$\lim_{n \to \infty} \xi_{\alpha,n} =: \xi_\alpha, \text{ in } L^2(H, \mu).$$

Moreover

(2.17) $$\int_H |\xi_\alpha(x)|^2 \mathcal{N}(0, S)(dx) = |\alpha|^2, \ \alpha \in H.$$

   Proof.   We have

$$\int_H |\xi_{\alpha,n+p}(x) - \xi_{\alpha,n}(x)|^2 \mu(dx) = \int_H \left| \sum_{k=n+1}^{n+p} \lambda_k^{-1/2} x_k \alpha_k \right|^2 \mu(dx)$$

$$= \sum_{k=n+1}^{n+p} (\lambda_h \lambda_k)^{-1/2} \alpha_h \alpha_k \int_H x_h x_k \mu(dx)$$

$$= \sum_{h=n+1}^{n+p} \frac{1}{\lambda_h} \alpha_h^2 \int_H x_h^2 \mu(dx) = \sum_{k=n+1}^{n+p} |\alpha_h|^2.$$

So $\{\xi_{\alpha,n}\}$ is a Cauchy sequence . Finally, letting $n \to \infty$ in the equalities

$$\int_H |\xi_{\alpha,n}(x)|^2 \mu(dx) = \sum_{h=1}^{n} \frac{1}{\lambda_h} \alpha_h^2 \int_H x_h^2 \mu(dx) = \sum_{h=1}^{n} |\alpha_h|^2,$$

(2.17) follows.   ∎

3. Heat Equation in Hilbert spaces

3.1 Notation. Let H be an infinite dimensional separable Hilbert space, $\{e_k\}$ a complete orthonormal system in H, and $\{\lambda_k\}$ a sequence of positive numbers. For any $x \in H$ we set $x_k =< x, e_k >, \ k \in \mathbb{N}$. A function $\varphi \in C_b(H)$, is said to be differentiable in the direction $e_k, \ k \in \mathbb{N}$, if there exists the limit in $C_b(H)$

$$D_k \varphi(x) =: \lim_{h \to 0} \frac{1}{h} (\varphi(x + he_k) - \varphi(x)).$$

We call $D_k \varphi$ the *partial derivative* of φ in the direction e_k. In a similar way we define partial derivatives of any order.

   We are here concerned with the Heat equation

(3.1) $$\begin{cases} u_t(t,x) = \sum_{k=1}^{\infty} \lambda_k D_k^2 u(t,x), \ t > 0, \ x \in H, \\ \\ u(0,x) = \varphi(x), \ x \in H, \ \varphi \in C_b(H), \end{cases}$$

and with its finite dimensional approximation

(3.2) $$\begin{cases} u_t(t,x) = \frac{1}{2} \sum_{k=1}^{n} \lambda_k D_k^2 u(t,x), \ t > 0, \ x \in H, \ \varphi \in C_b(H), \\ \\ u(0,x) = \varphi(x), \ x \in H. \end{cases}$$

The following result can be easily proved.

Proposition 3.1. *For all $\varphi \in C_b(H)$ problem (3.2) has a unique classical solution given by*

$$(3.3) \qquad u_n(t,x) = \frac{1}{\sqrt{(2\pi t)^n \lambda_1 \cdots \lambda_n}} \int_{\mathbb{R}^n} e^{-\sum_{k=1}^n \frac{\xi_k^2}{2t\lambda_k}} \varphi \left(x - \sum_{k=1}^n \xi_i e_i \right) d\xi,$$

for $t > 0$, $\xi \in \mathbb{R}^n$.

We define

$$(3.4) \qquad P_t^n \varphi(x) = u_n(t,x), \ t \geq 0 , \ \xi \in \mathbb{R}^n.$$

As easily seen P_t^n, $t \geq 0$ is a strongly continuous analytic semigroup in $C_b(H)$. Moreover

$$(3.5) \qquad \|P_t^n \varphi\|_0 \leq \|\varphi\|_0, \ t \geq 0, \ \varphi \in C_b(H).$$

Now we want to see whether there exists the limit

$$\lim_{n \to \infty} u_n(t,x),$$

in $C_b(H)$ for all initial datum $\varphi \in C_b(H)$. We first show that a necessary condition for the existence of the limit is the following

$$(3.6) \qquad \sum_{k=1}^{\infty} \lambda_k < +\infty.$$

Note in fact that choosing $\varphi(x) = e^{-\frac{|x|^2}{2}}$, $x \in H$, we have, as easily checked,

$$u_n(t,x) = \left[\prod_{k=1}^n (1 + \lambda_k t) \right]^{-1/2} e^{-\frac{1}{2} \sum_{k=1}^n \frac{x_k^2}{1+\lambda_k t}} e^{-\frac{1}{2} \sum_{k=n+1}^{\infty} x_k^2}.$$

Now we remark that the infinite product $\prod_{k=1}^{\infty} (1 + t\lambda_k)$, $t > 0$, is finite if and only if (3.6) holds. Thus if (3.6) holds we have

$$\lim_{n \to \infty} u_n(t,x) = u(t,x), \ t \geq 0, \ x \in H,$$

where

$$u(t,x) = \left[\prod_{k=1}^{\infty} (1 + \lambda_k t) \right]^{-1/2} e^{-\frac{1}{2} \sum_{k=1}^{\infty} \frac{x_k^2}{1+\lambda_k t}},$$

whereas if does not hold

$$\lim_{n \to \infty} u_n(t,x) = \begin{cases} 1 \text{ if } x = 0, \ t \geq 0 \\ \\ 0 \text{ if } x \neq 0, t > 0, \end{cases}$$

and u_n does not converge to a continuous function.

From now on we assume that the condition (3.6) holds. In this case the linear operator Q defined by

$$(3.7) \qquad Qx = \sum_{k=1}^{\infty} \lambda_k x_k, \ x \in H,$$

belongs to $\mathcal{L}_1(H)$, and problem (3.1) can be rewritten in a more compact way as

$$(3.8) \qquad \begin{cases} u_t(t,x) = \dfrac{1}{2} \ \mathrm{Tr}\, [QD^2u(t,x)], \ t > 0, \ x \in H, \\[2mm] u(0,x) = \varphi(x), \ x \in H. \end{cases}$$

3.2 Construction of the Heat semigroup.

Beside the semigroups P_t^n, $t \geq 0$, defined by (3.4) we will need also the following one

$$(3.9) \quad S_t^n \varphi(x) = \frac{1}{\sqrt{(2\pi t)^n \lambda_n}} \int_{-\infty}^{+\infty} e^{-\frac{\xi^2}{2t\lambda_n}} \varphi(x - \xi_n e_n) \, d\xi, \ \varphi \in C_b(H), \ t > 0,$$

and $S_0^n = I$. Clearly

$$(3.10) \qquad P_t^n \varphi = \prod_{k=1}^{n} S_t^n \varphi, \ t > 0, \ \varphi \in C_b(H).$$

The following theorem is proved in [3]. We sketch the proof for the reader convenience.

Theorem 3.1. *For all* $\varphi \in C_b(H)$ *there exists the limit*

$$(3.11) \qquad P_t\varphi := \lim_{n \to \infty} P_t^n \varphi$$

in $C_b(H)$, *uniformly in* t *on the bounded subsets of* $[0, +\infty[$. *Moreover* P_t *is a strongly continuous contraction semigroup in* $C_b(H)$ *and*

$$P_t\varphi = \prod_{k=1}^{\infty} S_t^k \varphi, \ \varphi \in C_b(H).$$

Moreover the semigroup P_t *is order preserving, that is for all* $\varphi, \psi \in C_b(H)$ *the following holds.*

$$(3.13) \qquad \varphi(x) \leq \psi(x) \ \forall \, x \in H \Leftrightarrow P_t\varphi(x) \leq P_t\psi(x) \ \forall \, x \in H, \ t \geq 0.$$

Proof. We first prove the existence of the limit when $\varphi \in C_b^{1,1}(H)$. Since

$$P_t^n \varphi - P_t^{n-1} \varphi = \prod_{k=1}^{n} S_t^k \varphi - \prod_{k=1}^{n-1} S_t^k \varphi = \prod_{k=1}^{n-1} S_t^k = (S_t^n \varphi - \varphi),$$

it follows

(3.14) $$\left\|P_t^n\varphi - P_t^{n-1}\varphi\right\|_0 \le \left\|S_t^n\varphi - \varphi\right\|_0.$$

Moreover

$$S_t^n\varphi(x) - \varphi(x) = \frac{1}{\sqrt{2\pi\lambda_n t}} \int_{-\infty}^{+\infty} e^{-\frac{\zeta^2}{2\lambda_n t}} (\varphi(x - \zeta e_n) - \varphi(x))d\zeta$$

$$= \frac{1}{\sqrt{2\pi\lambda_n t}} \int_{-\infty}^{+\infty} e^{-\frac{\zeta^2}{2\lambda_n t}} \int_0^1 <D\varphi(x - \zeta(1-\rho)e_n), \zeta e_n > d\zeta d\rho.$$

Since

$$\int_{-\infty}^{+\infty} e^{-\frac{\zeta^2}{2\lambda_n t}} <D\varphi(x), \zeta e_n > d\zeta = <D\varphi(x), e_n> \int_{-\infty}^{+\infty} e^{-\frac{\zeta^2}{2\lambda_n t}} \zeta d\zeta = 0,$$

it follows

$$S_t^n\varphi(x) - \varphi(x) = \frac{1}{\sqrt{2\pi\lambda_n t}}$$

$$\times \int_{-\infty}^{+\infty} e^{-\frac{\zeta^2}{2\lambda_n t}} \int_0^1 <D[\varphi(x - \zeta(1-\rho)e_n) - \varphi(x)], \zeta e_n > d\zeta d\rho,$$

from which

$$|S_t^n\varphi(x) - \varphi(x)| \le \frac{1}{\sqrt{2\pi\lambda_n t}} \int_{-\infty}^{-\infty} |\zeta|^2 e^{-\frac{\zeta^2}{2\lambda_n t}} \|\varphi\|_{1,1} = \|\varphi\|_{1,1}\lambda_n t.$$

Then from (3.14) it follows

$$\left\|P_t^{n+p}\varphi - P_t^n\varphi\right\|_0 \le t\|\varphi\|_{1,1} \sum_{k=n+1}^{n+p} \lambda_k$$

for all $n, p \ge 1$. Consequently, $\{P_t^n\varphi\}_{n\in\mathbb{N}}$ is a Cauchy sequence in $C_b(H)$, uniformly for t in the bounded subsets of $[0, +\infty[$. Therefore the limit (3.11) exists for all $\varphi \in C_b^{1,1}(H)$. Since by Theorem 2.1, $C_b^{1,1}(H)$ is dense in $C_b(H)$, (3.11) follows. The other statements are easy to check ∎

REMARK 3.1. Unless P_t^n the semigroup P_t is neither analytic, nor differentiable, see P. Guiotto [8].

3.3 Integral representation of the heat semigroup. In this section we want to find an integral representation of the semigroup $P_t, t \ge 0$.

We recall that, by Theorem 3.1, we have

$$\|P_t\varphi - P_t^n\varphi\|_0 \le t \sum_{k=n+1}^{\infty} \lambda_k\|\varphi\|_{1,1}, \forall \varphi \in C_b^{1,1}(H).$$

To prove a representation result we need the following well known result on the Daniell representation theorem.

Theorem 3.2. *Let E be a complete separable metric space, and let $\mathcal{R}$ be a Riesz space of real functions defined in E (^{2}) including the function identically equal to 1. Let $L : \mathcal{R} \mapsto \mathbb{R}$ be a linear mapping such that*
(i) $\varphi \in \mathcal{R}, \varphi \geq 0 \Rightarrow L(\varphi) \geq 0$.
(ii) $L(1) = 1$.
(iii) It holds

$$\{\varphi_n\} \subset \mathcal{R},\ \varphi_n(x) \uparrow \varphi(x),\ \forall\, x \in E,\ \varphi \in \mathcal{R} \Rightarrow L(\varphi_n) \uparrow L(\varphi).$$

Then there exists a unique probability measure μ on $(E, \mathcal{F}(\mathcal{R}))$ such that, for any Borel bounded function φ

$$L(\varphi) = \int_E \varphi(x)\mu(dx),$$

where $\mathcal{F}(\mathcal{R})$ is the σ–algebra generated by $\mathcal{R}$.

We prove now the result.

Proposition 3.2. *Let $\varphi \in C_b(H)$ and let $\{\varphi^k\}$ be a sequence in $C_b(H)$ such that*

$$\varphi^k(x) \uparrow \varphi(x),\ \text{as } k \to \infty,\ \forall\, x \in H.$$

Then for all $t \geq 0$ we have

$$P_t\varphi^k(x) \uparrow P_t\varphi(x),\ \text{as } k \to \infty,\ \forall\, x \in H.$$

Proof. Denote by φ_α and φ_α^k the regularized of φ and φ^k given by (2.1). We have

$$P_t\varphi(x) - P_t\varphi^k(x) = I_1 + I_2 + I_3,$$

where

$$I_1 = P_t\varphi(x) - P_t\varphi_\alpha(x),\ I_2 = P_t\varphi_\alpha(x) - P_t\varphi_\alpha^k(x),\ I_3 = P_t\varphi_\alpha^k(x) - P_t\varphi^k(x).$$

Notice that by (2.1) we have

$$\varphi_\alpha^k(x) \leq \varphi^k(x),\ \forall\, x \in H,$$

and so $I_3 \leq 0$. It follows

$$(3.16) \qquad\qquad P_t\varphi(x) - P_t\varphi^k(x) \leq I_1 + I_2.$$

For any $\varepsilon > 0$ there exists $\alpha_\varepsilon > 0$ such that

$$(3.17)) \qquad\qquad I_1 = P_t\varphi(x) - P_t\varphi_\alpha(x) \leq \frac{\varepsilon}{2},\ \forall\, x \in H,\ \forall\, \alpha \leq \alpha_\varepsilon.$$

From (3.16) it follows

$$(3.18) \qquad\qquad P_t\varphi(x) - P_t\varphi_{\alpha_\varepsilon}(x) \leq \frac{\varepsilon}{2} + |P_t\varphi_{\alpha_\varepsilon}(x) - P_t\varphi_{\alpha_\varepsilon}^k(x)|.$$

---

^{2}that is such that $\varphi, \psi \in \mathcal{R} \Rightarrow f \wedge g \in \mathcal{R},\ f \vee g \in \mathcal{R}$.

But by (2.3) we have

$$[\varphi_{\alpha_\varepsilon}]_{1,1} \leq \frac{1}{\alpha_\varepsilon}, \; [\varphi_{\alpha_\varepsilon}^k]_{1,1} \leq \frac{1}{\alpha_\varepsilon},$$

and so, for some constant C_ε

$$\|\varphi_{\alpha_\varepsilon}\|_{1,1} \leq C_\varepsilon, \; \|\varphi_{\alpha_\varepsilon}^k\|_{1,1} \leq C_\varepsilon.$$

Now we have

$$|P_t\varphi_{\alpha_\varepsilon}(x) - P_t\varphi_{\alpha_\varepsilon}^k(x)| \leq |P_t\varphi_{\alpha_\varepsilon}(x) - P_t^n\varphi_{\alpha_\varepsilon}(x)| + |P_t^n\varphi_{\alpha_\varepsilon}(x) - P_t^n\varphi_{\alpha_\varepsilon}^k(x)|$$

$$+ |P_t^n\varphi_{\alpha_\varepsilon}^k(x) - P_t\varphi_{\alpha_\varepsilon}^k(x)|.$$

By (3.18) it follows

$$|P_t\varphi_{\alpha_\varepsilon}(x) - P_t\varphi_{\alpha_\varepsilon}^k(x)| \leq 2t \sum_{k=n+1}^{\infty} \lambda_k + |P_t^n\varphi_{\alpha_\varepsilon}(x) - P_t^n\varphi_{\alpha_\varepsilon}^k(x)|.$$

Let finally $n_\varepsilon \in \mathbb{N}$ such that $2tC_\varepsilon \sum_{k=n+1}^{\infty} \lambda_k \leq \frac{\varepsilon}{4}$. By the Dominated Convergence Theorem it follows

$$\lim_{k\to\infty} |P_t^{n_\varepsilon}\varphi_{\alpha_\varepsilon}(x) - P_t^{n_\varepsilon}\varphi_{\alpha_\varepsilon}^k(x)| = 0,$$

and so, due to the arbitrariness of ε we can conclude that

$$P_t\varphi(x) - P_t\varphi_k(x) \to 0, \forall \, x \in H.$$

∎

We can prove now the result

Theorem 3.3. *We have*

(3.19) $$P_t\varphi(x) = \int_H \varphi(y) \, \mathcal{N}(x, tQ)(dy)$$

Proof. First we note that $C_b(H)$ is a Riesz space including the function identically equal to 1 and that the σ–algebra generated by $C_b(H)$ is $\mathcal{B}(H)$. Let $t > 0$ and $x \in H$ be fixed and denote by $L = L_{t,x}$ the mapping

$$L_{t,x} : C_b(H) \mapsto \mathbb{R}, \; \varphi \mapsto P_t\varphi(x).$$

In view of Proposition 3.2 the mapping $L_{t,x}$ fulfills the hypotheses of Theorem 3.2 and so there exists a Borel measure $\mu_{t,x}$ in $(H, \mathcal{B}(H))$ such that

$$P_t\varphi(x) = \int_H \varphi(y)\mu_{t,x}(dy), \; \forall \, \varphi \in C_b(H).$$

It remains to check that $\mu_{t,x} = \mathcal{N}(x, tQ)$. To prove this we first remark that for all $n \in \mathbb{N}$ we have

(3.20) $$P_t^n(e^{i<h,\cdot>})(x) = e^{i<h,x> - \frac{t}{2}<Q_nh,h>}, \; x, h \in H,$$

where

$$Q_n x = \sum_{k=1}^{n} \lambda_k < x, e_k > e_k, \ x \in H.$$

This can be seen easily by checking that $u(t,x) = e^{i<h,x>-\frac{t}{2}<Q_n h,h>}$ is the solution to (3.2) corresponding to initial datum $\varphi = e^{i<h,\cdot>}$. Letting now n tend to infinity in (3.21), we find

$$P_t(e^{i<h,\cdot>})(x) = e^{i<h,x>-\frac{t}{2}<Qh,h>}, \ x, h \in H.$$

Now we compute the characteristic function of the measure $\mu_{t,x}$. We have

$$\int_H e^{i<h,y>} \mu_{t,x}(dy) = e^{i<h,x>-\frac{t}{2}<Qh,h>}, \ x, h \in H.$$

This implies $\mu_{t,x} = \mathcal{N}(x, tQ)$ as required. ∎

3.4 Regularity properties of the heat semigroup. This section is devoted to study regularity properties of P_t, $t \geq 0$. We first prove that $P_t\varphi$ is differentiable in any direction e_k.

Theorem 3.4. *Let $\varphi \in C_b(H)$ and let $t > 0$. Then there exists $D_k P_t\varphi$ for all $k \in \mathbb{N}$ and*

$$(3.21) \qquad D_k P_t\varphi = \prod_{n=1, \ n\neq k}^{\infty} S_t^n D_k S_t^k \varphi.$$

Moreover

$$(3.22) \qquad D_k P_t\varphi(x) = \frac{1}{\lambda_k t} \int_H y_k \varphi(x+y) \mathcal{N}(0, tQ)(dy), \ x \in H.$$

Proof.   Setting $\eta = x_k - \xi$ in (3.9) (with n replaced by k) we find

$$S_t^k \varphi(x) = \frac{1}{\sqrt{2\pi\lambda_k t}} \int_{-\infty}^{+\infty} e^{-\frac{(x_k-\eta)^2}{2\lambda_k t}} \varphi(x + (\eta - x_k)e_k) d\eta.$$

It follows

$$S_t^k \varphi(x) = -\frac{1}{\lambda_k t} \frac{1}{\sqrt{2\pi\lambda_k t}} \int_{-\infty}^{+\infty} (x_k - \eta) e^{-\frac{(x_k-\eta)^2}{2\lambda_k t}} \varphi(x + (\eta - x_k)e_k) d\eta$$

$$(3.23) \qquad = -\frac{1}{\lambda_k t} \frac{1}{\sqrt{2\pi\lambda_k t}} \int_{-\infty}^{+\infty} \xi e^{-\frac{\xi^2}{2\lambda_k t}} \varphi(x - \xi e_k) d\xi$$

$$= \frac{1}{\lambda_k t} \frac{1}{\sqrt{2\pi\lambda_k t}} \int_{-\infty}^{+\infty} \xi e^{-\frac{\xi^2}{2\lambda_k t}} \varphi(x + \xi e_k) d\xi$$

We are now ready to prove (3.22). If $n > k$ we have

$$D_k P_t^n \varphi = \prod_{l=1, \ l\neq k}^{n} S_t^l D_k S_t^k \varphi.$$

By Theorem 3.1, with the Hilbert space H replaced by $e_k^\perp$, we find

$$\lim_{n \to \infty} D_k P_t^n \varphi = \prod_{l=1,\; l \neq k}^{\infty} S_t^l D_k S_t^k \varphi.$$

Since D_k is closed we see that (3.22) holds. Finally (3.23) follows from the representation formula (3.20) (with the Hilbert space H replaced by $e_k^\perp$), formula (3.24) and Fubini's Theorem. ■

In a similar way we can prove that $P_t \varphi$ is twice differentiable in any pair of directions e_h, e_k. The proof is left as an exercise to the reader.

Theorem 3.5. *Let $\varphi \in C_b(H)$ and let $t > 0$. Then there exists $D_h D_k P_t \varphi$ for all $h, k \in \mathbb{N}$ and*

$$(3.24) \qquad D_h D_k P_t \varphi = \prod_{n=1,\; n \neq h,k}^{\infty} S_t^n D_h D_k S_t^k \varphi,$$

and

$$D_h D_k P_t \varphi(x) = \frac{1}{\lambda_h \lambda_k t^2} \int_H y_h y_k \varphi(x + y) \mathcal{N}(0, tQ)(dy)$$

$$(3.25)$$

$$- \frac{\delta_{h,k}}{\lambda_h t} P_t \varphi(x), \; x \in H.$$

3.5. Global regularity properties of the heat semigroup. In this subsection we want to replace (3.23) and (3.26) by global identities, not involving coordinates y_k. For this we have to introduce new spaces $C_Q^1(H)$ and $C_Q^2(H)$.

Definition 3.1. *$C_Q^1(H)$ is the set of all $\varphi \in C_b(H)$ such that*
(i) There exists $D_k \varphi$ for all $k \in \mathbb{N}$.
(ii) $\displaystyle \sup_{x \in H} \sum_{k=1}^{\infty} \lambda_k |D_k \varphi(x)|^2 < +\infty.$
(iii) The mapping $D_Q \varphi$ defined as

$$D_Q \varphi : H \mapsto H, \; x \mapsto \sum_{k=1}^{\infty} \sqrt{\lambda_k} D_k \varphi(x) e_k,$$

is uniformly continuous.

It is easy to see that $C_b^1(H) \subset C_Q^1(H)$ and that if $\varphi \in C_b^1(H)$ then

$$D_Q \varphi(x) = Q^{1/2} D\varphi(x), \; x \in H.$$

Definition 3.2. $C_Q^2(H)$ *is the set of all* $\varphi \in C_b(H)$ *such that*
(i) *There exists* $D_h D_k \varphi$ *for all* $h, k \in \mathbb{N}$.
(ii) $\displaystyle\sup_{x \in H} \sum_{h=1}^{\infty} \left(\sum_{k=1}^{\infty} \sqrt{\lambda_h \lambda_k} D_h D_k \varphi(x) y_k \right)^2 \leq C|y|^2$, *for all* $y \in H$ *and some* $C > 0$.
(iii) *The mapping* $D_Q^2 \varphi$ *defined as*

$$D_Q^2 \varphi : H \mapsto \mathcal{L}(H), \; x \mapsto D_Q^2 \varphi,$$

where

$$< D_Q^2 \varphi(x) z, z > = \sum_{h,k=1}^{\infty} \sqrt{\lambda_h \lambda_k} D_h \varphi(x) D_k \varphi(x) z_h z_k, \; z \in H,$$

is uniformly continuous.

The proof of the following result is straightforward and it is left as an exercise to the reader.

Proposition 3.4. *Assume that here exists* $D_h D_k \varphi$ *for all* $h, k \in \mathbb{N}$, *and let* $D_{Q_n}^2 \varphi(x)$ *be the linear operator in* H *defined by*

$$< D_{Q_n}^2 \varphi(x) \alpha, \beta > = \sum_{h,k=1}^{n} \sqrt{\lambda_h \lambda_k} D_h D_k \varphi(x) \alpha_h \beta_k, \; \alpha, \beta \in H.$$

Assume that
(i) *There exists* $C > 0$ *such that*

$$\left| < D_{Q_n}^2 \varphi(x) \alpha, \beta > \right| \leq C|\alpha| \, |\beta|, \; \alpha, \beta, x \in H.$$

(ii) *There exists the* $\displaystyle\lim_{n \to \infty} < D_{Q_n}^2 \varphi(x) \alpha, \beta >$, *in* $C_b(H)$, *for all* $\alpha, \beta \in H$.
Then $\varphi \in C_Q^2(H)$ *and*

$$\lim_{n \to \infty} < D_{Q_n}^2 \varphi(x) \alpha, \beta > = < D_Q^2 \alpha, \beta >, \; \text{in } C_b(H), \; \alpha, \beta \in H.$$

We are now ready to prove

Theorem 3.6. *Let* $\varphi \in C_b(H)$ *and* $t > 0$. *Then* $P_t \varphi \in C_Q^1(H)$ *and*

$$(3.26) \quad < D_Q P_t \varphi(x), \alpha > = \frac{1}{t} \int_H < \alpha, Q^{-1/2} y > \varphi(x + y) \, \mathcal{N}(0, tQ)(dy), \; \alpha \in H.$$

Moreover the following estimate holds.

$$(3.27) \qquad |D_Q P_t \varphi(x)| \leq \frac{1}{\sqrt{t}} \, \|\varphi\|_0, \; x \in H.$$

Proof.   We have to check properties (i)–(iii) of Definition 3.1. Property (i) follows from Theorem 3.4, let us check (ii). Let $\alpha \in H$, then by (3.23) we have

$$\sum_{k=1}^{n} \sqrt{\lambda_k} D_k \varphi(x) \alpha_k = \sum_{k=1}^{n} \frac{1}{t \sqrt{\lambda_k}} \int_H y_k \alpha_k \varphi(x + y) \mathcal{N}(0, tQ)(dy).$$

It follows, by Hölder's inequality

$$\left| \sum_{k=1}^{n} \sqrt{\lambda_k} D_k \varphi(x) \alpha_k \right|^2 \leq \frac{\|\varphi\|_0^2}{t^2} \int_H \left(\sum_{k=1}^{n} \frac{y_k \alpha_k}{\sqrt{\lambda_k}} \right)^2 \mathcal{N}(0, tQ)(dy)$$

$$= \frac{\|\varphi\|_0^2}{t^2} \sum_{h,k=1}^{n} \frac{\alpha_h \alpha_k}{\sqrt{\lambda_h \lambda_k}} \int_H y_h y_k \, \mathcal{N}(0, tQ)(dy)$$

$$= \frac{\|\varphi\|_0^2}{t^2} \sum_{k=1}^{n} \frac{\alpha_k^2}{\lambda_k} \int_H y_k^2 \, \mathcal{N}(0, tQ)(dy) = \frac{\|\varphi\|_0^2}{t^2} \sum_{k=1}^{n} \alpha_k^2.$$

By the arbitrariness of α it follows

$$(3.28) \qquad \sum_{k=1}^{n} |D_k \varphi(x)|^2 \leq \frac{\|\varphi\|_0^2}{t^2}, \ n \in \mathbb{N},$$

and the assumption (ii) of Definition 3.1 is fulfilled.

Let us prove (iii). By (3.28) it follows that the series

$$D_Q \varphi(x) = \sum_{k=1}^{\infty} \sqrt{\lambda_k} D_k \varphi(x)$$

is convergent and we have, recalling Proposition 2.2,

$$< D_Q \varphi(x), \alpha > = \frac{1}{t} \int_H < \alpha, Q^{-1/2} y > \varphi(x+y) \, \mathcal{N}(0, tQ)(dy), \ \alpha \in H.$$

If $x, \overline{x} \in H$, we have

$$< D_Q \varphi(x) - D_Q \varphi(\overline{x}), \alpha > = \frac{1}{t} \int_H < \alpha, Q^{-1/2} y > (\varphi(x+y) - \varphi(\overline{x}+y)) \, \mathcal{N}(0, tQ)(dy).$$

It follows, by Hölder's inequality

$$|< D_Q \varphi(x) - D_Q \varphi(\overline{x}), \alpha >|^2$$

$$\leq \frac{\omega_\varphi(x - \overline{x})}{t^2} \int_H |< \alpha, Q^{-1/2} y >|^2 \, \mathcal{N}(0, tQ)(dy) \leq \frac{\omega_\varphi(x - \overline{x})}{t^2} |\alpha|^2.$$

This implies

$$|D_Q \varphi(x) - D_Q \varphi(\overline{x})| \leq \frac{\omega_\varphi(x - \overline{x})}{t^2},$$

so that $\varphi \in C_Q^1(H)$. ∎

Next we give a global version of formula (3.26).

Theorem 3.7. Let $\varphi \in C_b(H)$ and $t > 0$. Then $P_t\varphi \in C_Q^2(H)$ and
(3.29)

$$< D_Q^2 P_t\varphi(x)\alpha, \beta > \leq \frac{1}{t^2} \int_H < \alpha, Q^{-1/2}y >< \beta, Q^{-1/2}y > \varphi(x+y) \, \mathcal{N}(0, tQ)(dy)$$

$$-\frac{1}{t} < \alpha, \beta > P_t\varphi(x), \ \alpha, \beta \in H.$$

Moreover the following estimates hold

(3.30)
$$\|D_Q^2 P_t\varphi(x)\| \leq \frac{\sqrt{2}}{t} \|\varphi\|_0, \ x \in H,$$

(3.31)
$$\|D_Q^2 P_t\varphi(x)\| \leq \frac{1}{\sqrt{t}} \|D_Q\varphi\|_0, \ x \in H.$$

Proof. It is easy to check that, for any $n \in \mathbb{N}$

$$< D_{Q_n}^2 P_t\varphi(x)\alpha, \beta > = \frac{1}{t^2} \int_H < \alpha, Q_n^{-1/2}y >< \beta, Q_n^{-1/2}y > \varphi(x+y) \, \mathcal{N}(0, tQ_n)(dy)$$

$$-\frac{1}{t} < \alpha, \beta > P_t^n\varphi(x), \ \alpha, \beta \in H.$$

Now (3.29) follows from Proposition 3.3. Let us prove (3.30). By (3.29) we have

$$< D_Q^2 P_t\varphi(x)\alpha, \alpha > = \frac{1}{t^2} \int_H \left\{ | < \alpha, Q^{-1/2}y > |^2 - t|\alpha|^2 \right\} \varphi(x+y) \, \mathcal{N}(0, tQ)(dy).$$

By Hölder's inequality, it follows

$$| < D_Q^2 P_t\varphi(x)\alpha, \alpha > |^2 = \frac{\|\varphi\|^2}{t^4} \int_H | \left[< \alpha, Q^{-1/2}y > |^2 - t \right]^2 \mathcal{N}(0, tQ)(dy),$$

and estimate (3.30) follows by computing the integral above. Finally (3.31) is an easy consequence of Theorem 3.6. ∎

Let $\theta \in]0, 1[$. We denote by $C_Q^\theta(H)$ the space of all functions $\varphi \in C_b(H)$ such that

$$|\varphi|_{\theta,Q} = \sup_{x,y \in H} \frac{|\varphi(Q^{1/2}x) - \varphi(Q^{1/2}y)|}{|x - y|^\theta} < +\infty.$$

Space $C_Q^\theta(H)$, endowed with the norm

$$\|\varphi\|_{\theta,Q} = \|\varphi\|_0 + |\varphi|_{\theta,Q},$$

is a Banach space. The following characterization is proved in [3, Theorem 2.1]

(3.32)
$$(C_b(H), C_Q^1(H))_{\theta,\infty} = C_Q^\theta(H), \ \forall \, \theta \in]0, 1[,$$

where $(C_b(H), C_Q^1(H))_{\theta,\infty}$ represents a real interpolation space , see H. Triebel [15]. Using this result and interpolating between (3.30) and (3.32) we find

Proposition 3.4. Let $\varphi \in C_b(H)$, $\theta \in]0,1[$ and $t > 0$. Then we have

$$(3.33) \qquad \|D_Q^2 P_t\varphi(x)\| \leq \frac{2^{\theta/2}}{t^{\frac{\theta+1}{2}}} \|\varphi\|_{\theta,Q}.$$

We end this section by an additional regularity result, similar to one of Gross [7], that will be useful later. For this we need the following result proved in N. Dunford and J.T. Schwartz [6, Lemma 14 (a), page 1098].

Lemma 3.1. Let $T \in \mathcal{L}(H)$. Assume that there exists $K > 0$ such that for all finite rank linear bounded operators N in $\mathcal{L}(H)$ one has

$$(3.34) \qquad |\text{Tr } (NT)| \leq K\|N\|.$$

Then $T \in \mathcal{L}_1(H)$ and

$$(3.35) \qquad \|T\|_{\mathcal{L}_1(H)} \leq K.$$

We prove now the result

Theorem 3.8. Let $\varphi \in C_b^1(H)$ and $t > 0$. Then $D_Q^2 P_t\varphi(x) \in \mathcal{L}_1(H)$ and

$$(3.36) \qquad \text{Tr } [D_Q^2 P_t\varphi(x)] = \frac{1}{t} \int_H < D(\varphi(x+y)), y > \mathcal{N}(0, tQ)(dy), \ x \in H.$$

Moreover $\text{Tr } [D_Q^2 P_t\varphi(\cdot)] \in C_b(H)$ and

$$(3.37) \qquad \|D_Q^2 P_t\varphi\|_{\mathcal{L}_1(H)} \leq \frac{1}{\sqrt{t}} \|\varphi\|_0.$$

Proof.   Since $\varphi \in C_b^1(H)$ we have for any $\alpha \in H$,

$$< DP_t\varphi(x), Q^{1/2}\alpha >= \int_H < D\varphi(x+y), Q^{1/2}\alpha > \mathcal{N}(0, tQ)(dy) = P_t\psi(x),$$

where

$$\psi(x) =< DP_t\varphi(x), \alpha >, \ x \in H.$$

Let now $\beta \in H$. Then from Theorem 3.6 it follows

$$< D_Q P_t\psi(x), \beta >= \frac{1}{t} \int_H \psi(x+y) < \beta, Q^{-1/2}y > \mathcal{N}(0, tQ)(dy)$$

$$= \frac{1}{t} \int_H < D\varphi(x+y), Q^{1/2}\alpha >< \beta, Q^{-1/2}y > \mathcal{N}(0, tQ)(dy).$$

On the other hand it is easy to check that

$$< D_Q P_t\psi(x), \beta >=< D_Q^2 P_t\varphi(x)\alpha, \beta >,$$

so that we have

$$< D_Q^2 P_t \varphi(x)\alpha, \beta >$$

$$= \frac{1}{t} \int_H < D\varphi(x+y), Q^{1/2}\alpha >< \beta, Q^{-1/2}y > \mathcal{N}(0, tQ)(dy).$$

Let now $N \in \mathcal{L}(H)$ be of finite rank operator. Then we have

$$< N D_Q^2 P_t \varphi(x)\alpha, \beta >$$

$$= \frac{1}{t} \int_H < D\varphi(x+y), Q^{1/2}\alpha >< N^*\beta, Q^{-1/2}y > \mathcal{N}(0, tQ)(dy).$$

It follows

$$\text{Tr } [N D_Q^2 P_t] = \frac{1}{t} \int_H < D\varphi(x+y), Q^{1/2}N Q^{-1/2}y > \mathcal{N}(0, tQ)(dy).$$

By Hölder inequality we find

$$\left| \text{Tr } [N D_Q^2 P_t] \right|^2 \leq \frac{\|\varphi\|_1^2}{t^2} \int_H |Q^{1/2}N Q^{-1/2}y|^2 \, \mathcal{N}(0, tQ)(dy).$$

Since

$$\int_H |Q^{1/2}N Q^{-1/2}y|^2 \, \mathcal{N}(0, tQ)(dy) = t \, \text{Tr } (NQ) \leq t\|N\| \, \text{Tr } Q,$$

the conclusion follows from Lemma 3.1. ∎

4. Elliptic equations

We shall denote by $\mathcal{A}$ the infinitesimal generator of the strongly continuous semi-group P_t, $t \geq 0$. Since

$$\|P_t \varphi\|_0 \leq \|\varphi\|_0, \ t \geq 0,$$

by the Hille–Yosida Theorem we have that the spectrum of $\mathcal{A}$ is contained in $\{\lambda \in \mathbf{C} : \text{ Re } \lambda \leq 0\}$. Moreover for any $\lambda > 0$ the resolvent of $\mathcal{A}$ is given by

$$(4.1) \qquad R(\lambda, \mathcal{A})\varphi = \int_0^{+\infty} e^{-\lambda t} P_t \varphi \, dt.$$

We want now to describe the relationship between $\mathcal{A}$ and the differential operator $\mathcal{A}_0$ defined by

$$(4.2) \qquad \begin{cases} D(\mathcal{A}_0) = \{\varphi \in C_Q^2(H) : \ D_Q^2\varphi(x) \in \mathcal{L}_1(H), \ \text{Tr } [D_Q^2\varphi] \in C_b(H)\} \\[2mm] \mathcal{A}_0\varphi = \frac{1}{2}\text{Tr } [D_Q^2\varphi]. \end{cases}$$

Proposition 4.1. *The following statements hold.*

(i) $D(\mathcal{A}_0)$ *is dense in* $C_b(H)$.

(ii) $\mathcal{A}$ *is an extension of* $\mathcal{A}_0$.

(iii) $\mathcal{A}$ *is the closure of* $\mathcal{A}_0$.

Proof.   (i)–Let $\varphi \in C_b^1(H)$, then $P_t\varphi \in D(\mathcal{A}_0)$ by Theorem 3.8. Thus $D(\mathcal{A}_0)$ is dense in $C_b^1(H)$ and consequently in $C_b(H)$, since $C_b^1(H)$ is dense in $C_b(H)$ by Theorem 2.1.

(ii)–Let $\varphi \in D(\mathcal{A}_0)$. Define

$$g(t) = P_t\varphi, \quad g_n(t) = P_t^n\varphi, \ t \in [0,1].$$

By Theorem 3.1 we have

$$g_n \to g \ \text{in} \ C([0,1]; C_b(H)).$$

Moreover, since

$$\frac{d}{dt}g_n(t) = \frac{1}{2}\sum_{k=1}^{n}\lambda_k D_k^2 g_n(t) = \frac{1}{2}\sum_{k=1}^{n}\lambda_k D_k^2 P_t^n\varphi = P_t^n\left(\frac{1}{2}\sum_{k=1}^{n}\lambda_k D_k^2\varphi\right),$$

we have

$$\frac{d}{dt}g_n \to P_t(\mathcal{A}_0\varphi) \ \text{in} \ C([0,1]; C_b(H)).$$

This implies

$$\frac{d}{dt}g(t) = P_t(\mathcal{A}_0\varphi).$$

Setting now $t = 0$, we see that $\varphi \in D(\mathcal{A})$ and $\mathcal{A}\varphi = \mathcal{A}_0\varphi$.

(iii)– Let $\varphi \in D(\mathcal{A})$. We have to show that there exists a sequence included in $D(\mathcal{A}_0)$ such that

(4.3) $\varphi_n \to \varphi, \ \mathcal{A}_0\varphi_n \to \mathcal{A}\varphi \ \text{in} \ C_b(H),$

as $n \to \infty$.

Fix $\lambda > 0$ and set

$$\psi = \lambda\varphi - \mathcal{A}\varphi.$$

By Theorem 2.1 there exists a sequence $\{\psi_n\} \subset C_b^1(H)$ such that

$$\psi_n \to \psi \ \text{in} \ C_b(H),$$

as $n \to +\infty$. We are going to show that, setting $\varphi_n = R(\lambda, \mathcal{A})\psi_n$, the sequence $\{\varphi_n\}$ fulfills (4.3). Clearly $\varphi_n \to \varphi$ as $n \to +\infty$. Moreover by Theorem 3.8, $\varphi_n \in D(\mathcal{A}_0)$ and

$$\|\mathcal{A}_0\varphi_n\|_0 \le \int_0^{+\infty}\frac{dt}{\sqrt{t}}\|\psi_n\|_1.$$

Now (4.3) follows easily. ∎

Let us consider now the equation

(4.4) $\lambda\varphi - \mathcal{A}\varphi = g$

with $\lambda \ge 0$ fixed.

The function $\varphi = \int_0^{+\infty} e^{-\lambda t}P_t g$, is called a *strong solution* to (4.4). If $\varphi \in D(\mathcal{A}_0)$ we say that it is a *strict solution* to equation (4.4).

We study now some regularity properties of the strong solution φ of (4.4).

Theorem 4.1. Let $\lambda > 0$, $g \in C_b(H)$ and let φ the strong solution to (4.4). Then the following statements hold.

(i) If $g \in C_Q^\theta(H)$, $\theta \in]0,1[$, then φ belongs to $C_Q^2(H)$ and

$$(4.5) \qquad\qquad D_Q^2 \varphi = \int_0^{+\infty} e^{-\lambda t} D_Q^2 P_t g \, dt.$$

(ii) If $g \in C_b^1(H)$, $\theta \in]0,1[$, then $D_Q^2 \varphi(x) \in \mathcal{L}_1(H)$ for all $x \in H$ and

$$(4.6) \qquad\qquad \mathcal{A}\varphi(x) = \frac{1}{2} \, \mathrm{Tr} \, [D_Q^2 \varphi(x)].$$

Moreover the solution φ is strict.

Proof. (i) follows from 3.33 and (ii) from Theorem 3.8. ∎

To state Schauder estimates we need a last notation

We denote by $C_Q^{2+\theta}(H)$ the space of all functions $f \in C_Q^2(H)$ such that

$$|f|_{2+\theta,Q} := \sup_{x,x',y\in H, y\neq 0} \frac{\|D_Q^2 f(Q^{1/2}x) - D_Q^2 f(Q^{1/2}x')\|_{\mathcal{L}(H)}}{|x - x'|^\theta} < +\infty.$$

Space $C_Q^{2+\theta}(H)$, endowed with the norm

$$\|f\|_{2+\theta,Q} = \|f\|_{1,Q} + |f|_{2,Q} + |f|_{2+\theta,Q},$$

is a Banach space.

The following result in proved in [3]

Theorem 4.2. Let $\lambda > 0$, $\theta \in]0,1[$ and $g \in C_b^\theta(H)$. Then $\varphi \in C_Q^{2+\theta}(H)$.

5. Potential

If $H = \mathbb{R}^n$ and $g \in C_b(H)$ has a bounded support then it is well-known that there exists a unique $\varphi \in C_b^{1+\theta}(H)$ such that

$$-\frac{1}{2}\Delta\varphi = g.$$

Moreover, if $n \geq 3$ one has

$$\varphi(x) = C_n \int_{\mathbb{R}^n} \frac{g(y)}{|x - y|^{n-2}} \, dy = \int_0^\infty P_t g(x) dt.$$

We will show now that in the infinite dimensional case we can still define the potential of any function $g \in C_b(H)$ having a bounded support. For this we need a result on the asymptotic behaviour in time of $P_t g$.

Lemma 5.1. *Let* $g \in C_b(H)$ *with a bounded support* K. *Then for all* $n \in \mathbb{N}$ *there exists a constant* $C_N = C_N(K)$ *such that*

(i) It holds

$$(5.1) \qquad |P_t g(x)| \leq C_N \, t^{-N/2} \|g\|_0,$$

$$(5.2) \qquad |D_Q P_t g(x)| \leq C_N \, t^{-(N+1)/2} \|g\|_0,$$

$$(5.3) \qquad \|D_Q^2 P_t g(x)\| \leq C_N \, t^{-(N+2)/2} \|g\|_0.$$

(ii) If $g \in C_b^1(H)$, *we have*

$$\|P_t g\|_1 \leq C_N \, t^{-N/2} \|D_Q g\|_0.$$

(iii) If $g \in C_b^\theta(H)$, $\theta \in]0,1[$, *we have*

$$\|P_t g\|_\theta \leq C_N \, t^{-N/2} \|g\|_{\theta,Q}.$$

Proof. We first prove (5.1), following H. Kuo [9, Lemma 4.5]. We have

$$P_t g(x) = \int_H g(x+y) \mathcal{N}(0, tQ)(dy) = \int_K g(y) \mathcal{N}(x, tQ)(dy),$$

where $\mathcal{N}(x, tQ)$ is the Gaussian measure with mean x and covariance operator tQ. It follows

$$P_t g(x) \leq \|g\|_0 \int_K \mathcal{N}(x, tQ)(dy).$$

Set

$$L = \sup_{z \in K}\{| < z, e_k > |, \; k = 1, ..., N\}$$

and

$$\Lambda = \{x \in H : \; |x_k| \leq L\}.$$

We have clearly $K \subset \Lambda$. Since Λ is a cylindrical set, one can compute easily the integral

$$\int_K \mathcal{N}(x, tQ)(dy) = (2\pi t)^{-N/2} (\lambda_1 ... \lambda_N)^{-1/2} \prod_{k=1}^N \int_{-L}^L e^{-\frac{|y_k - x_k|^2}{2\lambda_k t}} \, dy_k.$$

Thus

$$|P_t g(x)| \leq \|g\|_0 \frac{(2L)^N}{(2\pi)^{N/2} \lambda_1 \ldots \lambda_N} t^{-N/2}.$$

The proof of (5.2), and (5.3), are similar. Also (5.4) follows in the same way, starting from the formula

$$DP_t g(x) = \int_H Dg(x+y) \mathcal{N}(0, tQ)(dy) = \int_K Dg(y) \mathcal{N}(x, tQ)(dy).$$

Finally (5.5) follows by interpolation. ∎

Theorem 5.1. *Let $g \in C_b^\theta(H)$ with bounded support and with $\theta \in]0, 1[$, and let*

$$(5.6) \qquad \varphi(x) = \int_0^{+\infty} P_t g(x) \, dt.$$

Then the following statements hold
(i) $\varphi \in D(\mathcal{A})$ and $\mathcal{A}\varphi = g$.
(ii) φ is the unique strong solution to equation (5.1).
(iii) $\varphi \in C_Q^{2+\theta}(H)$.

Proof. The first statement follows by Lemma 5.1 computing the limit

$$\lim_{s \to 0} \frac{1}{s}(P_s \varphi - \varphi).$$

To prove *(ii)* fix $\lambda > 0$, then φ is the strong solution to the equation

$$(5.7) \qquad \lambda\varphi - \mathcal{A}\varphi = \lambda\varphi - g =: g_1.$$

Moreover it is easy to see that this fact implies that φ is the strong solution to (5.1). The last statement follows from Theorem 4.2 since, by Lemma 5.1, $g_1 \in C_b^\theta(H)$. ∎

REMARK 5.1. In [7] and in [9] are essentially proved, in the setting of Wiener spaces, statements *(i)* and *(ii)*. Statement *(iii)* is new.

REFERENCES

1. Cannarsa P. & Da Prato G., *A semigroup approach to Kolmogoroff equations in Hilbert spaces*, Appl. Math. Letters **4**, 1 (1991), 49–52.
2. Cannarsa P. & Da Prato G., *On a functional analysis approach to parabolic equations in infinite dimensions*, J. Funct. Anal. **118**,1 (1993), 22–42..
3. Cannarsa P. & Da Prato G., *Infinite Dimensional Elliptic Equations with Hölder continuous coefficients*, Advances in Differential equations (to appear).
4. Cannarsa P. & Da Prato G., *Schauder estimates for Kolmogorov equations in Hilbert spaces*, Capri.
5. Daleckij Ju. L., *Differential equations with functional derivatives and stochastic equations for generalized random processes*, Dokl. Akad. Nauk SSSR **166** (1966), 1035–1038.
6. Dunford N. & Schwartz J. T., *Linear Operators*, vol. II, 1956.
7. Gross L, *Potential Theory in Hilbert spaces*, J.Funct. Analysis **1** (1965), 123-189.
8. Guiotto P., *Nondifferentiability of the Heat semigroup in infinite dimensional Hilbert spaces*, preprint.
9. Kuo H.H, *Gaussian Measures in Banach Spaces*, Springer-Verlag, 1975.
10. Lasry J. M. & Lions P. L., *A remark on regularization in Hilbert spaces*, Israel J. Math. **55** 3 (1986), 257–266.
11. Lunardi A., *Analytic semigroups and optimal regularity in parabolic problems*, Birkhauser Verlag, Basel, 1995.
12. Nemirovski A.S. & Semenov S.M., *The polynomial approximation of functions in Hilbert spaces*, Mat. Sb. (N.S) **92** 134 (1973), 257-281.
13. Piech A., *Regularity of the Green's operator on Abstract Wiener Space*, J. Diff. Eq. **12** (1969), 353–360..
13. Piech A., *A fundamental solution of the prabolic equation on Hilbert space*, J. Funct. An. **3** (1972), 85–114.
14. Triebel H., *Interpolation Theory, Function Spaces, Differential Operators*, North-Holland publaddr Amsterdam, 1986.

PIERMARCO CANNARSA
DIPARTIMENTO DI MATEMATICA, UNIVERSITÀ DI ROMA TOR VERGATA
VIA DELLA RICERCA SCIENTIFICA, 00133, ROMA, ITALY
E–mail: Cannarsa@axp.mat.utovrm.it

GIUSEPPE DA PRATO
SCUOLA NORMALE SUPERIORE DI PISA
PIAZZA DEI CAVALIERI 6, 56126 PISA, ITALY
E–mail: Daprato@sabsns.sns.it

Proceedings of Symposia in Applied Mathematics
Volume **54**, 1998

RECENT DEVELOPMENTS IN THE THEORY OF THE
BOLTZMANN EQUATION

CARLO CERCIGNANI

ABSTRACT. In the last few years the theory of the nonlinear Boltzmann equation has witnessed a veritable torrent of contributions, spurred by the basic result of DiPerna and Lions. This lecture is an attempt to survey the recent developments in various mathematical aspects of kinetic theory with particular attention to the theory of existence of solutions, the problem of trend to equilibrium and numerical methods of solution.

1. INTRODUCTION.

This beautiful meeting is intended to celebrate the 70th birthday of Peter Lax and Louis Nirenberg, two distinguished mathematicians of Courant. This circumstance brought my memory back to the days when Harold Grad, another widely known mathematician of the same Institute, was still alive. I thus trust that Peter and Louis will not resent it, if I dedicate this particular paper in the volume in their honor, to the memory of Harold.

Harold Grad was the man who brought mathematical clarity into kinetic theory as well as the apostle of the use of the Boltzmann equation in rarefied gas dynamics. He advocated the use of all the available mathematical techniques in order to uncover the information that is hidden in kinetic theory. In particular he formulated the problem of validity for the Boltzmann equation in clear terms, developed systematically the theory of the linearized Boltzmann equation, set the basis for the treatment of kinetic layers and introduced the first basic steps required to enter into the theory of exact nonlinear perturbations of equilibrium solutions. He was also deeply interested in the applications of kinetic theory to practical problems. He attended several of the early Symposia on Rarefied Gas Dynamics, where he also gave illuminating lectures that one can still read in the proceedings. Perhaps it is not irrelevant to recall that in the early 1960's there were people well qualified in continuum gas dynamics who doubted the relevance of the Boltzmann equation for rarefied gas dynamics.

Unfortunately, in 1972 he abandoned research in kinetic theory to devote himself to his studies on plasma physics. The field lost its recognized leader, but his work was the solid basis on which new accomplishments were built. In fact the mathematical theory of the Boltzmann equation developed even beyond his hopes.

In the last few years of his life he showed again interest in kinetic theory and I still remember his opening lecture at the 15th Symposium on Rarefied Gas Dynamics held not too far from here, in Grado, in 1986. This was his last scientific talk and

---

1991 *Mathematics Subject Classification.* 82C40, 76P05.

he was never able to give the editors of the Proceedings his manuscript. His health deteriorated and, when a few months later he was awarded the Maxwell prize, he was not able to deliver a speech for that occasion. He passed away in the fall of 1986.

The aim of this lecture is to survey the known results on the Cauchy and the mixed problems which arise when we consider the time evolution of a rarefied gas either in the entire space $\Re^3$ or in a subset of it which may be either a vessel Ω (Ω is a bounded open set of $\Re^3$ with a sufficiently smooth boundary $\partial\Omega$), endowed with unit normal $n(x)$ (pointing into Ω) or the region outside a solid body.

The evolution equation for the distribution function $f(x,\xi,t)$ is the Boltzmann equation [1, 2, 3]

$$\Lambda f = Q(f,f) \qquad \text{in } \mathcal{D} \quad (\mathcal{D} = \Omega \times \Re^3 \times (0,T)) \tag{1.1}$$

In Eq. (1.1) Λf is the free streaming operator defined by

$$\Lambda f = \frac{\partial f}{\partial t} + \xi \cdot \frac{\partial f}{\partial x} \tag{1.2}$$

and $Q(f,f)$ the collision term

$$Q(f,f)(x,\xi,t) = \int_{\Re^3} \int_{\mathcal{B}^+} (f'f'_* - ff_*)B(V,n)d\xi_* dn \quad (V = \xi - \xi_*) \tag{1.3}$$

Here $B(V,n)$ is a kernel containing the details of the molecular interaction, f', f'_*, f_* are the same as f, except for the fact that the velocity argument ξ ($\in \Re^3$) is replaced by ξ', ξ'_*, ξ_*, respectively, ξ_* being an integration variable (having the meaning of the velocity of a molecule colliding with the molecule of velocity ξ, whose path we are following). ξ' and ξ'_* are the velocities of two molecules entering a collision that will bring them to have velocities ξ and ξ_*, whereas the unit vector n describes a hemisphere $\mathcal{B}^+$ of the unit sphere $\mathcal{B}$. The relations between ξ', ξ'_*, on one hand, and ξ, ξ_*, on the other hand, read as follows:

$$\xi' = \xi - n[(\xi - \xi_*) \cdot n]$$
$$\xi'_* = \xi_* + n[(\xi - \xi_*) \cdot n] \tag{1.4}$$

We shall assume that the gain and loss parts of $Q(f,f)$, denoted by $Q^\pm$, are separately meaningful (a.e.). Please remark that the loss term $Q^-(f,f)$ equals f times an expression linear in f that we shall denote by $\nu(f)$. Occasionally we simply write Q or $Q^\pm$ to denote $Q(f,f)$ or $Q^\pm(f,f)$.

Eq. (1.1) must be solved with an initial condition

$$f(x,\xi,0) = f_0(x,\xi) \tag{1.5}$$

and, in the case of mixed problems, with the boundary condition

$$\gamma^+ f(x,\xi,t) = K\gamma^- f \qquad \text{on } E^+ \tag{1.6}$$

(where, as explained below, K is a linear integral operator and $\gamma^\pm f$ denote the traces of f on $E^\pm = \{(t,x,\xi) \in \partial\Omega \times \Re^3 \times (0,T) \mid \pm \xi \cdot n(x) > 0\}$)

The purely mathematical theory of the Boltzmann equation is concerned with such topics as existence and uniqueness of solutions and is regarded by most people interested in the applications as a waste of time invented to amuse mathematicians and without any interest for the more practical aspects of rarefied gas dynamics. For this reason, it is to be stressed that, although this chapter is, among those of the theory, the most abstract and remote from applications, there is a relation between contiguous chapters and progress in one of them will soon or later have a beneficial influence on the others. In particular the theorems which are proved by mathematicians insure that there is no internal contradiction in the theory and the techniques used to prove them throw light on other aspects, such as numerical techniques, estimates of errors in approximate solutions, etc.

The purely mathematical aspects of the Boltzmann equation began to be investigated in the thirties by the famous Swedish mathematician T. Carleman [41], who provided an existence proof for the purely initial value problem with homogeneous data (*i.e.* data independent of position x). The same problem was revisited by Arkeryd [42] in 1972; he provided solutions in a (weighted) L^1 space, rather than in a (weighted) L^∞ space. Solutions depending on the space variables are much more difficult to handle, if we do not restrict our attention to solutions existing only locally in time but look for solutions existing for an arbitrarily long time interval; the first steps in this direction were taken by Harold Grad in a famous paper [43], where he obtained some basic estimates for the rigorous theory of the solutions close to an absolute Maxwellian. These results were essential for the important theorems obtained later by several Japanese authors [44, 45, 46].

The idea is the following: if we work in spaces of sufficiently regular solutions, we can introduce the semigroup $U(t)$ (actually a group) associated with the collisionless evolution ($\Lambda f = 0$). Then the above initial-boundary value problem reduces to solving the following integral equation

$$f(t) = U(t)f_0 + \int_0^t U(t-s)Q(f(s), f(s))ds \tag{1.7}$$

It is easy to prove local existence of this equation in several spaces (first introduced by Grad), typically:

$$X_{\alpha,\beta} = \{f| \ (1 + |\xi|^2)^{\alpha/2} \exp(-\beta|\xi|^2)f \in L^\infty(\Omega \times \Re^3)\} \tag{1.8}$$

with norm

$$\| f \|_{\alpha,\beta} = \| (1 + |\xi|^2)^{\alpha/2} \exp(-\beta|\xi|^2)f \|_{L^\infty(\Omega \times \Re^3)} \tag{1.9}$$

A case in which one can say a lot more about the solutions of initial boundary value problems is the case when the data are compatible with a solution close to a uniform Maxwellian distribution M, given by

$$M = A \exp(-\beta|\xi - u|^2) \tag{1.10}$$

where β is the inverse temperature and u the average velocity (frequently assumed to be zero). Then techniques from rigorous perturbation theory [4] can be used. To this end, let us introduce the perturbation h such that

$$f = M + M^{1/2}h \tag{1.11}$$

and assume (for simplicity) that, in the case of mixed problems, M coincides with the wall Maxwellian, so that Eq. (1.6) is satisfied by the restrictions of M to $E^\pm$. Eqs. (1.1-3) can then be rewritten in the following way:

$$\Lambda h = Lh + \Gamma(h, h) \qquad \text{in } \mathcal{D} \tag{1.12}$$

$$\gamma^+ h(x, \xi, t) = \hat{K}\gamma^- h \qquad \text{on } E^+ \tag{1.13}$$

$$h(x, \xi, 0) = h_0(x, \xi) \tag{1.14}$$

where L and Γ are two suitable operators, linear and quadratic, respectively, whereas

$$\hat{K} = M^{-1/2} K M^{1/2} \tag{1.15}$$

Now the global solution can be found by perturbation techniques, provided the linearized operator

$$B = -\xi \cdot \partial/\partial x + L \tag{1.16}$$

with the boundary conditions (1.13) generates a semigroup $T(t)$ with a nice decay.

One finds two types of result: the first of them with a decay like $t^{-\alpha}$ ($\alpha > 0$) applies to unbounded domains (in particular, to the Cauchy problem [44, 45]), using estimates first provided by Grad [43], whereas the second, with an exponential decay, applies to bounded domains.

There are several papers dealing with the proofs of the behaviors indicated above. The case of a bounded domain has been considered by Guiraud [5] in the case of diffuse reflection and by Shizuta and Asano [6] in the case of specular reflection, both assuming that Ω is convex. The case of unbounded domains exterior to a bounded convex obstacle was treated by several Japanese authors [7, 8].

Illner and Shinbrot [47] provided an existence proof for a global solution close to vacuum; their assumptions were later relaxed by Bellomo and Toscani [48], while Toscani [49] considered solutions close to a nonhomogeneous Maxwellian of a special class.

Cercignani proved [50] existence for a very particular case with data arbitrarily far from equilibrium, the so-called affine homoenergetic flows. Arkeryd, Esposito and Pulvirenti [51] proved existence for solutions close to a homogeneous solution (different from a Maxwellian). One should also mention the important paper of Arkeryd [52] who proved an existence theorem in the context of non-standard analysis.

It should be realized that any equation similar to the Boltzmann equation but having a little more of compactness in the dependence upon the space variables is rather easy to deal with; this was shown by Morgenstern [53] in the 1950's and Povzner [54] in the 1960's; they introduced mollifying kernels in the collision term of the Boltzmann equation, producing 8-fold and 6-fold integrations, respectively, in place of the original 5-fold integration. In the 1970's Cercignani, Greenberg and Zweifel [55] indicated that a theorem of existence and uniqueness can be proved if the particles can sit only at discrete positions on a lattice.

When we want to deal with the more difficult case of arbitrarily large data, we can hope for some significant results only if we use the L^1 framework and the techniques introduced by DiPerna and Lions [9] to deal with the Cauchy problem, as discussed with more detail in the next few sections.

2. The results of DiPerna and Lions.

A turning point in the existence theory of the Boltzmann equation occurred in 1987, when Golse et al. [10] were able to prove certain results that have become known as "velocity averaging lemmas". Also in 1987, DiPerna and Lions used these lemmas and other estimates to prove the first general global existence theorem for the Boltzmann equation [9]. Their result, with proof, is given in the latter paper, in a review article by Gérard [11] and in a recent book [3]. Here we shall merely outline the ideas of the proof as well as the meaning and limitations of the result.

We begin by fixing some notation. If $\Omega \subset \Re^3$ is open, $L^p_{\text{loc}}(\Omega) = \{f : \Omega \to \Re, \ f\mid_U \in L^p(U)$ for all $U \subset \Omega$ which are open and relatively compact$\}$. If $\Omega \subset \Re^3$, $\Omega' \subset \Re^l$, the space of all measurable functions on $\Omega' \times \Omega$ whose restrictions to $\Omega' \times U$ is in $L^p(\Omega' \times U)$ for each open and relatively compact $U \subset \Omega$ will be denoted by $L^p(\Omega' \times \Omega_{\text{loc}})$. $S(\Re^3)$ denotes the Schwartz space of rapidly decreasing C^∞-functions on $\Re^3$. For each $s \in \Re_+$, $H^s(\Re^3)$ is the usual Sobolev space, i.e. the completion of $S(\Re^3)$ with respect to the norm

$$\|f\|_{H^s} := \left(\int (1 + |z|^{s/2})^2 |\hat{f}(z)|^2 \, dz \right)^{1/2}. \tag{2.1}$$

As in Sect.1, we shall use Λ as an abbreviation for the transport operator given by Eq. (1.2).

In fact, we first discuss some results about a generalized Boltzmann equation

$$(\partial_t + \xi \cdot \nabla_x) f = \iint\limits_{\Re^3 \, S^2} q(x, V, n) [f' f'_* - f f_*] \, dn \, d\xi_* \tag{2.2}$$

with the convention $f_* = f(\cdot, \xi_*, \cdot)$, $f' = f(\cdot, \xi', \cdot)$, etc. Of course, for Eq. (1.1) $q(\dots) = B(V, n)$.

Lemma 2.1. *Suppose that q is a nonnegative measurable function in $L^\infty_{\text{loc}}(\Re^3 \times \Re^3 \times S^2)$, which depends only on x, V and $|V \cdot n|$ and grows at most polynomially with respect to x and V. Then, if $f \in C^1\left(\Re_+, S(\Re^3 \times \Re^3)\right)$ is a positive solution of (1.3) such that $|\ln f|$ grows at most polynomially in (x, ξ), uniformly on compact time intervals in $\Re_+$, we have*

$$\iint f \, dx \, d\xi = \iint f_0 \, dx \, d\xi, \tag{2.4}$$

$$\iint f |\xi|^2 \, dx \, d\xi = \iint f_0 |\xi|^2 \, dx \, d\xi, \tag{2.5}$$

$$\iint f |x - t\xi|^2 \, dx \, d\xi = \iint f_0 |x|^2 \, dx \, d\xi, (2.6) \tag{2.6}$$

$$\iint f \ln f \, dx \, d\xi + \int_0^t \iint e(f)(\cdot, s) \, dx \, d\xi \, ds = \iint f_0 \ln f_0 \, dx \, d\xi, \tag{2.7}$$

where

$$e(f)(x, \xi, t)$$
$$= \frac{1}{4} \iint (f' f'_* - f f_*) \ln\left(\frac{f' f'_*}{f f_*}\right)(x, \xi, \xi_*, n, t) \cdot q(x, \xi - \xi_*, n) \, d\xi_* \, dn.$$

These identities imply the estimates

$$\iint f(1 + |x|^2 + |\xi|^2) \, dx \, d\xi \le \iint f_0(1 + 2|x|^2 + (2t^2 + 1)|\xi|^2) \, dx \, d\xi \tag{2.8}$$

and

$$\iint f|\ln f| \, dx \, d\xi + \int_0^\infty \iint e(f)(\cdot, s) \, ds \, dx \, d\xi$$

$$\le \iint f_0(|\ln f_0| + 2|x|^2 + 2|\xi|^2) \, dx \, d\xi + C \tag{2.9}$$

where C is a purely numerical constant.

For the proof we refer to the papers and book quoted above. We also remind the reader that $e(f)$ is always nonnegative (Boltzmann inequality) [1-3] . It also well-known that weight functions ψ which will lead to conservation equations are $1, \xi_i, 1 \le i \le 3$ (momentum), $|\xi|^2 \, x_i\xi_j - x_j\xi_i, 1 \le i \le j \le 3$ (angular momentum), $x_i - t\xi_i, 1 \le i \le 3$, (center of mass) and $|x - t\xi|^2$ (moment of inertia).

Let us now specify the assumptions on the collision kernel $q(V, n)$ for which a general existence result will be proved. Notice that we assume no dependence of q on x; such a dependence only enters when we construct approximate solutions a little later.

Suppose that $q \in L^1_{\text{loc}}(\Re^3 \times S^2)$, $q \ge 0$, and that q depends only on $|V|$ and $|V \cdot n|$. Let

$$A(V) = \int_{S^2} q(V, n) \, dn. \tag{2.10}$$

Suppose, furthermore, that for every $R > 0$

$$\frac{1}{1 + |\xi|^2} \int_{|\xi_*| \le R} A(\xi - \xi_*) \, d\xi_* \to 0 \qquad \text{as} \quad |\xi| \to \infty \tag{2.11}$$

and that

$$A \in L^\infty_{\text{loc}}(\Re^3) \tag{2.12}$$

(this last assumption was not made by DiPerna and Lions [9], but, as noticed by Gérard [11], it simplifies certain technicalities of the proof).

We now split $Q(f, f) = Q_+(f, f) - Q_-(f, f)$ and write $Q_-(f, f) = f\nu(f)$. Note that $\nu(f) = A*f$, where $*$ denotes a convolution product in velocity space.

We do not know whether a global classical solution of the Boltzmann equation exists for general initial data. One of the crucial steps in the paper by DiPerna and Lions [9] is to introduce weaker solution concepts which lighten the burden of proof, but are still strong enough to guarantee that the collision terms are defined. As before, we write $U(t)$ $(t \in \Re)$ for the one-parameter family of operators defined by

$$U(t)g(x_i, \xi_i) = g(x - \xi t, \xi) \tag{2.13}$$

for each measurable g on $\Re^3 \times \Re^3$. First we reformulate the mild solution concept, with minimal integrability constraints on the collision terms.

Definition 2.2. *A measurable function $f = f(x, \xi, t)$ on $[0, \infty) \times \Re^3 \times \Re^3$ is a mild solution of the Boltzmann equation to the (measurable) initial value $f_0(x, \xi)$ if for almost all (x, ξ) $U(-s)Q_\pm(f, f)(x, \xi, s)$ are in $L^1_{\text{loc}}[0, \infty)$, and if for each $t \ge 0$ Eq. (1.7) holds.*

One of the key ideas of DiPerna and Lions was to introduce a new concept of solution, such that the bounds (2.8) and (2.9) could be put to best use, and then to regain mild solutions via a limiting procedure. They called the relaxed solution concept "renormalized solution".

Definition 2.3. *A function $f = f(x, \xi, t) \in L^1_+(\Re^+_{\text{loc}} \times \Re^3 \times \Re^3)$ is called a renormalized solution of the Boltzmann equation if*

$$\frac{Q_\pm(f, f)}{1 + f} \in L^1_{\text{loc}}(\Re_+ \times \Re^3 \times \Re^3) \tag{2.14}$$

and if for every Lipschitz continuous function $\beta : \Re_+ \to \Re$ which satisfies $|\beta'(t)| \le \frac{C}{1+t}$ for all $t \ge 0$ one has

$$\Lambda\beta(f) = \beta'(f)Q(f, f) \tag{2.15}$$

in the sense of distributions.

We remark that the division by $1 + f$ is natural inasmuch as it leads to a "quasi-linearization" of $Q(f, f)$.

Lemma 2.4. *Let $f \in (L^1_{\text{loc}} \times \Re^3 \times \Re^3)$.*
i) If f satisfies (2.14) and (2.15) with $\beta(t) = \ln(1 + t)$, then f is a mild solution of the Boltzmann equation.
ii) If f is a mild solution of the Boltzmann equation and if $\frac{Q_{\pm}(f, f)}{1 + f} \in L^1_{\text{loc}}(\Re_+ \times \Re^3 \times \Re^3)$, then f is a renormalized solution.

Proof. See Refs. [3,9,10].

The basic result of DiPerna and Lions is given by

Theorem 2.5. *[9,3,10] Suppose that $f_0 \in L^1_+(\Re_+ \times \Re^3 \times \Re^3)$ is such that $\int \int f_0(1 + |x|^2 + |\xi|^2)\, dx\, d\xi$ and $\int \int f_0 |\ln f_0|\, dx\, d\xi$ are bounded. Then there is a renormalized solution of the Boltzmann equation such that $f \in C(\Re_+, L^1(\Re^3 \times \Re^3))$, $f\big|_{t=0} = f^0$, and (2.8), (2.9) hold.*

The renormalized solution f is found as a limit of functions solving truncated equations. For some $\delta > 0$ and some modified nonnegative collision kernel $\bar{q} \in C^\infty_0(\Re^3 \times S^2)$ such that $\bar{q}$ vanishes for $|V \cdot n < \delta|$, let

$$\bar{Q} = \int \int \bar{q}(g'g'_* - gg_*)\, dn\, d\xi_* \tag{2.16}$$

and

$$\tilde{Q} = (1 + \delta \int |g|\, d\xi)^{-1} \bar{Q} \tag{2.17}$$

Lemma 2.6. *Let $f_0 \in S(\Re^3 \times \Re^3)$ be nonnegative such that $|\ln f_0|$ grows at most polynomially. Then the Cauchy problem*

$$\Lambda f = \tilde{Q}(f, f), \qquad f\big|_{t=0} = f_0 \tag{2.18}$$

has a unique global solution f which satisfies the hypotheses of Lemma 2.1. It also satisfies the estimates (2.8) and (2.9).

This assertion can be proved by contraction mapping.

Let now $q_n \in C^\infty_{0,+}(\Re^3 \times S^2)$ satisfy (2.11) and (2.12) (uniformly for all n) and suppose that $q_n \to q$ a.e. Furthermore, we approximate f_0 in $L^1_+(\Re^3 \times \Re^3)$ by a sequence $\{f_0^n\}_n \subset S(\Re^3 \times \Re^3)$ such that

$$\forall n \quad f_0^n \geq \mu_n e^{-|x|^2 - |\xi|^2} \qquad (\mu_n > 0), \tag{2.19}$$

$$\int \int f_0^n (1 + |x|^2 + |\xi|^2)\, dx\, d\xi \longrightarrow \int \int f_0(1 + |x|^2 + |\xi|^2)\, dx\, dv, \tag{2.20}$$

$$\int \int f_0^n |\ln f_0^n|\, dx\, d\xi \longrightarrow \int \int f_0 |\ln f_0|\, dx\, d\xi. \tag{2.21}$$

Let $\delta_n \searrow 0$, and let Q^n be $\tilde{Q}$ (from (2.17)) with $\delta = \delta_n$, $\bar{q} = q_n$. Then, Lemma 2.6 assures us that there is a sequence $\{f^n\}$ such that $Tf^n = Q^n(f^n, f^n)$, $f^n|_{t=0} = f_0^n$, and (by (2.8) and (2.9))

$$\forall T > 0 \ \sup_{t \in [0,T]} \sup_n \int \int f^n(1 + |x|^2 + |\xi|^2)\, dx\, d\xi < \infty, \tag{2.22}$$

$$\forall T > 0 \ \sup_{t \in [0,T]} \sup_n \int \int f^n |\ln f^n|\, dx\, d\xi < \infty, \tag{2.23}$$

$$\sup_n \int_0^\infty \int \int e_n(f^n)\, dx\, d\xi\, dt < \infty, \tag{2.24}$$

where

$$e_n(f^n) = \frac{1}{4}(1 + \delta_n \int f^n \, d\xi)^{-1} \cdot \int \int (f^{n'} f_*^{n'} - f^n f_*^n) \ln\left(\frac{f^{n'} f_*^{n'}}{f^n f_*^n}\right) q_n \, d\xi_* \, dn \quad (2.25)$$

We recall now the Dunford-Pettis criterion for weak compactness in L^1:

Let $\{f_n\}_{n \in \mathcal{N}}$ be a sequence of functions in $L^1(\Re^3)$. Then the following i) and ii) are equivalent.

i) $\{f_n\}$ is contained in a weakly sequentially compact set of $L^1(\Re^3)$.

iia) $\{f_n\}$ is bounded in $L^1(\Re^3)$.

iib) $\forall \epsilon > 0 \quad \exists \delta > 0$ such that $\forall E \subset \Re^3$ (E measurable) such that $\lambda(E) < \delta$, we also have $\sup_n \int_E |f_n| \, dx \le \epsilon$.

iic) $\forall \epsilon > 0 \exists K$ compact, $K \subset \Re^3$, such that $\sup_n \int_{\Re^3 - K} |f_n| \, dx \le \epsilon$. We will apply the criterion to the following situation. If $h \in C(\Re_+, \Re_+)$ and $w \in L^\infty_{\text{loc}}(\Re^3, \Re_+)$ are such that $h(t)/t \to \infty$ ($t \to \infty$) and $w(x) \to \infty$ ($|x| \to \infty$), then the inequality

$$\sup_n \int_{\Re^3} [h(|f_n|) + |f_n|(1 + w)] \, dx < \infty \quad (2.26)$$

implies that $\{f_n\}_{n \in \mathcal{N}}$ satisfies ii).

A major problem with weak convergence is the well-known fact that nonlinear functions are in general not weakly continuous. A useful property is, however, the fact that convex functions are at least lower semi-continuous: If $F : \Re \to \Re$ is convex and if $f_n \to wf$ in L^1, then

$$\int F \circ f \, dx \le \liminf_{n \to \infty} \int F \circ f_n \, dx \quad (2.27)$$

Also, if one of the factors in a product converges a.e. and the other factor converges weakly, then the product is compact in the weak topology. Specifically, let $f_n \to wf$ in L^1, let $\{g_n\} \subset L^\infty$ be bounded and let $g_n \to g$ a.e., then

$$f_n g_n \to fg \text{ in } L^1. \quad (2.28)$$

This follows because for every $\epsilon > 0$ there is a compact set K such that the supremum with respect to n of $\int_{\Re^3 \setminus K}(|f_n g_n| + |fg|) \, dx$ is not larger than ϵ, and by Egorov's theorem, there is a set $E \subset K$ such that $\sup_n \int_E |f_n| \, dx \le \epsilon$ and such that $g_n \to g$ uniformly on $K \setminus E$.

We can now work with the "approximating sequence of solutions to modified equations" given above. Q_+^n, Q_-^n, A_n all refer to this situation. The collision kernel in $Q_\pm^n$ is really x- (and t-) dependent, and given by

$$q_n(x, V, n) = \frac{1}{1 + \delta_n \int f_n \, d\xi} q_n(V, n).$$

Lemma 2.7. *For all $T > 0$, $R > 0$, the sequences*

$$\frac{Q_+^n(f_n, f_n)}{1 + f_n} \quad \text{and} \quad \frac{Q_-^n(f_n, f_n)}{1 + f_n} \quad (2.29)$$

are contained in weakly compact subsets of $L^1\big((0, T) \times \Re^3 \times B_R\big)$, where $B_R = \{\xi \in \Re^3; \|\xi\| \le R\}$.

Since $\{f^n\}$ has uniformly bounded entropy and second moments, (2.26) implies that we can extract a subsequence (again denoted by $\{f^n\}$) which converges weakly in $L^1((0,T) \times \Re^3 \times \Re^3)$,

$$f^n \rightharpoonup f. \tag{2.30}$$

Let $g_\delta^n := \frac{1}{\delta} \ln(1 + \delta f^n)$. The uniform bounds on entropy and second moments for f^n easily imply that

$$\sup_{t \in [0,T]} \sup_n \|f^n - g_\delta^n\|_{L^1(\Re^3 \times \Re^3)} \longrightarrow 0 \qquad \text{as} \quad \delta \to 0. \tag{2.31}$$

Also, since

$$\Lambda g_\delta^n = \frac{1}{1 + \delta f^n} Q^n(f^n, f^n), \tag{2.32}$$

then

$$U(-t-h)g_\delta^n - U(-t)g_\delta^n = \int\limits_t^{t+h} \frac{U(-s)Q^n(f^n, f^n)(s)}{1 + \delta U(-s)f_n(s)} \, ds. \tag{2.33}$$

By the compactness insured by Lemma 2.7, then $\forall \delta > 0 \; \forall T > 0 \; \forall R > 0$

$$\sup_{t \in [0,T]} \sup_n \|U(-t-h)g_\delta^n - U(-t)g_\delta^n(t)\|_{L^1(\Re^3 \times B_R)} \longrightarrow 0$$

as $h \to 0$. We next estimate, by (2.31) and (2.8),

$$\sup_t \|U(-t-h)f^n - U(-t)f^n\|_{L^1(\Re^3 \times \Re^3)}$$

$$\leq 0(\delta) + \sup_t \|U(-t-h)g_\delta^n - U(-t)g_\delta^n\|_{L^1(\Re^3 \times B_R)}. \tag{2.34}$$

This easily entails

$$\sup_{t \in [0,T]} \sup_n \|U(-t-h)f^n - U(-t)f^n\|_{L^1(\Re^3 \times \Re^3)} \to 0 \quad (h \to 0), \tag{2.35}$$

and a standard equicontinuity argument shows that the (weak) limit f must then satisfy

$$U(-t)f \in C(\Re_+; L^1(\Re^3 \times \Re^3)) \tag{2.36}$$

and, for all $T > 0$

$$\sup_{t \in [0,T]} \|U(-t-h)f - U(-t)f\|_{L^1} \to 0 \quad (\to 0). \tag{2.37}$$

Actually, by using an elementary argument from integration theory,

$$f \in C(\Re_+; L^1(\Re^3 \times \Re^3)). \tag{2.38}$$

Also, by using the convexity of the function $x \cdot \max(\ln x, 0)$,

$$\forall t \quad \int\int f |\ln f| \, d\xi \, dx + \limsup_n \int\limits_0^t \int\int e_n(f^n) \, d\xi \, dx$$

$$\leq \int\int f_0(|\ln f_0| + 2|x|^2 + 2|\xi|^2) \, d\xi \, dx + C, \tag{2.39}$$

and

$$\forall t \quad \int\int f(1 + |x|^2 + |\xi|^2) \, d\xi \, dx \leq \int\int f_0(1 + 2|x|^2 + (2t^2 + 1)|\xi|^2) \, d\xi \, dx. \tag{2.40}$$

By now, we have a weakly convergent sequence $f_n \rightharpoonup f$, and the limit f is in $C([0,T]; L^1)$. Subsequences of $Q_{+,-}^n(f^n, f^n)/(1 + f^n)$ will also converge weakly (by

Lemma 2.7), but we cannot say *a priori* whether the limits will by $Q_{+,-}(f,f)/(1 + f)$, because nonlinear functionals are in general not weakly continuous. This problem was first overcome by DiPerna and Lions by a skillful use of results known as "velocity averaging lemmas" (see [10]). We present these below (actually, we confine our discussion to a simplified situation, which is all we need).

Lemma 2.8. *Let $u \in L^2(\Re \times \Re^3 \times \Re^3)$ have compact support, and suppose that $\Lambda u \in L^2(\Re \times \Re^3 \times \Re^3)$. Then $\int u \, d\xi \in H^{1/2}(\Re \times \Re^3)$, and the $H^{1/2}$-norm of $\int u \, d\xi$ is bounded in terms of $\|u\|_{L^2}$, $\|\Lambda u\|_{L^2}$ and the support of u.*

We will use Lemma 2.8 to pass from weak to strong convergence in L^1-settings. The next lemma is the crucial step.

Lemma 2.9. *Suppose that $\{g_n\} \subset L^1((0,T) \times \Re^3 \times \Re^3)$ is weakly relatively compact, and that $\{\Lambda g_n\}$ is weakly relatively compact in $L^1_{\text{loc}}((0,T) \times \Re^3 \times \Re^3)$. Then, if $\{\psi_n\}$ is a bounded sequence in $L^\infty((0,T) \times \Re^3 \times \Re^3)$ which converges a.e., then $(\int g_n \psi_n \, d\xi)$ is compact in the norm topology in $L^1((0,T) \times \Re^3)$.*

We note an immediate

Corollary. *Under the hypotheses of Lemma 2.9, if $g_n \to g$ weakly in $L^1((0,T) \times \Re^3 \times \Re^3)$ and $\psi_n \to \psi$ a.e., then*

$$\left\| \int g_n \psi_n \, d\xi - \int g\psi \, d\xi \right\|_{L^1((0,T) \times \Re^3)} \to 0. \tag{2.41}$$

We also have:

Lemma 2.10. *Let $\{f_n\}$ be a relatively compact sequence in $L^1((0,T) \times \Re^3 \times \Re^3)$, and suppose that there is a family of real-valued uniformly Lipschitz continuous functions $\{\beta_\delta\}_{\delta>0}$, $\beta_\delta(0) = 0$ for all δ, such that*
i) $\beta_\delta(s) \to s$ as $\delta \to 0$, uniformly on compact subsets of $\Re_+$,
ii) the sequence $\{T(\beta_\delta(f^n))\}$ is, $\forall \delta$, weakly relatively compact in $L^1_{\text{loc}}((0,T) \times \Re^3 \times \Re^3)$. Then, if $f^n \to f$ weakly in L^1_+, $\{\psi_n\}_n$ is bounded in $L^\infty((0,T) \times \Re^3 \times \Re^3)$ and $\psi_n \to \psi$ a.e.,

$$\lim_{n \to \infty} \left\| \int f^n \psi_n \, d\xi - \int f\psi \, d\xi \right\|_{L^1} = 0. \tag{2.42}$$

We return now to our sequence. We have:

Lemma 2.11. *Let $\{f^n\}$ be the sequence of solutions to approximating problems as above. There is a subsequence such that for each $T > 0$*
i) $\int f^n \, d\xi \to \int f \, d\xi$ a.e. and in $L^1((0,T) \times \Re^3)$,
ii) $A_n * f^n \to A * f$ in $L^1((0,T) \times \Re^3 \times B_R)$ for all $R > 0$, and a.e.,
iii) for each compactly supported function $\varphi \in L^\infty((0,T) \times \Re^3 \times \Re^3)$,

$$\left(\frac{\int Q^n_\pm(f^n, f^n) \varphi \, d\xi}{1 + \int f^n \, d\xi} \right) \longrightarrow \left(\frac{\int Q_\pm(f, f) \varphi \, d\xi}{1 + \int f \, d\xi} \right) \tag{2.43}$$

in $L^1((0,T) \times \Re^3)$.

Consider now Λ^{-1}, defined by $u = \Lambda^{-1} g$, i.e. $\Lambda u = g$ with $u|_{t=0} = 0$:

$$\Lambda^{-1} g(x, \xi, t) = \int_0^t g(x - (t - s)\xi, \xi, s) \, ds. \tag{2.44}$$

Λ^{-1} is, as one can check immediately, a continuous and weakly continuous operator from $L^1((0,T) \times \Re^3 \times \Re^3_{\text{loc}})$ into $C([0,T])$; $L^1(\Re^3 \times \Re^3_{\text{loc}})$, and if $g \geq 0$, also $\Lambda^{-1} g \geq 0$. We use Λ^{-1} to rewrite the Boltzmann equation in yet another form.

Suppose that $F \in C\left([0,T]; L^1(\Re^3 \times \Re^3_{\text{loc}})\right)$, $\Lambda F \geq 0$. The operator defined as follows: $\Lambda_F^{-1} := e^{-F}\Lambda^{-1}e^F$ is then continuous (and weakly continuous) from $L^1((0,T) \times \Re^3 \times \Re^3_{\text{loc}})$ into $C([0,T]; L^1(\Re^3 \times \Re^3_{\text{loc}}))$.

If $\{F_n\}$ is a bounded sequence in $C\left([0,T]; L^1(\Re^3 \times \Re^3_{\text{loc}})\right)$ such that $\Lambda F_n \geq 0$, $F_n(x, \xi, t) \to F(x, \xi, t)$ for all t and almost all (x,ξ), and if $g_n \to g$ weakly in $L^1\left((0,T) \times \Re^3 \times \Re^3_{\text{loc}}\right)$, then, for all $t \in [0,T]$,

$$\Lambda_{F_n}^{-1}g_n(t) \longrightarrow \Lambda_F^{-1}g(t) \tag{2.45}$$

in $L^1(\Re^3 \times \Re^3_{\text{loc}})$. (This is easily proved by using the explicit solution formula for Λ^{-1}, given above.)

To use (2.45), let $F_n = \Lambda^{-1}(A_n * f^n)$, where f^n, A_n are from the modified Boltzmann equation, as used above. This equation can be written as

$$\Lambda f^n + (A_n * f^n)f^n = Q_+^n(f^n, f^n) \tag{2.46}$$

or (after multiplication with e^{F_n} and observing that

$$\Lambda(f^n e^{F_n}) = (\Lambda f^n)e^{F_n} + f^n(\Lambda F_n)e^{F_n} = e^{F_n}Q_+^n(f^n, f^n),$$

$$f^n = f_0^n e^{-F_n} + \Lambda_{F_n}^{-1}Q_+^n(f^n, f^n). \tag{2.47}$$

By Lemma 2.11 ii) and the above remarks, $\{F_n\}$ turns out to be a bounded sequence in $C\left((0,T); L^1(\Re^3) \times \Re^3_{\text{loc}}\right)$ and for all $t \in \Re_+$

$$F_n \to F = \Lambda^{-1}(A * f) \qquad \text{a.e.} \tag{2.48}$$

Starting from these remarks a careful argument gives [9,11,3]:

Lemma 2.12. *For all $t \in \Re_+$, we have $\Lambda_F^{-1}Q_+(f, f) \in L^1(\Re^3 \times \Re^3_{\text{loc}})$ and*

$$f = U(t)f_0 e^{-F} + \Lambda_F^{-1}Q_+(f, f). \tag{2.49}$$

Eq. (2.49) is already saying that f satisfies the Boltzmann equation in some sense. We will now simply check that it satisfies the criteria for a renormalized solution (as given before).

First, it is easy to show that for every $T < \infty$

$$\frac{Q_-(f,f)}{1+f} \in L^1([0,T] \times \Re^3 \times \Re^3_{\text{loc}}) \tag{2.50}$$

(just use the condition on A and that

$$\sup_{t\in[0,T]} \sup_n \int\int f^n(1 + |x|^2 + |\xi|^2)\, dx\, d\xi < \infty). \tag{2.51}$$

As for $Q_+(f,f)/(1+f)$, by an elementary inequality [3,9,10], we have

$$Q_\pm^n(f^n, f^n)\left(1 + \delta\int f^n\, d\xi\right)^{-1}$$

$$\leq 2Q_\mp^n(f^n, f^n)\left(1 + \delta\int f^n\, d\xi\right)^{-1} + \frac{4e_n}{\ln 2}\left(1 + \delta\int f^n\, d\xi\right)^{-1} \tag{2.52}$$

and

$$\sup_n \int_0^\infty\int\int e_n(f^n)\, dx\, d\xi\, ds < \infty. \tag{2.53}$$

Because of the nonnegativity of $e_n(f^n)$ and (2.53), we can assume that $e_n(f^n)$ converges weakly (in $\mathcal{D}'$, or in the vague topology on the bounded measures) to a bounded, nonnegative measure μ by Lemma 2.11; we also know that the other two terms in (2.52) converge weakly in L^1, and so

$$\frac{Q_\pm(f,f)}{1+\delta\int f\,d\xi} \leq \frac{2Q_\mp(f,f)}{1+\delta\int f\,d\xi} + \frac{4}{\ln 2}(1+\delta\int f\,d\xi)\mu. \tag{2.54}$$

(2.54) remains true if we replace μ by its absolutely continuous part $e \in L^1\big((0,T)\times \Re^3 \times \Re^3\big)$, and by taking $\delta \to 0$, it follows that

$$Q_\pm(f,f) \leq 2Q_\mp(f,f) + E \tag{2.55}$$

with $E \in L^1\big((0,T)\times\Re^3\times\Re^3\big)$. (2.50) and (2.55) now entail that

$$\frac{Q_+(f,f)}{1+f} \in L^1\big([0,T]\times\Re^3\times\Re^3_{\mathrm{loc}}\big). \tag{2.56}$$

To show that $Q_+(f,f)(x,\xi,\cdot) \in L^1(0,T)$ for almost all (x,ξ), we use that by Lemma 2.12 for all t, $\Lambda_F^{-1}Q_+(f,f) \in L^1(\Re^3\times\Re^3_{\mathrm{loc}})$ and $F \in L^1(\Re^3\times\Re^3_{\mathrm{loc}})$. Explicitly, we see that

$$\int_0^t U(-s)Q_+(f,f)\exp-(U(-t)F-U(-s)F)\,ds \tag{2.57}$$

is in $L^1(\Re^3\times\Re^3_{\mathrm{loc}})$ for all t, and because $U(-t)F$ is nonnegative, _increasing_ with respect to t and in $L^1(\Re^3\times\Re^3_{\mathrm{loc}})$ with respect to (x,ξ), it follows that $U(-t)Q_+(f,f) \in L^1(0,T)$ for almost all (x,ξ). For Q_-, the same assertion follows from (2.55). Now we can use Lemma 2.12 to conclude that f is a mild solution of the Boltzmann equation in the sense defined above.

The only remaining step is the verification of the entropy estimate (2.9) from (2.33). This is a consequence of the proof of Lemma 2.11, which, for all $\delta > 0$, entails

$$\frac{f^n f^n_*}{1+\delta\int f^n\,d\xi} \to \frac{ff_*}{1+\delta\int f\,d\xi}, \qquad \frac{f^{n'} f^{n'}_*}{1+\delta\int f^n\,d\xi} \to \frac{f'f'_*}{1+\delta\int f\,d\xi},$$

weakly in $L^1\big((0,T)\times\Re^3_x\times\Re^3_\xi\times S^2\big)$. Now, by using the convexity of the function

$$(x,y) \mapsto (x-y)\ln\frac{x}{y}$$

on $\Re_+ \times \Re_+$, we see that for all $T > 0$

$$\int_0^T\int\int \frac{e(f)}{1+\delta\int f\,d\xi}\,dx\,d\xi\,dt \leq \liminf_{n\to\infty}\int_0^T\int\int\frac{e_n(f^n)}{1+\delta\int f^n\,d\xi}\,dx\,d\xi\,dt.$$

The entropy estimate (2.9) follows from this and the monotone convergence theorem in the limit $\delta \to 0$.

Once the lemmas above are taken for granted, this completes the proof of the theorem of DiPerna and Lions.

This result is of the greatest importance for the theory of the Boltzmann equation, yet leaves a lots of problems open. Is the "renormalized solution" a weak solution in the usual sense as well? A positive answer in a particular case will be given in Sect. 7. Is the solution unique? (this question is open even for the case just mentioned). How regular is the solution? Thus, as any important new result in mathematics, the theorem of DiPerna and Lions solved a problem but opened a new chapter of research.

3. More on boundary conditions.

As we indicated in Sect. 1 on $\partial\Omega$ we impose a linear boundary condition of a rather standard form [1-3]:

$$\gamma^+ f(t,x,\xi) = \int_{\xi'\cdot n<0} K(\xi' \to \xi; x,t)\gamma^- f(t,x,\xi')d\xi' \equiv K\gamma^- f$$

$$(x \in \partial\Omega,\ \xi \cdot n > 0) \tag{3.1}$$

where $K(\xi' \to \xi; x,t)$ is a kernel (the boundary scattering kernel) such that:

$$K(\xi' \to \xi; x,t) \geq 0 \tag{3.2}$$

$$\int_{\xi\cdot n>0} K(\xi' \to \xi; x,t)|\xi \cdot n|d\xi = |\xi' \cdot n| \tag{3.3}$$

$$M_w(\xi) = \int_{\xi'\cdot n<0} K(\xi' \to \xi; x,t)M_w(\xi')d\xi' \tag{3.4}$$

where M_w is the wall Maxwellian given by Eq. (1.10) with $u = 0$ and $\beta = \beta_w$ (β_w being the inverse temperature of the wall), whereas $\gamma^\pm$ are the trace operators introduced in Sect. 1.

The case of an isothermal boundary has been treated by Hamdache [12]. In the case of non-isothermal data along $\partial\Omega$ the initial-boundary value problem possesses boundary data which are compatible, not with a Maxwellian, but rather with one of those steady solutions, whose theory is still in its infancy (for an example see the paper by Arkeryd et al. [13]; thus one cannot expect the solution to tend toward a Maxwellian when $t \to \infty$ as has been recently shown for other kinds of boundary conditions [14, 17]. The main difficulties in tackling this problem seem to lie with large velocities. For this reason, Arkeryd and Cercignani [18] introduced a modified Boltzmann equation in which they cut off all the collisions such that the sum of the squares of two colliding molecules is larger than m^2 where m is an assigned positive constant. The only place where they used this cutoff was the entropy estimate and the need for the cutoff disappears when the temperature is constant. Thus their paper contains also a slightly different proof of Hamdache's theorem, with an extension to more general boundary conditions, to a more detailed study of the boundary behavior, and for the full class of collision operators of the DiPerna-Lions existence context [9].

A central observation for these proofs was an inequality introduced by Darrozès and Guiraud [19] in 1966 and subsequently discussed by several authors. We shall state this inequality in the form of a lemma

Lemma 3.1. [1, 19, 20, 21, 22] *If Eqs. (3.1), (3.2), (3.3) and (3.4) hold, then:*

$$\int \xi \cdot n\gamma f \log \gamma f d\xi \leq -\beta_w \int \xi \cdot n|\xi|^2 \gamma f d\xi \qquad (a.e.\ in\ t\ and\ x \in \partial\Omega) \tag{3.5}$$

where β_w is the inverse temperature evaluated at the point $x \in \partial\Omega$. Unless the kernel in Eq. (3.1) is a delta function, equality holds in Eq. (3.9) if and only if the trace γf of f on $\partial\Omega$ coincides with M_w (the wall Maxwellian).

For a proof see Refs. [21, 2, 22].

If the wall is moving, the above relations hold in the reference frame of the wall. Then, since the Maxwellian M_w has a drift velocity u_w, if we want to adopt a reference frame, with respect to which the wall moves, then ξ must be replaced by $\xi - u_w$.

In a paper by Arkeryd and Maslova [23] the work of Arkeryd and Cercignani [18] was extended to the noncutoff case at the price of introducing some restrictions on the kernel $K(\xi' \to \xi; x, t)$. Subsequently it was shown [24] that one of the conditions can be replaced by a more natural one thanks to the above inequality.

As for external problems there appears to be just one paper [25] dealing with the existence problem at the level of generality of the DiPerna and Lions paper [9].

All these results will be surveyed in the subsequent sections. Attention will be paid to some further results restricted to the case when the solution just depends on one space coordinate and to the problem of the trend to equilibrium which was mentioned above.

4. The results of Arkeryd and Cercignani.

Before discussing the existence theorems for initial-boundary value problems, we need to recall some trace results giving the L^1 regularity of the trace of f on the boundary and study the semigroup generated by the free streaming operator, including a sort of Green's formula, that will be used in Sect. 6. This is done in detail by Arkeryd and Cercignani [18] and will not be repeated here.

In order to deal with the existence theorem in a vessel at rest, with a temperature that varies from one point to another, it is convenient to remark that there is a Maxwellian naturally associated with the problem at each point of the boundary, *i.e.* the wall Maxwellian M_w; an exception is offered by specularly reflecting boundaries, which will not be considered here because they have no temperature associated with the boundary. Eq.(3.5), gives (for smooth solutions):

$$\int \xi \cdot n \gamma f \log \gamma f d\xi + \beta_w \int \xi \cdot n |\xi|^2 \gamma f d\xi \leq 0 \quad \text{(a.e. in } t \text{ and } x \in \partial\Omega) \qquad (4.1)$$

For this reason Arkeryd and Cercignani [18] consider an inverse temperature $\beta(x)$ with $\inf \beta(x) > 0$ which reduces to β_w at each point of the boundary and otherwise depends smoothly on x and the modified H-functional:

$$H = \int f \log f d\xi dx + \int \beta(x) |\xi|^2 f d\xi dx \qquad (4.2)$$

In general H will not decrease in time, as a consequence of the Boltzmann equation and inequality (4.1), because a simple calculation shows that

$$\frac{dH}{dt} \leq - \int \xi \cdot \frac{\partial \beta}{\partial x} |\xi|^2 f d\xi dx \qquad (4.3)$$

Thanks to the truncation for large speeds [18] $|\xi| \leq m$ and the right hand side of Eq. (4.3) is bounded by a constant C given by

$$C = m^3 \parallel \frac{\partial \beta}{\partial x} \parallel_{L^\infty} \int f_0 d\xi dx \qquad (4.4)$$

Thus H is bounded by $H_0 + CT$ on $[0, T]$ if bounded initially by H_0.

This is the only point where the truncation for large speeds is needed; further, the truncation can be dispensed with, if β is constant because the left hand side of Eq. (4.3) is bounded by 0.

Eq. (4.2) implies [18] that both $\int f|\log f|d\xi dx$ and $\int |\xi|^2 f d\xi dx$ are separately bounded in terms of the initial data.

In the following we shall use the following notation:

$$< f, g >= \int_{\mathcal{D}} f g \, dx d\xi dt \qquad (4.5)$$

$$< f, g >_{\pm}= \int_{\partial E^{\pm}} f g \, d\sigma^{\pm} \qquad (4.6)$$

$$< f, g >_t= \int_{\Omega \times R^3} f(t, x, \xi) g(t, x, \xi) dx d\xi \qquad (4.7)$$

$$(f, g)_{\pm} = \int_{\pm \xi \cdot n > 0} f g |\xi \cdot n(x)| d\xi. \qquad (4.8)$$

We also define the backward and forward stay times as

$$t^+ = t^+(x, \xi, t) = \inf\{s > 0; x - s\xi \in \partial\Omega\} \qquad (4.9)$$

$$t^- = t^-(x, \xi, t) = \inf\{s > 0; x + s\xi \in \partial\Omega\} \qquad (4.10)$$

with the related quantities

$$s^+(x, \xi, t) = \min\{t, t^+(x, \xi, t)\}, \qquad (4.11)$$

$$s^-(x, \xi, t) = \min\{T - t, t^-(x, \xi, t)\}. \qquad (4.12)$$

We also use the mappings

$$R^s : \overline{\mathcal{D}} \to \overline{\mathcal{D}}$$

with

$$R^s(x, \xi, t) = (t + s, x + s\xi, \xi) \qquad (4.13)$$

to define

$$f^{\#}(s, x, \xi, t) = f^{\#}(s) = f(R^s((x, \xi, t))). \qquad (4.14)$$

As hinted at in Sect.1, in this and the next sections use will be made of the equivalent concepts of exponential, mild, and renormalized solutions as defined by DiPerna and Lions [9] for the Cauchy problem and such solutions will be found as limits of functions solving truncated equations.

The definitions of these solutions require some comment because of the boundary conditions, which are not satisfied exactly but only in the form of an inequality, as Hamdache [12] first pointed out. This aspect of the matter has been discussed by Arkeryd and Maslova [23] in some detail. The basic point is that, when approximating a solution with a sequence, we partially lose control upon the traces, that can only be shown to tend to measures $\mu^{\pm}$ of the spaces $\mathcal{M}^{\pm}$ of σ-finite measures defined on the σ-algebras $\mathcal{B}^{\pm}$ of Borel sets from $E^{\pm}$. Each of these measures can be decomposed into a part completely continuous $\mu_c^{\pm}$ with density $f^{\pm}$ with respect to the Lebesgue measure $d\sigma^{\pm}$, and a singular part $\mu_s^{\pm}$. The measures $\mu_s^{\pm}$ satisfy the boundary conditions:

$$\mu^+ = K\mu^- \qquad (4.15)$$

where K can be defined for measures via

$$< \varphi, K\mu^- >_+=< K^*\varphi, \mu_- >_- \qquad (4.16)$$

where the adjoint operator K^* is defined by:

$$K^*\varphi(x,\xi',t) = \int_{\xi\cdot n(x)>0} \varphi(x,\xi,t)K(\xi' \to \xi; t, x)|\xi \cdot n(x)|/|\xi' \cdot n(x)|d\xi. \quad (4.17)$$

and assumed to carry $C_0(E^+)$ into $C_b(E^-)$.

The traces of a solution $\gamma^\pm f$ will only satisfy

$$\frac{d\mu^\pm}{d\sigma^\pm} \geq \gamma^\pm f \quad (4.18)$$

Then we can introduce the following definitions:

Definition 4.1. f *is a mild solution of (1.1-3) if*

$$f \in L^1(\mathcal{D}), \quad f \geq 0, \quad (Q^\pm)^\# \in L^1([0,s_-(x,\xi,t)]) \quad (4.19)$$

$$f^\#(s,x,\xi,t) = f^\#(\tau,x,\xi,t) + \int_\tau^s Q^\#(z,x,\xi,t)dz, \quad 0 \leq s \leq \tau \leq s^-(x,\xi,t) \quad (4.20)$$

$$f = f_0(x,\xi) \quad in \ E^0 \quad (E^0 = (x,\xi,t \in \overline{\mathcal{D}}; t=0) \quad (4.21)$$

for a.e. $(x,\xi,t) \in E^\pm \cup E^0$*, and there are* $\mu^\pm \in \mathcal{M}^\pm$ *satisfying (4.15) and (4.18).*

Definition 4.2. f *is a solution in exponential multiplier form (or exponential solution for short) of (1.1-3) if*

$$f \in L^1(\mathcal{D}), \quad f \geq 0, \quad \nu(f) \in L^1_{loc}(\mathcal{D}) \quad (4.22)$$

$$f^\#(s,x,\xi,t) = [f_0(x,\xi)\chi^0 + \gamma_+ f\chi^+]\exp(-\int_0^s (\nu(f))^\#(z,x,\xi,t)dz)+$$
$$\int_0^s (Q^+)^\#(z,x,\xi,t)\exp(-\int_z^s (\nu(f))^\#(z',x,\xi,t)dz')dz, \quad 0 \leq s \leq s^-(x,\xi,t)$$
$$\quad (4.23)$$

for a.e. $(x,\xi,t) \in E^\pm \cup E^0$*, and there are* $\mu^\pm \in \mathcal{M}^\pm$ *satisfying (4.15) and (4.18). Here* χ^+ *and* χ^0 *denote the characteristic functions of* E^+ *and* E^0*.*

Definition 4.3. f *is a renormalized solution of (1.1-3) if*

$$f \in L^1(\mathcal{D}), \quad f \geq 0, \quad \nu(f) \in L^1_{loc}(\mathcal{D}) \quad (4.24)$$

and f *is a weak solution (with test functions vanishing in the neighborhood of* $E^\pm$*) of*

$$\Lambda \log(1+f) = \frac{Q(f,f)}{1+f} \quad in \ \mathcal{D} \quad (4.25)$$

and there are $\mu^\pm \in \mathcal{M}^\pm$ *satisfying (4.15) and (4.18).*

The existence theorem proved by Arkeryd and Cercignani [18] reads as follows

Theorem 4.1. *Let* $f^0 \in L^1(\Omega \times \Re^3)$ *be such that*

$$\int f^0(1+|\xi|^2)d\xi dx < \infty; \quad \int f^0|\log f^0|d\xi dx < \infty. \quad (4.26)$$

Then there is a solution $f \in C(\Re_+, L^1(\Omega \times \Re^3))$ *of the Boltzmann equation such that* $f(.,0) = f^0$*, which also satisfies mass conservation and has an H-functional with a bounded time derivative.*

For the proof we refer to the original paper [18] .

It is interesting to study the boundary condition satisfied by these solutions and prove

Theorem 4.2. *There is a solution as in Theorem 4.1, which satisfies*

$$\gamma^+(f) \geq K(\gamma^- f) \quad a. \ e. \ on \ E^+ \quad , \quad (4.27)$$

For the proof, we refer again to the original paper. We remark that the fact that we obtain an inequality is a consequence of the fact that we can only expect convergence of the traces of the approximating sequence to measures, as discussed above. In fact Eq. (4.27) follows by taking the completely continuous part of Eq. (4.16) .

Theorems 4.1 and 4.2 contain Hamdache's result [12] and extend it. The extension is of interest for the study of the solutions of the Boltzmann equation when the boundaries drive the gas out of equilibrium. In order to obtain a realistic result one needs to remove the cutoff, as done for the first time by Arkeryd and Maslova [23] and discussed in the next section.

5. THE RESULTS OF ARKERYD AND MASLOVA.

In this section we study the results presented by Arkeryd and Maslova in a recent paper [23] . They introduce a class of boundary operators for which (1.6) (3.1-3) hold, by restricting the adjoint operator K^*, but are able to avoid the cutoff for large velocities.

A better control of mass, energy and entropy for the distributions emerging from the wall are provided by the following conditions:

There exists $K_2 > 0$ such that $K^* |\xi \cdot n(x)| \geq K_2$ (spreading condition)   (5.2)

There exists $K_3 \leq \infty$ such that $K^* |\xi|^2 \leq K_3$ (energy condition)   (5.3)

There exists $K_4 < \infty$ and $\alpha \in [0, 1)$ such that, for every $f \in L^1(\Gamma^-)$ with $f \geq 0$,

it holds $< Kf, \log(Kf/(f, 1)_-) >_+ -\alpha \hat{H}^- \leq K_4(q_2^- + q)$ (entropy condition)
(5.4)

Here

$$\hat{H}^- = < f, \log(f/(f, 1)_-) >_-$$
$$q_j^\pm = < f, |\xi|^j >_\pm \qquad q = < f, |\xi \cdot n| >_+ + < f, |\xi \cdot n| >_-$$ (5.5)

These conditions are reasonable for a linear operator, except for (5.4), which appears a bit unusual, since it is nonlinear, albeit homogeneous of first degree, in f. In the next section we shall discuss how to dispense with that condition by using Lemma 3.1 in a suitable way.

The other conditions have the following role (Lemma 4.1 of Arkeryd and Maslova [23]):

1) Eq. (5.2) (together with a proper use of momentum balance) gives a control on the mass flow hitting the boundary.

2) Eq. (5.3) (together with 1)) gives a control on q_2^+.

3) Using 1) and 2) together with energy balance one obtains an a priori bound upon the energy without using the entropy estimates.

At this point Arkeryd and Maslova [23] use Eq. (5.4) to bound entropy and entropy source. They also obtain bounds on $< f, |\log(f/(f, 1)_-)| >_\pm$ which are related to entropy flows. To bound the latter, however, one should remove the denominator $(f, 1)_-$, which does not appear to be an easy matter.

The following lemma holds [23]:

Lemma 5.1. *Assume Eqs. (3.2-3) and (5.2-4), together with*

$$(f_0, \log f_0) \in L^1(\Omega) \tag{5.6}$$

$$(Q, \log f) \le 0, \quad (Q, \psi) = 0 \quad for \ \psi = 1, \xi, |\xi|^2. \tag{5.7}$$

Then f satisfies the inequality

$$- < Q, \log f > + H(T) + < f, |\log(f/(f,1)_-)| >_\pm \le C(T), \tag{5.8}$$

with $C(T) > 0$ depending only on f_0 and on K_2, K_3, K_4.

Having these a priori bounds they proceed more or less as in the paper by Arkeryd and Cercignani [18] , the main change being that they prefer to avoid the semigroups that were used there, and finally arrive at

Theorem 5.2. *Assume that*

$$(1 + |\xi|^2)f_0, \quad f_0 \log f_0 \in L^1(\Omega \times R^3), \qquad f_0 \ge 0.$$

and Eqs. (5.2-4). Then there exists an exponential solution of (1.1-3) satisfying

$$f \in C([0,T], L^1(\Omega \times R^3)), \quad f \ge 0, \quad < f, 1 >_t = < f_0, 1 >_0;$$

$$(1 + |\xi|^2)\gamma^\pm f \in L^{1\pm} ; \tag{5.9}$$

$$\sup_{t \le T} [< f, \ln f >_t + < f, |\xi|^2 >_t] + < e(f), 1 > \le C(T). \tag{5.10}$$

Here, in agreement with the notation in Sect. 2:

$$e(f) = \frac{1}{4} \int_{R^3} \int_{B^+} (f'f'_* - ff_*) \log \frac{f'f'_*}{ff_*} B(\xi - \xi_*, n) d\xi_* dn, \tag{5.11}$$

6. Another result and a generalization

In this section, following a paper of the author [24] , we shall describe a result which relaxes one of the assumptions of Arkeryd and Maslova [23] as well as a generalization to the case of moving boundaries.

The result alluded to is that Lemma 5.1 and (as a consequence) Theorem 5.2 hold without assuming (5.4), but only the compatibility with a Maxwellian, Eq. (3.4). According to Lemma 3.1, the inequality of Darrozès and Guiraud [19] then holds. Using Green's formula [6, 15, 20], approximation and Lemma 3.1 we can prove [24]

Lemma 6.1. *Assume Eqs. (3.2-4) and (5.2-3) , together with $\beta_w \le C_o(T)$ and*

$$(f_0, \log f_0) \in L^1(\Omega) \tag{6.1}$$

$$(Q, \log f) \le 0, \quad (Q, \psi) = 0 \quad for \ \psi = 1, \xi, |\xi|^2. \tag{6.2}$$

Then f satisfies the inequality

$$- < Q, \log f > + H(T) \le C(T), \tag{6.3}$$

with $C(T) > 0$ depending only on f_0 and on K_2, K_3.

The only part of the thesis of Lemma 5.1, which does not follow from the new assumptions is the boundedness of the entropy flows $< f, |\log(f/(f,1)_-)| >_\pm$. This part of the lemma is never used in the proof of theorem (5.2) and thus we can prove

Theorem 6.2. *Assume that*

$$(1 + |\xi|^2)f_0, \quad f_0 \log f_0 \in L^1(\Omega \times R^3), \qquad f_0 \ge 0. \tag{6.4}$$

and Eqs. (3.2-4) and (5.2-3). Then there exists an exponential solution of (1.1-3) satisfying

$$f \in C([0,T], L^1(\Omega \times R^3)), \quad f \geq 0, \quad < f, 1 >_t = < f_0, 1 >_0; \tag{6.5}$$

$$(1 + |\xi|^2)\gamma^{\pm} f \in L^{1\pm} ; \tag{6.6}$$

$$\sup_{t \leq T}[< f, \ln f >_t + < f, |\xi|^2 >_t] + < e(f), 1 > \leq C(T). \tag{6.7}$$

where $e(f)$ is given by Eq. (5.11).

Remark. The above result applies also to the inhomogeneous boundary condition

$$\gamma^+ f(x, \xi, t) = \alpha f_+ + (1 - \alpha) K \gamma^- f \quad (0 \leq \alpha \leq 1) \tag{6.8}$$

where $f_+ \geq 0$ is assigned.

We pass now to problems with moving boundaries. These are of some interest and do not seem to have been considered so far. The main difference is that in Eq. (1.6) E^+ varies with time, because $\partial\Omega$ does. As remarked in Sect. 2, all the relations concerning the kernel K hold in the reference frame of the wall. Then, when the Maxwellian M_w has a drift velocity u_w, if we want to adopt a reference frame, with respect to which the wall moves, then ξ must be replaced by $\xi - u_w$. In particular, the indices $+$ and $-$ refer now to $(\xi - u_w) \cdot n > 0$ and $(\xi - u_w) \cdot n < 0$, respectively.

When we integrate the Boltzmann equation to obtain a priori inequalities, we obtain a factor $(\xi - u_w) \cdot n$ in place of $\xi \cdot n$, so that most of the changes compensate. The main difference arises in the entropy inequality, where a factor $|\xi - u_w|^2$ appears in place of simply ξ^2. The extra terms can be easily controlled, however, by means of the momentum balance equation (after scalar multiplication by a smooth vector-valued function $u(x, t)$, which reduces to u_w on the wall)

Then we have the following

Corollary 6.3. *Theorem 6.2 holds in the presence of moving walls as well.*

7. IMPROVED RESULTS IN THE CASE OF A SLAB

In a recent paper [26] , R. Illner and the author proved a new result on the initial-boundary value problem for the nonlinear Boltzmann equation in the interval $\Omega = [0, 1]$ in one-dimensional spatial geometry, with general diffusive boundary conditions at $x = 0$ and $x = 1$. Thus in this section $x \in \Re$; in addition, the x-, y- and z- component of the velocity $\xi \in \Re^3$ will be denoted by ξ, η and ζ respectively; in order to avoid confusion we shall replace, when needed, the notation for the velocity vector $\xi \in \Re^3$ by v. The Boltzmann equation reads as follows

$$\frac{\partial f}{\partial t} + \xi \frac{\partial f}{\partial x} = Q(f, f) \tag{7.1}$$

the remaining part of the notation being as before.

Cercignani and Illner [26] needed some truncations on the collision kernel B, in order to obtain more advanced results on the solution, in particular to replace the renormalized solutions of DiPerna and Lions by the more powerful result that the solution is a weak one in the usual sense. This line was started by the author [27, 28, 29]. In order to present their result, we assume that there is an $\epsilon > 0$ such that

$$B(\dots) = 0 \text{ if } |v - v_*| \leq \epsilon. \tag{7.2}$$

$$B \text{ is bounded.} \tag{7.3}$$

A third and less serious assumption on B is that the ratio r between $\int_{\mathbf{S}}[n \cdot (v - v_*)]^2 B(n \cdot (v - v*), |v - v_*|)dn$ and $|v - v_*|^2 \int_{\mathbf{S}} B(n \cdot (v - v_*), |v - v_*|)dn$ is bounded from below.

The assumption (7.2) can be summarized as saying that "collisions with small relative speed are disregarded" and is therefore physically more reasonable than the assumptions made by Arkeryd in Ref. [30].

For $x \in \partial\Omega$, i.e., $x \in \{0,1\}$, and $\omega = (-1)^x$, we impose the usual boundary conditions discussed in Sects. 1 and 3.

The objective of Ref. [26] was to show that under reasonable assumptions on the diffuse boundary condition, and with the truncations on the collision kernel B made in (7.2) and (7.3), the initial-boundary value problem for the Boltzmann equation has a global weak solution in the usual sense. The main step in Ref. [26] was a proof that the gain and loss terms of the collision term $Q(f, f)$ are in $L^1([0, 1] \times \Re^3 \times [0, T])$ for any positive time $T > 0$. Cercignani and Illner [26] also showed that the boundary conditions are satisfied as identities in the weak sense, and obtained uniform bounds (for a given time interval) on the second moment (the kinetic energy) of f.

The assumptions on the boundary kernels are the same made in the previous section and thus exclude specular and bounce-back reflection.

In order to discuss the results of Ref. [26], we need some additional notation. For each $x \in [0, 1]$ and $t \geq 0$, let

$$\rho(x, t) = \int f(x, v, t)\, dv$$

$$m(t) = \int \rho(x, t)\, dx$$

$$j(x, t) = \int \xi f(x, v, t)\, dv$$

$$p(x, t) = \int \xi^2 f(x, v, t)\, dv$$

$$q(x, t) = \int \xi v^2 f(x, v, t)\, dv. \tag{7.4}$$

We call ρ the mass density, $m(t)$ the total mass, j the mass flux (or momentum) in x−direction, p the momentum flux, and q the energy flux. At the boundaries we will need the ingoing and outgoing parts of these quantities. We use the abbreviations

$$\rho_+ = \int_{\xi>0} f\, dv, \ \rho_- = \int_{\xi<0} f\, dv, j_+ = \int_{\xi>0} \xi f\, dv, \ j_- = \int_{\xi<0} |\xi| f\, dv, \tag{7.5}$$

etc., such that $\rho = \rho_+ + \rho_-$, $j = j_+ - j_-$, $p = p_+ + p_-$ and $q = q_+ - q_-$.

Introducing an extension of the work of Bony [31] for discrete velocity models to the continuous velocity case, the following functional was considered [26-29]:

$$I[f](t) = \underbrace{\int \int}_{x<y} \int_v \int_{v_*} (\xi - \xi_*) f(x, v, t) f(y, v_*, t)\, dv_* dv dx dy \tag{7.6}$$

where the first double integral is over the triangle $0 \leq x < y \leq 1$. One can then prove the following relation

$$\int_0^T \int_0^1 \int_v \int_{v_*} (\xi - \xi_*)^2 f(x, v_*, t) f(x, v, t) \, dv dv_* dx dt$$

$$= \quad I[f](0) - I[f](T) + \int_0^T \left(p(0,t) \int_0^1 \rho(x,t) \, dx + p(1,t) \int_0^1 \rho(x,t) \, dx \right) dt \; ; \quad (7.7)$$

and show that the left-hand side of (7.7) is bounded for any finite time interval, though it may grow exponentially in time.

As remarked in Ref. [26], boundedness of the left-hand side of (7.7) follows, if we can obtain bounds on

$$\int_0^1 j_\pm(x,t) \, dx, \quad \int_0^T p(0,t) \, dt, \text{ and on } \int_0^T p(1,t) \, dt.$$

Such bounds were obtained in Ref. [26] by a series of estimates, analogous to those used in Refs. [23] and [24], which lead to

Lemma 7.1. *If f is a sufficiently smooth solution of the initial-boundary value problem for Eq. (7.1) and the boundary conditions of the kind considered in the previous section, with initial value f_0, then*

$$E(t), \quad \int_0^t (p(1,\tau) + p(0,\tau)) \, d\tau \quad , \int_0^1 (j_+ + j_-)(x,t) \, dx$$

and

$$\int_0^t \int_0^1 \int_v \int_{v_*} (\xi - \xi_*)^2 f(x, v_*, \tau) f(x, v, \tau) \, dv dv_* dx d\tau$$

can grow at most exponentially in time.

The next objective is then to show that the collision terms themselves can also grow at most exponentially in time. The method used in Ref. [26] to this end is the same as in Refs. [27, 28, 29].

Following largely the notation of Refs. [26-29], let

$$d\mu = dn dv_* dv dx \quad (7.8)$$

and, for $0 \leq \tau \leq T$,

$$\Delta(\tau, T) =$$

$$\int_{[0,1] \times \Re^6 \times S^2 \times [\tau, T]} B(n \cdot (v - v_*), |v - v_*|) f(x, v, t) f(x, v_*, t) \, d\mu dt. \quad (7.9)$$

Lemma 7.2. *If the solution of the initial-boundary value problem for (7.1), defined as above exists as a classical solution for $t \in (0, \infty)$, and if the initial value f_0 has a finite H-functional $H[f_0]$ and finite energy $E(0) = \int_0^1 \int v^2 f_0(x,v) \, dv dx$, then there is a constant K (depending on the initial data, and ϵ and growing at most exponentially with T) such that*

$$\Delta(\tau, T) \leq K. \quad (7.10)$$

The proof of Lemma 7.2 is a simple consequence of the next two lemmas.

Lemma 7.3. *Let u_1 be the x-component of the bulk velocity*

$$u_1 = \frac{\int \xi f d\xi}{\int f d\xi} \quad (7.11)$$

Then

$$\int_{\Re^3 \times \Re^3 \times [0,T] \times \Re} (\xi - u_1)^2 f(x,v,t) f(x,,v_*,t) dx dt dv dv_*$$

$$< K_0 \tag{7.12}$$

where K_0 is a constant, which only depends on the initial data, and can grow at most exponentially with T.

In fact, the integral in Eq.(7.12) is nothing else than the integral in Lemma 7.1 (except for a factor 2) suitably rearranged. It is enough to expand the squares in both integrals and replace $\int \xi f d\xi$ by $u_1 \int f d\xi$.

Using an argument from Refs. [28] and [29], one can prove:

Lemma 7.4. *Under the above assumptions*

$$\int_{\Re^3 \times \Re^3 \times \mathbf{S}^2 \times [0,T] \times \Re} |v - u|^2 f(x,v,t) f(x,v_*,t) B(n \cdot (v - v*), |v - v_*|) dt dx dv dv_* dn$$

$$< K_0 \tag{7.13}$$

where K_0 is a constant, which only depends on the initial data and can grow at most exponentially with T.

A simple rearrangement leads to:

Lemma 7.5. *Under the assumptions of Lemma 7.3, we have, for smooth solutions:*

$$\int_{\Re^3 \times \Re^3 \times \mathbf{S}^2 \times [0,T] \times \Re} |v - v_*|^2 f(x,v,t) f(x,v_*,t) B(n \cdot (v - v*), |v - v_*|) d\mu dt$$

$$< K_0 \tag{7.14}$$

where K_0 is the same constant as in Lemma 7.4 and hence can grow at most exponentially with T.

Lemma 7.2 now follows thanks to Eq. (7.14) and the fact that $B(.,.)$ is zero for $|v - v_*| \leq \epsilon$. Then we have

$$\int_{\Re^3 \times \Re^3 \times \mathbf{S}^2 \times [0,T] \times \Re} f(x,v,t) f(x,v_*,t) B(n \cdot (v - v*), |v - v_*|) dt dx dv dv_* dn$$

$$< K_0/\epsilon^2 \tag{7.15}$$

As in Ref. [26] the above estimates imply the existence of a global weak solution for the initial-boundary value problem, with the boundary conditions satisfied as equalities and not as inequalities. This can be stated in the form of a theorem as follows

Theorem 7.6. *Let $f_0 \in L^1([0,1] \times \Re^3)$ be such that*

$$\int f_0(\cdot)(1 + |x|^2 + |v|^2) dv dx < \infty; \qquad \int f_0 |\ln f_0(.)| dv dx < \infty. \tag{7.16}$$

Also, assume that the collision kernel B and the boundary conditions satisfy the conditions made above. Then there is a global weak solution $f(x,v,t)$ of the initial-boundary value problem for Eq. (7.1) such that $f \in C(\Re_+, L^1([0,1] \times \Re^3))$, $f(.,0) = f_0$. This solution also satisfies the boundary conditions (7.4) a.e. *Proof.* See Ref. [26].

8. TREND TO EQUILIBRIUM

Recently, important progress has been achieved in the old problem of studying the trend to equilibrium. Since the foundations of kinetic theory were laid down by Maxwell and Boltzmann in the last century, one was led to believe that the solution of the Boltzmann equation describing the evolution of the gas in a domain bounded by solid walls kept at a constant temperature, should tend to a Maxwellian distribution. In 1965 H. Grad commented[18] on the difficulty of proving the trend to equilibrium in a rigorous way. Later, doubts even on the possibility of proving this result in a mathematically rigorous way were cast by Truesdell and Muncaster[19], who claimed that the H-theorem which follows from the Boltzmann inequality "is not a sufficient condition for the strict trend to equilibrium"; this is of course correct, if one disregards boundary conditions (which might be incompatible with the trend to equilibrium, as is case, e. g., when the temperature of the boundary is not uniform), or there are no boundaries. If, on the other hand, the boundary conditions are taken into account and are compatible with equilibrium, then the popular belief that the solution should tend to a Maxwellian appears to be well justified[20].

Clearly, several situations may arise. Among the most typical ones, we quote:

1. Ω is a box with periodicity boundary conditions (flat torus). Then there is no boundary and there is a natural (space homogeneous) Maxwellian associated with the total mass, momentum and energy (which are of course conserved). The proof that f tends to this Maxwellian is more complicated than in the space homogeneous case.

2. Ω is a compact domain with specular reflection. There might seem to be a difficulty for the choice of the natural Maxwellian because momentum is not conserved. In the case of a box a simple argument shows that the total momentum must vanish when $t \to \infty$. If $\partial\Omega$ is not a surface of revolution (in which case the gas might rotate as a solid body) the same conclusion should hold. Thus if rotationally invariant domains are excluded, then M is a non-drifting Maxwellian with constant density and temperature. A simpler case is that of the reverse reflection, or "bounce-back" boundary condition; in this case the rotationally invariant domains do not constitute an exception.

3. Ω is the entire space. Then the asymptotic behavior of the initial values at space infinity is of paramount importance. If the gas is initially concentrated at finite distances from the origin, one physically expects and can mathematically prove that the gas escapes through infinity and the asymptotic state is a vacuum.

4. Ω is a compact domain but the boundary conditions on $\partial\Omega$ are different from specular reflection. Then the asymptotic state might be completely different from a Maxwellian, unless the temperature is constant along the boundary.

In the papers discussed in the previous sections, it was generally assumed that there is a boundary Maxwellian M_w, which may vary along the boundary itself. This Maxwellian is uniquely determined except for very special boundary conditions, such as specular and reverse reflection, which we have frequently excluded. If we assume that the Maxwellian is the same at each point of the boundary, as in the paper by Hamdache [12], we can conjecture that the solution will tend asymptotically in time toward the nondrifting Maxwellian M_w. A proof of this was provided in several papers. In fact, the circumstance that there is a weak limit when time tends to infinity and that this limit is a Maxwellian was discussed by Desvillettes

[14] and Cercignani [15] (see also Ref. 3), starting from a remark by DiPerna and Lions [32]. Subsequently L. Arkeryd [16] proved that f actually tends to a Maxwellian in a strong sense for a periodic box, but his argument works in other cases as well; his proof uses techniques of nonstandard analysis and, as such, is outside the scope of this paper. Then P.-L. Lions [17] obtained the same result without resorting to nonstandard analysis. The author [33], following the approach of Ref. [17] gave a proof that is particularly suitable to deal with the solutions in a slab discussed in the previous section. The main differences from Lions's proof [17] are: a) his assumption that $B > 0$ a.e. is not needed; b) the Maxwellian is uniquely determined, thanks to the boundary conditions which allow a unique Maxwellian.

We also point out that recently Arkeryd and Nouri [34] have sketched a proof of the fact that for boundary conditions satisfying the restriction of Ref. 23 and $B > 0$, the Maxwellian is uniquely determined (for renormalized solutions). This had already been pointed out for the weak limit in Ref. [15] (see also Ref. [3]).

We remark that the new point with respect to the general case is that we have constants in our estimates in place of functions which may grow exponentially in time [12, 33]. We then obtain:

Theorem 8.1. *Let f be a solution of the initial boundary value problem (1.3), (1.6), where a suitably vanishing B is allowed. Then, when t tends to infinity, $f(.,.,t)$ converges strongly to the global Maxwellian $n_0 M_w$ where the constant factor n_0 is uniquely fixed by mass conservation.*

Proof. It is enough to show that for every sequence t_n tending to ∞ there exists a subsequence t_{n_k} such that $f_{n_k}(x,v,t) = f(x,v,t+t_{n_k})$ converges in $L^1((0,1) \times \Re^3 \times [0,T])$ to $n_0 M_w$ for any $T > 0$. The weak convergence of this sequence follows from the uniform boundedness of mass, energy and entropy.

Thus $f_n(x,v,t) = f(x,v,t+t_n)$ is weakly compact in $L^1(\Omega \times \Re^3 \times [0,T])$ for any sequence t_n of nonnegative numbers and any $T > 0$. If $t_n \to \infty$, then there exist a subsequence t_{n_k} and a renormalized solution $M(x,v,t)$ in $L^1(\Omega \times \Re^3 \times [0,T])$ such that f_{n_k} converges weakly to $M(x,v,t)$ in $L^1(\Omega \times \Re^3 \times [0,T])$ for any $T > 0$; in addition, the gain term $Q^+(f_n, f_n)$ converges a. e. to $Q^+(M, M)$. In order to prove that M is a Maxwellian, we remark that, since the integral $\int \int e(f) dv dt$ as given by (5.11) is finite, then

$$\int_{t_{n_k}}^{T+t_{n_k}} \int_{\Re^3} \int_\Omega \int_{S^2} \int_{\Re^3} [f(x,v',t)f(x,v'_*,t) - f(x,v,t)f(x,v_*,t)]$$

$$\times \log \frac{f(x,v',t)f(x,v'_*,t)}{f(x,v,t)f(x,v_*,t)} B(n \cdot (v - v_*), |v - v_*|) d\mu dt \to 0 \qquad (k \to \infty)$$

and thus

$$\int_0^T \int_{\Re^3} \int_\Omega \int_{S^2} \int_{\Re^3} [f_{n_k}(x,v',t)f_{n_k}(x,v'_*,t) -$$

$$f_{n_k}(x,v,t)f_{n_k}(x,v_*,t)] \log \frac{f_{n_k}(x,v',t)f_{n_k}(x,v'_*,t)}{f_{n_k}(x,v,t)f_{n_k}(x,v_*,t)}$$

$$\times B(n \cdot (v - v_*), |v - v_*|) d\mu dt \to 0 \qquad (8.1)$$

$$(k \to \infty)$$

And, according to an argument by DiPerna and Lions [32] (see also Ref. 3) we can pass to the limit and obtain

$$\int_0^T \int_0^1 \int_{\Re^3} \int_{S^2} \int_{\Re^3} [M(x,v',t)M(x,v'_*,t) - M(x,v,t)M(x,v_*,t)]$$

$$\times \log \frac{M(x,v',t)M(x,v_*,t)}{M(x,v,t)M(x,v_*,t)} B(n \cdot (v - v_*), |v - v_*|) d\mu dt = 0 \quad . \tag{8.2}$$

This implies

$$M(x,v',t)M(x,v'_*,t) = M(x,v,t)M(x,v_*,t)$$

$$\text{(a. e. in } v_*, n, x, v_*, t \quad \text{for } B(V,n) \geq 0) \tag{8.3}$$

In the case of the slab discussed in the previous section we have the unusual restriction on the relative speed which produces a vanishing kernel for $V < \epsilon$. We use, however, the fact that one can use local arguments (in v, v_*) to deduce that $M(x,v,t$ is a local (in x and t) Maxwellian. This was clear to Boltzmann [35, 36, 37] for twice differentiable functions and has been extended to the case when f is only assumed to be a distribution by Wennberg [38]. Then we conclude that $M(x,v,t)$ is a local Maxwellian.

But we have for all $K > 1$

$$|f'_{n_k} f'_{n_k*} - f_{n_k} f_{n_k}| \leq (K-1) f_{n_k}, f_{n_k} + \frac{1}{\log K} (f_{n_k}(x,v',t) f_{n_k}(x,v'_*,t)$$

$$-f_{n_k}(x,v,t) f_{n_k}(x,v_*,t)) \log \frac{f_{n_k}(x,v',t) f_{n_k}(x,v'*,t)}{f_{n_k}(x,v,t) f_{n_k}(x,v_*,t)} \quad . \tag{8.4}$$

Then, since $e(f_{n_k})$ converges to 0 a.e. and $Q^+(f_{n_k}, f_{n_k})$ converges to $Q^+(M,M)$ a. e. we deduce that the loss term $Q^-(f_{n_k}, f_{n_k})$ converges a. e. to $Q^-(M,M)$. Now, the loss term is of the form $f\nu(f)$, where $\nu(f)$ is a convolution product in velocity space. Then $f_{n_k}\nu(f_{n_k}) \to M\nu(M)$ a.e.. Then either ρ_M is zero, in which case f_{n_k} converges strongly to zero (a. e. in v), or is nonzero. In the second case $\nu(M)$ is also nonzero and if we let $u_{n_k} = \nu(f_{n_k})/L(M)$ we have that $u_{n_k} \to 1$ a.e. (by the averaging lemma). Then since $f_{n_k} u_{n_k}$ tends to $M(x,v,t)$ a. e., we conclude that $f_{n_k} \to M$ a.e.

But $M(x,v,t)$ must be a (renormalized and hence weak) solution of the Boltzmann equation or, since the collision term vanishes:

$$\Lambda f = 0 \tag{8.5}$$

In addition M must satisfy the boundary condition (1.6) [26].

Thus the solutions of the Boltzmann equation in a slab with the boundary conditions (1.6) tend (in the case of a boundary at constant temperature) to Maxwellians satisfying the free transport equation, $\Lambda f = 0$. These Maxwellians are well known since Boltzmann and are, e. g., discussed in Chapter III of Ref. [3]. Now if we specialize this general solution to the case when $M(x,.,t)$ is an L^1 function for any $t \geq 0$, and satisfies the boundary conditions, we see that M is a Maxwellian with no drift and constant temperature; this immediately implies that M is a uniform Maxwellian, which must coincide with M_w (which is an absolute nondrifting Maxwellian) except for a factor, which is fixed by mass conservation. Thus we have proved Theorem 8.1.

9. EXTERNAL PROBLEMS

For applications to aerospace problems, the external case is of paramount importance. This aspect had been treated [6,7] only in the perturbation framework described in Sect. 1, till P.-L. Lions [25] provided the relevant tools for dealing with external problems for large data in an L^1 framework. .

In this situation $\Omega = \overline{\mathcal{O}}^c$, where $\mathcal{O}$ is a smooth bounded set in $\Re^3$. One provides an initial condition (1.5), boundary condition (1.6) on the boundary $\partial\Omega \times \Re^3 \times (0,T)$ and a condition at infinity

$$f \to M \quad \text{as} \quad |x| \to \infty \tag{9.1}$$

where M is any assigned Maxwellian.

The results proved by Lions [25] are based on the following a priori estimates

$$\sup_{t\in[0,T]} \int_{\Omega\times\Re^3} dx d\xi (f \log \frac{f}{M} + M - f) \leq C \tag{9.2}$$

$$\sup_{t\in[0,T]} \sup_{x_0\in\Omega} \int_{\Omega\cap(x_0+B^1)} dx \int_{\Re^3} d\xi f(1 + |v|^2 + |\log f|) \leq C \tag{9.3}$$

whenever f_0 satisfies

$$\int_{\Omega\times\Re^3} dx d\xi (f_0 \log \frac{f_0}{M} + M - f_0) \leq C_0 \tag{9.4}$$

for some positive constant $C = C(C_0, T)$. In the above B^1 is the unit ball.

Lions [25] gives a sketch of the proofs of the following results in the case when $\partial\Omega$ is specularly reflecting:

Theorem 9.1. *Let $f_0 \geq 0$ satisfy (9.4). Then there exists a global renormalized solution satisfying Eq. (9.1).*

This solution is also a weak solution in a sense introduced in Ref. [17].

The result can be extended to other boundary conditions with the usual difficulty of the trace control. In fact we can apply the results of previous sections to $\Omega^R \times \Re^3$, where $\Omega^R = B^R \cap \Omega$ with B^R's radius R large enough and the following boundary condition on the artificial boundary:

$$\gamma^+ f = M \quad \text{on} \quad (\partial B^R \cap \Omega) \times \Re^3 \times (0,T) \tag{9.5}$$

which is a particular case of (6.8) for $\alpha = 1$ and $f^+ = M$. Then using the compactness properties of sequences of approximated solutions (see Theorem IV.2 in Lions's paper [21], with obvious changes) we can pass to the limit as R goes to ∞. This can be easily performed because the proof of the a priori bound (9.2) is only modified because of an additional boundary term of the form

$$\int_{\partial B^R\cap\Omega} d\sigma \int d\xi \xi \cdot n(f \log \frac{f}{M} + M - f) =$$

$$-\int_{\partial B^R\cap\Omega} d\sigma \int_{\xi\cdot n<0} d\xi |\xi \cdot n|(f \log \frac{f}{M} + M - f) \tag{9.6}$$

which is obviously negative, contributes to the time derivative. Then the estimates remain valid with constants independent of R.

10. NUMERICAL METHODS.

Although analytical techniques have been very useful in obtaining approximate solutions and forming qualitative ideas on the solutions of practical problems, in general they do not provide us with detailed and precise answers to the sort of questions that are posed by the space engineer. Various numerical procedures exist which either attempt to solve for f by conventional techniques of numerical analysis or efficiently by-pass the formalism of the integrodifferential equation and simulate the physical situation that the equation describes (Monte Carlo methods). Only recently proofs have been given that these partly deterministic, partly stochastic games provide solutions that converge (in a suitable sense) to solutions of the Boltzmann equation. There appear to be very few limitations to the complexity of the flow fields that this approach can deal with. Chemically reacting and ionized flows can and have been analysed by these methods.

Numerical solutions of the Boltzmann equation based on finite difference methods meet with severe computational requirements due to the large number of independent variables. In practice the only method that can be used for practical calculations is the technique of Hicks, Yen and Nordsiek [56, 57], which is based on a Monte Carlo quadrature method to evaluate the collision integral. This method was further developed by Aristov and Tcheremissine [58, 59] and has been subsequently applied to a few two-dimensional flows [60, 61].

An additional difficulty for these methods is the fact that chemically reacting and thermally radiating flows are hard to describe with theoretical models having the same degree of accuracy as the Boltzmann equation for monatomic nonreacting and nonradiating gases. These considerations paved the way to the development of simulation methods, which started with the work of Bird [62, 63] on the so called Direct Simulation Monte Carlo (DSMC) method and have become a powerful tool for practical calculations.

In the DSMC method, the intermolecular collisions are considered on a probabilistic rather than a deterministic basis. Furthermore, the gas of real life is modeled by some thousands of simulated molecules on a computer. For each of them the space coordinates and velocity components (as well as the variables describing the internal state, if we deal with a polyatomic molecule) are stored in the memory and are modified with time as the molecules are simultaneously followed through representative collisions and boundary interactions in the simulated region of space. The calculation is unsteady and the steady solutions are obtained as asymptotic limits of unsteady ones. The flow field is subdivided into cells, which are taken to be small enough for the solution to be approximately constant through the cell. The time variable is advanced in discrete steps of size Δt, small with respect to the mean free time, *i. e.* the time between two subsequent collisions of a molecule. This permits a separation of the inertial motion of the molecules from the collision process: one first moves the molecules according to collision-free dynamics and subsequently the velocities are modified according to the collisions occurring in each cell.

Some variations of Bird's method have appeared, due to Koura [67], Belotserkovskii and Yanitskii [68], Deshpande [69]. They differ from Bird's method because of the method used to sample the time interval between two subsequent collisions.

A different method was proposed by Nanbu [73]. What is new in his method is the circumstance that it does not try to simulate the N-body dynamics, but the description of the system given by the Boltzmann equation. Nanbu's method is now well understood, from both a physical and mathematical standpoint and has been rigorously proven to yield approximations to solutions of the Boltzmann equation, provided the number of test particles is sufficiently large [70, 71]. Subsequently convergence of Bird's method was proved [64, 66]. The first of these proofs is nonconstructive, but the second one is constructive and is based on the strategy used in the validation proof of O. Lanford [65].

The computing task of a simulation method varies with the molecular model. For models other than Maxwell's it is proportional to N for Bird's method, while it is proportional to N^2 for Nanbu's method. Babovsky [70, 71], however, found a procedure to reduce the computing task of Nanbu's method and make it proportional to N. His modification is based on the idea of subdividing the interval [0,1] into N equal subintervals.

The Monte Carlo method discussed above is not only a practical tool for engineers, but also a good method for probing into uncovered areas of the theory of the Boltzmann equation, such as stability of the solutions of this equation.

While the problems related to the instability of laminar flows and their transition to turbulence have been studied for a long time in classical hydrodynamics, the corresponding problems in kinetic theory have been paid attention only recently. This circumstance is clearly related to the extremely complex character of such problems. It is clear, however, that the study of such problems might be of great importance for the purpose of understanding fundamental phenomena of instability and self-organization in molecular dynamics as well as computing hypersonic rarefied flows. Typical examples in this area are Bénard's instability [74], Taylor vortices [75, 77] and channel flow [76].

In the last few years , objections have been raised against the ability of the DSMC technique to faithfully resolve vortex motion, due to the lack of accurate conservation of angular momentum in collisions. It seems that the first paper to cast doubts on this ability is due to Meiburg [78], who applied both molecular dynamics and direct simulation Monte Carlo to the case of a rarefied flow past an impulsively started inclined flat plate. A clear vortex structure was obtained in the wake region of the molecular dynamics calculation, but the wake in the direct simulation Monte Carlo was relatively devoid of structure. As pointed out by Bird [79], the approximations associated with either method cannot be tested for the unsteady flow past a plate because this problem places excessive demands on computer time. He therefore introduced the forced vortex flow produced by a moving wall in a two-dimensional cavity as an alternative test case and came to the conclusion that as long as the cell size requirement are met, the lack of conservation of angular momentum in collisions does not appear to have a significant effect on the results of direct simulation Monte Carlo calculations. Bird [79] also examined the values of the parameters in Meiburg's calculations [78] and showed that the density of the gas was too high to employ the direct simulation Monte Carlo method, because the mean free path was of the order of the molecular diameter and the size of the cell is too large to analyse the vortical wake structure. The issue of lack of conservation of angular momentum was addressed by Nanbu et al. [80], who showed that if the cell size is sufficiently small, the total angular momentum of the molecules in a cell is almost conserved in the Monte Carlo calculations.

The results obtained by S. Stefanov and C. Cercignani [74, 75, 77] have a preliminary character, but clearly indicate that the formation of such vortex patterns arising from the aforementioned instabilities are possible. The calculations refer to Knudsen numbers of order 10^{-2} and different choices of the other parameters (including high values of the Mach number).

Encouraged by the positive result of these investigations, the same authors also considered [76] the fluctuations of the macroscopic quantities in a rarefied gas flowing in a channel under the action of a constant external force in a direction parallel to the walls (which are assumed to be at rest with the same temperature). Their final aim is the study of the transition to turbulence by means of the Boltzmann equation, but they remained far from achieving this goal, since their calculations are restricted to a two-dimensional geometry. The main aim of their calculations was to investigate the time evolution of both the macroscopic quantities and their fluctuations. The numerical experiments refer to Knudsen numbers between .005 and .1 and three different values of the body force. The analysis of the results indicates an increase in the macroscopic fluctuation for $Kn \leq .05$ and certain magnitudes of the force. In order to recognize the possible formation of vortex patterns and to estimate the macroscopic fluctuations, a data analysis was performed. If one takes into account the results of this analysis, he is tempted to conclude that he observes a transition from laminar to two-dimensional small-scale turbulence.

A common statement in the books and papers dealing with turbulence is that the scales of turbulence and molecular motions are widely separated and thus one may well ask what is the purpose of calculations of the kind discussed above. To answer this objection, one may remark that the common statement alluded to in the previous sentence may always be true for a liquid; but what about a gas? Let us consider the ratio between the mean free path ℓ and the dissipative scale of turbulence ℓ_D:

$$\ell/\ell_D = (\ell/L)(Re)^{3/4} = Kn(Re)^{3/4} = (Ma)^{3/4}(Kn)^{1/4} \tag{17}$$

where L is a macroscopic scale and the relation $Re = Ma/Kn$ has been taken into account. If we take $Re = 10^4$ for fully developed turbulence, we obtain that ℓ and ℓ_D are of the same order for $Kn = 10^{-3}$ and $Ma = 10$. Thus, in hypersonic flow, for moderate rarefaction, turbulence scale and mean free path are of the same order of magnitude!

These estimates are, of course, to be taken with caution, since there might be sizable factors "of order unity". These considerations should, however, not to be ignored when considering high Mach number flows. This field is completely unexplored and is certainly worth more attention.

11. Concluding remarks

We have surveyed the initial and initial-boundary value problems for the Boltzmann equation, with particular attention to recent results and the trend to equilibrium.

It appears that the subject has reached a certain maturity. Only difficult problems (such as smoothness, uniqueness, asymptotic behavior for long times or small mean free paths, steady solutions), which seem to require significantly new ideas, appear to remain open.

Another problem, which might be easier to solve, is to find conditions under which equality holds in (4.18). A simple but significant case, discussed in Sect. 7, is provided in the recent paper by Cercignani and Illner [26] .

Acknowledgement. This survey was prepared when the author was visiting the University of Kaiserslautern (Germany) as a Humboldt Prize awardee.

REFERENCES

[1] C. Cercignani, *Mathematical Methods in Kinetic Theory*. 2nd edition, Plenum Press, New York 1990.

[2] C. Cercignani, *The Boltzmann Equation and its Applications*. Springer, New York 1987.

[3] C. Cercignani, R. Illner, and M. Pulvirenti, *The Mathematical Theory of Dilute Gases*. Springer-Verlag, New York (1994)

[4] T. Kato, *Perturbation Theory of Linear Operators*. Springer, New York 1966.

[5] J. P. Guiraud, *An H theorem for a gas of rigid spheres in a bounded domain. Théories cinétiques classiques et rélativistes*, G. Pichon, Ed., CNRS, Paris, 1975, 29-58.

[6] Y. Shizuta and K. Asano, *Global solutions of the Boltzmann equation in a bounded convex domain*. Proc. Japan Acad., Vol. 53A, 1977, 3-5.

[7] S. Ukai, *Solutions of the Boltzmann equation*. Pattern and Waves-Qualitative Analysis of Nonlinear Differential Equations, 1986, 37-96.

[8] S. Ukai and K. Asano, *On the initial boundary value problem of the linearized Boltzmann equation in an exterior domain*. Proc. Japan Acad., Vol. 56, 1980, 12-17.

[9] R. DiPerna and P. L. Lions, *On the Cauchy problem for Boltzmann equations*. Ann. of Math., Vol. 130, 1989, 321-366.

[10] F. Golse, B. Perthame, P. L. Lions, R. Sentis, *Regularity of the moments of the solution of a transport equation*. J. Funct. Anal., Vol. 76, 1988, 110-125.

[11] P. Gérard, *Solutions globales du problème de Cauchy pour l'équation de Boltzmann*. Seminaire Bourbaki , nr. 699, 1987-88.

[12] K. Hamdache, *Initial boundary value problems for Boltzmann equation. Global existence of weak solutions*. Arch. Rat. Mech. Analysis, Vol. 119, 1992, 309-353.

[13] L. Arkeryd, C. Cercignani and R. Illner, *Measure solutions of the steady Boltzmann equation in a slab*. Commun. Math. Phys., Vol. 142, 1991, 285-296.

[14] L. Desvillettes, *Convergence to equilibrium in large time for Boltzmann and BGK equations*. Arch. Rat. Mech. Analysis, Vol. 110, 1990, 73-91.

[15] C. Cercignani, *Equilibrium states and trend to equilibrium in a gas according to the Boltzmann equation*. Rend. Mat. Appl., Vol. 10, 1990, 77-95,

[16] L. Arkeryd, *On the strong L^1 trend to equilibrium for the Boltzmann equation*. Studies in Appl. Math., Vol. 87, 1992, 283-288.

[17] P. L. Lions, *Compactness in Boltzmann's equation via Fourier integral operators and applications. I*. Journal of Mathematics of Kyoto University, Vol. 34, 1994, 391-427.

[18] L. Arkeryd and C. Cercignani, *A global existence theorem for the initial-boundary value problem for the Boltzmann equation when the boundaries are not isothermal*. Arch. Rat. Mech. Anal., Vol. 125, 1993, 271-288.

[19] J. Darrozès and J.-P. Guiraud, *Généralisation formelle du théorème H en présence de parois. Applications*, C.R.A.S. Paris, Vol. A262, 1966, 1368-1371.

[20] C. Cercignani and M. Lampis, *Kinetic models for gas-surface interactions*. Transp. Th. Stat. Phys., Vol. 1, 1971, 101-114.

[21] C. Cercignani, *Scattering kernels for gas-surface interactions*. Transp. Th. Stat. Phys., Vol. 2, 1972, 27-53.

[22] C. Truesdell and R. G. Muncaster, *Fundamentals of Maxwell's Kinetic Theory of a Simple Monatomic Gas*. Academic Press, New York 1980.

[23] L. Arkeryd and N. Maslova, *On diffuse reflection at the boundary for the Boltzmann equation and related equations*. Preprint ISSN 03407-2809, University of Göteborg, January 1994.

[24] C. Cercignani, *On global existence theorems for the initial value problem for the Boltzmann Equation*. Asymptotic Modelling in Fluid Mechanics, P.A. Bois *et al.*, Eds., LNP 442, 1995, 165-177.

[25] P. L. Lions, *Conditions at infinity for Boltzmann's equation*. Cahiers de Mathématiques de la décision n^0 9334, CEREMADE 1993.

[26] C. Cercignani, and R. Illner, *Global weak solutions of the Boltzmann equation in a slab with diffusive boundary conditions*, Arch. Rat. Mech. Anal., 1995, Vol. 134, 1996, 1-16.

[27] C. Cercignani, *Weak solutions of the Boltzmann equation and energy conservation*, Appl. Math. Lett., Vol. 8, 1995, 53-59.

[28] C. Cercignani, *A remarkable estimate for the solutions of the Boltzmann equation*. Appl. Math. Lett., Vol. 5, 1992, 59-62.

[29] C. Cercignani, *Errata: Weak solutions of the Boltzmann equation and energy conservation*, Appl. Math. Lett., Vol. 8, 1995, 95-99.

[30] L. Arkeryd, *Existence theorems for certain kinetic equations and large data*. Arch. Rat. Mech. Anal., Vol. 103, 1988, 139-149.

[31] M. Bony, *Existence globale et diffusion en théorie cinétique discrète*. In *Advances in Kinetic Theory and Continuum Mechanics*, R. Gatignol and Soubbarameyer, Eds., Springer-Verlag, Berlin 1991, 81-90.

[32] R. DiPerna, and P. L. Lions, *Global solutions of Boltzmann's equation and the entropy inequality*. Arch. Rat. Mech. Anal., Vol. 114, 47–55, 1991)

[33] C. Cercignani, *Trend to equilibrium of weak solutions of the Boltzmann equation in a slab with diffusive boundary conditions*, J. Stat. Phys., Vol. 84, 1996, 875-888.

[34] L. Arkeryd, and A. Nouri, *Asymptotics of the Boltzmann equation with diffuse reflection boundary conditions*. Submitted to Monatshefte für Mathematik, 1995.

[35] C. Cercignani, *Are there more than five linearly independent collision invariants for the Boltzmann equation?*. J. Stat. Phys., Vol. 58, 1990, 817-823.

[36] L. Boltzmann, *Über das Wärmegleichgewicht von Gasen, auf welche äussere Kräfte wirken*. Sitzungsberichte der Akademie der Wissenschaften Wien, Vol. 72, 1875, 427-457.

[37] B. Boltzmann, *Über die Aufstellung und Integration der Gleichungen, welche die Molekularbewegungen in Gasen bestimmen*. Sitzungsberichte der Akademie der Wissenschaften Wien, Vol. 74, 1876, 503-552.

[38] B. Wennberg, *On an entropy dissipation inequality for the Boltzmann equation*. C. R. A. S. (Paris), Vol. 315, I, 1992, 1441-1446.

[39] G.A. Bird, *Molecular Gas Dynamics and the Direct Simulation of Gas Flows*. Clarendon Press, Oxford 1994.

[40] C. Cercignani, *Aerodynamical Applications of the Boltzmann Equation*. La Rivista del Nuovo Cimento, Vol. 18, 7, 1995, 1-40.

[41] T. Carleman, *Sur la théorie de l'équation intégro-differentielle de Boltzmann*, Acta Mathematica Vol. 60, 1933, 91-146.

[42] L. Arkeryd, *On the Boltzmann equation. Part I: Existence", "On the Boltzmann equation. Part II: The full initial value problem*, Archive for Rational Mechanics and Analysis Vol. 45, 1972, 1-16 and 17-34.

[43] H. Grad, *Asymptotic equivalence of the Navier-Stokes and non-linear Boltzmann equation*, in: Proc. AMS Symposia on Applied Math. Vol. 17, 1965, 154-183.

[44] S. Ukai, *On the existence of global solutions of mixed problem for non-linear Boltzmann equation*, Proceedings of the Japan Academy Vol. 50, 1974, 179-184.

[45] T. Nishida and K. Imai, *Global solutions to the initial value problem for the nonlinear Boltzmann equation*, Publications of the Research Institute for Mathematical Sciences, Kyoto University Vol. 12, 1977, 229-239.

[46] Y. Shizuta and K. Asano, *Global solutions of the Boltzmann equation in a bounded convex domain*, Proceedings of the Japan Academy Vol. 53, 1974, 3-5.

[47] R. Illner and M. Shinbrot, *The Boltzmann equation: Global existence for a rare gas in an infinite vacuum*, Comm. Math. Phys. Vol. 95, 1984, 217-226.

[48] N. Bellomo and G. Toscani, *On the Cauchy problem for the nonlinear Boltzmann equation. Global existence, uniqueness and asymptotic stability*, J. Math. Phys. Vol. 26, 1985, 334-338.

[49] G. Toscani, *Global solution of the initial value problem for the Boltzmann equation near a local Maxwellian*, Arch. Rat. Mech. Anal Vol. 102, 1988, 231-241.

[50] C. Cercignani, *Existence of homoenergetic affine flows for the Boltzmann equation*, Arch. Rational Mech. Anal. Vol. 105, 1989, 377-387.

[51] L. Arkeryd, R. Esposito and M. Pulvirenti, *The Boltzmann equation for weakly inhomogeneous data*, Commun. Math. Phys. Vol. 111, 1988, 393-407.

[52] L. Arkeryd, *Loeb solutions of the Boltzmann equation*, Arch. Rat. Mech. Anal. Vol. 86, 1984, 85-97.

[53] D. Morgenstern, *Analytical studies related to the Maxwell-Boltzmann equation* J. Rational Mech. Anal. Vol. 4, 1955, 533-543.

[54] A. Ya. Povzner,*The Boltzmann equation in the kinetic theory of gases*, Mat. Sbornik, Vol. 58, 65-86; translated in American Math. Soc. Translations Vol. 47, 1962, 193-216.

[55] C. Cercignani, W. Greenberg, and P. Zweifel, *Global solutions of the Boltzmann equation on a lattice*, J. Stat. Phys. Vol. 20, 1979, 449-462.

[56] A. Nordsiek and B. Hicks: *Monte Carlo evaluation of the Boltzmann collision integral*, in: Rarefied Gas Dynamics, C. L. Brundin, ed., Vol. II, 695-710, Academic Press, New York 1967.

[57] S. M. Yen, B. Hicks, and R. M. Osteen: *Further development of a Monte Carlo method for the evaluation of Boltzmann collision integral*, in: Rarefied Gas Dynamics, M. Becker and M. Fiebig, eds., Vol. I, A.12-1-A.12-10, DFVLR-Press, Porz-Wahn 1974

[58] V. Aristov, F. G. Tcheremissine: *The conservative splitting method for solving the Boltzmann equation*, U. S. S. R. Computational Mathematics and Mathematical Physics, Vol. 20, 1980, 208-225.

[59] F. G. Tcheremissine: *Numerical methods for the direct solution of the kinetic Boltzmann equation"*: U. S. S. R. Computational Mathematics and Mathematical Physics, Vol. 25, 1985, 156-166.

[60] F. G. Tcheremissine: *Advancement of the method of direct numerical solving of the Boltzmann equation*, in: Rarefied Gas Dynamics: Theoretical and Computational Techniques, E. P. Muntz, D. P. Weaver and D. H. Campbell, eds., 343-358, AIAA, Washington 1989

[61] Cercignani, C., and A. Frezzotti: *Numerical simulation of supersonic rarefied gas flows past a flat plate: effects of the gas-surface interaction model on the flow-field*, in: Rarefied Gas Dynamics: Theoretical and Computational Techniques, E. P. Muntz, D. P. Weaver and D. H. Campbell, eds., 552-566, AIAA, Washington 1989

[62] G. A. Bird: *Direct simulation and the Boltzmann equation*, Physics of Fluids, Vol. 13, 1970, 2676-2687.

[63] G. A. Bird: *Monte-Carlo Simulation in an Engineering Context*, in: Rarefied Gas Dynamics, S. S. Fischer, Editor, Vol. I, 239-255, AIAA, New York 1981

[64] W. Wagner: *A convergence proof for Bird's direct simulation Monte Carlo method for the Boltzmann equation*, J. Stat. Phys., Vol. 66, 1992, 1011-1044.

[65] O. Lanford III, *Time evolution of large classical systems,* in: Dynamic Systems, Theory and Applications, E. J. Moser, ed.. Lecture Notes in Physics Vol. 38, 1-111. Springer Verlag, Berlin 1975.

[66] M. Pulvirenti, W. Wagner, M. B. Zavelani, *Convergence of particle schemes for the Boltzmann equation* Euro. J. of Mechanics B, Vol. 7, 1994, 339-351.

[67] K. Koura: *Transient Couette flow of rarefied binary gas mixtures*, Physics of Fluids, Vol. 13, 1970, 1457-1466.

[68] O. M. Belotserkovskii, and V. Yanitskii: *Statistical particle-in-cell method for solving rarefied gas dynamics problems*, Zhurnal Vychitelnik Matematiki i Matematicheskii Fiziki, Vol. 15, 1975, 1195-1201. (in Russian)

[69] S. M. Deshpande: Department of Aeronautical Engineering of the Indian Institute for Science Report no. 78, FM4, 1978

[70] H. Babovsky, *A convergence proof for Nanbu's Boltzmann simulation scheme,* European J. Mech. B: Fluids Vol. 8(1), 1989, 41-55.

[71] H. Babovsky, and R. Illner: *A convergence proof for Nanbu's simulation method for the full Boltzmann equation*, SIAM Journal of Numerical Analysis, Vol. 26, 1989, 45-65.

[72] I. Kuščer: *Phenomenological aspects of gas-surface interaction* in: Fundamental Problems in Statistical Mechanics, Vol. IV, E. G. D. Cohen and W. Fiszdon, Eds., 441-467, Ossolineum, Warsaw, 1978

[73] K. Nanbu: *Direct simulation scheme derived from the Boltzmann equation*, Journal of the Physical Society of Japan, Vol. 49, 1980, 2042-2049.

[74] S. Stefanov, C. Cercignani: *Monte Carlo Simulation of Bénard's Instability in a Rarefied Gas*, Euro. J. of Mechanics B, Vol. 5, 1992, 543-552.

[75] S. Stefanov, C. Cercignani: *Monte Carlo Simulation of Taylor-Couette Instability in a Rarefied Gas*, J. Fluid Mech., Vol. 256, 1993, 199-213.

[76] S. Stefanov, C. Cercignani: *Monte Carlo Simulation of Channel Flow of a Rarefied Gas*, Euro. J. of Mechanics B, Vol. 13, 1994, 93-114.

[77] S. Stefanov, C. Cercignani, *Taylor-Couette flow of a rarefied gas*, in: Proceedings of the International Symposium on Aerospace and Fluid Science, p. 490-500, Institute of Fluid Science, Tohoku University, Sendai 1994

[78] E. Meiburg: *Comparison of the molecular dynamics method and the direct simulation Monte Carlo technique for flows around simple geometries*, Phys. Fluids, Vol. 29, 1986, 3107-3113.

[79] G. A. Bird: *Direct simulation of high-vorticity gas flows*, Phys. Fluids, Vol. 30, 1987, 364-366.

[80] K. Nanbu, Y. Watanabe, S. Igarashi: *Conservation of angular momentum in the direct simulation Monte Carlo method*, Jour. Phys. Soc. Japan, Vol. 57, 1988, 2877-2880.

DIPARTIMENTO DI MATEMATICA, POLITECNICO DI MILANO, MILANO, ITALY
E-mail address: CARCER@MATE.POLIMI.IT

Proceedings of Symposia in Applied Mathematics
Volume **54**, 1998

New Results for the Asymptotics of Orthogonal Polynomials and Related Problems via the Lax-Levermore Method

P. Deift, T. Kriecherbauer, and K. T-R McLaughlin

Dedicated to Peter Lax and Louis Nirenberg in honor of their 70th birthdays.

ABSTRACT. We present new results on the equilibrium measure for logarithmic potentials on a finite interval of the real line in the presence of an external field.

1. Introduction

It is a remarkable fact, arising from the work of many authors, in many different areas, and over many years, that a certain broad class of classical problems in analysis are intimately related. We begin by describing some of these problems and their interconnections, but first we need some notation and some definitions. Let

$$(1) \quad \mathcal{A} = \left\{ Positive\ Borel\ measures\ \mu,\ \int d\mu = 1,\ supp(\mu) \subseteq \Sigma = [-1,1] \right\}.$$

By an *external field* V we mean simply a map from Σ to $\mathbf{R}$. Henceforth, and throughout the paper, we assume that

$$(2) \qquad all\ external\ fields\ V\ lie\ in\ C^2(\Sigma),$$

i.e. $V \in C^2(-1,1)$, and $V(x)$, $V'(x)$, and $V''(x)$ have continuous extensions to $[-1,1]$. Although many of the definitions and results presented below are true for general closed subsets Σ of $\mathbf{R}^2$, and also for more general external fields V, we restrict our attention to the case $\Sigma = [-1,1]$, with fields V that are twice continuously differentiable as in (2) above. Throughout the paper we use the following

1991 *Mathematics Subject Classification.* 31A99, 41A99, 60K99.

Percy Deift was supported in part by NSF grant # DMS-9500867.

Thomas Kriecherbauer was supported in part by the DFG.

Kenneth T-R McLaughlin was supported in part by NSF postdoctoral fellowship grant # DMS-9508946.

notation: if $S = \{x_1, \dots, x_n\}$ is a set of n distinct points in $\mathbf{R}$, and δ_{x_i} is the Dirac delta measure concentrated at x_i, then

$$(3) \qquad \rho_S = \frac{1}{n} \sum_{i=1}^{n} \delta_{x_i}$$

is the *normalized counting measure* (NCM) for S.

Problem 1. Electrostatic Equilibrium

Consider the following maximization problem:

$$(4) \qquad E = E_V = \sup_{\mu \in \mathcal{A}} \left[\frac{1}{2} \int \int \log{(x-y)^2} d\mu(x) d\mu(y) + \int V(y) d\mu(y) \right].$$

Here $\frac{1}{2} \int \int \log{(x-y)^2} d\mu(x) d\mu(y)$ is the equilibrium energy for a charged distribution $d\mu(x)$ on the *conductor* $\Sigma = [-1, 1]$, and $V d\mu$ is the energy of the charges in the external field $V(y)$. It is well known that the supremum in (4) is attained at a unique Borel measure $\mu = \mu_{ES}^V$ in $\mathcal{A}$; the measure μ_{ES}^V is called the *equilibrium measure* for Σ and V. Moreover, the equilibrium measure $d\mu_{ES}^V$ is characterized by the following variational conditions: for some constant ℓ,

$$(5) \qquad V(x) + \int \log{(x-y)^2} d\mu_{ES}^V(y) \le \ell,$$

with equality on the support of $d\mu_{ES}^V$.

Problem 2. Weighted Transfinite Diameter

For each positive integer n, let

$$(6) \qquad d_{V,n} = \left[\max_{\{x_1, \dots, x_n\} \subset [-1,1]} \prod_{i<j} |x_j - x_i| e^{\frac{V(x_i)}{2}} e^{\frac{V(x_j)}{2}} \right]^{\frac{2}{n(n-1)}}$$

and set

$$(7) \qquad d_V = \lim_{n \to \infty} d_{V,n}.$$

The quantity d_V is the *weighted transfinite diameter* for the interval $\Sigma = [-1, 1]$. Observe that in the unweighted case, $V = 0$, the quantity $d_{0,n}$ has the geometric interpretation as the maximum of the geometric means of the distances between n points located in Σ.

For each n, a set $\{x_1, \dots, x_n\}$ which realizes the maximum of (6) is called an *nth weighted Fekete set*, and the points $x_1, \dots, x_n$ are called *weighted Fekete points*. As is well known, Fekete points play a distinguished role in Lagrange interpolation. Also, in another direction, in the unweighted case and for a general closed set $\Sigma \subset \mathbf{R}^2$, Fekete [9] showed that the transfinite diameter d_0 controls the number N_Σ of polynomials with integer coefficients, and leading coefficient fixed, all of whose roots are simple and lie in Σ: if $d_0 < 1$, then $N_\Sigma < \infty$.

If $S^{(n)} = \{x_1^{(n)}, \dots, x_n^{(n)}\}$ is an nth weighted Fekete set for $\Sigma = [-1, 1]$, consider

$$(8) \qquad \mu_{TD}^V = \lim_{n \to \infty} \rho_{S^{(n)}}.$$

This limit is known to exist in the weak - $*$ topology for measures.

Problem 3. Weighted Chebyshev Polynomials

For each $n \geq 0$, set

$$(9) \quad T_{V,n} = \inf \left\{ \sup_{|x| \leq 1} \left| e^{\frac{nV(x)}{2}} p(x) \right| \ : \ \begin{array}{c} p(x) = x^n + \cdots \\ is\ a\ monic\ polynomial\ of\ degree\ n \end{array} \right\}.$$

For each n, the infimum is attained at a unique polynomial, called the *nth weighted Chebyshev polynomial*. Furthermore,

$$(10) \qquad\qquad T_V = \lim_{n \to \infty} (T_{V,n})^{1/n}$$

exists. Let $S^{(n)} = \{x_1^{(n)}, \ldots, x_n^{(n)}\}$ denote the zeros of the nth weighted Chebyshev polynomials, and consider

$$(11) \qquad\qquad \mu_{WC}^V = \lim_{n \to \infty} \rho_{S^{(n)}}.$$

Once again, this limit is known to exist in the weak - $*$ topology of measures.

The Chebyshev polynomials, which play a distinguished role in approximation theory, are obtained as above by minimizing the supremum norm: to obtain the nth orthogonal polynomial corresponding to the measure $e^{nV(x)} dx$ on $[-1, 1]$, one must of course minimize the L^2 norm,

$$\left\{ \int |p(x)|^2 e^{nV(x)} dx \ : \ p(x) = x^n + \cdots \right\}.$$

Problem 4. Zero distribution of Orthogonal Polynomials.

Let $P(x) = x^m + \cdots$ be a monic polynomial of order m. For $t > 0$, let $d\alpha_n^{t, \pm P}$ denote the measures $e^{\pm P(x)} X_{[-(tn)^{1/m}, (tn)^{1/m}]} dx$ obtained by restricting the measures $e^{\pm P(x)} dx$ to the interval $[-(tn)^{1/m}, (tn)^{1/m}]$. (Here X_B is the indicator function of the set B.) Let $p_n = p_n^{t, \pm P} = x^n + \cdots$ denote the nth monic orthogonal polynomial obtained by applying the Gram - Schmidt orthogonalization procedure to the sequence $1, x, \ldots, x^n$, with respect to the measure $d\alpha_n^{t, \pm P}$, in the usual way.

Let $y_1^{(n)}, \ldots, y_n^{(n)}$ denote the zeros of $p_n(x)$, and let $S^{(n)} = \{x_1^{(n)}, \ldots, x_n^{(n)}\}$, where $x_i^{(n)} = \frac{y_i^{(n)}}{(tn)^{1/m}}$ and consider the limit

$$(12) \qquad\qquad \mu_{OP}^{t, \pm P} = \lim_{n \to \infty} \rho_{S^{(n)}}.$$

This limit is known to exist in the weak - $*$ topology, and in fact, is independent of the lower order terms of $P(x)$.

In the case where m is an even integer, the above problem arises naturally in the following way. Consider the nth monic orthogonal polynomial $\tilde{p}_n(x)$ obtained by applying the Gram - Schmidt procedure to $1, x, \ldots, x^n$ with respect to the measure $e^{-P(x)} dx$ on **R**. *Question*: what is the smallest interval $[-A, A]$ to which one can restrict $e^{-P(x)} dx$ in such a way that the nth polynomial p_n for the restricted measure $e^{-P(x)} X_{[-A, A]}(x) dx$ is a close approximation to $\tilde{p}_n$? See **Remark (2) regarding Problem 4**, below.

Problem 5. Fast Decreasing Polynomials

A set of polynomials $\{p_n(x), degree(p_n) \leq n, \ n = 0, 1, \ldots\}$ is called a *set of fast decreasing polynomials* for a weight $e^{\frac{V}{2}}$, $V(0) = 0$, if there exists a constant C independent of n such that

$$(13) \quad p_n(0) = 1 \quad and \quad |p_n(x)| \leq C e^{-\frac{nV(x)}{2}} \ for\ x \in [-1, 1]\ and\ n \geq 0.$$

Such polynomials arise in approximation theory, and provide a polynomial approximation to the Dirac delta function at the origin.

Question: given $V(x)$, do such polynomials exist?

Problem 6. Hankel Determinants.

Let $d\alpha_n^V(x) = e^{nV(x)}dx$ be a measure on $[-1, 1]$. Let $c_j^{(n)} = \int x^j d\alpha_n^V(x)$, $j = 0, 1, \ldots$ denote the moments of $d\alpha_n^V(x)$, and for each $k \geq 1$, construct the *Hankel determinant*

$$(14) \qquad H_k^{(n)} = det\left(c_{i+j}^{(n)}\right)_{0 \leq i,j \leq k}.$$

Consider

$$(15) \qquad h_V = \lim_{n \to \infty}\left[H_n^{(n)}\right]^{1/n^2}.$$

This limit is known to exist.

Problems 1-6 are related in the following way.

Denote the support of the equilibrium measure $\mu = \mu_{ES}^V$ for the electrostatic maximization problem (4) by σ_{ES}^V. Then

$$(16) \qquad \mu_{ES}^V = \mu_{TD}^V = \mu_{WC}^V$$

and in the case where $V(x) = \pm tx^m$, we have

$$(17) \qquad \mu_{ES}^V = \mu_{TD}^V = \mu_{WC}^V = \mu_{OP}^{t,\pm P}$$

for any polynomial $P(x) = x^m + \cdots$. Furthermore, fast decreasing polynomials of type (13) exist if and only if

$$(18) \qquad 0 \in \sigma_{ES}^V.$$

Define the *weighted logarithmic capacity* of $\Sigma = [-1, 1]$ by

$$(19) \qquad K = K_V = e^{E_V},$$

where E_V is the supremum of the electrostatic energy in (4). Then, in addition,

$$(20) \qquad K_V = d_V = T_V e^{\int \frac{V(y)}{2} d\mu_{ES}^V(y)} = h_V.$$

It is clear from the above results and considerations that the key problem is to determine the equilibrium measure μ_{ES}^V. In fact, all that is needed is σ_{ES}^V, the support of μ_{ES}^V. Indeed, as we see below, μ_{ES}^V can be computed explicitly via the solution of a scalar Riemann - Hilbert problem, once σ_{ES}^V is known.

A general reference for results in potential theory, and in particular for the properties of the equilibrium measure of **Problem 1** is [14]. The proof that the limit (7) exists in the unweighted case is due to Fekete [10], and in the weighted case is apparently due to Leja [16]. The proof that $d_V = K_V$ (see (20)) is due to Fekete [10] and Szegő [25], and in the weighted case to Leja [16]. The existence of limit (8), and the relation $\mu_{TD}^V = \mu_{ES}^V$ (see (16)) is apparently due to Leja [16]. The proof of the existence of the limit in (10), and the relation $K_V = T_V e^{\int \frac{V}{2} d\mu_{ES}^V}$ (see (20)) is due to Gonchar and Rachmanov [12] and Mhaskar-Saff [19], and the proof of the existence of the limit in (11) and the relation $\mu_{WC}^V = \mu_{ES}^V$ (see (16)) was given by Rachmanov [22] and Mhaskar-Saff [19]. The proof that the limit in (12) exists, and that $\mu_{OP}^V = \mu_{ES}^V$ (see (17)) was given by Rachmanov [22] and Mhaskar-Saff [19]. Fast decreasing polynomials were considered, for example, by DeVore [5] and Nevai and Totik [**20**]. The statement that fast decreasing polynomials exist if and only if (18) holds is due to Totik [**27**]. The proof that the limit in (15) exists and that $h_V = K_V$ is due to Rachmanov [**22**] and Mhaskar-Saff [**19**].

An extremely useful reference containing all the above material, together with historical notes, is the recent book by Saff and Totik [**23**]. Historically, it seems that Gauss was the first to begin the analysis of the electrostatic equilibrium problem with external fields. Working in 3 dimensions (so that the logarithmic potential is replaced by the $\frac{1}{r}$ Coulombic potential), Gauss [**11**] derived the Euler Lagrange equations for the analog of (4), and solved these equations in simple cases. Gauss recognized early the analytical difficulties in computing the equilibrium measure for such problems, as he writes, ([**11**], Sec. 35) "Die wirkliche Bestimmung der Vertheilung der Masse auf einer gegebenen Fläche für jede vorgeschriebene Form von U übersteigt in den meisten Fällen die Kräfte der Analyse in ihrem gegenwärtigen Zustande. (The actual determination of the distribution of the mass on a given surface for an arbitrary function U lies, in most cases, beyond the powers of present day analysis.)"

At this stage, what indeed is known about μ_{ES}^V and, in particular, σ_{ES}^V? It turns out that in recent years considerable progress on the determination of the general properties of σ_{ES}^V for one dimensional conductors Σ has come from yet another direction, viz., random matrix theory in the spirit of Wigner and Dyson (see [**18**]).

Consider, in particular, the space of $n \times n$ Hermitean matrices M with probability distribution

$$(21) \qquad \Pi_n(M)dM = \tilde{Z}_n^{-1} e^{-n\ Tr(V(M))} dM$$

where $V(x) = a_{2m}x^{2m} + a_{2m-1}x^{2m-1} + \cdots + a_0$, $a_{2m} > 0$, is a polynomial which is real valued and bounded below,

$$(22) \qquad dM = \left(\prod_{k=1}^n dM_{kk} \right) \prod_{k<j} d\left(ReM_{kj}\right) d\left(ImM_{kj}\right)$$

is "Lebesgue" measure for Hermitean matrices, and $\tilde{Z}_n$ is the normalization factor. Using the spectral theorem $M = U\Lambda U^*$, $\Lambda = diag(\lambda_1(M), \ldots, \lambda_n(M))$, U unitary, as a change of variables, the distribution (21) takes the form (see [**18**])

$$(23) \qquad \Pi_n(M)dM = Z_n^{-1} e^{-n\sum_{i=1}^n V(\lambda_i)} \prod_{i<j} (\lambda_i - \lambda_j)^2 \prod_{i=1}^n d\lambda_i \ dU$$

where dU is the restriction to $U(n)/T(n)$, the quotient of the space of $n \times n$ unitary matrices by the n-torus, of Haar measure on $U(n)$, and Z_n is the normalization constant (partition function)

$$(24) \qquad Z_n = \int_{-\infty}^{\infty} \cdots \int_{-\infty}^{\infty} e^{-n\sum_{i=1}^n V(\lambda_i)} \prod_{i<j} (\lambda_i - \lambda_j)^2 \, d\lambda_1 \cdots d\lambda_n.$$

As is well known, a familiar calculation due to Heine (see [**24**]) shows that Z_n is precisely the Hankel determinant H_{n-1}^{n-1} which arises in the study of polynomials orthogonal with respect to the weight $d\alpha_n(x) = e^{-nV(x)}dx$ (see (14) above).

Let $N_n(\Delta)$ denote the integral of the NCM $\frac{1}{n}\sum_{i=1}^n \delta_{\lambda_i}$ for the eigenvalues $\lambda_1, \ldots, \lambda_n$ of a random Hermitean matrix M, over an interval $\Delta = (a, b)$: thus

$N_n(\Delta) = n^{-1} \# \{\lambda_i \in \Delta\}$. Now consider $E(N_n(\Delta))$, the expectation of $N_n(\Delta)$ with respect to the probability measure (23). A simple calculation shows that

$$(25) \qquad E(N_n(\Delta)) = \int_\Delta u_n(\lambda_1) d\lambda_1$$

where $u_n(\lambda_1)$ is the $(n-1)$-fold integral,

$$(26) \quad u_n(\lambda_1) = Z_n^{-1} \int_{-\infty}^{\infty} \cdots \int_{-\infty}^{\infty} e^{-n \sum_{i=1}^n V(\lambda_i)} \prod_{i<j} (\lambda_i - \lambda_j)^2 \, d\lambda_2 \cdots d\lambda_n,$$

the so-called one point function for (23). The fact of the matter is the following: as $n \to \infty$, $E(N_n(\Delta))$ converges, and in addition to the connections (16) - (20) above, we have (see, in particular [21])

$$(27) \qquad N(\Delta) \equiv \lim_{n \to \infty} E(N_n(\Delta)) = \mu_{ES}^{-V}(\Delta),$$

where μ_{ES}^{-V} is the equilibrium measure for $-V$, but now with conductor $\Sigma = (-\infty, \infty)$. Moreover the following is true.

LEMMA 28 (See, for example, [21]). *N is absolutely continuous with respect to Lebesgue measure, $N(\Delta) = \int_\Delta \rho(\lambda) d\lambda$, with Radon - Nikodym derivative $\rho(\lambda)$, the so-called density of states, given by a Hölder continuous function with compact support consisting of a finite union of intervals.*

The proof of (28) follows from a remarkable identity for the Borel transform

$$(29) \qquad U(z) = \int_{\mathbf{R}} \frac{N(d\lambda)}{\lambda - z}, \quad z \notin \mathbf{R}$$

of the limiting measure $N(\cdot)$. Indeed for $z \notin \mathbf{R}$, one has

$$(30) \qquad U(z)^2 + V'(z)U(z) + L(z) = 0$$

where

$$(31) \qquad L(z) = \int_{\mathbf{R}} \frac{V'(z) - V'(\lambda)}{z - \lambda} N(d\lambda).$$

Solving for $U(z)$ from (30), and using well known properties of the Borel transform, the proof of (28) is immediate. Equation (30) was apparently first introduced by Bessis, Itzykson and Zuber in [2] in 1980, in the special case that V is quadratic (see equation (A.4.35) ibid.). We refer the reader to [21] for a review of recent analytical results in random matrix theory, and also to the very interesting papers [3] and [13] and the many references therein.

Our goal in this paper is to announce a variety of new results for equilibrium measures and the above related problems. Details will be presented in a subsequent publication. Our results are of two types. Firstly, we have proved a generalization of (28) which is appropriate for the finite conductor case, $\Sigma = [-1, 1]$, which allows V to be an arbitrary C^2 function, and which uses techniques that are perhaps more familiar to the orthogonal polynomial / approximation theory community. Indeed we work directly with the weighted Fekete points arising in **Problem 2** above. Secondly, we have analyzed σ_{ES}^V in detail in the monomial case $V(x) = \pm t x^m$, $t > 0$, both for m even and for m odd, as t varies. Here we have used techniques introduced in [17] and [4] to analyze the continuum limit of the Toda lattice, and also the small dispersion limit of the Korteweg de Vries equation (see below).

In a sequel to this announcement / paper, we will show how to use the equilibrium measure μ_{ES}^V to compute the asymptotics of orthogonal polynomials via the steepest descent / stationary phase method for Riemann - Hilbert problems introduced in [7], and further developed in [8] and [6]. In related work, we refer the reader to the recent paper [1], in which the authors use Riemann - Hilbert techniques together with the theory of isomonodromy deformations to compute the asymptotics of orthogonal polynomials, and also to address the so-called "universality conjecture" arising in random matrix theory.

Our results are the following. Define

$$(32)\quad L_{\frac{1}{2},\infty} = \left\{ h(x)\ mble.\ on\ [-1,1] : \|h\|_{\frac{1}{2},\infty} \equiv \sup_{|x|\leq 1} \sqrt{1-x^2}|h(x)| < \infty \right\}.$$

Define

$$(33)\quad q^{(0)}(x) = \left(\frac{V'(x)}{2}\right)^2 + \int \frac{V'(x)-V'(y)}{x-y} d\mu_{ES}^V(y) +$$
$$+ \frac{1}{x^2-1}\left(1 + \int V'(y)(x+y)d\mu_{ES}^V(y)\right),$$

and let $q^{(0)}(x) = q_+^{(0)}(x) - q_-^{(0)}(x)$, $q_\pm^{(0)} \geq 0$, denote the decomposition of $q^{(0)}$ into positive and negative parts. The function $q^{(0)}$ arises naturally in the analysis of the weighted Fekete points $x_1^*, \ldots, x_n^*$.

THEOREM 34. *The equilibrium measure μ_{ES}^V for (4) with V in C^2 is absolutely continuous with respect to Lebesgue measure,*

$$(35)\quad d\mu_{ES}^V = \psi(x)dx, \quad \psi(x) \geq 0,$$

where $\psi \in L_{\frac{1}{2},\infty}$ and is given by

$$(36)\quad \psi = \frac{1}{\pi}\sqrt{q_-^{(0)}(x)} \leq \frac{c}{\sqrt{1-x^2}},$$

for some $0 < c < \infty$. In particular ψ is continuous in $(-1,1)$: if V is C^k, $k \geq 3$, then ψ is Hölder continuous of order $\frac{1}{2}$, but in general no better.

Inserting (35) into (33) above, we obtain

$$(37)\quad q^{(0)}(x) = \left(\frac{V'(x)}{2}\right)^2 + \int \frac{V'(x)-V'(y)}{x-y}\psi(y)dy +$$
$$+ \frac{1}{x^2-1}\left(1 + \int V'(y)(x+y)\psi(y)dy\right).$$

THEOREM 38. *Suppose that $V(x)$ is real analytic in a neighborhood of $\Sigma = [-1,1]$. Then in addition to the results of Theorem 34, ψ is supported on a **finite** number of closed subintervals in $[-1,1]$.*

Suppose $supp(\mu_{ES}^V) = J = \cup_{j=0}^N (\alpha_j, \beta_j)$, where $-1 \leq \alpha_0 < \beta_0 < \alpha_1 \cdots < \alpha_N < \beta_N \leq 1$. There are five cases:

    (i) $N = 0$, $\alpha_0 = -1$, $\beta_0 = 1$
    (ii) $N > 0$, $\alpha_0 = -1$, $\beta_N = 1$

(iii) $N \geq 0$, $\alpha_0 = -1$, $\beta_N < 1$
(iv) $N \geq 0$, $\alpha_0 > -1$, $\beta_N = 1$
(v) $N \geq 0$, $\alpha_0 > -1$, $\beta_N < 1$

In case (i), if $\int_{-1}^{1} \frac{iV'}{(y^2-1)_+^{1/2}} dy = 0$, then for $z \in \mathbf{C} \backslash \mathbf{R}$, define

$$(39) \qquad F(z) = \left[z^2 - 1\right]^{1/2} \frac{1}{\pi i} \int_{-1}^{1} \frac{-iV'(y)/2\pi}{(y^2-1)_+^{1/2}} \frac{dy}{y-z} + \frac{i\gamma}{\left[z^2-1\right]^{1/2}};$$

if $\int_{-1}^{1} \frac{iV'}{(y^2-1)_+^{1/2}} dy \neq 0$, then again for $z \in \mathbf{C} \backslash \mathbf{R}$, define

$$(40) \quad F(z) = \left[\frac{z+1}{z-1}\right]^{1/2} \frac{1}{\pi i} \int_{-1}^{1} \frac{-iV'(y)}{2\pi} \left(\frac{y-1}{y+1}\right)_+^{1/2} \frac{dy}{y-z} + \frac{i\gamma}{\left[z^2-1\right]^{1/2}},$$

where in both (39) and (40), γ is chosen so that $\int_{-1}^{1} Re F_+(z) dz = 1$. In cases (ii) - (v), set

$$(41) \qquad F(z) = \frac{R(z)^{1/2}}{\pi i} \int_J \frac{-iV'(y)/2\pi}{R(y)_+^{1/2}} \frac{dy}{y-z}, \quad z \in \mathbf{C} \backslash \bar{J},$$

where

$$(42) \quad R(z) = \begin{cases} \frac{(z-\beta_0)(z-\alpha_N)}{(z^2-1)} \prod_{j=1}^{N-1} (z-\alpha_j)(z-\beta_j) & \text{in case (ii)} \\ \frac{(z-\beta_0)}{(z+1)} \prod_{j=1}^{N} (z-\alpha_j)(z-\beta_j) & \text{in case (iii)} \\ \frac{(z-\alpha_N)}{(z-1)} \prod_{j=0}^{N-1} (z-\alpha_j)(z-\beta_j) & \text{in case (iv)} \\ \prod_{j=0}^{N} (z-\alpha_j)(z-\beta_j) & \text{in case (v)} \end{cases}$$

Then in all cases

$$(43) \qquad \psi(x) = Re F_+(x), \quad x \in supp(\mu_{ES}^V).$$

Remark 1 The radicals in Theorem 38 are chosen such that

$$\left(z^2 - 1\right)^{1/2} > 0,$$
$$R(z)^{1/2} > 0$$

for $z > \beta_N$, and the subscript "+" denotes the limit from above, $F_+(z) = \lim_{\epsilon \downarrow 0} F(z + i\epsilon)$, etc. In (42), in case (ii) the product is taken to be 1 if $N = 1$, and in cases (iii) and (iv), the products are taken to be 1 if $N = 0$.

Remark 2 In [23] and [28] the authors identify classes of weights which they prove lead to case (i), and case (v) with $N = 0$, or case (ii) with $N = 1$ and $\beta_0 = -\alpha_1$.

THEOREM 44. *In the case that $V(x)$ is a polynomial of degree m, let the number of subintervals comprising $supp(\mu_{ES}^V)$ be given by $N + 1 = N_V$. Then*

$$(45) \qquad N_V \leq m + 1.$$

As noted above, our final results provide more detailed information in the case that $V(x)$ is a monomial, $V(x) = \pm tx^m$, $t > 0$.

THEOREM 46. *If the external field is*

$$(47) \qquad V(x) = -tx^{2q},$$

then there exists precisely one critical value $t_{-,2q}$, $t_{-,2q} = \frac{1}{q} \prod_{\ell=1}^{q} \frac{2\ell}{2\ell-1}$.

For $t_{-,2q} \le t$, the solution of the maximization problem falls into case (v) of Theorem 38 with $\beta_0 = -\alpha_0$, i.e. supp $\mu_{ES}^V = (-\beta_0, \beta_0)$. Moreover, $\psi = ReF_+$, with F defined in Theorem 38, case (v), and β_0 is determined by the following equation:

$$(48) \qquad \int_{-1}^{1} ReF_+(x)dx = 1,$$

or, equivalently,

$$(49) \qquad \beta_0^{2q} tq \prod_{\ell=1}^{q} \frac{2\ell - 1}{2\ell} = 1 \;.$$

For $0 < t \le t_{-,2q}$, the solution of the maximization problem is described by case (i) of Theorem 38 above.

THEOREM 50. *Suppose the external field is*

$$(51) \qquad V(x) = tx^{2q}.$$

(A) *Suppose further that $q = 1$. Then there exists one critical value $t_{+,2} = 2$.*

> **A_1 : $t_{+,2} < t$**
> *For $t_{+,2} < t$, the solution of the maximization problem falls into case (ii) of Theorem 38, with $N = 1$, and $\beta_0 = -\alpha_1$, i.e.*

$$(52) \qquad supp(\mu_{ES}^V) = (-1, -\alpha_1) \cup (\alpha_1, 1).$$

> *The parameter α_1, and hence, through Theorem 38, $\psi = ReF_+(x)$ are determined by the equation*

$$(53) \qquad \int_{-1}^{1} ReF_+(x)dx = 1.$$

> *As in Theorem 46 above, this condition may be evaluated explicitly, and yields an algebraic equation for α_1.*

> **A_2 : $0 < t \le t_{+,2}$**
> *For $0 < t \le t_{+,2}$, the solution of the maximization problem is described by case (i) of Theorem 38, i.e.*

$$(54) \qquad supp(\mu_{ES}^V) = (-1, 1).$$

(B) *Now suppose $q = 2$. Then there are two critical values $t_{+,4}^{(2)} = \frac{8}{3} < t_{+,4}^{(1)}$.*

> **B_1 : $t_{+,4}^{(1)} \le t$**
> *For $t_{+,4}^{(1)} \le t$, the solution of the maximization problem falls into case (ii) of Theorem 38, with $N = 1$, as in part (A) above. That is,*

$$(55) \qquad supp(\mu_{ES}^V) = (-1, -\alpha_1) \cup (\alpha_1, 1),$$

with α_1 and (again) $\psi = ReF_+(x)$ determined by condition (53), with F defined in Theorem 38, with $V(x) = tx^4$.

B$_2$: $t_{+,4}^{(2)} < t < t_{+,4}^{(1)}$

For $t_{+,4}^{(2)} < t < t_{+,4}^{(1)}$, the solution of the maximization problem falls into case (ii) of Theorem 38, but now $N = 2$. More specifically, we have

(56) $$supp(\mu_{ES}^V) = (-1, -\alpha_2) \cup (-\beta_1, \beta_1) \cup (\alpha_2, 1).$$

The parameters β_1 and α_2 are determined by the condition (53) together with a condition arising from the fact that for a variational problem such as (4), the Lagrange multiplier must be the same in all intervals comprising the complement of the support of μ_{ES}^V (see below).

B$_3$: $0 < t \leq t_{+,4}^{(2)}$

For $0 < t \leq t_{+,4}^{(2)}$, the solution of the maximization problem falls into case (i) of Theorem 38. That is,

(57) $$supp(\mu_{ES}^V) = (-1, 1),$$

and $\psi(x) = ReF_+(x)$, with F defined in Theorem 38, case (i), with $V(x) = tx^4$.

(C) *Suppose $q > 2$. Then there are two critical values $t_{+,2q} < t_{+,2q}^{(1)}$ (note there is no superscript on the first critical value).*

C$_1$: $t_{+,2q}^{(1)} \leq t$

For $t_{+,2q}^{(1)} \leq t$, the maximization problem is described by case (ii) of Theorem 38, with $N = 1$.

C$_2$: $t_{+,2q}^{(1)} - \epsilon < t < t_{+,2q}^{(1)}$

For $t_{+,2q}^{(1)} - \epsilon < t < t_{+,2q}^{(1)}$, the solution is described by case (ii) of Theorem 38, but again in this regime $N = 2$. For a further description of the determination of the endpoints of the three intervals comprising $supp(\mu_{ES}^V)$, see below.

C$_3$: $t_{+,2q} < t < t_{+,2q} + \epsilon$

For $t_{+,2q} < t < t_{+,2q} + \epsilon$, the maximization problem is described by case (ii) of Theorem 38, with $N = 2$.

C$_4$: $0 < t \leq t_{+,2q}$

For $0 < t \leq t_{+,2q}$, the solution of the maximization problem is described by case (i) of Theorem 38.

Remark Note that in the general case $q > 2$, we do not have a global description of the maximizer ψ for all values of t. Of course, by Theorem 44, we know that the number of intervals is bounded by $m + 1 = 2q + 1$. We believe, however, that in the entire region $t_{+,2q} < t < t_{+,2q}^{(1)}$ $supp(\mu_{ES}^V)$ contains precisely three intervals.

Remark Some of the quantities in Theorem 50 can easily be evaluated explicitly (e.g. $t_{+,2} = 2$, $t_{+,4} = \frac{8}{3}$, $\alpha_1 = (1 - \frac{2}{t})^{1/2}$ in case $\mathbf{A}_1$), whereas other quantities are determined by more complicated relations. We have omitted most of this information in the statement of the Theorem, but it will be presented in the proof of the Theorem which will be given in the subsequent publication mentioned above. This remark also applies to the following Theorem.

THEOREM 58. *Suppose the external field is*

(59)
$$V(x) = tx^{2q+1}.$$

(A) *Suppose further that $q = 0$. Then there exists one critical value $t_{+,1} = 2$.*

$\mathbf{A}_1 : t_{+,1} \leq t$
    For $t_{+,1} \leq t$, the solution of the maximization problem falls into case (iv) of Theorem 38, with $N = 0$, i.e.

(60)
$$supp(\mu_{ES}^V) = (\alpha_0, 1).$$

$\mathbf{A}_2 : 0 < t \leq t_{+,1}$
    For $0 < t \leq t_{+,1}$, the solution of the maximization problem is described by case (i) of Theorem 38.

(B) *Now suppose $q = 1$. Then there are three critical values $t_{+,3}^{(3)} = \frac{4}{3} < t_{+,3}^{(2)} = \frac{25}{6} < t_{+,3}^{(1)}$.*

$\mathbf{B}_1 : t_{+,3}^{(1)} \leq t$
    For $t_{+,3}^{(1)} \leq t$, the solution of the maximization problem falls into case (iv) of Theorem 38, with $N = 0$:

(61)
$$supp(\mu_{ES}^V) = (\alpha_0, 1).$$

$\mathbf{B}_2 : t_{+,3}^{(2)} < t < t_{+,3}^{(1)}$
    For $t_{+,3}^{(2)} < t < t_{+,3}^{(1)}$, the solution of the maximization problem falls into case (iv) of Theorem 38, but now $N = 1$. More

specifically, we have

$$(62) \qquad supp(\mu_{ES}^V) = (\alpha_0, \beta_0) \cup (\alpha_1, 1).$$

$\mathbf{B_3}:\ t_{+,3}^{(3)} \le t \le t_{+,3}^{(2)}$

For $t_{+,3}^{(3)} \le t \le t_{+,3}^{(2)}$, *the solution of the maximization problem falls into case (iv) of Theorem 38, with $N = 0$. That is,*

$$(63) \qquad supp(\mu_{ES}^V) = (\alpha_0, 1).$$

$\mathbf{B_4}:\ 0 < t \le t_{+,3}^{(3)}$

For $0 < t \le t_{+,3}^{(3)}$, *the solution of the maximization problem is described by case (i) of Theorem 38, i.e.*

$$(64) \qquad supp(\mu_{ES}^V) = (-1, 1).$$

(C) *Suppose $q > 1$. Then there are two critical values $t_{+,2q+1} < t_{+,2q+1}^{(1)}$ (note there is no superscript on the first critical value).*

$\mathbf{C_1}:\ t_{+,2q+1}^{(1)} \le t$

For $t_{+,2q+1}^{(1)} \le t$, *the maximization problem is described by case (iv) of Theorem 38, with $N = 0$:*

$$(65) \qquad supp(\mu_{ES}^V) = (\alpha_0, 1).$$

$\mathbf{C_2}:\ t_{+,2q+1}^{(1)} - \epsilon < t < t_{+,2q+1}^{(1)}$

For $t_{+,2q+1}^{(1)} - \epsilon < t < t_{+,2q+1}^{(1)}$, *the solution is described by case (iv) of Theorem 38, but again in this regime $N = 1$:*

$$(66) \qquad supp(\mu_{ES}^V) = (\alpha_0, \beta_0) \cup (\alpha_1, 1).$$

$\mathbf{C_3}:\ t_{+,2q+1} \le t < t_{+,2q+1} + \epsilon$

For $t_{+,2q+1} \le t < t_{+,2q+1} + \epsilon$, *the maximization problem is described by case (iv) of Theorem 38, with $N = 0$. That is,*

$$(67) \qquad supp(\mu_{ES}^V) = (\alpha_0, 1).$$

$\mathbf{C_4}:\ 0 < t \le t_{+,2q+1}$

For $0 < t \le t_{+,2q+1}$, *the solution of the maximization problem is described by case (i) of Theorem 38, i.e.*

$$(68) \qquad supp(\mu_{ES}^V) = (-1, 1).$$

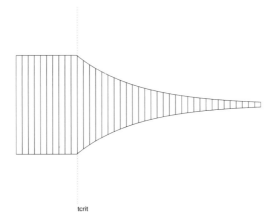

tcrit

FIGURE 1. Plot of the support J as a function of t, for the case $V = -tx^{2q}$, $t > 0$.

Remark As in Theorem 50 above, note that for general $q > 1$, we do not have a global description of the maximizer ψ for all values of $t > 0$. Again by Theorem 44, we know that the number of intervals is bounded by $m + 1 = 2q + 2$. However, we believe that for all $t \in (t_{+,2q+1}, t_{+,2q+1}^{(1)})$, the support of ψ contains precisely three intervals.

Remark: For Theorems 46, 50, 58, for t in any open interval (a, b) on which the number N_V is constant, the endpoints $\{\alpha_j\}$ and $\{\beta_j\}$ of the intervals comprising the support of μ_{ES}^V are analytic functions of t. On the other hand $J = \cup_{j=0}^{N}(\alpha_j, \beta_j)$ is continuous in the natural sense, even at the critical values of t, and hence μ_{ES}^V is weak-* continuous for all $t > 0$.

A picture of the support of μ_{ES}^V for these cases facilitates visualization of the results contained in Theorems 46 - 58. We fix m, and simultaneously plot all endpoints $\{\alpha_j(t)\}$ and $\{\beta_j(t)\}$ as functions of t, for $t > 0$. Connecting α_j and β_j with vertical lines at any fixed t, we have a snapshot of the support of μ_{ES}^V. As t varies, we see how the support evolves (see Figures 1 - 7).

Remark 1 regarding Problem 4: In **Problem 4** above, it was mentioned that the limit in (12) is in fact independent of all lower order terms in the polynomial $P(x)$. Indeed, in [**12**], Gonchar and Rachmanov show that

(69) $$\mu_{OP}^{t,\pm P} = \mu_{ES}^V$$

with

(70) $$V(x) = \pm tx^m,$$

where m is the degree of the weight - polynomial $P(x)$ in **Problem 4**. Thus Theorems 46 - 58 describe the asymptotic distribution of zeros of the polynomials of **Problem 4**.

Remark 2 regarding Problem 4: Rachmanov [**22**] has shown that for measures of the form $e^{-tx^{2q}} dx$ on all of **R**, the asymptotic distribution of zeros, when suitably scaled to the interval $[-1, 1]$ coincides with $\mu_{ES}^{-tx^{2q}}$, provided $t_{-,2q} \le t$. Thus *the answer to the question posed at the end of* **Problem 4** *is to take the* "cutoff parameter" $A = tn^{1/m}$, *with* $t \ge t_{-,2q}$.

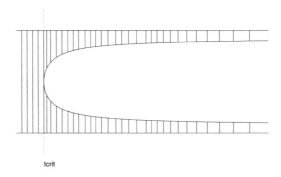

FIGURE 2. Plot of the support J as a function of t, for the case $V = tx^2$, $t > 0$.

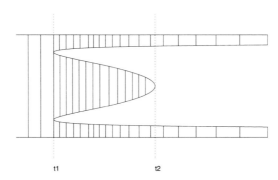

FIGURE 3. Plot of the support J as a function of t, for the case $V = tx^4$, $t > 0$.

Our approach to asymptotic problems for orthogonal polynomials is motivated by the work of Lax and Levermore [15], who considered the zero dispersion limit of the KdV equation with fixed initial data,

$$(71) \qquad\qquad y_t - 6yy_x + \epsilon^2 y_{xxx} = 0,$$

$$(72) \qquad\qquad y(x, 0) = y_0(x),$$

as $\epsilon \downarrow 0$. The authors used scattering and inverse scattering theory for the associated Lax operator

$$(73) \qquad\qquad L = -\epsilon^2 \frac{d^2}{dx^2} + y_0,$$

under the assumption that the effect of the reflection coefficient $r(z)$ is negligible, to obtain a formula in closed form for $y(x, t)$. In a remarkable calculation, they

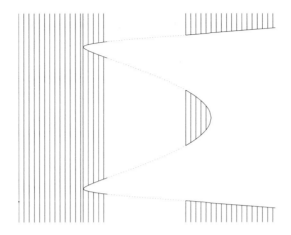

FIGURE 4. Plot of the support J as a function of t, for the general case $V = tx^{2q}$, $t > 0$, $q \geq 3$. The dotted lines denote our conjecture for the region $t_{+,2q} + \epsilon < t < t^{(1)}_{+,2q} - \epsilon$.

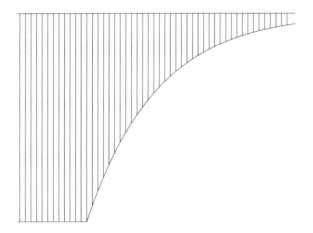

FIGURE 5. Plot of the support J as a function of t, for the case $V = tx$, $t > 0$.

then showed that asymptotically as $\epsilon \downarrow 0$ this formula is governed by an associated maximization problem, quite analogous to (4), but now with *two* external parameters x and t. They then proceeded to show that for all x and t this maximization problem possesses a unique solution, which is in fact an L^p function, for any $p \in [1, 2)$. Furthermore, they showed that at $t = 0$, the support of the maximizer for each x is one interval, whose endpoints are related in an elegant way to the initial data $y_0(x)$. Until a critical later time t_b, the shock time for Burger's equation $y_t - 6yy_x = 0$, $y(x, 0) = y_0(x)$, they showed that for all x the support of the maximizer is again precisely one interval, and that the relation between the endpoints and $y(x, t)$ which held at $t = 0$ remains true for $0 \leq t \leq t_b$.

Beyond the critical time t_b, they postulated that for each x the support becomes a finite union of subintervals. However, this remained a conjecture until the work of

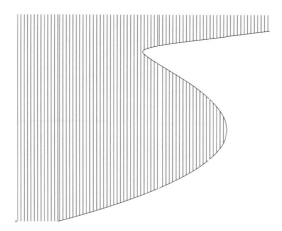

FIGURE 6. Plot of the support J as a function of t, for the case $V = tx^3$, $t > 0$.

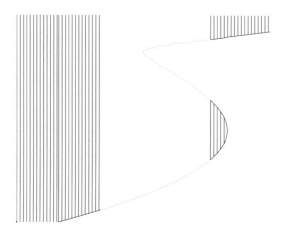

FIGURE 7. Plot of the support J as a function of t, for the general case $V = tx^{2q+1}$, $t > 0$, $q \geq 2$. The dotted lines denote our conjecture for the region $t_{+,2q+1} + \epsilon < t < t_{+,2q+1}^{(1)} - \epsilon$.

Tian [26] (see also Wright [32]), who showed that for a very general class of initial data, as t crosses t_b, the support of the maximizer experiences a transition from one interval to two for values of x in an interval $(x_-(t), x_+(t))$; for $x > x_+(t)$ and $x < x_-(t)$, the maximizer continues to be supported on a single interval. While this result is local in t, Tian further showed that for a restricted class of initial data, the zero dispersion limit of the KdV equation is governed for all $t > t_b$ by a maximizer whose support for each x is either one interval, or two.

Our work uses an approach to the "phase transition" problem that is different from that of Tian's, and is based on the calculations and results in [17], [4], and in [6], which in turn has provided a method to generalize the results of Tian (see [4]). Indeed, our methods establish the following result.

THEOREM 74. *For real analytic initial data $y_0(x)$, the number of intervals comprising the support of the maximizer remains finite for all x and t.*

We will provide a proof of this result in a later publication.

Lax - Levermore theory has been extended in a number of highly non-trivial directions, particularly in the work of Venakides [29] [30] [31]: whereas Lax and Levermore compute $y(x, t)$ to leading order as $\epsilon \downarrow 0$, in [31] Venakides shows how to compute the asymptotics to the next order. This paper of Venakides in turn leads on to [6], which then provides the model for our approach to the computation of the asymptotics of orthogonal polynomials. As indicated above, our results for the asymptotics of orthogonal polynomials will be presented in a sequel to this paper.

Acknowledgements The authors are particularly grateful to P. Nevai for introducing (one of) us to the problems considered in this paper, and for generous help and encouragement. The authors are also particularly grateful to V. Totik and E. Saff for providing them with an advance copy of their very useful new book [23] on equilibrium measures. Finally, the authors would like to thank many of their colleagues for very useful conversations, particularly W. Van Assche, A. Magnus, F. Marcellan, P. Nevai, P. Sarnak, V. Totik, S. Venakides, and X. Zhou. X. Zhou, in particular, provided great assistance in extending certain basic formulae (for example (30)) to general C^2 potentials.

References

1. P. Blecher and A. Its. Asymptotics of Orthogonal Polynomials and Universality in Matrix Models. *preprint* (1996).
2. E. Brezin, C. Itzykson, J.B. Zuber. Quantum Field Theory Techniques in Graphical Enumeration. *Adv. in Appl. Math.*, **1**, 109-157, (1980).
3. A. Boutet de Monvel, L.A. Pastur, and M. Shcherbina. On the statistical mechanics approach to random matrix theory: integrated density of states. *J. of Stat. Phys.*, **79**, 585-611, (1995).
4. P. Deift and K. T-R McLaughlin. A Continuum Limit of the Toda Lattice. *To appear, Memoirs of the AMS.*
5. R. DeVore. The approximation of continuous functions by positive linear operators. *Lecture notes in Math.*, **293**, Springer-Verlag, Berlin-Heidelberg, New York, (1972).
6. P. Deift, S. Venakides, and X. Zhou. New Results in Small Dispersion KdV by an Extension of the Steepest Descent Method for Riemann - Hilbert problems. *Submitted to Proc. Nat'l Acad. Sci.*, (1996).
7. P. Deift and X. Zhou. A steepest descent method for oscillatory Riemann - Hilbert problems. Asymptotics for the mKdV equation, *Ann. of Math.* **137**, 295-370, (1993).
8. P. Deift and X. Zhou. Asymptotics for the Painleve II equation. *Comm. Pure and Appl. Math.* **48**, 277-337, (1995).
9. M. Fekete. Über die Verteilung der Wurzeln bei gewissen algebraischen Gleichungen mit ganzzahligen Koeffizienten. Math Z., 17, 228-249, 1923.
10. M. Fekete. Über den transfiniten Durchmesser ebener Punktmengen. *Math. Z.*, **32**, 108-114, 1930.
11. C.F. Gauss. Allgemeine Lehrsätze in Beziehung auf die im verkehrten Verhältnisse des Quadrats der Entfernung wirkenden Anziehungs- und Abstossungs- Kräfte. Gesammelte Werke (V). Goettingen, 1877, 197-242.
12. Gonchar and Rachmanov. Equilibrium Measure and the Distribution of Zeros of Extremal Polynomials. *Math. USSR Sbornik*, **53**, 119-130, (1986).
13. K. Johansson. On Fluctuations of Eigenvalues of Random Hermitian Matrices. *preprint TRITA-MAT-1995-MA-19* (Sept. 1995).
14. N.S. Landkof. *Foundations of Modern Potential Theory.* Springer-Verlag, Berlin, (1972).
15. P. Lax, C.D. Levermore. The Small Dispersion Limit of the Korteweg -de Vries Equation I, II, III. *Comm. Pure Appl. Math.*, **36** (1983), 253-290, 571-593, 809-829.
16. F. Leja. Sur les moyennes arithmetiques, geometriques et harmoniques des distances mutuelles des points d'un ensembler. *Ann. Polon. Math.*, **9**, 211-218, (1961).

17. K. T-R McLaughlin. A Continuum Limit of the Toda Lattice. *PhD thesis, NYU* (1994).

18. M.L. Mehta. *Random Matrices,* 2nd ed., Academic Press, San Diego, 1991.

19. H.N. Mhaskar and E.B. Saff. Extremal problems for polynomials with exponential weights. *Trans. Amer. Math. Soc.*, **285**, 204-234, (1984).

20. P. Nevai and V. Totik. Weighted Polynomial Inequalities. *Constr. Approx.*, **2**, 113-127, (1986).

21. L.A. Pastur. Spectral and Probabilistic Aspects of Matrix Models. *preprint* (1995).

22. E.A. Rachmanov. On asymptotic properties of polynomials orthogonal on the real axis. *Mat. Sb.*, **119**, 163-203 (1982). English Transl.: *Math. USSR - Sb.*, **47**, (1984).

23. E.B. Saff and V. Totik. Logarithmic Potentials with External Fields. *Submitted to Springer - Verlag* (1996).

24. G. Szegő. *Orthogonal Polynomials.* AMS Colloquium Publications, vol.23, 1939.

25. G. Szegő. Bemerkungen zur einer Arbeit von Herrn Fekete "Über den transfiniten Durchmesser ebener Punktmengen". *Math. Zeitscher*, **21**, 203-208, (1924).

26. F.R. Tian. Oscillations of the Zero Dispersion Limit of the Korteweg de Vries Equation , (preprint). *Comm. Pure Appl. Math.*, (1993).

27. V. Totik. Fast decreasing polynomials via potentials. *J. D'Analyse Math.*, **62**, 131-154, (1994).

28. A.B.J. Kuijlaars and W. Van Assche. A Problem of Totik on Fast Decreasing Polynomials. *preprint* (1996).

29. S. Venakides. The zero dispersion limit of the Korteweg de Vries Equation with nontrivial reflection coefficient. *Comm. Pure and App. Math.*, **38**, 125-155, (1985).

30. S. Venakides. The generation of modulated wavetrains in the solution of the Korteweg de Vries Equation: *Comm. Pure and App. Math.*, **38**, 883-909, (1985).

31. S. Venakides. The Korteweg de Vries Equation with small dispersion: higher order Lax-Levermore theory. *Comm. Pure and App. Math.*, **43**, 335-361, (1990).

32. O. Wright. Korteweg de Vries Zero Dispersion Limit: A Restricted Initial Value Problem. *PhD Thesis, Princeton University*, (1991).

PERCY DEIFT

COURANT INSTITUTE OF MATHEMATICAL SCIENCES, NEW YORK, NY 10003
E-mail address: deift@cims.nyu.edu

THOMAS KRIECHERBAUER

UNIVERSITY OF MUNICH, MUNICH, GERMANY
E-mail address: tkriech@rz.matematik.uni-muenchen.de

KENNETH T-R McLAUGHLIN

DEPARTMENT OF MATHEMATICS, OHIO STATE UNIVERSITY, AND PRINCETON UNIVERSITY
E-mail address: mcl@math.princeton.edu

Proceedings of Symposia in Applied Mathematics
Volume **54**, 1998

Evolution of trajectory correlations in steady random flows

Albert Fannjiang, Leonid Ryzhik, and George Papanicolaou

To P.Lax and L. Nirenberg.

ABSTRACT. We analyze the behavior of the correlation for two nearby trajectories of motion in a random incompressible flow with nonzero mean and small
fluctuations. We show that the Fourier transform of the Richardson function
of a passive scalar advected by the flow satisfies, under certain conditions, a
radiative transport equation. We also study the stretching of curves advected
by the flow and show that their length grows algebraically in time, and not
exponentially as it does for time dependent, zero mean random flows.

1. INTRODUCTION

We consider motion in a steady, random, incompressible flow with constant
drift $\mathbf{u}$. The random part of the flow $\mathbf{v}(\mathbf{x})$, with $\nabla \cdot \mathbf{v} = 0$, is a smooth spatially
homogeneous random field with zero mean and rapidly decaying correlations. The
trajectory $\mathbf{X}^\varepsilon(t, \mathbf{x})$ advected by the flow satisfies the scaled equation

$$(1.1) \qquad \frac{d\mathbf{X}^\varepsilon}{dt} = \mathbf{u} + \sqrt{\varepsilon}\mathbf{v}\left(\frac{\mathbf{X}^\varepsilon}{\varepsilon}\right)$$

$$\mathbf{X}^\varepsilon(0, \mathbf{x}) = \mathbf{x},$$

where $\mathbf{x} \in R^d$, $d = 2, 3$, is the starting position. The small scaling parameter ε
determines the size of the random fluctuations and their correlation length. We are
interested in the behavior of the rescaled two-point difference function

$$(1.2) \qquad \mathbf{Y}^\varepsilon(t, \mathbf{y}, \mathbf{x}) = \frac{1}{\varepsilon}\{\mathbf{X}^\varepsilon(t, \mathbf{x}) - \mathbf{X}^\varepsilon(t, \mathbf{x} - \varepsilon\mathbf{y})\}$$

in the limit of $\varepsilon \to 0$. The individual trajectories $\mathbf{X}^\varepsilon(t)$ are close to the trajectories of
the deterministic flow $\mathbf{X} = \mathbf{x} + \mathbf{u}t$ for finite times t, when ε is small, and thus random
perturbations do not play an essential role in their behavior. The rescaled trajectory
differences $\mathbf{Y}^\varepsilon$ that start nearby are, however, affected and that is what we study
in this paper. We note that this is different from the analysis of the long time
correlations of trajectories that are initially a fixed distance apart, independently
of ε [**19, 22**].

1991 *Mathematics Subject Classification.* Primary 60F05, 76F05, 76R50; Secondary 58F25.

The paper is organized as follows. In Section 2 we review briefly the known results for the long time (of order ε^{-1}) behavior of single trajectories, or groups of trajectories that are not close initially. In Section 3 we show that the asymptotic behavior of $\mathbf{Y}^\varepsilon$ as $\varepsilon \to 0$ can be deduced from the limit theorems of [19, 22, 17], stated here in Appendix A. This allows us to derive in Sections 4 and 5 a singular diffusion equation for the rescaled Richardson function. Higher moments of the Richardson function are studied in Section 6. In Section 7 we show how the diffusion equation for the Richardson function can be transformed into a radiative transport equation for the Wigner distribution associated with the trajectory difference function $\mathbf{Y}^\varepsilon$. We study the duality between the limit two-point diffusion process and its analog in Fourier space in Section 8. A scaling more general than 1.1 is treated in Section 9. In Section 10 we show that the Jacobian matrix of the flow (1.1) also converges to a diffusion process in the limit $\varepsilon \to 0$, and we analyze in detail its law in two-dimensional potential flows in Section 11. In Section 12 we apply these results to study the stretching of curves transformed by (1.1). We show that their length grows algebraically in this case and not exponentially, as in the case of zero mean randomly time dependent flows [2, 8, 32].

2. Long time diffusive behavior

The long time diffusive behavior of trajectories of mean zero random velocity fields was first studied by G.I.Taylor [26, 33]. He considered the trajectories of the system

$$(2.1) \qquad \begin{aligned} d\mathbf{X}(t) &= \mathbf{v}(t, \mathbf{X}(t)) + \kappa d\mathbf{W}(t) \\ \mathbf{X}(0) &= 0, \end{aligned}$$

where $\mathbf{v}(t, \mathbf{x})$ is a mean zero random velocity field with rapidly decaying correlations, and $\mathbf{W}(t)$ is the standard d-dimensional Brownian motion. He argued that in the long time limit the trajectories behave like those of d-dimensional Brownian motion with covariance matrix

$$(2.2) \qquad \int_0^\infty E\left\{ v_i(t, \mathbf{X}(t)) v_j(0, 0) + v_j(t, \mathbf{X}(t)) v_i(0, 0) \right\} dt + \kappa^2 \delta_{ij}.$$

The first term in expression (2.2), which corresponds to the turbulent enhancement of the diffusion coefficient, depends also on the molecular diffusivity κ through the path $\mathbf{X}(t)$. It is convenient to scale equations (2.1) with a small parameter ε

$$(2.3) \qquad \begin{aligned} d\mathbf{X}^\varepsilon(t) &= \frac{1}{\varepsilon} \mathbf{v}\left(\frac{t}{\varepsilon^2}, \frac{\mathbf{X}^\varepsilon}{\varepsilon}\right) + \kappa d\mathbf{W}(t) \\ \mathbf{X}^\varepsilon(0) &= 0 \end{aligned}$$

and analyze the behavior of the trajectories as $\varepsilon \to 0$. When the molecular diffusivity κ is positive and the flow is incompressible, $\nabla \cdot \mathbf{v} = 0$, it is shown by homogenization methods [28, 27, 1, 24, 14] that as $\varepsilon \to 0$ the trajectories $X^\varepsilon \to 0$ converge weakly to Brownian motion with the covariance matrix given by the homogenization formula (2.2) with $\mathbf{X}(t)$ the solution of (2.1) corresponding to $\varepsilon = 1$ in (2.3). When the molecular diffusivity κ is positive, diffusive behavior holds for random velocity fields that, in the time independent case, are spatially homogeneous, incompressible and have a square integrable stream matrix [14]. Time dependent flows when $\kappa > 0$ are analyzed in [12] and when $\kappa = 0$ and there is rapid time decorrelation of $\mathbf{v}$ in [18], [13] .

The weak fluctuation scaling (described in detail in Appendix A) was intensively studied (with $\kappa = 0$ and $\kappa > 0$ [17, 19, 22]). A typical scaling in this case, with $\kappa = 0$, is

(2.4)
$$\frac{d\mathbf{X}^\varepsilon}{dt} = \frac{1}{\varepsilon}\mathbf{v}\left(\frac{t}{\varepsilon^2}, \frac{\mathbf{X}^\varepsilon(t)}{\varepsilon^\gamma}\right),$$
$$\mathbf{X}^\varepsilon(0) = \mathbf{x},$$

with $0 \leq \gamma < 1$, so that now the random velocity field is varying on spatial scales large compared to ε, but possibly small on the overall scale if $\gamma \neq 0$. Then, as before, the process $\mathbf{X}^\varepsilon(t)$ converges weakly as $\varepsilon \to 0$ to a diffusion process. Its generator is given by

$$L_{\mathbf{x}}f(\mathbf{x}) = \frac{1}{2}\sum_{i,j=1}^{d} a_{ij}(\mathbf{x})\frac{\partial^2 f}{\partial x_i \partial x_j}.$$

This operator is self-adjoint because the velocity field $\mathbf{v}(\mathbf{x})$ is incompressible. The corresponding generator for a compressible flow has an extra drift term [19, 17, 22]. The diffusion matrix is given by

(2.5)
$$a_{ij}(\mathbf{x}) = \int_0^\infty E\left\{v_i(0,\mathbf{x})v_j(t,\mathbf{x}) + v_j(0,\mathbf{x})v_i(t,\mathbf{x})\right\}dt.$$

Expression (2.5) is called the Kubo formula. Of particular interest here is the system

(2.6)
$$\frac{d\mathbf{X}^\varepsilon}{dt} = \frac{1}{\varepsilon}\mathbf{v}\left(\mathbf{X}^\varepsilon(t) + \frac{t}{\varepsilon^2}\mathbf{u}\right),$$
$$\mathbf{X}^\varepsilon(0) = \mathbf{x},$$

where $\mathbf{v}(\mathbf{x})$ is a space homogeneous, mean zero divergence free random field. Then $\mathbf{X}^\varepsilon(t,\mathbf{x})$ converges to Brownian motion with the covariance matrix given by the Kubo formula

(2.7)
$$c_{ij}(0) = \int_{-\infty}^\infty E\left\{v_i(t\mathbf{u})v_j(0)\right\}dt,$$

This corresponds to the Taylor prediction (2.2) with the path $\mathbf{X} = \mathbf{u}t$ being frozen and independent of the random medium. We shall see in Section 3 why the path in (2.7) is deterministic. The joint law for two trajectories of (2.6) starting at two points $\mathbf{x}, \mathbf{x}'$ was also obtained in [19, 22]. The joint process $(\mathbf{X}^\varepsilon, \mathbf{X}'^\varepsilon)$ converges weakly to a diffusion process $(\mathbf{X}, \mathbf{X}')$ with generator

$$\tilde{L}f(\mathbf{x},\mathbf{x}') = L_{\mathbf{x}}f + L_{\mathbf{x}'}f + L'f,$$

where the cross-term L' is

$$L'f = \frac{1}{2}c_{ij}(\mathbf{x},\mathbf{x}')\frac{\partial^2 f}{\partial x_i \partial x_j'}.$$

Here

$$c_{ij}(\mathbf{x} - \mathbf{x}') = \sum_{i,j=1}^{d}\int_{-\infty}^\infty E\left\{v_i(0)v_j(\mathbf{x}' - \mathbf{x} + t\mathbf{u})\right\}dt.$$

Thus the difference function $\mathbf{Y}^\varepsilon = \mathbf{X}^\varepsilon - \mathbf{X}'^\varepsilon$ converges weakly to a diffusion process $\mathbf{Y}$ with generator

(2.8)
$$L_{\mathbf{Y}}f = (c_{ij}(0) - c_{ij}(\mathbf{y}))\frac{\partial^2 f}{\partial y_i \partial y_j}.$$

The behavior of the difference function for a stochastic flow on a compact manifold was considered in [**3**], where the properties of the two-point motion for various Lyapunov exponents were investigated.

The scaling in (1.1) is different from both that of (2.3) and (2.4). As we noted above, the trajectories of (1.1) stay close to the deterministic trajectories $\mathbf{X} = \mathbf{x} + \mathbf{u}t$. We shall show, however, that after a rescaling the dynamics of many quantities of interest can be reduced to a form similar to (2.4). For example, the rescaled difference function (1.2) for the trajectories of (1.1) converges to a diffusion process with the same generator (2.8).

3. FLOW CORRELATIONS

Let $\mathbf{X}^\varepsilon(t, \mathbf{x})$ and $\mathbf{X}^\varepsilon(t, \mathbf{x} - \varepsilon\mathbf{y})$ be two trajectories of (1.1) that are $\varepsilon\mathbf{y}$ apart initially. We consider the rescaled difference variable

$$\mathbf{Y}^\varepsilon(t, \mathbf{y}, \mathbf{x}) = \frac{\mathbf{X}^\varepsilon(t, \mathbf{x}) - \mathbf{X}^\varepsilon(t, \mathbf{x} - \varepsilon\mathbf{y})}{\varepsilon},$$

which is the scaled separation of the particles at time t. Then $\mathbf{X}^\varepsilon(t, \mathbf{x})$, $\mathbf{Y}^\varepsilon(t, \mathbf{y}, \mathbf{x})$ satisfy

$$(3.1) \qquad \begin{aligned} \frac{d\mathbf{X}^\varepsilon}{dt} &= \mathbf{u} + \sqrt{\varepsilon}\mathbf{v}\left(\frac{\mathbf{X}^\varepsilon}{\varepsilon}\right), \quad \mathbf{X}^\varepsilon(0) = \mathbf{x} \\ \frac{d\mathbf{Y}^\varepsilon}{dt} &= \frac{1}{\sqrt{\varepsilon}}\left\{\mathbf{v}\left(\frac{\mathbf{X}^\varepsilon}{\varepsilon}\right) - \mathbf{v}\left(\frac{\mathbf{X}^\varepsilon}{\varepsilon} - \mathbf{Y}^\varepsilon\right)\right\}, \quad \mathbf{Y}^\varepsilon(0) = \mathbf{y}. \end{aligned}$$

As we noted above the trajectory $\mathbf{X}^\varepsilon(t, \mathbf{x})$ is close to the deterministic trajectory $\mathbf{X} = \mathbf{x} + \mathbf{u}t$ for finite t, and so we introduce its rescaled fluctuations

$$(3.2) \qquad \mathbf{Z}^\varepsilon(t, x) = \frac{\mathbf{X}^\varepsilon(t, \mathbf{x}) - \mathbf{u}t - \mathbf{x}}{\varepsilon}.$$

Then the system (3.1) becomes

$$(3.3) \qquad \begin{aligned} \frac{d\mathbf{Z}^\varepsilon}{dt} &= \frac{1}{\sqrt{\varepsilon}}\mathbf{v}\left(\frac{\mathbf{x} + \mathbf{u}t}{\varepsilon} + \mathbf{Z}^\varepsilon\right), \quad \mathbf{Z}^\varepsilon(0) = 0 \\ \frac{d\mathbf{Y}^\varepsilon}{dt} &= \frac{1}{\sqrt{\varepsilon}}\left\{\mathbf{v}\left(\frac{\mathbf{x} + \mathbf{u}t}{\varepsilon} + \mathbf{Z}^\varepsilon\right) - \mathbf{v}\left(\frac{\mathbf{x} + \mathbf{u}t}{\varepsilon} + \mathbf{Z}^\varepsilon - \mathbf{Y}^\varepsilon\right)\right\}, \quad \mathbf{Y}^\varepsilon(0) = \mathbf{y}. \end{aligned}$$

This system has the general form (14.1) to which the limit theorem of Appendix A applies. The full statement of this theorem and the necessary assumptions on the random velocity field in the general case are recalled in Appendix A. We shall assume in particular that $\mathbf{v}(\mathbf{y})$ is a mean zero, divergence-free, space homogeneous and strongly mixing velocity field bounded in $C^3(R^d)$. Then the limit theorem implies the following.

THEOREM 3.1. *The processes* $\mathbf{Z}^\varepsilon$, *and* $\mathbf{Y}^\varepsilon$ *converge weakly to the correlated diffusion processes* $\mathbf{Z}$ *and* $\mathbf{Y}$ *whose joint generator is*

$$\begin{aligned} Lf &= \frac{1}{2}\sum_{i,j=1}^{d}\left[\int_{-\infty}^{\infty}R_{ij}(\mathbf{u}t)dt\frac{\partial^2 f}{\partial z_i \partial z_j}\right. \\ &\quad + \int_{-\infty}^{\infty}\left(2R_{ij}(\mathbf{u}t) - R_{ij}(\mathbf{u}t + \mathbf{y}) - R_{ij}(\mathbf{u}t - \mathbf{y})\right)dt\frac{\partial^2 f}{\partial y_i \partial y_j} \\ (3.4) &\quad \left. + \int_{-\infty}^{\infty}\left(2R_{ij}(\mathbf{u}t) - 2R_{ij}(\mathbf{u}t - \mathbf{y})\right)dt\frac{\partial^2 f}{\partial z_i \partial y_j}\right]. \end{aligned}$$

Here the covariance tensor R_{ij} is defined by

(3.5) $$E\{v_i(y)v_j(y+h)\} = R_{ij}(h),$$

The individual generators for Z and Y are:

(3.6) $$L_Z f(z) = \frac{1}{2} \sum_{i,j=1}^{d} \int_{-\infty}^{\infty} R_{ij}(\mathbf{u}t)dt \frac{\partial^2 f}{\partial z_i \partial z_j},$$

and

(3.7) $$L_Y f(y) = \sum_{i,j=1}^{d} \int_{-\infty}^{\infty} \left[R_{ij}^s(\mathbf{u}t) - R_{ij}^s(\mathbf{y} + \mathbf{u}t) \right] dt \frac{\partial^2 f}{\partial y_i \partial y_j},$$

respectively, where

$$R_{ij}^s = \frac{R_{ij} + R_{ji}}{2}.$$

Thus the fluctuations $\mathbf{Z}^\varepsilon(t)$ converge to Brownian motion with covariance matrix given by the Kubo formula (2.7). It is clear in this case that the Kubo formula (2.7) is the expression (2.2) predicted by Taylor. The path $\mathbf{X}(t) = \mathbf{x} + \mathbf{u}t$ is now the deterministic trajectory of (1.1) around which $\mathbf{X}^\varepsilon(t,\mathbf{x})$ is fluctuating because these two paths are close to each other. The generator (3.7) for the separation process $\mathbf{Y}(t,\mathbf{y})$ coincides with the one given by (2.8), as expected. We note that the diffusion coefficient in (3.7) vanishes for $\mathbf{y} = c\mathbf{u}$, so that if two particles start at two nearby positions on the same deterministic trajectory then their separation is not changed by the flow in the limit. The generator L_Y is asymptotically close to L_Z for all $\mathbf{y}$ that have large component in the direction perpendicular to the mean flow $\mathbf{u}$. The reason for this is that when the two starting points are separated by a large distance in the direction normal to $\mathbf{u}$, their trajectories are almost independent, and the rescaled difference trajectory behaves like the fluctuations of each individual trajectory. The other end of the asymptotics, the small $\mathbf{y}$ behavior of L_Y, is related to the Jacobian of the map $\mathbf{x} \to \mathbf{X}^\varepsilon(t,\mathbf{x})$, and is described in detail in Section 10.

4. The Richardson function and its evolution

The Richardson function of a scalar $\phi(t,\mathbf{x})$ advected by a random flow

$$\frac{\partial \phi}{\partial t} + \mathbf{v}(\mathbf{x}) \cdot \nabla_\mathbf{x} \phi = 0$$
$$\phi(0,\mathbf{x}) = \phi_0(\mathbf{x})$$

is defined [26] by

$$Q(t,\mathbf{x},\mathbf{y}) = \phi(\mathbf{x})\phi(\mathbf{x}+\mathbf{y}).$$

It was predicted by Richardson [30] that the expectation of this function satisfies the usual diffusion equation in $\mathbf{y}$, in the long time limit. This problem was studied extensively by physicists (see [20] for an extensive review and references), especially for δ-correlated in time velocity fields, but to the best of our knowledge the only mathematical results are those in [19] and [25, 29]. Molchanov and Piterbarg [25] considered in particular a convection-diffusion equation of the form

$$(4.1) \qquad \frac{\partial T^\varepsilon}{\partial t} + \frac{1}{\varepsilon} \mathbf{v}(\frac{t}{\varepsilon^2}, \mathbf{x}) \cdot \nabla_{\mathbf{x}} T^\varepsilon = \kappa \Delta T^\varepsilon$$
$$T^\varepsilon(0, \mathbf{x}) = T_0(\mathbf{x}).$$

They argued that the expectation $E\{Q^\varepsilon(\mathbf{x}, \mathbf{y})\}$ of the Richardson function has a limit $Q(\mathbf{x}, \mathbf{y})$ as $\varepsilon \to 0$. If the initial density $T_0(\mathbf{x})$ is a space homogeneous isotropic random process with correlation function $Q_0(r)$, the limit Richardson function satisfies the diffusion equation

$$(4.2) \qquad \frac{\partial Q}{\partial t} = \frac{1}{r^{d-1}} \frac{\partial}{\partial r} r^{d-1}(2\kappa + F(r)) \frac{\partial Q}{\partial r}$$
$$Q(0, r) = Q_0(r).$$

Here

$$F(r) = \int_{-\infty}^{\infty} (R_L(t, 0) - R_L(t, r)) \, dt,$$

and $R_L(t, r)$ comes from the correlation matrix $R(t, \mathbf{x})$ of the isotropic velocity field $\mathbf{v}(\mathbf{x})$:

$$R_{ij} = \left(R_L(t, r) + \frac{r}{d-1} \frac{\partial R_L}{\partial r} \right) \delta_{ij} - \frac{x_i x_j}{r(d-1)} \frac{\partial R_L}{\partial r}.$$

Equation (4.2) is the generalization of (2.8) for $\kappa \neq 0$. We note that the additional term due to non-zero molecular diffusivity is additive. This is different from the homogenization formulas of the type (2.2) which come from the homogenization scaling of (2.3).

We study here the rescaled Richardson function of oscillatory initial densities advected by the random flow. Let $\phi^\varepsilon(\mathbf{x})$ be the solution of

$$(4.3) \qquad \frac{\partial \phi^\varepsilon}{\partial t} + \left(\mathbf{u} + \sqrt{\varepsilon} \mathbf{v}\left(\frac{\mathbf{x}}{\varepsilon} \right) \right) \cdot \nabla \phi^\varepsilon = 0$$
$$\phi^\varepsilon(0, \mathbf{x}) = \phi_0^\varepsilon(\mathbf{x}),$$

where the initial density $\phi_0^\varepsilon(\mathbf{x})$ is ε-oscillatory (see Appendix B for the precise definition of this notion). It may be either deterministic, or random but independent of $\mathbf{v}(\mathbf{x})$. An example of such random initial data is

$$(4.4) \qquad \phi_0^\varepsilon(\mathbf{x}) = \phi_0(\frac{\mathbf{x}}{\varepsilon}),$$

where $\phi_0(\mathbf{x})$ is a spatially homogeneous ergodic random process. Interesting deterministic initial data of oscillatory form are the localized initial densities

$$(4.5) \qquad \phi_0^\varepsilon(\mathbf{x}) = \frac{1}{\varepsilon^{d/2}} \phi_0(\frac{\mathbf{x}}{\varepsilon}),$$

where $\phi_0(\mathbf{x})$ is an $L^1(R^d) \cap L^2(R^d)$ function, and the inhomogeneous wave family

$$(4.6) \qquad \phi^\varepsilon(\mathbf{x}) = A(\mathbf{x}) e^{iS(\mathbf{x})/\varepsilon}$$

where $A(\mathbf{x})$ and $S(\mathbf{x})$ are smooth functions, and $S(\mathbf{x})$ is real valued. The latter family describes the distribution of tracers which have the form of high frequency waves propagating in the direction $\nabla S(\mathbf{x})$ with amplitude $A(\mathbf{x})$. A particular case of (4.6) is the high frequency plane wave

$$(4.7) \qquad \phi^\varepsilon(\mathbf{x}) = A e^{i\mathbf{x} \cdot \mathbf{p}/\varepsilon}.$$

The rescaled Richardson function of the family $\phi^\varepsilon(\mathbf{x})$ is defined by

$$(4.8) \qquad \hat{W}^\varepsilon(t, \mathbf{x}, \mathbf{y}) = Q(t, \mathbf{x} - \varepsilon\mathbf{y}, \varepsilon\mathbf{y}) = \phi^\varepsilon(t, \mathbf{x} - \varepsilon\mathbf{y})\phi^{\varepsilon*}(t, \mathbf{x}),$$

so that its expectation is the correlation function of the filed $\phi^\varepsilon(\mathbf{x})$ at two nearby points separated by $\varepsilon\mathbf{y}$. It has a weak limit $\hat{W}(t, \mathbf{x}, \mathbf{y})$ as $\varepsilon \to 0$ in the space $\mathcal{A}'$ dual to (see Appendix B)

$$\mathcal{A} = \left\{ f(\mathbf{x}, \mathbf{y}) : \int d\mathbf{y} \sup_{\mathbf{x}} |f(\mathbf{x}, \mathbf{y})| < \infty \right\},$$

introduced in [**23**]. The Richardson function of the family (4.4) is

$$(4.9) \qquad \hat{W}(\mathbf{x}, \mathbf{y}) = R_0(\mathbf{y}),$$

with probability one, by the ergodic theorem, and is independent of $\mathbf{x}$. Here $R_0(\mathbf{x})$ is the covariance function of the random process $\phi_0(\mathbf{x})$. The Richardson functions of the localized deterministic family (4.5) is

$$(4.10) \qquad \hat{W}(\mathbf{x}, \mathbf{y}) = \delta(\mathbf{x}) \int \hat{Q}_0(\mathbf{z}, \mathbf{y}) d\mathbf{z},$$

where

$$(4.11) \qquad \hat{Q}_0(\mathbf{x}, \mathbf{y}) = \phi_0(\mathbf{x} - \mathbf{y})\phi_0^*(\mathbf{x}).$$

The limit Richardson function of the WKB family (4.6) is

$$(4.12) \qquad \hat{W}(\mathbf{x}, \mathbf{y}) = |A(\mathbf{x})|^2 e^{-i\nabla S(\mathbf{x}) \cdot \mathbf{y}},$$

which reduces to

$$(4.13) \qquad \hat{W}(\mathbf{x}, \mathbf{y}) = |A|^2 e^{-i\mathbf{p} \cdot \mathbf{y}}$$

for the plane waves (4.7). The limits (4.10) and (4.12) correspond to localization in space and direction, respectively.

The rescaled Richardson function $\hat{W}^\varepsilon(t, \mathbf{x}, \mathbf{y})$ of the oscillatory field $\phi^\varepsilon(t, \mathbf{x})$ may be reduced to the form studied in [**19, 25**], when the initial data has the form (4.4) and $\phi_0(\mathbf{x})$ is Gaussian. In that case $\phi^\varepsilon(t, \mathbf{x}) = \psi^\varepsilon(\frac{\mathbf{x} - \mathbf{u}t}{\varepsilon}, t)$, where $\psi^\varepsilon(t, \mathbf{x})$ satisfies an equation of the form (4.1) with $\kappa = 0$

$$(4.14) \qquad \frac{\partial \psi^\varepsilon}{\partial t} + \frac{1}{\varepsilon} \mathbf{v}\left(\mathbf{x} + \mathbf{u}\frac{t}{\varepsilon^2}\right) \cdot \nabla_\mathbf{x} \psi^\varepsilon = 0$$

$$\psi^\varepsilon(0, \mathbf{x}) = \phi_0(\mathbf{x}).$$

The rescaled Richardson function $W^\varepsilon(t, \mathbf{x}, \mathbf{y})$ of $\phi^\varepsilon(\mathbf{x})$ and the unscaled Richardson function $Q^\varepsilon(t, \mathbf{x}, \mathbf{y})$ of $\psi^\varepsilon(\mathbf{x})$ are related by

$$W^\varepsilon(t, \mathbf{x}, \mathbf{y}) = Q^\varepsilon\left(t, \frac{\mathbf{x} - \mathbf{u}t}{\varepsilon}, \mathbf{y}\right).$$

When the initial data for (4.3) is random, space homogeneous of the form (4.4), both the scaled and unscaled Richardson functions are independent of $\mathbf{x}$ in the limit $\varepsilon \to 0$, hence they coincide in this limit, and the analysis for (4.3) can be reduced to the one in [**25**] and [**19**]. We consider here a more general class of ε-oscillatory initial data, not necessarily random and Gaussian, and also a more general scaling in Section 9. The following theorem shows that a version of (2.8) and (4.2) still holds.

THEOREM 4.1. *Let $\phi^\varepsilon(t, \mathbf{x})$ satisfy (4.3) with the initial data $\phi_0^\varepsilon(\mathbf{x})$ being either*

(i) *random of the form (4.4) with $\phi_0(\mathbf{x})$ spatially homogeneous and independent of $\mathbf{v}(\mathbf{x})$, or*

(ii) *deterministic of the form (4.5) with $\phi_0(\mathbf{x}) \in L^2(R^d)$, or*

(iii) *the plane wave (4.7).*

Let $\hat{W}(t, \mathbf{x}, \mathbf{y})$ be the weak limit of the expectation of the rescaled Richardson function $\hat{W}^\varepsilon(t, \mathbf{x}, \mathbf{y})$ in $\mathcal{A}'$ as $\varepsilon \to 0$, and $\hat{W}_0(\mathbf{x}, \mathbf{y})$ be the limit of the expectation of the scaled Richardson function for $\phi_0^\varepsilon(\mathbf{x})$. Then $\hat{W}(t, \mathbf{x}, \mathbf{y})$ satisfies the diffusion equation with drift:

$$(4.15) \qquad \frac{\partial \hat{W}}{\partial t} + \mathbf{u} \cdot \nabla_{\mathbf{x}} \hat{W} = L_Y \hat{W}$$

$$\hat{W}(0, \mathbf{x}, \mathbf{y}) = \hat{W}_0(\mathbf{x}, \mathbf{y}),$$

where the operator L_Y is given by (3.7). The initial data $\hat{W}_0(\mathbf{x}, \mathbf{y})$ in (4.15) are given by (4.9) in the random case, and by (4.10) and (4.13) in the two deterministic cases.

We shall give the proof of Theorem 4.1 for the deterministic case (ii) of a localized pulse, with the modification for random initial data (4.4) and the plane waves (4.7) being routine. Our proof given in Section 5 is in several steps. First, we obtain an equation for $\hat{W}^\varepsilon(t, \mathbf{x}, \mathbf{y})$ in the form (14.1) but with initial data $\hat{W}^\varepsilon(0, \mathbf{x}, \mathbf{y})$ depending on ε. In the second step we show that these initial data may be replaced by the limit rescaled Richardson function $\hat{W}_0(\mathbf{x}, \mathbf{y})$. Finally we apply the limit theorem of Appendix A to obtain equation (4.15) for $\hat{W}(t, \mathbf{x}, \mathbf{y}) = \lim_{\varepsilon \to 0} E\{W^\varepsilon(t, \mathbf{x}, \mathbf{y})\}$.

We believe that this theorem is true for a much wider class of initial data than (4.4), (4.5) and (4.7). The technical difficulty in generalizing this result arises in the second step of the proof, that is, replacing the initial data $\hat{W}_0^\varepsilon$ by the limit $\hat{W}_0$. The limit theorem of Appendix A does not provide any information on the uniformity of the convergence to the limit process with respect to the starting point in all of R^n. In particular, we do not have uniform bounds for the probability to visit a fixed set, which are needed in the general case when we cannot control the convergence of $\hat{W}_0^\varepsilon(\mathbf{x}, \mathbf{y})$ to $\hat{W}_0(\mathbf{x}, \mathbf{y})$. This does not allow us to include the physically important case of general wave data (4.6) in Theorem 4.1.

5. THE EXPECTATION OF THE RICHARDSON FUNCTION

Let the initial data for (4.3) be of the form (4.5). Then $\hat{W}_0^\varepsilon(\mathbf{x}, \mathbf{y}) = \frac{1}{\varepsilon^d} \hat{Q}_0(\frac{\mathbf{x}}{\varepsilon}, \mathbf{y})$, and equation (4.3) implies that $\hat{W}^\varepsilon$ satisfies the initial value problem

$$(5.1) \qquad \frac{\partial \hat{W}^\varepsilon}{\partial t} + \mathbf{u} \cdot \nabla_{\mathbf{x}} \hat{W}^\varepsilon + \sqrt{\varepsilon} \mathbf{v}\left(\frac{\mathbf{x}}{\varepsilon}\right) \cdot \nabla_{\mathbf{x}} \hat{W}^\varepsilon$$

$$+ \frac{1}{\sqrt{\varepsilon}} \left\{ \mathbf{v}\left(\frac{\mathbf{x}}{\varepsilon}\right) - \mathbf{v}\left(\frac{\mathbf{x}}{\varepsilon} - \frac{\mathbf{y}}{\varepsilon}\right) \right\} \cdot \nabla_{\mathbf{y}} \hat{W}^\varepsilon = 0$$

with

$$(5.2) \qquad \hat{W}^\varepsilon(0, \mathbf{x}, \mathbf{y}) = \frac{1}{\varepsilon^d} \hat{Q}_0(\frac{\mathbf{x}}{\varepsilon}, \mathbf{y}),$$

where $\hat{Q}_0$ is given by (4.11). The solution of (5.1) is given explicitly by

$$\hat{W}^\varepsilon(t, \mathbf{x}, \mathbf{y}) = \hat{W}_0^\varepsilon(\mathbf{x} - \mathbf{u}t + \varepsilon \mathbf{Z}^\varepsilon(-t, \mathbf{x}), \mathbf{Y}^\varepsilon(-t, \mathbf{y}, \mathbf{x})).$$

Let

$$\hat{W}'^{\varepsilon}(t, \mathbf{x}, \mathbf{y}) = \hat{W}_0(\mathbf{x} - \mathbf{u}t + \varepsilon \mathbf{Z}^{\varepsilon}(-t, \mathbf{x}), \mathbf{Y}^{\varepsilon}(-t, \mathbf{y}, \mathbf{x}))$$

be the solution of (5.2) with the initial data $\hat{W}_0^{\varepsilon}(\mathbf{x}, \mathbf{y})$ replaced by $\hat{W}_0(\mathbf{x}, \mathbf{y})$, which is given by (4.10). We claim that

$$(5.3) \qquad E\{\hat{W}^{\varepsilon}(t, \mathbf{x}, \mathbf{y}) - \hat{W}'^{\varepsilon}(t, \mathbf{x}, \mathbf{y})\} \to 0$$

weakly in $\mathcal{A}'$ as $\varepsilon \to 0$. Let $f(\mathbf{x}, \mathbf{y}) \in \mathcal{A}$ be a continuous non-random test function of compact support. Then, since the flow is incompressible

$$E\left\{\iint d\mathbf{x} d\mathbf{y} (\hat{W}^{\varepsilon}(t, \mathbf{x}, \mathbf{y}) - \hat{W}'^{\varepsilon}(t, \mathbf{x}, \mathbf{y})) f(\mathbf{x}, \mathbf{y})\right\}$$

$$= \iint d\mathbf{x} d\mathbf{y} (\hat{W}_0^{\varepsilon}(\mathbf{x}, \mathbf{y}) - \hat{W}_0(\mathbf{x}, \mathbf{y})) E\{f(\mathbf{x} + \mathbf{u}t + \varepsilon \mathbf{Z}^{\varepsilon}(t, \mathbf{x}), \mathbf{Y}^{\varepsilon}(t, \mathbf{y}, \mathbf{x}))\}$$

$$= \iint d\mathbf{x} d\mathbf{y} (\frac{1}{\varepsilon^d} \hat{Q}_0(\frac{\mathbf{x}}{\varepsilon}, \mathbf{y})$$

$$- \delta(\mathbf{x}) \int d\mathbf{p} \hat{Q}_0(\mathbf{p}, \mathbf{y})) E\{f(\mathbf{x} + \mathbf{u}t + \varepsilon \mathbf{Z}^{\varepsilon}(t, \mathbf{x}), \mathbf{Y}^{\varepsilon}(t, \mathbf{y}, \mathbf{x}))\}$$

$$= \iint d\mathbf{x} d\mathbf{y} \hat{Q}_0(\mathbf{x}, \mathbf{y}) E\{f(\varepsilon \mathbf{x} + \mathbf{u}t + \varepsilon \mathbf{Z}^{\varepsilon}(t, \varepsilon \mathbf{x}), \mathbf{Y}^{\varepsilon}(t, \mathbf{y}, \varepsilon \mathbf{x}))$$

$$- f(\mathbf{u}t + \varepsilon \mathbf{Z}^{\varepsilon}(t, 0), \mathbf{Y}^{\varepsilon}(t, \mathbf{y}, 0))\}.$$

Theorem 3.1 implies that

$$(5.4) \qquad E\{f(\mathbf{u}t + \varepsilon \mathbf{Z}^{\varepsilon}(t, 0), \mathbf{Y}^{\varepsilon}(t, \mathbf{y}, 0))\} \to \bar{f}(0, \mathbf{y}),$$

strongly in the uniform norm on compact sets in $\mathbf{y}$. Here the function $\bar{f}(\mathbf{x}, \mathbf{y})$ satisfies the diffusion equation with a drift

$$\frac{\partial \bar{f}}{\partial t} = \mathbf{u} \cdot \nabla_{\mathbf{x}} \bar{f} + L_Y \bar{f}$$

$$\bar{f}(0, \mathbf{x}, \mathbf{y}) = f(\mathbf{x}, \mathbf{y}),$$

with the operator L_Y given by (3.7). The limit theorem of Appendix A applies to the system

$$\frac{d\tilde{\mathbf{Z}}^{\varepsilon}}{dt} = \frac{1}{\sqrt{\varepsilon}} \mathbf{v}\left(\frac{\mathbf{u}t}{\varepsilon} + \tilde{\mathbf{Z}}^{\varepsilon}\right), \quad \tilde{\mathbf{Z}}^{\varepsilon}(0) = \mathbf{x}$$

$$\frac{d\tilde{\mathbf{Y}}^{\varepsilon}}{dt} = \frac{1}{\sqrt{\varepsilon}} \left[\mathbf{v}\left(\frac{\mathbf{u}t}{\varepsilon} + \tilde{\mathbf{Z}}^{\varepsilon}\right) - \mathbf{v}\left(\frac{\mathbf{u}t}{\varepsilon} + \tilde{\mathbf{Z}}^{\varepsilon} - \tilde{\mathbf{Y}}^{\varepsilon}\right)\right],$$

where $\tilde{\mathbf{Z}}^{\varepsilon}(t, \mathbf{x}) = \mathbf{x} + \mathbf{Z}^{\varepsilon}(t, \varepsilon \mathbf{x})$, and $\tilde{\mathbf{Y}}^{\varepsilon}(t, \mathbf{y}, \mathbf{x}) = \mathbf{Y}^{\varepsilon}(t, \mathbf{y}, \varepsilon \mathbf{x})$. It implies that the expectation

$$E\{f(\varepsilon \mathbf{x} + \mathbf{u}t + \varepsilon \mathbf{Z}^{\varepsilon}(t, \varepsilon \mathbf{x}), \mathbf{Y}^{\varepsilon}(t, \mathbf{y}, \varepsilon \mathbf{x}))\}$$

has the same limit as in (5.4). Then by the dominated convergence theorem (5.4) vanishes in the limit $\varepsilon \to 0$ since $\hat{Q}_0(\mathbf{x}, \mathbf{y}) \in L^1(R^d \times R^d)$. Thus (5.3) holds, and we may replace the initial data $\hat{W}_0^{\varepsilon}$ by $\hat{W}_0$ in (5.1), and

$$\hat{W}(t, \mathbf{x}, \mathbf{y}) = \lim_{\varepsilon \to 0} E\{\hat{W}^{\varepsilon}(t, \mathbf{x}, \mathbf{y})\} = \lim_{\varepsilon \to 0} E\{\hat{W}'^{\varepsilon}(t, \mathbf{x}, \mathbf{y})\}.$$

The limit theorem applies to the weak form of the resulting system, that is, for any test function $f(\mathbf{x}, \mathbf{y})$ we have

$$\iint E\left\{\hat{W}'^{\varepsilon}(t, \mathbf{x}, \mathbf{y})\right\} f(\mathbf{x}, \mathbf{y}) dxdy$$

$$= \iint \hat{W}_0(\mathbf{x}, \mathbf{y}) E\left\{f(\mathbf{x} + \mathbf{u}t + \varepsilon \mathbf{Z}^{\varepsilon}(t, \mathbf{x}), \mathbf{Y}^{\varepsilon}(t, \mathbf{x}, \mathbf{y}))\right\} dxdy$$

$$= \iint dxdy \hat{Q}_0(\mathbf{x}, \mathbf{y}) E\left\{f(\mathbf{u}t + \varepsilon \mathbf{Z}^{\varepsilon}(t, 0), \mathbf{Y}^{\varepsilon}(t, 0, \mathbf{y}))\right\} dxdy$$

$$\rightarrow \iint dxdy \hat{Q}_0(\mathbf{x}, \mathbf{y}) \bar{f}(\mathbf{u}t, \mathbf{y}) dxdy$$

by the dominated convergence theorem. Thus the function $\hat{W}(t, \mathbf{x}, \mathbf{y})$ is the weak solution of

$$(5.5) \qquad \frac{\partial \hat{W}}{\partial t} + \mathbf{u} \cdot \nabla_{\mathbf{x}} \hat{W} = L_Y \hat{W}$$

$$\hat{W}(0, \mathbf{x}, \mathbf{y}) = \hat{W}_0(\mathbf{x}, \mathbf{y}).$$

Here the operator L_Y is given by (3.7). This finishes the proof of Theorem 4.1.

6. Higher moment equations

We show how the joint behavior of trajectories starting at several points may be studied using the limit theorem of Appendix A. Consider first two pairs of trajectories, $\mathbf{X}^{\varepsilon}(t, \mathbf{x}_1)$, $\mathbf{X}^{\varepsilon}(t, \mathbf{x}_1 - \varepsilon \mathbf{y}_1)$, and $\mathbf{X}^{\varepsilon}(t, \mathbf{x}_2)$, $\mathbf{X}^{\varepsilon}(t, \mathbf{x}_2 - \varepsilon \mathbf{y}_2)$. Then the corresponding processes $\mathbf{Z}_j^{\varepsilon}$ and $\mathbf{Y}_j^{\varepsilon}$ satisfy the following system:

$$(6.1) \quad \frac{d\mathbf{Z}_1^{\varepsilon}}{dt} = \frac{1}{\sqrt{\varepsilon}} \mathbf{v}\left(\frac{\mathbf{x}_1 + \mathbf{u}t}{\varepsilon} + \mathbf{Z}_1^{\varepsilon}\right), \quad \mathbf{Z}_1^{\varepsilon}(0) = 0$$

$$\frac{d\mathbf{Y}_1^{\varepsilon}}{dt} = \frac{1}{\sqrt{\varepsilon}} \left\{\mathbf{v}\left(\frac{\mathbf{x}_1 + \mathbf{u}t}{\varepsilon} + \mathbf{Z}_1^{\varepsilon}\right) - \mathbf{v}\left(\frac{\mathbf{x}_1 + \mathbf{u}t}{\varepsilon} + \mathbf{Z}_1^{\varepsilon} - \mathbf{Y}_1^{\varepsilon}\right)\right\}, \quad \mathbf{Y}_1^{\varepsilon}(0) = \mathbf{y}_1$$

$$\frac{d\mathbf{Z}_2^{\varepsilon}}{dt} = \frac{1}{\sqrt{\varepsilon}} \mathbf{v}\left(\frac{\mathbf{x}_2 + \mathbf{u}t}{\varepsilon} + \mathbf{Z}_2^{\varepsilon}\right), \quad \mathbf{Z}_2^{\varepsilon}(0) = 0$$

$$\frac{d\mathbf{Y}_2^{\varepsilon}}{dt} = \frac{1}{\sqrt{\varepsilon}} \left\{\mathbf{v}\left(\frac{\mathbf{x}_2 + \mathbf{u}t}{\varepsilon} + \mathbf{Z}_2^{\varepsilon}\right) - \mathbf{v}\left(\frac{\mathbf{x}_2 + \mathbf{u}t}{\varepsilon} + \mathbf{Z}_2^{\varepsilon} - \mathbf{Y}_2^{\varepsilon}\right)\right\}, \quad \mathbf{Y}_2^{\varepsilon}(0) = \mathbf{y}_2.$$

The joint process $(\mathbf{Z}_1^{\varepsilon}, \mathbf{Y}_1^{\varepsilon}, \mathbf{Z}_2^{\varepsilon}, \mathbf{Y}_2^{\varepsilon})$ converge to the process $(\mathbf{Z}_1, \mathbf{Y}_1, \mathbf{Z}_2, \mathbf{Y}_2)$ with generator of the form

$$L^{(2)} = L_1 + L_2 + L_{12},$$

where the operators L_1 and L_2 are given by (3.4) in the variables $(\mathbf{z}_1, \mathbf{y}_1)$ and $(\mathbf{z}_2, \mathbf{y}_2)$, respectively. The cross term L_{12} involves terms of the type

$$c_{ij}(\mathbf{x}_1, \mathbf{x}_2, \mathbf{y}_1, \mathbf{y}_2) \frac{\partial^2}{\partial z_1^i \partial y_2^j}, \quad d_{ij}(\mathbf{x}_1, \mathbf{x}_2, \mathbf{y}_1, \mathbf{y}_2) \frac{\partial^2}{\partial y_1^i \partial y_2^j},$$

and other similar ones. The coefficients c_{ij}, d_{ij}, ... are non-zero only for points $\mathbf{x}_1$, $\mathbf{x}_2$ which lie on the same deterministic trajectory, that is, $\mathbf{x}_1 - \mathbf{x}_2 = \lambda \mathbf{u}$ for some $\lambda \in R$. Thus when the points $\mathbf{x}_1$ and $\mathbf{x}_2$ do not lie on the same deterministic trajectory, the pairs of the limit processes $\mathbf{Z}_1$, $\mathbf{Y}_1$ and $\mathbf{Z}_2$, $\mathbf{Y}_2$ are independent and their joint generator is the sum of the corresponding generators:

$$L^{(2)} = L_1 + L_2.$$

This phenomenon occurs also for all higher moments.

Let us consider now the case when the two trajectories start at the same point $\mathbf{x}_1 = \mathbf{x}_2 = \mathbf{x}$, so that $\mathbf{Z}_1^\varepsilon = \mathbf{Z}_2^\varepsilon$. Then the system (6.1) reduces to

$$(6.2) \qquad \frac{d\mathbf{Z}^\varepsilon}{dt} = \frac{1}{\sqrt{\varepsilon}}\mathbf{v}\left(\frac{\mathbf{x} + \mathbf{u}t}{\varepsilon} + \mathbf{Z}^\varepsilon\right), \quad \mathbf{Z}^\varepsilon(0) = 0$$

$$\frac{d\mathbf{Y}_1^\varepsilon}{dt} = \frac{1}{\sqrt{\varepsilon}}\left\{\mathbf{v}\left(\frac{\mathbf{x} + \mathbf{u}t}{\varepsilon} + \mathbf{Z}^\varepsilon\right) - \mathbf{v}\left(\frac{\mathbf{x} + \mathbf{u}t}{\varepsilon} + \mathbf{Z}^\varepsilon - \mathbf{Y}_1^\varepsilon\right)\right\}, \quad \mathbf{Y}_1^\varepsilon(0) = \mathbf{y}_1$$

$$\frac{d\mathbf{Y}_2^\varepsilon}{dt} = \frac{1}{\sqrt{\varepsilon}}\left\{\mathbf{v}\left(\frac{\mathbf{x} + \mathbf{u}t}{\varepsilon} + \mathbf{Z}^\varepsilon\right) - \mathbf{v}\left(\frac{\mathbf{x} + \mathbf{u}t}{\varepsilon} + \mathbf{Z}^\varepsilon - \mathbf{Y}_2^\varepsilon\right)\right\}, \quad \mathbf{Y}_2^\varepsilon(0) = \mathbf{y}_2.$$

The joint generator for the limit processes $\mathbf{Y}_1$, $\mathbf{Y}_2$ has the form:

$$L^{(2)}f(\mathbf{y}_1, \mathbf{y}_2) = \frac{1}{2}a_{ij}^{\alpha\beta}(\mathbf{y}_1, \mathbf{y}_2)\frac{\partial^2 f}{\partial y_i^\alpha \partial y_j^\beta},$$

where the Greek indices label the points and the Latin indices label the coordinates. The coefficients $a_{ij}^{\alpha\beta}(\mathbf{y}_1, \mathbf{y}_2)$ are given by

$$a_{ij}^{\alpha\beta}(\mathbf{y}_1, \mathbf{y}_2) = \int_{-\infty}^\infty [R_{ij}(\mathbf{u}t) - R_{ij}(\mathbf{u}t - \mathbf{y}_\alpha) - R_{ji}(\mathbf{u}t - \mathbf{y}_\beta) + R_{ij}(\mathbf{u}t - \mathbf{y}_\beta + \mathbf{y}_\alpha)]\, dt.$$

Let $\hat{G}(t, \mathbf{x}, \mathbf{y}_1, \mathbf{y}_2)$ be the second moment of the Richardson function for the solution of (4.3)

$$\hat{G}(t, \mathbf{x}, \mathbf{y}_1, \mathbf{y}_2) \quad = \lim_{\varepsilon \to 0} E\left\{\hat{W}^\varepsilon(t, \mathbf{x}, \mathbf{y}_1)\hat{W}^\varepsilon(t, \mathbf{x}, \mathbf{y}_2)\right\}$$

$$= \lim_{\varepsilon \to 0} E\left\{\phi^\varepsilon(t, \mathbf{x} - \varepsilon\mathbf{y}_1)\phi^\varepsilon(t, \mathbf{x})\phi^\varepsilon(t, \mathbf{x} - \varepsilon\mathbf{y}_2)\phi^\varepsilon(t, \mathbf{x})\right\}.$$

Assume that the initial data is of the form (4.4) with $\phi_0(\mathbf{x})$ being Gaussian. Then $\hat{G}$ satisfies the initial value problem:

$$(6.3) \qquad \frac{\partial \hat{G}}{\partial t} = L^{(2)}\hat{G},$$

$$\hat{G}(0, \mathbf{y}_1, \mathbf{y}_2) = 2R_0(\mathbf{y}_1)R_0(\mathbf{y}_2) + R_0(\mathbf{y}_1 - \mathbf{y}_2)R_0(0).$$

All the higher moment equations for the Richardson function can be derived in a similar way.

7. The Wigner distribution and connections with radiative transport theory

The Wigner distribution of an oscillatory family $\phi^\varepsilon(\mathbf{x})$ is defined as the inverse Fourier transform of the rescaled Richardson function

$$(7.1) \qquad W^\varepsilon(t, \mathbf{x}, \mathbf{k}) = \int \frac{d\mathbf{y}}{(2\pi)^d}e^{i\mathbf{k}\cdot\mathbf{y}}\phi^\varepsilon(t, \mathbf{x} - \varepsilon\mathbf{y})\phi^{\varepsilon*}(t, \mathbf{x}).$$

The Wigner transform has a weak limit $W(t, \mathbf{x}, \mathbf{k})$ as $\varepsilon \to 0$, which is a non-negative measure [15] (see Appendix B). The limit Wigner distribution may be interpreted as the limit energy (or particle) density of an ε-oscillatory family $\phi^\varepsilon(\mathbf{x})$, travelling in the direction $\mathbf{k}$ at position $\mathbf{x}$. In particular, we have

$$(7.2) \qquad \lim_{\varepsilon \to 0}|\phi^\varepsilon(\mathbf{x})|^2 = \int W(t, \mathbf{x}, \mathbf{k})d\mathbf{k}$$

in the weak sense if and only if $\phi^\varepsilon(\mathbf{x})$ is ε-oscillatory. The limit Wigner distribution also determines the limit values as $\varepsilon \to 0$ of various others quantities of interest, for example, the correlation functions of the derivatives of ϕ_ε. The limit of the correlation matrix of the gradient

$$\tilde{W}^\varepsilon_{ij}(t, \mathbf{x}, \mathbf{y}) = \varepsilon \frac{\partial \phi^\varepsilon}{\partial x_i}(\mathbf{x} - \varepsilon \mathbf{y}) \varepsilon \frac{\partial \phi^{*\varepsilon}}{\partial x_j}(\mathbf{x})$$

is the Hessian in $\mathbf{y}$ of $\hat{W}(t, \mathbf{x}, \mathbf{k})$:

$$(7.3) \qquad \tilde{W}^\varepsilon_{ij}(t, \mathbf{x}, \mathbf{y}) \to \frac{\partial^2 \hat{W}}{\partial y_i \partial y_j}.$$

The limits of the higher order derivatives can be similarly expressed via $W(t, \mathbf{x}, \mathbf{k})$. Thus, in the high frequency limit the Wigner distribution gives a complete description both of the field $\phi^\varepsilon(t, \mathbf{x})$ itself and its derivatives. Relations of the type (7.2) and (7.3) are the main reason for the recent studies of the Wigner distribution for waves in random media [**31**].

Let us briefly recall how radiative transport equations for waves arise [**31**]. Let $\mathbf{w}^\varepsilon(t, \mathbf{x}) \in C^N$ be the solution of a symmetric hyperbolic system (this scaling corresponds to $\beta = 1$ in (4.3))

$$(7.4) \qquad \left\{ A(\mathbf{x}) + \sqrt{\varepsilon} V\left(\frac{\mathbf{x}}{\varepsilon}\right) \right\} \frac{\partial \mathbf{w}^\varepsilon}{\partial t} + D^j \frac{\partial \mathbf{w}^\varepsilon}{\partial x^j} = 0$$
$$\mathbf{w}^\varepsilon(0, \mathbf{x}) = \mathbf{w}^\varepsilon_0(\mathbf{x}).$$

Here the matrix $A(\mathbf{x})$ is positive definite and the matrices D^j are symmetric. The initial data $\mathbf{w}^\varepsilon_0(\mathbf{x})$ is assumed to be ε-oscillatory and deterministic, and $V(\mathbf{y})$ is a matrix valued, space homogeneous random process. Then the $N \times N$ limit Wigner matrix $W(t, \mathbf{x}, \mathbf{k})$ has a special form

$$(7.5) \qquad W(t, \mathbf{x}, \mathbf{k}) = \sum_{\alpha, i, j} W^\alpha_{ij}(t, \mathbf{x}, \mathbf{k}) \mathbf{b}^{\alpha, i} \mathbf{b}^{*\alpha, j}.$$

Here the vectors $\mathbf{b}^{\alpha, j} \in C^N$ form a basis for the eigenspace of the dispersion matrix

$$L(\mathbf{x}, \mathbf{k}) = A^{-1}(\mathbf{x}) k_j D^j$$

of the system (7.4), corresponding to the eigenvalue ω_α. The size of the square matrices W^α is equal to the degeneracy of the eigenvalue. They satisfy a system of radiative transport equations. This system is decoupled when the eigenvalues are simple and $\omega_\alpha(\mathbf{k}) \neq \omega_\beta(\mathbf{p})$ for all β, $\mathbf{k}$ and $\mathbf{p}$. Then the radiative transport equation for the scalar W^α has the form

$$(7.6) \qquad \frac{\partial W^\alpha}{\partial t} + \nabla_\mathbf{k} \omega_\alpha \cdot \nabla_\mathbf{x} W^\alpha - \nabla_\mathbf{x} \omega_\alpha \cdot \nabla_\mathbf{k} W^\alpha$$
$$= \int d\mathbf{k}' \sigma(\mathbf{x}, \mathbf{k}, \mathbf{k}') \delta(\omega_\alpha(\mathbf{k}) - \omega_\alpha(\mathbf{k}'))(W^\alpha(\mathbf{k}') - W^\alpha(\mathbf{k})).$$

The function $\sigma(\mathbf{x}, \mathbf{k}, \mathbf{k}')$ is the differential scattering cross-section and is determined by the power spectrum of $V(\mathbf{y})$. Equation (7.6) has a long history. It was proposed phenomenologically by Rayleigh in the beginning of this century and then studied extensively by physicists [**10**]. Various derivations of this equation were given in 1960's (see [**31**] for references) when $\mathbf{w}^\varepsilon$ is a solution of the wave equation or Maxwell's equations. A general case was treated in [**31**] but the results were derived only formally. The only case when the radiative transport equation was obtained rigorously, to the best of our knowledge, is for the Schrödinger equation with $V(\mathbf{y})$ Gaussian and only for small t [**11, 16**].

For the Wigner distribution of the density of a passive scalar in a random flow we have the following theorem.

THEOREM 7.1. *Let $\phi^\varepsilon(t, \mathbf{x})$ be the solution of (4.3), with the initial data of the form (4.4), (4.5) or (4.7). Let $W_0(\mathbf{x}, \mathbf{k})$ be the expectation of the limit Wigner distribution of $\phi_0^\varepsilon(\mathbf{x})$. Then $E\{W^\varepsilon(t, \mathbf{x}, \mathbf{k})\}$ converges to $W(t, \mathbf{x}, \mathbf{k})$ weakly in $\mathcal{A}'$, where $W(t, \mathbf{x}, \mathbf{k})$ satisfies the radiative transport equation*

$$(7.7) \quad \frac{\partial W}{\partial t} + \mathbf{u} \cdot \nabla_{\mathbf{x}} W = \int \frac{d\mathbf{k}'}{(2\pi)^{d-1}} k_i k_j \hat{R}_{ij}^s(\mathbf{k}' - \mathbf{k}) \delta((\mathbf{k}' - \mathbf{k}) \cdot \mathbf{u})(W(\mathbf{k}') - W(\mathbf{k}))$$

with the initial condition

$$W(0, \mathbf{x}, \mathbf{k}) = W_0(\mathbf{x}, \mathbf{k}).$$

This theorem follows immediately from Theorem 4.1 by applying the inverse Fourier transform to (5.5). Equation (7.7) has the usual form of a radiative transport equation (7.6). The dispersion law of (4.3) is

$$\omega(\mathbf{k}) = \mathbf{u} \cdot \mathbf{k}.$$

The scattering operator on the right side of (7.7) is symmetric since $(\mathbf{k}, \hat{R}(\mathbf{k} - \mathbf{k}')\mathbf{k}) = (\mathbf{k}', \hat{R}(\mathbf{k}' - \mathbf{k})\mathbf{k}')$ because of the incompressibility of the random field $\mathbf{v}(\mathbf{y})$. The transport equation is valid globally in time and for random velocity fields that are not necessarily Gaussian. This tells us that in more general cases the radiative transport equation should also be valid globally in time and for non-Gaussian random fluctuations. It should also be valid for general inhomogeneous high frequency waves of the form (4.6), the restriction to plane wave initial data is technical as explained in the remarks after Theorem 4.1.

8. DIFFUSION–TRANSPORT DUALITY

The radiative transport equations (7.6) for waves have a nice interpretation in terms of a certain Markovian jump process [5]. Consider the backward characteristics of (7.6)

$$(8.1) \quad \dot{\mathbf{X}} = -\nabla_{\mathbf{k}} \omega(\mathbf{X}, \mathbf{K}),$$
$$\dot{\mathbf{K}} = \nabla_{\mathbf{x}} \omega(\mathbf{X}, \mathbf{K}),$$

starting at $\mathbf{x}(0) = \mathbf{x}$ and $\mathbf{K}(0) = \mathbf{k}$. Let a particle move along a trajectory of (8.1) for a random time τ_1 and then switch wave vector randomly from $\mathbf{K}$ to $\mathbf{K}'$. After that it follows trajectories of (8.1) for a random time τ_2, at which moment it switches it direction again. The process is continued in an obvious manner. The probability distribution of the j-th jump time τ_j is

$$P\{\tau_j > t\} = \exp\left\{-\int_0^t \Sigma(\mathbf{x}(s, \mathbf{x}_{j-1}, \mathbf{k}_{j-1}), \mathbf{k}(s, \mathbf{x}_{j-1}, \mathbf{k}_{j-1}))ds\right\}$$

Here $\Sigma(\mathbf{x}, \mathbf{k})$ is the total scattering cross-section:

$$\Sigma(\mathbf{x}, \mathbf{k}) = \int d\mathbf{k}' \sigma(\mathbf{x}, \mathbf{k}', \mathbf{k}) \delta(\omega_\alpha(\mathbf{k}) - \omega_\alpha(\mathbf{k}')),$$

and $\mathbf{x}(s, \mathbf{x}_{j-1}, \mathbf{k}_{j-1})$, $\mathbf{k}(s, \mathbf{x}_{j-1}, \mathbf{k}_{j-1})$ is the trajectory of (8.1) starting at the position and wave number of the previous jump. The probability density that the wave number jumps from direction $\mathbf{k}$ into direction $\mathbf{k}'$ is given by

$$p(\mathbf{x}, \mathbf{k}, \mathbf{k}') = \frac{\sigma(\mathbf{x}, \mathbf{k}, \mathbf{k}')}{\Sigma(\mathbf{x}, \mathbf{k})}.$$

The resulting process is well defined if the total scattering cross-section is bounded from above, so that only a finite number of jumps occur during any given time interval, with probability one. The Kolmogorov equation for the resulting process is (7.6), that is, given any test function $f(\mathbf{x}, \mathbf{k})$, the function $\bar{f}(t, \mathbf{x}, \mathbf{k}) = E\{f(\mathbf{X}(t, x, \mathbf{k}), \mathbf{K}(t, \mathbf{x}, \mathbf{k}))\}$ is the solution of (7.6) with the initial data $\bar{f}(0, \mathbf{x}, \mathbf{k}) = f(\mathbf{x}, \mathbf{k})$.

The integral operator on the right side of (7.7)

$$(8.2) \qquad \mathcal{Q}f = \int \frac{d\mathbf{k}'}{(2\pi)^{d-1}} k_i k_j \hat{R}_{ij}^s(\mathbf{k}' - \mathbf{k})\delta((\mathbf{k}' - \mathbf{k}) \cdot \mathbf{u})(f(\mathbf{k}') - f(\mathbf{k}))$$

is of the same form as the integral operator in (7.6). It is the generator of a jump process for which the particle is moving along the deterministic trajectory $\mathbf{x}(t) = \mathbf{x}(0) + \mathbf{u}t$, while the wave vector is the jump process described above with the differential scattering cross-section

$$\sigma(\mathbf{k}, \mathbf{k}') = \frac{1}{(2\pi)^{d-1}} k_i k_j \hat{R}_{ij}^s(\mathbf{k}' - \mathbf{k})\delta((\mathbf{k}' - \mathbf{k}) \cdot \mathbf{u})$$

and the total scattering cross-section

$$(8.3) \qquad \Sigma(\mathbf{k}) = \int d\mathbf{k}\, \frac{1}{(2\pi)^{d-1}} k_i k_j \hat{R}_{ij}^s(\mathbf{k}' - \mathbf{k})\delta((\mathbf{k}' - \mathbf{k}) \cdot \mathbf{u}).$$

Thus the projection of the wave vector on the mean velocity direction $\mathbf{u}$ is not changed after the jump. The total scattering cross-section (8.3) is an unbounded function of $\mathbf{k}$, and so the standard argument [5] that $\mathbf{K}(t, \mathbf{k})$ starting at $\mathbf{k}$ does not go to infinity in a finite time does not apply. We note that the process $\mathbf{K}(t, \mathbf{k})$ is dual to the diffusion process $\mathbf{Y}(t, \mathbf{x}, \mathbf{y})$ with generator L_Y given by (3.7). In fact we have that

$$L_Y(\mathbf{y})e^{i\mathbf{k}\cdot\mathbf{y}} = \mathcal{Q}(\mathbf{k})e^{i\mathbf{k}\cdot\mathbf{y}}$$

and hence

$$(8.4) \qquad E_Y\left\{e^{i\mathbf{k}\cdot\mathbf{Y}(t,\mathbf{y})}\right\} = E_K\left\{e^{i\mathbf{K}(t,\mathbf{k})\cdot\mathbf{y}}\right\}.$$

This means that the process $\mathbf{K}(t, \mathbf{k})$ is well defined as long as the diffusion process $\mathbf{Y}(t, \mathbf{y})$ is well defined, and so it exists for all time. The duality (8.4) has an interesting qualitative implication. Given any function $f(t, \mathbf{x}, \mathbf{y})$ we may view

$$\bar{f}(t, \mathbf{y}) = E\{f(\mathbf{Y}(t, \mathbf{y}))\}$$

either as a solution of

$$\frac{\partial \bar{f}}{\partial t} + \mathbf{u} \cdot \nabla_{\mathbf{x}} \bar{f} = L_Y \bar{f}$$
$$f(0, \mathbf{x}, \mathbf{y}) = f(\mathbf{x}, \mathbf{y})$$

or as the Fourier transform of the solution of

$$\frac{\partial \hat{f}}{\partial t} + \mathbf{u} \cdot \nabla_{\mathbf{x}} \hat{f} = \mathcal{Q}\hat{f}$$
$$\hat{f}(0, \mathbf{x}, \mathbf{k}) = \frac{\hat{f}(\mathbf{x}, -\mathbf{k})}{(2\pi)^d}.$$

The limit process $\mathbf{K}(t, \mathbf{k})$ has a unique invariant measure, which is Lebesgue measure on the plane $\mathbf{k} \cdot \mathbf{u} = \text{const}$. This means that in the long time limit solutions of (8.5) are nearly functions of $\mathbf{k} \cdot \mathbf{u}$ only, that is, their support fills out the whole plane, and the function is close to a small constant on this set. This implies that solutions of (8.5) tend to a delta function in the directions orthogonal to the mean flow $\mathbf{u}$, and so the in the long time limit $\mathbf{Y}(t)$ tends to be parallel to the mean velocity $\mathbf{u}$. In terms of the flow, this means that particles starting nearby tend to be aligned with the flow no matter what their relative position is initially.

9. GENERAL SCALING OF THE FLUCTUATIONS

The results described above may be generalized to the case when random fluctuations are oscillating on a scale finer than ε, but their strength is also scaled appropriately. The trajectories $\mathbf{X}^\varepsilon(t, \mathbf{x})$ satisfy the scaled equation

$$(9.1) \qquad \frac{d\mathbf{X}^\varepsilon}{dt} = \mathbf{u} + \varepsilon^\alpha \mathbf{v}\left(\frac{\mathbf{X}^\varepsilon}{\varepsilon^\beta}\right),$$

$$\mathbf{X}^\varepsilon(0, \mathbf{x}) = \mathbf{x}.$$

The trajectory fluctuation, defined as before by

$$(9.2) \qquad \mathbf{Z}^\varepsilon(t, \mathbf{x}) = \frac{\mathbf{X}^\varepsilon(t, \mathbf{x}) - \mathbf{u}t - \mathbf{x}}{\varepsilon}$$

has a non-trivial limit if $\alpha = 1 - \dfrac{\beta}{2}$, and $1 \le \beta < 2$. The strength of the fluctuations $\alpha = 1 - \beta/2$, which increases as the scale of the fluctuations decreases, is chosen so as to make the effect of the fluctuations on scale ε of order one. The case $\beta = 1$ corresponds to (1.1), while $\beta = 2$ corresponds to the homogenization scaling (2.3), which we do not consider here. In that case $\alpha = 0$ and fluctuations of the velocity field are no longer weak, which leads to entirely different results.

The fluctuations $\mathbf{Z}^\varepsilon$, defined by (9.2), and $\mathbf{Y}^\varepsilon$, defined by

$$\mathbf{Y}^\varepsilon(t, \mathbf{y}, \mathbf{x}) = \frac{\mathbf{X}^\varepsilon(t, \mathbf{x}) - \mathbf{X}^\varepsilon(t, \mathbf{x} - \varepsilon\mathbf{y})}{\varepsilon},$$

satisfy a scaled system similar to (3.3)

$$(9.3) \qquad \frac{d\mathbf{Z}^\varepsilon}{dt} = \frac{1}{\sqrt{\varepsilon^\beta}}\mathbf{v}\left(\frac{\mathbf{x} + \mathbf{u}t}{\varepsilon^\beta} + \frac{\mathbf{Z}^\varepsilon}{\varepsilon^{\beta-1}}\right)$$

$$(9.4) \qquad \frac{d\mathbf{Y}^\varepsilon}{dt} = \frac{1}{\sqrt{\varepsilon^\beta}}\left\{\mathbf{v}\left(\frac{\mathbf{x} + \mathbf{u}t}{\varepsilon^\beta} + \frac{\mathbf{Z}^\varepsilon}{\varepsilon^{\beta-1}}\right) - \mathbf{v}\left(\frac{\mathbf{x} + \mathbf{u}t}{\varepsilon^\beta} + \frac{\mathbf{Z}^\varepsilon}{\varepsilon^{\beta-1}} - \frac{\mathbf{Y}^\varepsilon}{\varepsilon^{\beta-1}}\right)\right\}$$

with the initial data $\mathbf{Z}^\varepsilon(0) = 0$ and $\mathbf{Y}^\varepsilon(0) = \mathbf{y}$. The limit theorem of Appendix A applies also to this system [17] as does the limit theorem for the process $\mathbf{Z}^\varepsilon$ of Section 3 in this scaling. The limit theorem for the process $\mathbf{Y}^\varepsilon$ has to be modified because the two terms in (9.4) become decorrelated in the limit $\varepsilon \to 0$, as opposed to (3.3) where they are not. That means that the $\mathbf{y}$-dependent cross term in (3.7) vanishes in the scaling of (9.1). The two particles become asymptotically independent of each other as ε tends to zero. This is because the initial separation $\varepsilon\mathbf{y}$ is large compared to the scale of the randomness and the two particles sample nearly uncorrelated parts of the random medium. The results of Theorems 4.1 and 7.1 for the limit Richardson and Wigner functions are modified because of this. Thus, the regime of validity of the radiative transport theory is restricted to the case of inhomogeneities that are comparable to the wavelength [31]. Inclusions that are smaller than the wavelength but with higher contrast than in the scaling (4.3) will cause dissipation of energy on smaller scales. The contrast still has to be small, which corresponds to $\alpha > 0$ in (9.1).

Let $\phi^\varepsilon(t, \mathbf{x})$ be the solution of the rescaled version of (4.3):

$$(9.5) \qquad \frac{\partial \phi^\varepsilon}{\partial t} + \left(\mathbf{u} + \varepsilon^{1-\beta/2}\mathbf{v}(\frac{\mathbf{x}}{\varepsilon^\beta})\right) \cdot \nabla_\mathbf{x} \phi^\varepsilon = 0.$$

The limit equation for $W(t, \mathbf{x}, \mathbf{k})$ in this scaling has the form

$$(9.6) \quad \frac{\partial W}{\partial t} + \mathbf{u} \cdot \nabla_\mathbf{x} W = -\int \frac{d\mathbf{k}'}{(2\pi)^{d-1}} k_i k_j \hat{R}_{ij}^s(\mathbf{k}' - \mathbf{k})\delta((\mathbf{k}' - \mathbf{k}) \cdot \mathbf{u})W(\mathbf{k}),$$

and

$$\iint W(t, \mathbf{x}, \mathbf{k})d\mathbf{x}d\mathbf{k} < \iint W_0(\mathbf{x}, \mathbf{k})d\mathbf{x}d\mathbf{k} = \lim_{\varepsilon \to 0}\int |\phi_0^\varepsilon(\mathbf{x})|^2 d\mathbf{x}$$

$$= \lim_{\varepsilon \to 0}\int |\phi^\varepsilon(t, \mathbf{x})|^2 d\mathbf{x}.$$

Thus, Proposition 1 of Appendix B implies that if the family $\phi_0^\varepsilon(\mathbf{x})$ is ε-oscillatory, then the solution $\phi^\varepsilon(t, \mathbf{x})$ does not remain ε-oscillatory for all times $t > 0$, and randomness generates oscillations on its own scale when $\beta > 1$.

10. Evolution of the Jacobian matrix

The behavior of the Jacobian matrix of the map induced by a mean zero time dependent random flow is the subject of recent study, both numerical [7] and theoretical [6, 8, 20, 21, 32]. One of the main quantities of interest in such a flow is the Lyapunov exponent responsible for the exponential growth in time of the norm of the Jacobian matrix. Its positivity was established in [8] for velocity fields that are a finite-dimensional Ornstein-Uhlenbeck process. The Jacobian and the evolution of curves in isotropic stochastic flows with zero drift is studied in [4]. The trajectories of (1.1) behave very differently from mean zero time dependent flows. In particular, as we shall see in Section 8, the length of the curves advected by this flow does not grow exponentially in time.

Let $J^\varepsilon(t, \mathbf{x})$ be the Jacobian matrix of the transformation $\mathbf{x} \to \mathbf{X}^\varepsilon(t, \mathbf{x})$:

$$(10.1) \qquad J_{ik}^\varepsilon(t, \mathbf{x}) = \frac{\partial X_i^\varepsilon(t, \mathbf{x})}{\partial x_k}$$

This map is volume preserving because the flow is incompressible, and thus, $\det J^\varepsilon = 1$. We are interested in the limit of $J^\varepsilon(t, \mathbf{x})$ as $\varepsilon \to 0$. The main result of this section is the following theorem.

THEOREM 10.1. *The Jacobian $J^\varepsilon(t, \mathbf{x})$ of the map $\mathbf{x} \to \mathbf{X}^\varepsilon(t, \mathbf{x})$ converges weakly to the diffusion process with generator*

$$(10.2) \qquad L_J f = -\frac{1}{2}\sum_{i,k,m,n=1}^{d}\int_0^\infty \frac{\partial^2 R_{im}(\mathbf{u}t)}{\partial y_j \partial y_l}dt\, J_{jk}J_{ln}\frac{\partial^2 f}{\partial J_{ik}\partial J_{mn}},$$

starting at $J = I$.

We derive (10.2) at the end of this section. First we note that the generator (10.2) for the limit diffusion process $J(t)$ may be interpreted naturally in terms of the generator (3.7) for the limit process $\mathbf{Y}(t)$. We recall that

$$\mathbf{Y}^\varepsilon(t, \mathbf{y}, \mathbf{x}) = \frac{\mathbf{X}^\varepsilon(t, \mathbf{x}) - \mathbf{X}^\varepsilon(t, \mathbf{x} - \varepsilon \mathbf{y})}{\varepsilon} \approx J^\varepsilon(t, \mathbf{x}) \mathbf{y}$$

for small $\mathbf{y}$. Thus the processes $\mathbf{Y}$ and $\mathbf{S} = J\mathbf{y}$ should behave in a similar way for small $\mathbf{y}$. In particular we should have that

(10.3) $$E\{f(\mathbf{Y}(t, \mathbf{y}))\} \approx E\{f(\mathbf{S}(t, \mathbf{y}))\}$$

for small $\mathbf{y}$. To show that (10.3) holds we expand the generator (3.7) of $\mathbf{Y}$ for small $\mathbf{y}$ and obtain

$$
\begin{aligned}
L_Y f &= \frac{1}{2} \int_{-\infty}^{\infty} \left(R_{im}^s(\mathbf{u}t) - R_{im}^s(\mathbf{u}t + \mathbf{y}) \right) dt \frac{\partial^2 f}{\partial y_i \partial y_m} \\
&= \frac{1}{2} \int_{-\infty}^{\infty} \left(R_{im}(\mathbf{u}t) - \frac{1}{2} R_{im}(\mathbf{u}t + \mathbf{y}) - \frac{1}{2} R_{im}(\mathbf{u}t - \mathbf{y}) \right) dt \frac{\partial^2 f}{\partial y_i \partial y_m} \\
(10.4) \qquad &\approx -\frac{1}{2} \int_0^\infty \frac{\partial^2 R_{im}(\mathbf{u}t)}{\partial y_j \partial y_l} y_j y_l \, dt \frac{\partial^2 f}{\partial y_i \partial y_m} = c_{imjl} y_j y_l \frac{\partial^2 f}{\partial y_i \partial y_m},
\end{aligned}
$$

where

(10.5) $$c_{imjl} = -\frac{1}{2} \int_0^\infty \frac{\partial^2 R_{im}(\mathbf{u}t)}{\partial y_j \partial y_l} \, dt.$$

Let $J(t, A, \omega)$ be the diffusion with generator L_J starting at $J(0) = A$. We claim that the generator for the process $\mathbf{S}(t, \mathbf{y}, \omega) = J(t, A, \omega)\mathbf{y}$, starting at $\mathbf{s} = A\mathbf{y}$ is given by (10.4). Theorem 10.1 implies that given any function $f(\mathbf{s})$, the expectation

$$\bar{f}(t, \mathbf{y}, A) = E\{f(\mathbf{S}(t))\} = E\{f(J(t)\mathbf{y})\}$$

satisfies the diffusion equation

$$
\begin{aligned}
\frac{\partial \bar{f}}{\partial t} &= c_{imjl} A_{jk} A_{ln} \frac{\partial^2 \bar{f}}{\partial A_{ik} \partial A_{mn}}, \\
\bar{f}(0, y, A) &= f(Ay).
\end{aligned}
$$

Then a simple calculation shows that

$$\bar{f}(t, A, \mathbf{y}) = \bar{g}(t, A\mathbf{y}),$$

where the function $\bar{g}(t, \mathbf{s})$ satisfies

$$
\begin{aligned}
\frac{\partial \bar{f}}{\partial t} &= c_{imjl} s_j s_l \frac{\partial^2 \bar{g}}{\partial s_i \partial s_m} \\
\bar{g}(0, \mathbf{s}) &= f(\mathbf{s}).
\end{aligned}
$$

Thus, the generator for $\mathbf{S}$ coincides with the small $\mathbf{y}$ asymptotics (10.4) of the generator L_Y. We recall that the large $\mathbf{y}$ asymptotics of L_Y gives rise to the generator L_Z for the limit process $\mathbf{Z}(t)$. Thus the limit difference function $\mathbf{Y}(t)$ contains information about both limit processes $\mathbf{Z}(t)$ and $J(t)$. In physical terms the behavior of the Jacobian matrix can be recovered from the large $\mathbf{k}$ behavior of the Wigner distribution which corresponds to the small $\mathbf{y}$ behavior of the Richardson function. The behavior of the fluctuations $\mathbf{Z}(t)$ may be recovered from small $\mathbf{k}$ limit of the Wigner distribution or, equivalently, large $\mathbf{y}$ behavior of the Richardson function.

We note further that the generator (10.2) is formally of the form

(10.6) $$L_J f = <cJ, J> \frac{\partial^2 f}{\partial J \partial J}.$$

Such diffusion processes have typically positive non-zero Lyapunov exponents, that is, the limit

$$\lambda = \lim_{T \to \infty} \frac{\ln \|J(t)\|}{T}$$

exists with probability one, and $\lambda > 0$. The computation of λ for some time dependent flows was done in [4, 8, 32, 20, 21]. We shall show in Section 11 that in our case $\lambda = 0$. The reason for this is the strong degeneracy of (1.1) in the direction of the mean flow $\mathbf{u}$.

We derive now formula (10.2) for the generator of the limit process $J(t)$ and prove Theorem 10.1. The evolution equation for J^ε is obtained by differentiating (1.1) with respect to x_k:

(10.7)
$$\frac{dJ^\varepsilon_{ik}}{dt} = \frac{1}{\sqrt{\varepsilon}} \frac{\partial v_i}{\partial x_j} \left(\frac{\mathbf{X}^\varepsilon}{\varepsilon} \right) J^\varepsilon_{jk}$$
$$J(0, \mathbf{x}) = I.$$

We rewrite this equation using the fluctuations $\mathbf{Z}^\varepsilon(t, x)$, defined by (3.2), so as to put it in a form suitable for the limit theorem of Appendix A

(10.8)
$$\frac{d\mathbf{Z}^\varepsilon}{dt} = \frac{1}{\sqrt{\varepsilon}} \mathbf{v} \left(\frac{\mathbf{x} + \mathbf{u}t}{\varepsilon} + \mathbf{Z}^\varepsilon \right), \quad \mathbf{Z}^\varepsilon(0) = 0$$
$$\frac{dJ^\varepsilon}{dt} = \cdot \frac{1}{\sqrt{\varepsilon}} \nabla \mathbf{v} \left(\frac{\mathbf{x} + \mathbf{u}t}{\varepsilon} + \mathbf{Z}^\varepsilon \right) J^\varepsilon, \quad J^\varepsilon(0) = I.$$

¿From this theorem, the joint generator for the limit diffusion processes Z and J has the form

(10.9)
$$L = L_Z + L_J + L_{JZ} + L_{ZJ}.$$

The operator L_Z is given by (3.6) and the operator L_J is

(10.10) $L_J f$
$$= \frac{1}{2} \sum_{i,k,m,n=1}^{d} \int_0^\infty E \left\{ \frac{\partial v_i}{\partial x_j} \left(\frac{\mathbf{x}}{\varepsilon} + \mathbf{z} \right) J_{jk} \right.$$
$$\times \frac{\partial}{\partial J_{ik}} \left(\frac{\partial v_m}{\partial x_l} \left(\frac{\mathbf{x}}{\varepsilon} + \mathbf{u}t + \mathbf{z} \right) J_{ln} \frac{\partial f(\mathbf{z}, J)}{\partial J_{mn}} \right) \right\} dt$$
$$= \frac{1}{2} \sum_{i,k,m,n=1}^{d} \int_0^\infty E \left\{ \frac{\partial v_i}{\partial x_j} \left(\frac{\mathbf{x}}{\varepsilon} + \mathbf{z} \right) \frac{\partial v_m}{\partial x_l} \left(\frac{\mathbf{x}}{\varepsilon} + \mathbf{u}t + \mathbf{z} \right) \right\} dt$$
$$\times J_{jk} \frac{\partial}{\partial J_{ik}} \left(J_{ln} \frac{\partial f(\mathbf{z}, J)}{\partial J_{mn}} \right).$$

To simplify (10.10) we note that if we let

$$F_{ijml}(\mathbf{y}) = E \left\{ \frac{\partial v_i}{\partial x_j} (\mathbf{x}) \frac{\partial v_m}{\partial x_l} (\mathbf{x} + \mathbf{y}) \right\},$$

then

$$\frac{\hat{F}_{ijml}(\mathbf{q})}{(2\pi)^d} \delta(\mathbf{p} + \mathbf{q}) = ip_j iq_l E \left\{ \hat{v}_i(\mathbf{p}) \hat{v}_m(\mathbf{q}) \right\} = \frac{q_j q_l \hat{R}_{im}(\mathbf{q})}{(2\pi)^d} \delta(\mathbf{p} + \mathbf{q}),$$

and thus

$$F_{ijml}(\mathbf{y}) = -\frac{\partial^2 R_{im}(\mathbf{y})}{\partial y_j \partial y_l}.$$

Therefore the generator L_J has the form

$$(10.11) \quad L_J f = -\frac{1}{2} \sum_{i,k,m,n=1}^{d} \int_0^\infty \frac{\partial^2 R_{im}(\mathbf{u}t)}{\partial y_j \partial y_l} dt J_{jk} \frac{\partial}{\partial J_{ik}} \left(J_{ln} \frac{\partial f}{\partial J_{mn}} \right)$$

$$= -\frac{1}{2} \sum_{i,k,m,n=1}^{d} \int_0^\infty \frac{\partial^2 R_{im}(\mathbf{u}t)}{\partial y_j \partial y_l} dt J_{jk} J_{ln} \frac{\partial^2 f}{\partial J_{ik} \partial J_{mn}},$$

the last equality being due to incompressibility. Note that L_J is independent of $\mathbf{z}$. Next we compute L_{ZJ}:

$$L_{ZJ} f = \frac{1}{2} \sum_{k,i,j=1}^{d} \int_0^\infty E \left\{ v_k \left(\frac{\mathbf{x}}{\varepsilon} + \mathbf{z} \right) \frac{\partial}{\partial z_k} \left(\frac{\partial v_i}{\partial x_n} \left(\frac{\mathbf{x}}{\varepsilon} + \mathbf{u}t + \mathbf{z} \right) J_{nj} \frac{\partial f}{\partial J_{nj}} \right) \right\} dt$$

$$(10.12) \quad = \frac{1}{2} \sum_{k,i,j=1}^{d} \int_0^\infty E \left\{ v_k \left(\frac{\mathbf{x}}{\varepsilon} + \mathbf{z} \right) \frac{\partial v_i}{\partial x_n} \left(\frac{\mathbf{x}}{\varepsilon} + \mathbf{u}t + \mathbf{z} \right) \right\} dt J_{nj} \frac{\partial^2 f}{\partial z_k \partial J_{ij}}.$$

The coefficients of L_{ZJ} are also independent of $\mathbf{z}$. The first order derivative term vanishes after taking the expectation because of the incompressibility condition. The operator L_{JZ} is

$$L_{JZ} f = \frac{1}{2} \sum_{i,j,k=1}^{d} \int_0^\infty E \left\{ \frac{\partial v_i}{\partial x_n} \left(\frac{\mathbf{x}}{\varepsilon} + \mathbf{z} \right) J_{nj} \frac{\partial}{\partial J_{ij}} \left(v_k \left(\frac{\mathbf{x}}{\varepsilon} + \mathbf{u}t + \mathbf{z} \right) \frac{\partial f}{\partial z_k} \right) \right\} dt$$

$$(10.13) \quad = \frac{1}{2} \sum_{i,j,k=1}^{d} \int_0^\infty E \left\{ \frac{\partial v_i}{\partial x_n} \left(\frac{\mathbf{x}}{\varepsilon} + \mathbf{z} \right) v_k \left(\frac{\mathbf{x}}{\varepsilon} + \mathbf{u}t + \mathbf{z} \right) \right\} dt J_{nj} \frac{\partial^2 f}{\partial z_k \partial J_{ij}},$$

it does not have first order derivative terms as well, and has coefficients independent of $\mathbf{z}$. This shows that J^ε converges by itself to a diffusion process J with generator (10.11), and thus Theorem 10.1 holds.

11. APPLICATION TO TWO-DIMENSIONAL FLOWS

We apply the results of Section 10 to two dimensional flows. Any two dimensional incompressible flow in R^2 has the form

$$(11.1) \qquad \mathbf{v}(x) = \left(\frac{\partial \psi}{\partial x_2}, -\frac{\partial \psi}{\partial x_1} \right),$$

where $\psi(\mathbf{x})$ is the stream function. We assume that the random field $\psi(\mathbf{x})$ is space homogeneous, zero mean and isotropic with covariance function

$$R_\psi(x) = E \left\{ \psi(\mathbf{y}) \psi(\mathbf{x} + \mathbf{y}) \right\} = F \left(\frac{r^2}{2} \right),$$

where $r^2 = x_1^2 + x_2^2$. The Jacobian matrix of two-dimensional time dependent flows with zero mean $\mathbf{u}$ was studied in detail in [32], in the diffusion approximation. The results there are entirely different but may be formally recovered from our calculations as we explain below. The covariance matrix R_{ij} for the flow (11.1) is

$$(11.2) \qquad R = \begin{pmatrix} -1 & 0 \\ 0 & -1 \end{pmatrix} F' + \begin{pmatrix} -x_2^2 & x_1 x_2 \\ x_1 x_2 & -x_1^2 \end{pmatrix} F''.$$

The tensor c_{imjl} (10.5) in the generator (10.11) has now the form

$$
\begin{aligned}
c_{1111} &= -\frac{1}{|\mathbf{u}|}\left(I_2 + I_3 + \hat{u}_1^2\hat{u}_2^2 I_4\right)\\[4pt]
c_{1112} &= -\frac{1}{|\mathbf{u}|}\left(3\hat{u}_1\hat{u}_2 I_3 + \hat{u}_1\hat{u}_2^3 I_4\right)\\[4pt]
c_{1122} &= -\frac{1}{|\mathbf{u}|}\left(3I_2 + 6\hat{u}_2^2 I_3 + \hat{u}_2^4 I_4\right)\\[4pt]
c_{1211} &= \frac{1}{|\mathbf{u}|}\left(3\hat{u}_1\hat{u}_2 I_3 + \hat{u}_1^3\hat{u}_2 I_4\right)\\[4pt]
(11.3) \qquad c_{1212} &= \frac{1}{|\mathbf{u}|}\left(I_2 + I_3 + \hat{u}_1^2\hat{u}_2^2 I_4\right)\\[4pt]
c_{1222} &= \frac{1}{|\mathbf{u}|}\left(3\hat{u}_1\hat{u}_2 I_3 + \hat{u}_1\hat{u}_2^3 I_4\right)\\[4pt]
c_{2211} &= -\frac{1}{|\mathbf{u}|}\left(3I_2 + 6\hat{u}_1^2 I_3 + \hat{u}_1^4 I_4\right)\\[4pt]
c_{2212} &= -\frac{1}{|\mathbf{u}|}\left(3\hat{u}_1\hat{u}_2 I_3 + \hat{u}_1^3\hat{u}_2 I_4\right)\\[4pt]
c_{2222} &= -\frac{1}{|\mathbf{u}|}\left(I_2 + I_3 + \hat{u}_1^2\hat{u}_2^2 I_4\right).
\end{aligned}
$$

Here $\mathbf{u} = (u_1, u_2)$ is the mean flow, $|\mathbf{u}| = \sqrt{u_1^2 + u_2^2}$, $\hat{u}_j = \dfrac{u_j}{|\mathbf{u}|}$, and the constants

$$
I_2 = \int_0^\infty F''\left(\frac{t^2}{2}\right)dt, \; I_3 = \int_0^\infty t^2 F'''\left(\frac{t^2}{2}\right)dt, \; I_4 = \int_0^\infty t^4 F^{(iv)}\left(\frac{t^2}{2}\right)dt
$$

are related by

$$(11.4) \qquad\qquad I_2 = -I_3, \; I_4 = 3I_2.$$

The other entries are determined by the symmetries $c_{imjl} = c_{mijl} = c_{imlj}$. The generator (11.5) with c_{imjl} as above is very different form the generator the limit of the Jacobian J^ε for mean zero, time dependent flows [20, 21, 32]. We may not set even formally $\mathbf{u} = 0$ in (11.3) because the resulting expression diverges as it always happens with the Kubo formula. However, the results of [20, 21, 32] can be recovered in our calculation by setting formally $I_3 = I_4 = 0$ in (11.3). Then the generator (10.2) given by Theorem 10.1 reduces formally to the one obtained in [20, 21, 32].

Let us choose a coordinate system so that $u_2 = 0$, and $u_1 = |\mathbf{u}|$. Then the only non-zero entry is $c_{1122} = -\dfrac{3I_2}{|\mathbf{u}|}$ and the generator L_J is

$$
\begin{aligned}
L_J &= \frac{I_2}{2|\mathbf{u}|}\left[3J_{21}^2\frac{\partial^2}{\partial J_{11}^2} + 6J_{21}J_{22}\frac{\partial^2}{\partial J_{11}\partial J_{12}} + 3J_{22}^2\frac{\partial^2}{\partial J_{12}^2}\right]\\[6pt]
&= \frac{3I_2}{2|\mathbf{u}|}\left[J_{21}\frac{\partial}{\partial J_{11}} + J_{22}\frac{\partial}{\partial J_{12}}\right]^2.
\end{aligned}
$$

The matrix $J(0) = I$, and thus the only entry which is changed during the evolution in the limit is J_{12}. The generator L_J has the simple form

$$(11.5) \qquad\qquad L_J f = \frac{3I_2}{2|\mathbf{u}|}\frac{\partial^2 f}{\partial J_{12}^2}.$$

The Lyapunov exponent of the matrix valued process $J(t)$ is manifestly zero. The reason that only J_{12} is changed by the dynamics can be seen from the generator L_Y. Its coefficients depend only on y_2 and are independent of the $\mathbf{y}$-coordinate along the mean velocity $\mathbf{u}$. This is reflected in (11.5).

12. Deformation of the length

Theorem 10.1 allows us to study stretching of curves by the flow (1.1) in R^2. If the initial curve has length of order one, and is parameterized by $\mathbf{x} = \mathbf{x}(s)$, $0 \leq s \leq 1$, its length is given by

$$l = \int_0^1 \left| \frac{d\mathbf{x}}{ds} \right| ds.$$

The length of the curve $\mathbf{X}^\varepsilon(t, s) = \mathbf{X}^\varepsilon(t, \mathbf{x}(s))$ is

$$l^\varepsilon(t) = \int_0^1 \left| \frac{d\mathbf{X}^\varepsilon}{ds} \right| ds = \int_0^1 \left| J^\varepsilon(t, \mathbf{x}(s)) \frac{d\mathbf{x}}{ds}(s) \right| ds.$$

The limit theorem of Appendix A implies that for any two distinct points $\mathbf{x}$ and $\mathbf{y}$, the processes $J^\varepsilon(t, \mathbf{x})$ and $J^\varepsilon(t, \mathbf{y})$ are not only identically distributed, but also independent in the limit $\varepsilon \to 0$, unless $\mathbf{x} - \mathbf{y} = c\mathbf{u}$ for some c, which means that the points lie on the same deterministic trajectory. This is similar to the joint behavior of trajectories of (1.1) starting at points lying on different deterministic trajectories as described in Section 6. The same is true for any number of fixed initial points. Thus $l^\varepsilon(t) \to l(t)$, so that $E\{l^n(t)\} = (E\{l(t)\})^n$, and the length $l^\varepsilon(t)$ becomes deterministic in the limit.

Let us consider stretching of an interval of length one which is initially at angle θ with respect to the mean flow $\mathbf{u} = (|\mathbf{u}|, 0)$, so that $\dfrac{d\mathbf{x}}{ds} = (\cos\theta, \sin\theta)$, under the potential flow (11.1). We introduce the variables

$$J_1(\theta) = J_{11} \cos\theta + J_{12} \sin\theta = \cos\theta + J_{12} \sin\theta$$
$$J_2(\theta) = J_{21} \cos\theta + J_{22} \sin\theta = \sin\theta.$$

The behavior of the joint process $(J_1(\theta), J_2(\theta))$ in our case is trivial and degenerate. The point (J_1, J_2) performs an ordinary Brownian motion along the horizontal lines $J_2 = \text{const}$ with diffusion coefficient that depends only on J_2. In the time dependent case [32] there is no such degeneracy, the process (J_1, J_2) is not restricted to a line, and its norm grows exponentially in time.

We have $J_1(0) = \cos\theta$, and

$$\begin{aligned}
\bar{l}(t) &= E\{l(t)\} = \lim_{\varepsilon \to 0} E\{l^\varepsilon(t)\} = \int_0^1 E\left\{ \sqrt{J_1^2(\theta) + J_2^2(\theta)} \right\} ds \\
&= E\left\{ \sqrt{J_1^2(\theta) + \sin^2\theta} \right\},
\end{aligned}$$

because the law for the limit process J is independent of the starting point. Then, using (11.5) we see that the limit length $\bar{l}(t) = g(t, \cos\theta)$, where the function $g(t, x)$ satisfies the initial value problem:

$$(12.1) \qquad \frac{\partial g}{\partial t} = \frac{3I_2}{|\mathbf{u}|} \sin^2\theta \frac{\partial^2 g}{\partial x^2}$$

$$g(0, x) = \sqrt{\sin^2\theta + x^2}.$$

This means that unless $\theta = 0$ the length $\bar{l}(t) \sim C(\theta)\sqrt{t}$, and does not grow exponentially. The length does not change at all when $\theta = 0$, that is, when the interval is parallel to the mean flow. The algebraic growth is similar to the algebraic growth of the length in a shear flow [6]. The flow (1.1) is dominated by the mean flow, and randomness is not strong enough to generate the kind of mixing needed for exponential growth of the length.

13. SUMMARY AND CONCLUSIONS

We have shown that weak, time independent, incompressible fluctuations of a uniform flow produce certain non-trivial effects on time scales of order one. These are important in advection of passive scalars when the initial density varies on scales comparable to that of the inhomogeneities of the flow. This regime is similar to the one for which the radiative transport theory [31] holds, and we show that the Wigner distribution of the passive scalar satisfies the radiative transport equation (7.7). The limit Richardson function, or the two point correlation function, satisfies the degenerate diffusion equation (4.15). This result does not require that the velocity field or the initial tracer distribution be Gaussian.

The non-zero mean flow has a strong effect on the two-point motion introducing a degeneracy in its direction. As we explain in Section 8, using the duality between the limit diffusion process $\mathbf{Y}(t)$ and the corresponding jump process in Fourier space, the two point difference vector tends to be aligned with the mean flow in the long time limit.

We also study the evolution of the Jacobian of the flow map $\mathbf{x} \to \mathbf{X}^\varepsilon(t)$ in the limit $\varepsilon \to 0$ and show that its limit is a diffusion process. We show that because of the degeneracy caused by the mean flow the corresponding Lyapunov exponent vanishes. This implies in particular that the length of curves moving with the flow is growing only algebraically, which is quite different from strong, time dependent, mean zero flows studied in [8, 32]. Physically this is because of the sweeping effect of the mean flow and the fact that we are not looking at the long time limit but rather at finite time effects.

Our results can be generalized to the case of non-zero but small molecular diffusivity. The results regarding the two point motion remain essentially the same allowing for an additive term in the diffusion equation for the Richardson function, and an absorption term in the radiative transport equation.

14. APPENDIX A. A LIMIT THEOREM FOR TURBULENT DIFFUSION

The limit as $\varepsilon \to 0$ of the trajectories of the dynamical systems with the scaling as in (2.4) is described by a limit theorem, which was proven by Kesten and Papanicolaou [19, 22] for $\gamma = 1$, and later by Komorowski [17] for $0 \le \gamma < 1$. They considered equations of the form

$$(14.1) \qquad \frac{d\mathbf{Q}^\varepsilon}{dt} = \frac{1}{\varepsilon} \mathbf{G}\left(\frac{t}{\varepsilon^2}, \frac{\mathbf{Q}^\varepsilon(t)}{\varepsilon^\gamma}, \varepsilon\right),$$
$$\mathbf{Q}^\varepsilon(0) = 0,$$

where the function $\mathbf{G}$ satisfies the following conditions.

(A1) The function $\mathbf{G}(t, \mathbf{q}, \varepsilon, \omega)$ is jointly measurable in all its arguments and a.s. in $C^3(R^d)$ as a function of $\mathbf{q}$ for each t, ε.

(A2) The process $\{\mathbf{G}(t, \cdot, \varepsilon, \omega)\}$ is stationary in t for each fixed ε.

(A3) Let

$$\mathcal{G}_s^t(\varepsilon, M) = \sigma\{\mathbf{G}(u, \mathbf{q}, \varepsilon, \cdot) | s \le u \le t, |\mathbf{q}| \le M\},$$

and

$$\beta(t, M) = \sup_{s \ge 0, 0 < \varepsilon \le 1} \sup_{A \in \mathcal{G}_0^s(\varepsilon, M), B \in \mathcal{G}_{s+t}^\infty(\varepsilon, M)} |P(AB) - P(A)P(B)|.$$

Then

$$\int_0^\infty |\beta(t, M)|^{1/p} dt < \infty$$

for $p = 6 + 2d$.

(A4) $E\{\mathbf{G}(t, \mathbf{q}, \varepsilon)\} = 0$.

(A5) For each $M < \infty$ there exists a constant $C(M)$ independent of t, $\mathbf{k}$, and ε such that

$$(14.2) \qquad E\left\{\sup_{|\mathbf{q} - \mathbf{k}| \le M} |D^\beta \mathbf{G}(t, \mathbf{q}, \varepsilon)|^{\max(8, d)}\right\} \le C, \ 0 \le |\beta| \le 3,$$

and these integrals converge uniformly in $\mathbf{k}$. All the derivatives in (14.2) are with respect to $\mathbf{q}$.

(A6) The following limits exist uniformly on compact sets and are bounded functions of $\mathbf{q}$:

$$(14.3) \qquad A_{ij}(\mathbf{q}) = \lim_{\varepsilon \to 0} \int_0^\infty E\{G_i(0, \mathbf{q}, \varepsilon)G_j(t, \mathbf{q}, \varepsilon)\} \, dt$$

$$c_{ij}(\mathbf{q}) = \lim_{\varepsilon \to 0} \int_0^\infty E\left\{G_i(0, \mathbf{q}, \varepsilon)\frac{\partial}{\partial q_i}G_j(t, \mathbf{q}, \varepsilon)\right\} dt.$$

In addition the matrix $a_{ij} = A_{ij} + A_{ji}$ is twice continuously differentiable.

(A7) The velocity field is incompressible: $\operatorname{div}\mathbf{G}^\varepsilon = 0$.

By a solution of (14.1) we mean a continuous function $\mathbf{Q}^\varepsilon(t)$ which satisfies

$$(14.4) \qquad \mathbf{Q}^\varepsilon(t, \omega) = \mathbf{q} + \frac{1}{\sqrt{\varepsilon}}\int_0^t \mathbf{G}\left(\frac{\sigma}{\varepsilon}, \mathbf{Q}^\varepsilon(\sigma, \omega), \varepsilon, \omega\right) d\sigma.$$

THEOREM 14.1. [**19**, **17**] *Let* $\mathbf{G}$ *be a random field that satisfies conditions (A1-A7). Then with probability one (14.4) has a unique solution for all* $t \ge 0$. *Let* R^ε *be the measure on* $C([0, \infty); R^d)$ *induced by this solution. For each* $f \in C^2(R^d)$ *put*

$$(14.5) \qquad Lf(\mathbf{q}) = \frac{1}{2}\sum_{i,j=1}^d a_{ij}(\mathbf{q})\frac{\partial^2 f(\mathbf{q})}{\partial q_i \partial q_j}$$

Then R^ε *converges weakly to the probability measure* R *on* $C([0, \infty); R^d)$, *which corresponds to the diffusion with infinitesimal generator* L *and starting at the point* $\mathbf{q}$, *that is,* $R\{\mathbf{Q}(0) = 0\} = 1$.

The most general statement of this theorem, including more general time dependence and compressible velocity fields, may be found in [**17**].

15. Appendix B. Oscillatory functions and the Wigner distribution

We review briefly some definitions and basic facts about oscillatory families of functions and the Wigner distribution [15]. Let $f^\varepsilon(\mathbf{x})$ be a bounded family of functions in $L^2_{loc}(R^d)$.

DEFINITION 1. The family $f^\varepsilon(\mathbf{x})$ is ε-oscillatory if for any smooth test function of compact support $\phi(\mathbf{x})$

$$(15.1) \qquad \limsup_{\varepsilon \to 0} \int_{|\mathbf{k}| \geq R/\varepsilon} \left| \widehat{\phi f^\varepsilon}(\mathbf{k}) \right|^2 d\mathbf{k} \to 0 \text{ as } R \to +\infty.$$

A sufficient condition for (15.1) to hold is

$$\exists k > 0, \text{ such that } \varepsilon^k \left\| \frac{\partial^k f^\varepsilon}{\partial x^k} \right\|_{L^2_{loc}} \leq C.$$

DEFINITION 2. The family $f^\varepsilon(\mathbf{x})$ is said to be compact at infinity if

$$(15.2) \qquad \limsup_{\varepsilon \to 0} \int_{|\mathbf{x}| \geq R/\varepsilon} |f^\varepsilon(\mathbf{x})|^2 d\mathbf{x} \to 0 \text{ as } R \to +\infty.$$

A sufficient condition for (15.2) to hold is that there exists a set K such that $\operatorname{supp} f^\varepsilon(\mathbf{x}) \subset K$.

DEFINITION 3. The Wigner distribution is defined by

$$W^\varepsilon(\mathbf{x}, \mathbf{k}) = \int \frac{d\mathbf{y}}{(2\pi)^d} e^{i\mathbf{k} \cdot \mathbf{y}} f^\varepsilon(\mathbf{x} - \frac{\varepsilon \mathbf{y}}{2}) f^{\varepsilon *}(\mathbf{x} + \frac{\varepsilon \mathbf{y}}{2}).$$

It has a weak limit point $W(\mathbf{x}, \mathbf{k})$ in the space $\mathcal{S}'$ of Schwartz distributions provided that $f^\varepsilon(\mathbf{x})$ is bounded in L^2_{loc}. We shall assume that such a limit point is unique. The limit Wigner distribution may be interpreted as a phase space energy density. For instance, the limit Wigner distribution of the WKB family $f^\varepsilon(\mathbf{x}) = A(\mathbf{x}) e^{iS(\mathbf{x})/\varepsilon}$ is

$$W(\mathbf{x}, \mathbf{k}) = |A(\mathbf{x})|^2 \delta(\mathbf{k} - \nabla S(\mathbf{x})).$$

The limit Wigner distribution is mostly relevant for ε-oscillatory and compact at infinity families $f^\varepsilon(\mathbf{x})$.

PROPOSITION 1. [15] Let f_ε be a bounded family in L^2_{loc} with limit Wigner measure $W(\mathbf{x}, \mathbf{k})$. Then

(i) The limit Wigner distribution $W(\mathbf{x}, \mathbf{k}) \in \mathcal{S}'$ is non-negative (a measure).
(ii) If two families $f^\varepsilon(\mathbf{x})$ and $g^\varepsilon(\mathbf{x})$ coincide on an open ball $B \subset R^d$ then $W_f(\mathbf{x}, \mathbf{k}) = W_g(\mathbf{x}, \mathbf{k})$ for $\mathbf{x} \in B$ and all $\mathbf{k} \in R^d$.
(ii) For any smooth function of compact support $\theta(\mathbf{x})$

$$(15.3) \qquad \iint |\theta(\mathbf{x})|^2 W(d\mathbf{x}, d\mathbf{k}) \leq \limsup_{\varepsilon \to 0} \int_{R^d} |\theta(\mathbf{x}) f_\varepsilon(\mathbf{x})|^2 d\mathbf{x}$$

with equality holding if and only if f_ε is ε-oscillatory. In this case $\limsup$ can be replaced by $\lim$ on the right side of (15.3).

ACKNOWLEDGMENTS

This work was partially sponsored by the NSF under grant numbers DMS-9622854, DMS-9600119, and by the Air Force Office of Scientific Research, Air Force Materials Command, USAF, under grant number F49620-95-1-0315. The US Government is authorized to reproduce and distribute reprints for governmental purposes notwithstanding any copyright notation thereon. The views and conclusions contained herein are those of the authors and should not be interpreted as necessarily representing the official policies or endorsements, either expressed or implied, of the Air Force Office of Scientific Research or the US Government.

We are grateful to Joseph B. Keller for numerous discussions on the subject.

References

[1] M. Avellaneda and A. J. Majda, An Integral Representation And Bounds On The Effective Diffusivity In Passive Advection By Laminar And Turbulent Flows, Commun. Math.Phys. **138**, 1991, 339-391.

[2] G.K. Batchelor and A.A. Townsend, Turbulent diffusion, **Surveys in Applied Mechanics**, Cambridge University Press, 1956, 352-399.

[3] P. Baxendale, Statistical equilibrium and two-point motion for a stochastic flow of diffeomorphisms, Progress in Probab., **19**, **Spatial Stochastic Processes**, eds. K. Alexander and J. Watkins, Birkhäuser, Boston, 1991, 189-218.

[4] P.Baxendale and T. Harris, Isotropic stochastic flows, Annals of Probability, **14**, 1986, 1155-1179.

[5] A. Bensoussan, J.L. Lions and G. Papanicolaou, Boundary layers and homogenization of transport processes, Publ. RIMS, **15**, 1979, 54-157.

[6] R.Carmona, Transport properties of Gaussian velocity fields, **Stochastic Analysis**, ed. M.M. Rao, 1996, CRC Press.

[7] R.Carmona, S.Grishin, and S.Molchanov, Massively parallel simulations of motions in a Gaussian velocity field, Progress in Probab. **39**, **Stochastic Modeling in Physical Oceanography**, eds R. Adler, P. Mueller and B. Rozovskii, Birkhäuser, New York, 1996, 47-68.

[8] R.Carmona, S.Grishin, L.Xu and S.Molchanov, Surface stretching for Ornstein-Ulehnbeck velocity fields, Electron. Comm. Probab., 1996.

[9] R. Carmona and L. Xu, Homogenization for time-dependent two-dimensional incompressible Gaussian flows, The Annals of Applied Probability, **7**, 1997, 265-279.

[10] S.Chandrasekhar, **Radiative transfer**, Dover, New York, 1960.

[11] G. Dell'Antonio, Large time, small coupling behavior of a quantum particle in a random field, Ann. Inst. H. Poincare Sect. A (N.S), **39**, 1983, 339-384.

[12] A. Fannjiang and T. Komorowski, An invariance principle for diffusion in time dependent turbulence, Preprint, 1997.

[13] A. Fannjiang and T. Komorowski, Turbulent Diffusion in Markovian Flows. I, Preprint, 1997.

[14] A. Fannjiang and G. Papanicolaou, Diffusion In Turbulence, Probability Theory and Related Fields, **105**, 1996, 279-334.

[15] P. Gérard, Microlocal defect measures, Comm. PDEs, **16**, 1991, 1761-1794.

[16] T. Ho, L. Landau and A. Wilkins, On the weak coupling limit for a Fermi gas in a random potential, Rev. Math. Phys., **5**, 1993, 209-298.

[17] T. Komorowski, Diffusion approximation for the advection of particles in a strongly turbulent random environment, The Annals of Probability, **24**, 1996, 346-376.

[18] T. Komorowski and G. Papanicolaou, Motion in a Gaussian incompressible flow, The Annales of Applied Probability, **7**, 1997, 229-264.

[19] H. Kesten and G. Papanicolaou, A limit theorem in turbulent diffusion, Comm. Math. Phys., **65**, 1979, 97-128.

[20] V. Klyatskin, W. Woyczynski and D. Gurarie, Short-time correlation approximations for diffusing scalars in random velocity fields: a functional approach, in **Stochastic modeling in physical oceanography**, R. Adler, P. Müller and B. Rozovskii, eds., Birkhäuser, 1996, 221-270.

[21] V. Klyatskin, W. Woyczynski and D. Gurarie, Diffusive passive tracers in random incompressible flows: Statistical topography aspects, Journal Stat. Phys., **84**, 1996, 797-836.

[22] H. Kunita, **Stochastic flows and stochastic differential equations**, Cambridge University Press, Cambridge, 1990.

[23] P.L. Lions and T. Paul, Sur les Mesures de Wigner, Revista Mat. Iberoamericana, **9**, 1993, 553-618.

[24] S. Molchanov, Lectures on Random Media, in: P. Bernard, ed., Lectures On Probability Theory Ecole d'Eté de Probabilité de Saint-Flour XXII, 1992, 242-411, Springer Lecture Notes in Mathematics 1581.

[25] S.A. Molchanov and L.I. Piterbarg, Heat propagation in random flows, Russian Jour. Math. Phys., **1**, 1994, 353-376.

[26] A.Monin and A.Yaglom, **Statistical Fluid Mechanics: Mechanics of Turbulence**, MIT Press, Cambridge, Mass., 1975

[27] K. Oeschlager, Homogenization Of A Diffusion Process In A Divergence Free Random Field, Ann. Of Probability, **16**, 1988, 1084-1126.

[28] G.C. Papanicolaou and S.R.S. Varadhan, Boundary value problems with rapidly oscillating coefficients, **Random Fields Coll. Math. Soc Janos Bolyai**, J. Fritz, J. Lebowitz, eds., 1982, North-Holland, 835-873.

[29] L. Piterbarg, Short-correlation approximation in models of turbulent diffusion **Stochastic models in geosystems**, IMA Proc., vol 85, 1997, 313-352.

[30] L.F. Richardson, Atmospheric diffusion shown on a distance-neighbor graph, Proc. Roy. Soc., **110**, 1926, 709-727.

[31] L. Ryzhik, G. Papanicolaou and J.B. Keller, Energy transport for elastic and other waves in a random medium, Wave Motion, **24**, 1996, 327-370.

[32] A. Saichev and W. Woyczynski, Probability distributions of passive tracers in randomly moving media, **Stochastic models in geosystems**, IMA Proc., vol. 85, 1997 , 359-399.

[33] G.I. Taylor, Diffusions by continuous movements, Proc. London math. Soc. Ser. 2, **20**, 1923, 196-211.

DEPARTMENT OF MATHEMATICS, UNIVERSITY OF CALIFORNIA AT DAVIS, DAVIS CA 95616
E-mail address: fannjian@math.ucdavis.edu

DEPARTMENT OF MATHEMATICS, STANFORD UNIVERSITY, STANFORD CA, 94305
E-mail address: ryzhik@math.stanford.edu, papanico@math.stanford.edu

Proceedings of Symposia in Applied Mathematics
Volume **54**, 1998

Integrability: from d'Alembert to Lax

A.S. Fokas

ABSTRACT. The inverse spectral method is a nonlinear Fourier transform method for solving *initial value problems* for certain nonlinear PDE's in 2 and 3 dimensions. It is based on the solution of the so-called Riemann-Hilbert and DBAR problems. After reviewing this method we will present a new transform method for solving *initial boundary value problems* for both linear and integrable nonlinear PDE's in 2 dimensions. This method provides a unified approach to solving linear equations with simple boundary conditions, linear equations with complicated boundary conditions (Wiener-Hopf type problems) and nonlinear integrable equations. For linear equations the new method is in a sense the antithesis of separation of variables introduced in 1750 by d'Alembert.

1. Introduction

There exist nonlinear equations which can be written as the compatibility condition of linear equations. Such equations are called *integrable* [1] and the associated linear equations are called *Lax pairs* [2]. There exists a great variety of nonlinear integrable equations: ODE's, PDE's, singular integrodifferential equations, difference equations, algebraic equations, functional equations, cellular type equations, etc. Some of these integrable equations are physically significant. Calogero [3] has elucidated the reasons why certain nonlinear PDE's are both integrable and widely applicable. Due to space limitations I will only discuss ODE's and PDE's. I first give a brief overview of the main analytical results obtained for such equations.

Notation

$S(\mathbb{R})$ will denote the space of Schwartz functions on the line. $H^m(\mathbb{R}^+)$ will denote the space of square integrable functions in $(0, \infty)$, whose first m generalized derivatives are also square integrable.

Integrable ODE's

For integrable ODE's there exist rather complete results. A typical such equation is the classical Painlevé II equation,

$$(1.1) \qquad \frac{d^2 q}{dt^2} = 2q^3 + tq - \alpha,$$

where $q(t)$, t and the constant α are complex. Equation (1.1) is a nonlinear version of the Airy equation. The most important recent analytical results about Painlevé II

1991 *Mathematics Subject Classification*. Primary 35; Secondary 46.

are [4]: (a) The general solution can be expressed through the solution of a Fredholm linear integral equation of the second kind which is uniquely parameterized by two complex constants called monodromy data. (b) $q(t)$ is a meromorphic function of t. (c) For certain choices of the monodromy data and for t restricted on certain rays (or in certain sectors) of the complex t-plane, $q(t)$ is free of poles. (d) There exists a complete description of the asymptotic behavior of $q(t)$ as $t \to \infty$ in the complex plane. The particular form of this asymptotic behavior depends on α, on the monodromy data and on $\arg t$. Particular cases of these asymptotic formulae are nonlinear versions of the classical connection formulae of the Airy equation.

Painlevé II appears in many applications. For example it was recently derived in [5] in connection with a Hele-Shaw type problem. In this particular case the physical problem reduces to the mathematical problem of showing that there exists a global solution in $\frac{\pi}{3} < \arg t < \frac{5\pi}{3}$, such that $q \to \alpha/t$ as $t \to \infty$ in $\frac{\pi}{3} < \arg t < \frac{5\pi}{3}$. Using the integrability machinery it is possible to show that indeed, if

$$(1.2a) \qquad q(t) = \frac{\alpha}{t}\left(1 + O(\frac{1}{t^3})\right), t \to \infty \quad \text{in} \quad S: \frac{\pi}{3} < \arg t < \frac{5\pi}{3},$$

then $q(t)$ is unique and $q(t)$ is free of poles in S. Furthermore, it is possible to describe the asymptotic behavior of this solution in the entire complex t-plane: $q(t)$ behaves like certain elliptic functions in the sectors $0 < \arg t < \frac{\pi}{3}$ and in $\frac{5\pi}{3} < \arg t < 2\pi$ (the parameters of these elliptic functions are certain explicit functions of α). Also

$$(1.2b) \qquad q(t) = -e^{\frac{i\pi}{12}}\frac{\sin \pi\alpha}{\sqrt{\pi}|t|^{\frac{1}{4}}}e^{-\frac{2i}{3}|t|^{\frac{3}{2}}} + \frac{\alpha}{t} + O\left(\frac{1}{t^{\frac{7}{4}}}\right), \quad t \to \infty, \quad \arg t = \frac{5\pi}{3},$$

$$(1.2c) \qquad q(t) = -e^{-\frac{i\pi}{12}}\frac{\sin \pi\alpha}{\sqrt{\pi}|t|^{\frac{1}{4}}}e^{\frac{2i}{3}|t|^{\frac{3}{2}}} + \frac{\alpha}{t} + O\left(\frac{1}{t^{\frac{7}{4}}}\right), \quad t \to \infty, \quad \arg t = \frac{\pi}{3},$$

$(1.2d)$

$$q(t) = \hat{\alpha}\frac{d}{dt}\ln\left\{\sin\left[\frac{\sqrt{2}}{3}t^{\frac{3}{2}} + \frac{3}{\pi}\ln(2|\sin \pi\alpha|)\ln t + \varphi_0\right]\right\} + O\left(\frac{1}{t}\right), t \to \infty, \arg t = 0,$$

where $\hat{\alpha} = sign(\sin \pi\alpha)$ and φ_0 is a certain unique expression of α.

Integrable Evolution Equations in One Spatial Variable

For equations in one temporal and one spatial dimensions, i.e. for equations in $1 + 1$, there exist only partial results. A typical such equation is the nonlinear Schrödinger equation

$$(1.3) \qquad\qquad iq_t + q_{xx} + 2\lambda|q|^2q = 0, \lambda = \pm 1,$$

where $q(x,t) \in \mathbb{C}$ and and $x,t \in \mathbb{R}$. The best known analytical results for this equation are: (a) If $\lambda = -1$ then the Cauchy problem on the infinite line for decaying initial data (i.e. $x \in (-\infty, \infty)$, $q(x,0) = q_0(x) \in S(\mathbb{R})$) can be expressed through the solution of a Fredholm linear integral equation of the second kind which is uniquely defined in terms of $q(x,0)$. This equation has always a unique solution. Similar considerations apply to the case of $\lambda = 1$, but now the above linear integral equation must be supplemented by a system of linear algebraic equations.

For a particular choice of $q(x,0)$, the solution consists only of the part obtained by the above system of algebraic equations. This explicit solution describes the nonlinear interaction of certain N-particular entities, called *solitons*. (b) There exists a complete description of the asymptotic behavior of $q(x,t)$ as $t \to \infty$. In particular, if $\lambda = -1$, then as $t \to \infty$ and $x/t = O(1)$ the solution disperses away like $t^{-\frac{1}{2}}$. This behavior is similar with the behavior of the associated linear equation. However, if $\lambda = 1$, then as $t \to \infty$, $x/t = O(1)$, the solution decomposes into N-solitons (the number N depends on $q(x,0)$). (c) The general solution of the Cauchy problem with periodic data (i.e. $x \in [0,1]$, $q(0,t) = q(1,t)$, $q(x,0) = q_0(x)$ given) can be approximated by an explicit solution which describes the nonlinear interaction of certain N-particular solutions. These solutions, which are the periodic analogues of solitons, can be expressed in terms of certain Riemann theta functions.

The well-posedness of the above problems can also be established by PDE techniques. However, the important question of describing the long time behavior of the solution of the Cauchy problem on the infinite line has not been answered up to now by PDE techniques. Similarly, the approximation of the solution of the periodic problem in terms of theta functions has not been established up to now by PDE techniques. It is my opinion that solitons are important, not because they are explicit solutions, but because they describe the asymptotic behavior of certain nonlinear equations.

Integrable Evolution Equations in Two Spatial Variables

For equations in one temporal and two spatial dimensions, i.e. for equations in $2 + 1$, the situation is even less satisfactory than for equations in $1 + 1$. It should be pointed out that there exists a false general perception that there exist only a few integrable equations in $2 + 1$. Actually, to each $1 + 1$ integrable equation, there correspond *several* $2 + 1$ integrable generalizations. For example the nonlinear Schrödinger equation (1.3) generalizes to several integrable $2 + 1$ equations, which include the Davey-Stewartson I (DSI) and DSII equations. The simplest $2 + 1$ integrable generalization of equation (1.3) is the equation [6]

$$(1.4) \qquad iq_t + q_{xx} + 2\lambda q \int_{-\infty}^{y} |q(x,y',t)|_x^2 dy' = 0, \quad \lambda = \pm 1.$$

The best known analytical results for integrable equations in $2 + 1$ are: (a) The Cauchy problem on the infinite plane for decaying initial data (i.e. $x, y \in (-\infty, \infty)$, $q(x,y,0) = q_0(x,y) \in S(\mathbb{R}^2)$) can be expressed through the solution of a Fredholm linear integral equation of the second kind which is defined in terms of $q(x,y,0)$. For some nonlinear equations (e.g. DSI, defocusing DSII, N-wave interactions) this equation is uniquely defined in terms of $q(x,y,0)$ for any $q(x,y,0) \in S(\mathbb{R}^2)$. For other nonlinear equations (e.g. focusing DSII, Kadomtsev-Petviashvilli equation I (KPI), KPII), this is the case provided that a certain norm of $q(x,y,0)$ is sufficiently small. Also for some nonlinear equations (e.g. DSI, defocusing DSII, N-wave interactions, KPII), the above linear integrable equation has always a unique solution. For other nonlinear equations (e.g. focusing DSII, KPI) this is the case provided that a certain norm of $q(x,y,0)$ is sufficiently small. (b) If the $1 + 1$ reduction of a given $2 + 1$ equation possesses solitons, then the given $2 + 1$ equation possesses *line solitons*. These solutions decay in every direction except on certain lines. There also exist two types of localized solutions in $2 + 1$ which do not have $1 + 1$ analogues:

(i) Some $2 + 1$ equations (e.g. DSI) possess *dromions*. The origin of these solutions is the linear part of the equation. These solutions were called dromions in [7] to emphasize that they travel on tracks (in greek: dromos) dictated by the boundary conditions. (ii) Some $2 + 1$ equations (e.g. KPI, focusing DSII) possess lumps. These localized solutions are genuine nonlinear structures and thus, in some sense, are the proper $2 + 1$ analogues of solitons. (c) For some $2 + 1$ equations it is possible to describe the long time behavior of the solution. Rigorous results exist only for the cases that line solitons, dromions and lumps are absent [8]. There exist some formal arguments to suggest that dromions and lumps are the natural asymptotic states of the equations that possess them. (d) For periodic initial data there exist complete results only for KPII.

For some of the integrable $2 + 1$ equations well posedness cannot be established so far by PDE techniques. Also there exist $2 + 1$ integrable equations for which PDE techniques imply well posedness without a small norm assumption, while this assumption is still required for the application of the integrability techniques.

Most of the details of the results mentioned above can be found in [25].

Riemann-Hilbert and DBAR Problems

The above brief overview already indicates an important unifying feature of integrable equations: The solution of the Cauchy problem can be expressed through the solution of a linear integral equation. Actually, there is something deeper underlying integrability. All the linear integral equations associated with nonlinear integrable equations have the same origin: each of these equations expresses the solution of a DBAR ($\frac{\partial}{\partial \bar{k}}$) problem or of a *Riemann-Hilbert problem*[1] [9].

It was mentioned earlier that an integrable PDE is the compatibility condition of a pair of linear equations. However, it is only one of these equations, namely the one involving only spatial derivatives, that determines the associate DBAR or Riemann-Hilbert problem. For example the nonlinear Schrödinger equation (1.3) possesses the Lax pair [10].

$$(1.5) \quad \mu_x + ik[\sigma_3, \mu] = Q\mu, \qquad Q \doteq \begin{pmatrix} 0 & q \\ \lambda\bar{q} & 0 \end{pmatrix}, \qquad \lambda = \pm 1, \sigma_3 = \mathrm{diag}(1, -1),$$

$$(1.6) \quad \mu_t + 2ik^2[\sigma_3, \mu] = \tilde{Q}\mu, \qquad \tilde{Q} \doteq 2kQ - iQ_x\sigma_3 - \lambda i|q|^2\sigma_3,$$

where $\mu(x, t, k)$ is a 2×2 matrix and $[,]$ denotes the usual matrix commutation. The *spectral analysis* of equation (1.5) for $\lambda = -1$ yields the Riemann-Hilbert (RH) problem

$$(1.7a) \qquad \mu^+(x, t, k) = \mu^-(x, t, k)e^{-ikx\sigma_3}S(k, t)e^{ikx\sigma_3}, k \in \mathbb{R}$$

$$(1.7b) \qquad \mu = I + O\left(\frac{1}{k}\right), k \to \infty,$$

where $\mu^+(\mu^-)$ is a 2×2 matrix solution of equation (1.5) which is analytic in the upper (lower) half complex k-plane and the 2×2 matrix $S(k, t)$ contains the

---

[1] A Riemann-Hilbert problem is a particular case of a DBAR problem

so-called *spectral data*. Equation (1.6) simply determines the time dependence of $S(k,t)$, which turns out to be

$$(1.7c) \qquad S(k,t) = e^{-2ik^2t\sigma_3}S(k,0)e^{2ik^2t\sigma_3}, k \in \mathbb{R}.$$

The matrix $S(k,0)$ can be determined from $q(x,0)$. The situation for other integrable equations in $1+1$ and $2+1$ is similar. The only difference is that for equations in $2+1$ the RH problem (1.7a) is replaced by a DBAR problem.

In summary: *The integrability of the Cauchy problem on the infinite line or the infinite plane for equations in $1+1$ and $2+1$ respectively, involves the following. The spectral analysis of one of the equations forming the associated Lax pair yields a RH problem, or a DBAR problem.*[2] *The other equation of the Lax pair determines the evolution of the spectral data. The spectral data at $t=0$ can be evaluated from* $q(x,0)$ *or* $q(x,y,0)$.[3]

The above method of solving the Cauchy problem of nonlinear integrable equations appears very different than the method of solving the corresponding linear PDE's, which is based on Fourier transforms. However, it was emphasized by Gel'fand and the author [11] that linear PDE's can actually be solved by a method very similar to the above. Namely: *Linear PDE's in $1+1$ and $2+1$ possess a Lax pair formulation. The spectral analysis of one of these equations yields a RH problem or a DBAR problem. The other equation forming the Lax pair determines the evolution of the spectral data. The 1- and 2-dimensional Fourier transforms appear naturally in the formulation of the associated RH and DBAR problems.*

For example, the equation

$$(1.8) \qquad iq_t + q_{xx} = 0$$

possesses the Lax pair

$$(1.9a) \qquad \mu_x + ik\mu = q,$$

$$(1.9b) \qquad \mu_t + ik^2\mu = iq_x + kq, \qquad k \in \mathbb{C},$$

$\mu(x,t,k)$ scalar. Similarly, the equation

$$(1.10) \qquad iq_t + \frac{1}{2}q_{xx} - \frac{1}{2}q_{yy} = 0,$$

which is the linearized version of the DSII equation, possess the Lax pair

$$(1.11a) \qquad \mu_{\bar{z}} - k\mu = q, \qquad z = x + iy,$$

$$(1.11b) \qquad \mu_t = i(\mu_{zz} + k^2\mu + kq + q_{\bar{z}}),$$

---

[2]In some cases one obtains a nonlocal RH problem instead of a DBAR problem.

[3]For brevity of presentation I only consider the case that solitons are absent. The solitonic case can be reduced to a solitonless case supplemented by a system of algebraic equations. I also assume that dromions and lumps are absent.

$\mu(x, y, t, k)$ scalar. It will be shown in §2 that the spectral analysis of equations (1.9a) and (1.11a) yield a RH and a DBAR problem respectively. The 1- and 2-dimensional Fourier transforms will appear naturally in the analysis of these problems.

What is Integrability?

The above discussion indicates that linear PDE's and integrable nonlinear PDE's in $1+1$ and $2+1$, share the distinguished property that they possess a Lax pair formulation. If we accept this property as the defining property of integrability, then in order to understand "what is integrability" for nonlinear equations we shall first scrutinize further the integrability of linear equations. Let us return to equation (1.8). Since equation (1.8) is the compatibility condition of equations (1.9), it follows that (1.8) and (1.9a) imply (1.9b). Indeed, equations (1.8) and (1.9a) imply immediately

$$(1.12) \qquad\qquad i\mu_t + \mu_{xx} = 0.$$

It is straightforward to verify that the compatibilities of equations (1.9a) and (1.12) is equation (1.8): Applying the operator $i\partial_t + \partial_{xx}$ to equation (1.9a) and using that $i\partial_t + \partial_{xx}$ and $\partial_x + ik$ commute, equation (1.8) follows. Equations (1.9a) and (1.12) imply (1.9b). Similarly, the evolution equation

$$(1.13) \qquad\qquad q_t + \sum_0^N \alpha_j \partial_x^j q = 0,$$

α_j arbitrary functions of t, possesses the Lax pair

$$(1.14) \qquad\qquad \mu_x + ik\mu = q, \mu_t + \sum_0^N \alpha_j \partial_x^j \mu = 0.$$

Thus the integrability of linear PDE's in $1+1$ is intimately related to the equation $\mu_x + ik\mu = q$. The defining property of this equation is that it is the *simplest* equation whose spectral analysis gives rise to the 1-dimensional Fourier transform.

Following our understanding of integrability of linear equations it is straightforward not only to understand the integrability of nonlinear equations but also to algorithmically *construct* them. Indeed, *nonlinear integrable equations should arise as some deformation of linear equations. However, while it appears very difficult to perform this deformation at the level of the equations themselves, it is elementary to perform this deformation at the level of the Lax pairs.*

Consider for example the problem of *constructing* an integrable deformation of equation (1.8). Instead of studying equation (1.8) we concentrate on its Lax pair (1.9). It is important to realize that the essential part of this Lax pair is only the lhs of the relevant equations. Indeed, the lhs together with

$$(1.15) \qquad\qquad \mu = \frac{\mu_1(x, t)}{k} + O\left(\frac{1}{k^2}\right), k \to \infty,$$

imply that the rhs of equation (1.9a) consists of $i\mu_1$, while the rhs of equation (1.9b) consists of $-\mu_{1x} + ik\mu_1$. Thus we look for proper deformations of the lhs of equations (1.9), i.e. of equations

$$\phi_x + ik\phi, \qquad \phi_t + ik^2\phi.$$

The complex conjugation of these expressions yields

$$\bar\phi_x - ik\bar\phi, \qquad \bar\phi_t - ik^2\bar\phi.$$

Rewriting the above in a matrix form, one finds

$$\mu_x + ik\sigma_3\mu, \mu_t + ik^2\sigma_3\mu, \sigma_3 = diag(1, -1),$$

where $\mu(x, t, k)$ is a 2×2 matrix valued function. This suggests the introduction of the *dressing operators* [12]

$$D_x\mu = \mu_x + ik\mu\sigma_3, D_t\mu = \mu_t + \alpha ik^2\mu\sigma_3,$$

α constant. Starting from these operators and using $\mu = I + O(\frac{1}{k})$, $k \to \infty$, the so called dressing method [12], *algorithmically* implies equations (1.5), (1.6).

We note that the deformation procedure is not unique. For example if one starts with more "copies" of equations (1.9), one obtains more complicated nonlinear integrable versions of equation (1.3).

In summary: *The integrability of linear PDE's in $1+1$ is a consequence of the equation $\mu_x + ik\mu = q$. This is the simplest equation whose spectral analysis gives rise to the 1-dimensional Fourier transform. The integrability of nonlinear PDE's in $1+1$ is a consequence of the equations obtained by "dressing" this equation.*

The above discussion suggests that in order to extend the notion of integrability to $2+1$ we must find the simplest equation whose spectral analysis gives rise to the 2-dimensional Fourier transform. This remarkable equation is $\mu_x + i\mu_y + ik\mu = q$. The integrability of linear PDE's in $2 + 1$ is intimately related to this equation. Again, proper deformations of this equation can be used to *algorithmically construct* integrable nonlinear PDE's in $2 + 1$.

In summary: *The integrability of linear PDE's in $2 + 1$ is a consequence of the equation $\mu_x + i\mu_y + ik\mu = q$. This is the simplest equation whose spectral analysis gives rise to the 2-dimensional Fourier transform. The integrability of nonlinear PDE's in $2 + 1$ is a consequence of the equations obtained by "dressing" this equation.*

It should be emphasized that so far I have discussed only evolution equations. I expect that the astute reader has already realized that similar considerations are valid for equations in any 2 and 3 dimensions. Examples of Lax pairs of linear non-evolutionary equations will be given in §3.

Initial-Boundary Value Problems

Certain initial-boundary value (IBV) problems for linear PDE's can be solved by certain transforms which include the sine, the cosine and the Laplace transforms. Given a 2-dimensional PDE with constant coefficients, separation of variables gives rise to two linear ordinary differential operators; the *independent* spectral analysis of these operators yields two different integral transforms for the solution of a given IBV problem. Consider for example equation (1.8) in the quarter plane, $x, t \in [0, \infty)$, with $q(x, 0) = q_1(x) \in S(\mathbb{R}^+)$, $q(0, t) = q_2(t) \in S(\mathbb{R}^+)$. Separation of variables implies that the proper transforms are the sine transform in x or the Laplace transform in t. Using the sine transform it follows that

$$(1.16) \qquad q(x,t) = \frac{2}{\pi} \int_0^\infty \sin kx \left[e^{-ik^2 t} \hat{q}_1(k) + ik \int_0^t e^{-ik^2(t-t')} q_2(t')dt' \right] dk,$$

where $\hat{q}_1(k)$ is the sine transform of the initial data. The Laplace transform yields a more complicated expression.

It is the author's opinion that the separation of variables approach has severe limitations: (a) The second term of the rhs of equation (1.16) is *not* separable in t and k. Thus the rhs of equation (1.16) does *not* provide the complete spectral decomposition of $q(x,t)$. (b) Changing the boundary conditions changes drastically the relevant transform. For example if the combination $q_x(0,t) + \alpha q(0,t)$ is given, then the appropriate x-transform is a combination of the sine, the cosine and a certain discrete transform. (c) For higher order PDE's separation of variables *fails* to give an appropriate x-transform. For example there does *not* exist an appropriate x-transform for $q_t + q_{xxx} = 0$ in the quarter plane. (d) For more complicated boundary conditions, such as discontinuous mixed boundary conditions, separation of variables fails to generate proper transforms. Some of these problems have been solved by the Wiener-Hopf method [13], or by using certain ad hoc integral representations (see for example the problem of waves on sloping beaches [14]). (e) The problem of extending the transform methods to more complicated domains, such as moving boundaries, remains essentially open.

It appears that there exists a new method, based on the Lax pair formulation of linear PDE's, which overcomes most of the limitations of the method of separation of variables. The new method, developed by the author in [15] and [16], is in a sense the antithesis of separation of variables. Indeed, it uses the *simultaneous spectral analysis of both linear operators defining the Lax pair to construct a unifying transform for the solution of certain IBV problems.* For example, the new method implies that the solution of *any* IBV problem of equation (1.8) in the quarter plane, such that $q(x,t) \to 0$ as $x \to \infty$ and $q(x,t) \to 0$ as $t \to \infty$, is given by

$$(1.17) \qquad q(x,t) = \frac{1}{2\pi} \int_{-\infty}^\infty e^{-ikx-ik^2 t} \hat{q}_1(k)dk + \frac{1}{2\pi} \int_{L_1} e^{-ikx-ik^2 t} \nu(k)dk.$$

In this equation, $\hat{q}_1(k)$ is the Fourier transform of the initial data $q(x,0) = q_1(x)$,

$$(1.18) \qquad \hat{q}_1(k) = \int_0^\infty e^{ikx} q_1(x)dx,$$

L_1 consists of the negative imaginary axis and of the negative real axis (see Figure 1) and $\nu(k)$ is some function analytic and bounded for $\pi \le \arg k \le 3\pi/2$. The particular form of $\nu(k)$ depends on the particular boundary conditions. For example if any of the following conditions is given

$$(1.19a) \qquad\qquad q(0,t) = q_2(t),$$

$$(1.19b) \qquad\qquad q_x(0,t) = q_2(t),$$

$$(1.19c) \qquad\qquad q_x(0,t) + i\alpha q(0,t) = q_2(t),$$

where the given function $q_2(t) \in S(\mathbb{R}^+)$ and $\arg \alpha$ is outside the interval $[\pi, \frac{3\pi}{2}]$, then $\nu(k)$ for (a) or (b) or (c) is given by

$$(1.20a) \qquad \nu(k) = 2k\hat{q}_2(k) + \hat{q}_1(-k),$$

$$(1.20b) \qquad \nu(k) = 2i\hat{q}_2(k) - \hat{q}_1(-k),$$

$$(1.20c) \qquad \nu(k) = \frac{2ik}{k-\alpha}\hat{q}_2(k) - \frac{k+\alpha}{k-\alpha}\hat{q}_1(-k),$$

where

$$\hat{q}_2(k) = \int_0^\infty e^{ik^2 t} q_2(t) dt.$$

It should be emphasized that the representation (1.17) is consistent with the beautiful result known as Ehrenpreis principle [17].

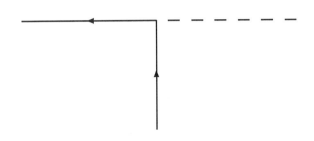

Figure 1
The contour L_1

An important advantage of the new method is that it provides explicit global relations between the boundary values of q. Because of these global relations it is possible to apply the new method even in the cases that the Shapiro-Lopatinsky [18] conditions imply that the given problem is in general ill-posed. For example, suppose that $\pi < \arg \alpha < \frac{3\pi}{2}$. Then the Shapiro-Lopatinsky conditions imply that $q(x,t)$ blows up exponentially as $t \to \infty$. However, the new method implies that even in this case there exists a solution $q(x,t) \to 0$ as $t \to \infty$, provided that the initial data $q_1(x)$ and the boundary data $q_2(t)$ satisfy the global constraint $\hat{q}_1(-\alpha) = i\hat{q}_2(\alpha)$.

It should be noted that the particular transforms obtained by separation of variables, can be rederived using the Lax pair formulation. For example the spectral analysis of equation (1.9a) can be used to derive the sine or the cosine transforms, while equation (1.9b) determines the evolution of the relevant transform data. Similarly, the traditional approach of using a t-transform to solve (1.8) can be rederived as follows. Equation (1.8) also possesses the Lax pair

$$(1.21a) \qquad \mu_t + ik\mu = q,$$

$(1.21b)$ $\mu_{xx} + k\mu = -iq.$

The spectral analysis of equation (1.21a) yields the Laplace transform in t, while equation (1.21b) determines the x-dependence of the Laplace data. These approaches should be contrasted with the new approach which involves performing a simultaneous spectral analysis of *both* equations (1.9), or *both* equations (1.21). The spectral analysis of the former equations is slightly simpler than the spectral analysis of the latter equations. In general, our experience indicates that for the implementation of the spectral analysis, the most convenient Lax pair is the one involving only first order derivatives of μ.

A great advantage of the new method is that it nonlinearizes. Indeed, the conceptual steps for solving an IBV problem for a given integrable PDE are the same with those for solving its linear version. It is the author's opinion that, the reason why we had failed for a long time to make essential progress in the solution of general IBV problems for nonlinear equations, is because we were trying to nonlinearize the wrong transforms. For example, the IBV problem of the nonlinear Schrödinger on the quarter plane should be studied not by nonlinearizing the sine transform or the Laplace transform, but by nonlinearizing equation (1.17).

Organization of the Paper

Section 2 contains details of the spectral analysis of the fundamental equations $\mu_x + ik\mu = q$ and $\mu_x + i\mu_y + ik\mu = q$. For simplicity, it is assumed that q is a Schwartz function. It is straightforward to replace this class of functions by a more restrictive class. In Section 3 the Lax pair formulation of linear PDE's in two variables is discussed. The Lax pair of the Laplace equation and of the linearized version of the celebrated Ernst equation are given as illustrative examples. Section 4 illustrates how the simultaneous spectral analysis of the Lax pair can be used to solve IBV problems for linear equations. The general method is illustrated by using the heat equation with $t \in [0, \infty)$, $x \in [0, l]$, $l > 0$. It is remarkable that the general solution involves only integrals (see Theorem 4.1), i.e. *there is no discrete spectrum* (the usual terms involving summation over sinuses will appear if one deforms the relevant contours along the real axis). The solution of certain IBV problems on the quarter plane for the equation $q_t + q_{xxx} = 0$ is given in [19]; other examples can be found in [15] and [16]. In Section 5 it is shown how to extend the new method to inhomogeneous linear equations. The equation $iq_t + q_{xx} = f(x, t)$ is used as an illustrative example. Section 6 discusses IBV problems for the nonlinear Schrödinger equation and for the equations of the transient stimulated Raman scattering [20]. The solution of the IBV problem for the Ernst equation is given in [21].

Further Discussion

It is the author's expectation that the new method developed in [15], [16] and discussed above, will be useful for studying a great number of open problems. Some of these problems are the following:

1. Progress has already been made for the solution of boundary value problems with changing type boundary conditions [19], as well as for IBV problems with moving boundaries. The latter problem yields a beautiful extension of the Ehrenpreis principle: the contours appearing in the integral representation of the solution are not fixed but they move with time. (I expect that for more complicated domains

the contours will depend on both x and t). Obviously, there exists a great number of similar problems that should be investigated.

2. It is possible to apply this method to any PDE with variable coefficients provided that this PDE is separable. Indeed, if such a PDE can be solved by some transform, then there exists an equation analogous to $\mu_x + ik\mu = q$ or $\mu_x + i\mu_y + ik\mu = q$, whose spectral analysis gives rise to this transform. Thus the given PDE will have a Lax pair formulation. The Lax pair of every such physical linear equation should be constructed, and proper deformations of these Lax pairs should be investigated. I conjecture that some of the nonlinear integrable equations constructed in this way will be of physical significance.

3. The application of the new method to IBV problems in 3-dimensions for both linear and integrable nonlinear PDE's will be presented elsewhere. The extension of the method to difference equations is straightforward [22].

4. The discussion in "what is integrability" suggests that in order to extend the notion of integrability to $3+1$ we must find the simplest equation whose spectral analysis gives rise to the 3-dimensional Fourier transform (i.e. the proper 3-dimensional generalization of the fundamental equation $\mu_x + i\mu_y + ik\mu = q$). This problem is under investigation.

5. The implications of this method for the development of new numerical spectral methods should be investigated.

2. The Spectral Analysis of the Fundamental Equations

The spectral analysis of a given linear equation means: *Define a solution $\mu(x, k)$, or $\mu(x, y, k)$, which is bounded for all complex values of k, and then compute $\frac{\partial \mu}{\partial k}$.* The general procedure for implementing this analysis will be illustrated using the two fundamental equations (1.9a) and (1.11a) [11].

2.1 The Spectral Analysis of $\mu_x + ik\mu = q$
Let the scalar function $\mu(x, k)$ satisfy

$$(2.1) \qquad \mu_x + ik\mu = q, \qquad k \in \mathbb{C},$$

where $q(x) \in S(\mathbb{R})$ and $x \in \mathbb{R}$. We will define a solution of this equation which is bounded for all $k \in \mathbb{C}$. This solution is defined by

$$(2.2a) \qquad \mu(x, k) = \begin{cases} \mu^+(x, k), & k_I \geq 0 \\ \mu^-(x, k), & k_I \leq 0, \end{cases}$$

where $k = k_R + ik_I$ and μ^+, μ^- are the following particular solutions of equation (2.1)

$$(2.2b) \qquad \mu^+(x, k) = -\int_x^\infty e^{-ik(x-\xi)} q(\xi) d\xi, k_I \geq 0,$$

$$(2.2c) \qquad \mu^-(x, k) = \int_{-\infty}^x e^{-ik(x-\xi)} q(\xi) d\xi, k_I \leq 0.$$

The superscript $+(-)$ indicates that $\mu^+(\mu^-)$ is addition to being bounded for $k \in \mathbb{C}^+(\mathbb{C}^-)$, it is also holomorphic for $k \in \mathbb{C}^+(\mathbb{C}^-)$. In the region of their overlap (i.e.

for $k \in \mathbb{R}$) μ^+ and μ^- are simply related. Indeed, if Δ is the difference of any two solutions of equation (2.1) then $\Delta = e^{-ikx}\rho(k)$. Actually subtracting (2.2b) and (2.2c) we find

$$(2.3a) \qquad \mu^+(x,k) - \mu^-(x,k) = -e^{-ikx}\hat{q}(k), k \in \mathbb{R},$$

where the function $\hat{q}(k)$ is defined by

$$(2.4a) \qquad \hat{q}(k) = \int_{-\infty}^{\infty} e^{ikx}q(x)dx.$$

Equations (2.2) and integration by parts imply

$$(2.3b) \qquad \mu(x,k) = \frac{-iq(x)}{k} + O\left(\frac{1}{k^2}\right), k \to \infty.$$

Equations (2.3) define an elementary Riemann-Hilbert (RH) problem [9]. Its unique solution is given by

$$(2.5) \qquad \mu(x,k) = -\frac{1}{2i\pi} \int_{-\infty}^{\infty} \frac{e^{-ilx}\hat{q}(l)dl}{l-k}, \qquad k \in \mathbb{C}.$$

Comparing the large k behavior of equation (2.5) with equation (2.3b), it follows that

$$(2.4b) \qquad q(x) = \frac{1}{2\pi} \int_{-\infty}^{\infty} e^{ikx}\hat{q}(k)dk.$$

Equations (2.4) are the classical formulae of the direct and inverse 1-dimensional Fourier transform.

2.2 The Spectral Analysis of $\mu_x + i\mu_y + 2ik\mu = 2q$.

Let the scalar function $\mu(x,y,k)$ satisfy

$$(2.6) \qquad \mu_x + i\mu_y + 2ik\mu = 2q,$$

where $q(x,y) \in S(\mathbb{R}^+)$ and $k \in \mathbb{C}$. We will define a solution of this equation which is bounded for all $k \in \mathbb{C}$. This solution is given by

$$(2.7a) \qquad \mu(x,y,k,\bar{k}) = \frac{1}{\pi} \int_{\mathbb{R}^2} \frac{e^{-ik(\bar{z}-\bar{\zeta})-i\bar{k}(z-\zeta)}q(\xi,\eta)d\xi d\eta}{z-\zeta},$$

where

$$(2.7b) \qquad z = x + iy, \qquad \zeta = \xi + i\eta.$$

The derivation of equation (2.7) is based on the inversion of the $\partial_{\bar{z}}$ operator. Indeed, using $z = x + iy$, equation (2.6) becomes

$$\mu_{\bar{z}} + ik\mu = q,$$

or

$$(2.8) \qquad \frac{\partial}{\partial \bar{z}} \left(\mu e^{ik\bar{z}+i\bar{k}z} \right) = q e^{ik\bar{z}+i\bar{k}z},$$

where the factor $e^{i\bar{k}z}$ was used in order to ensure that the exponential is bounded. Inverting the $\partial_{\bar{z}}$ operator [9], equation (2.8) yields equation (2.7). We emphasize that μ is holomorphic nowhere in the complex k-plane. Computing $\partial \mu / \partial \bar{k}$ it follows that

$$(2.9a) \qquad \frac{\partial}{\partial \bar{k}} \mu(x,y,k,\bar{k}) = \frac{-i}{\pi} e^{-ik\bar{z}-i\bar{k}z} \hat{q}(k_R, k_I),$$

where the function $\hat{q}$ is defined by

$$(2.10a) \qquad \hat{q}(k_R, k_I) = \int_{\mathbb{R}^2} e^{ik\bar{z}+i\bar{k}z} q(x,y) dx dy.$$

Equation (2.7) implies that

$$(2.9b) \qquad \mu(x,y,k) = \frac{-iq(x,y)}{k} + O\left(\frac{1}{k^2}\right), \; k \to \infty.$$

Equations (2.9) define an elementary DBAR problem. Its unique solution is given by

$$(2.11) \qquad \mu(x,y,k,\bar{k}) = -\frac{i}{\pi^2} \int_{\mathbb{R}^2} \frac{e^{-il\bar{z}-i\bar{l}z} \hat{q}(l_R, l_I)}{k-l} dl_R dl_I, \quad l = l_R + il_I.$$

Comparing the large k behavior of equation (2.11) with equation (2.9b), it follows that

$$(2.10b) \qquad q(x,y) = \frac{1}{\pi^2} \int_{\mathbb{R}^2} e^{-ik\bar{z}-i\bar{k}z} \hat{q}(k_R, k_I) dk_R dk_I.$$

Equations (2.10) are the classical formulae of the direct and inverse 2-dimensional Fourier transform.

3. Lax Pairs for Linear PDE's in 2 Variables

Proposition 3.1. Let $q(x,y)$ satisfy the linear PDE with x-independent coefficients

$$(3.1) \qquad M(\partial_x, \partial_y)q = 0,$$

where $M(\partial_x, \partial_y)$ is a linear operator of ∂_x and ∂_y with x-independent coefficients. This equation possesses the Lax pair

$$(3.2a) \qquad \mu_x + ik\mu = q, \qquad k \in \mathbb{C},$$

$$(3.2b) \qquad M(\partial_x, \partial_y)\mu = 0,$$

where $\mu(x,y,k)$ is a scalar function.

Proof Applying the operator M on equation (3.2a) and using the fact that M and $\partial_x + ik$ commute, it follows that the compatibility condition of equations (3.2) implies equation (3.1).   **QED**

The Lax pair (3.2) can be rewritten in a form which is more convenient for the implementation of its spectral analysis. We first discuss evolution equations.

Proposition 3.2. Let $q(x,t)$ satisfy the evolution equation

$$\left(\partial_t + \sum_0^N \alpha_j \partial_x^j\right) q = 0, \qquad N \in \mathbb{Z}^+,$$

where α_j, $0 \le j \le N$, are arbitrary functions of t. This equation possesses the following Lax pair

(3.3) $$\mu_x + ik\mu = q, \qquad k \in \mathbb{C},$$

$$\mu_t + \sum_0^N \alpha_j(-ik)^j \mu$$

(3.4) $$= -\sum_{j=1}^N \alpha_j \left(\partial_x^{j-1} + (-ik)\partial_x^{j-2} + (-ik)^2\partial_x^{j-3} + \cdots + (-ik)^{j-1}\right) q.$$

Proof. Equation (3.2b) takes the form

(3.5) $$\mu_t + \sum_0^N \alpha_j \partial_x^j \mu = 0.$$

Using equation (3.2a) to eliminate $\partial_x^j \mu$, $j = 1, ..., N$, this equation becomes equation (3.4).

Remark 3.1. Consider the case of linear PDE's with constant coefficients. The expression

$$e^{-ikx - w(k)y}$$

is a solution of equation (3.1) iff

(3.6) $$M(-ik, -w(k)) = 0.$$

In the particular case of evolution equations, equation (3.6) has one root $w(k)$, and the lhs's of the equations forming its Lax pair are given by

$$\mu_x + ik\mu, \qquad \mu_t + w(k)\mu.$$

In general, if the symbol of the differential operator M can be factorized into n roots, there exist n Lax pairs; the lhs's of the equations forming the lth Lax pair are given by

$$\mu_x + ik\mu, \qquad \mu_t + w_l(k)\mu, \qquad l = 1, ..., n.$$

Examples The symbol of the Laplace operator is $w^2 = k^2$. Laplace's equation can be written as

(3.7)
$$(\partial_x^2 + k^2)q = -(\partial_y^2 - w^2)q,$$

or

(3.8)
$$(\partial_x + ik)(\partial_x - ik)q = -(\partial_y - w)(\partial_y + w)q.$$

This implies that the Laplace equation

(3.9)
$$q_{xx} + q_{yy} = 0,$$

possesses the two Lax pairs

(3.10a)
$$\mu_x + ik\mu = q_y - wq,$$

(3.10b)
$$\mu_y + w\mu = -q_x + ikq,$$

where $w = \pm k$.

An alternative Lax pair is given by

$$\mu_x + ik\mu = Q, \ Q = q_x + iq_y,$$

$$\mu_y + k\mu = -iQ.$$

Remark 3.2. It is also possible to construct Lax pairs for linear PDE's with nonconstant coefficients. For example, the equation

$$(x - y)q_{xy} + \beta q_y - \alpha q_x = 0, \qquad \alpha, \beta \ \text{constants}$$

possesses the Lax pairs

$$\mu_x - \frac{\alpha}{x - k}\mu = \frac{q_x}{x - k},$$

$$\mu_y - \frac{\beta}{y - k}\mu = \frac{q_y}{y - k}, \qquad k \in \mathbb{C}.$$

The case of $\alpha = \beta = \frac{1}{2}$ is the linearized Ernst equation.

4. IBV Problems for Homogeneous Linear PDE's in 2 Variables

We will illustrate the general method by solving an IBV problem for the heat equation in a finite domain [22].

Theorem 4.1 Let $q(x,t)$ satisfy

(4.1)
$$q_t - q_{xx} = 0, \quad t \in [0, \infty), \ x \in [0, l], \ l > 0,$$

(4.2)
$$q(x, 0) = g(x), \quad q(0, t) = h_0(t), \quad q(l, t) = h_l(t),$$

where $g(x) \in C^2([0, l])$ and $h_0(t), h_l(t) \in S(\mathbb{R}^+)$. The unique solution of this IBV problem is given by

(4.3)
$$q(x, t) = \frac{1}{2\pi} \int_L e^{-ikx - k^2 t} \rho(k) dk,$$

where the directed contour L is depicted in Figure 2 and $\rho(k)$ is uniquely determined in terms of $g(x), h_0(t), h_l(t)$ as follows:

$$\rho(k) = \hat{g}(k), \quad k \in \mathbb{R}; \quad \rho(k) = \nu(k), \quad \arg k = \frac{5\pi}{4}, \; \frac{7\pi}{4};$$

(4.4) $$\rho(k) = e^{ikl}\lambda(k), \quad \arg k = \frac{\pi}{4}, \; \frac{3\pi}{4},$$

where

(4.5) $$\hat{g}(k) = \int_0^l e^{ikx} g(x) dx, \; 0 \le \arg k \le \pi,$$

(4.6)
$$\nu(k) = \frac{1}{1 - e^{-2ikl}} \left[\hat{g}(-k) - e^{-2ikl}\hat{g}(k) + 2ik(e^{-ikl}\hat{h}_l(k) - \hat{h}_0(k)) \right], \quad \frac{5\pi}{4} \le \arg k \le \frac{7\pi}{4},$$

(4.7)
$$\lambda(k) = \frac{1}{1 - e^{2ikl}} \left[e^{ikl}(\hat{g}(k) - \hat{g}(-k)) + 2ik(e^{ikl}\hat{h}_0(k) - \hat{h}_l(k)) \right], \quad \frac{\pi}{4} \le \arg k \le \frac{3\pi}{4},$$

and

(4.8) $$\hat{h}_0(k) = \int_0^\infty e^{k^2 t} h_0(t) dt, \quad \hat{h}_l(k) = \int_0^\infty e^{k^2 t} h_l(t) dt.$$

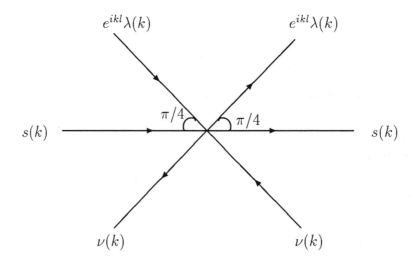

Figure 2
The contour L of Theorem 4.1

Proof   The conditions on $g(x), h_0(t), h_l(t)$ imply that the integral in equation (4.3) is well defined. Furthermore, differentiation and integration can be interchanged, which implies that $q(x,t)$ solves equation (4.1). In order to verify that $q(x,t)$ satisfies $q(x,0) = g(x)$ we note that

$$(4.9) \quad q(x,0) = \frac{1}{2\pi} \Big(\int_{-\infty}^{\infty} e^{-ikx} \hat{g}(k)dk + \int_{L_1} e^{-ikx}\nu(k)dk + \int_{L_2} e^{ik(l-x)}\lambda(k)dk \Big),$$

where the contours L_1 and L_2 are shown in Figure 3.

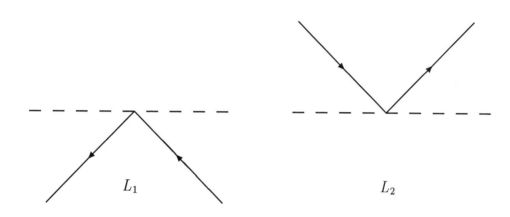

Figure 3
The contours L_1 and L_2

The integrand of the second and third integrals in equation (4.9) are analytic in $5\pi/4 \le \arg k \le 7\pi/4$ and $\pi/4 \le \arg k \le 3\pi/4$ respectively. Thus, Cauchy's theorem implies that these integrals give zero contributions and equation (4.9) reduces to $q(x,0) = g(x)$.

In order to verify that $q(x,t)$ satisfies $q(0,t) = h_0(t)$ we note that

$$q(0,t) = \frac{1}{2\pi} \int_{-\infty}^{\infty} e^{-k^2 t} \hat{g}(k)dk$$

$$+\frac{1}{2\pi} \int_{L_1} \frac{e^{-k^2 t}}{1 - e^{-2ikl}} \Big[\hat{g}(-k) - e^{-2ikl}\hat{g}(k) + 2ik(e^{-ikl}\hat{h}_l(k) - \hat{h}_0(k)) \Big] dk$$

$$(4.10) \quad +\frac{1}{2\pi} \int_{L_2} \frac{e^{-k^2 t}}{1 - e^{2ikl}} \Big[e^{2ikl}(\hat{g}(k) - \hat{g}(-k)) + 2ik(e^{2ikl}\hat{h}_0(k) - e^{ikl}\hat{h}_l(k)) \Big] dk.$$

Letting $k \to -k$ in the third integral, the above simplifies to

$$(4.11) \quad q(0,t) = \frac{1}{2\pi} \Big(\int_{-\infty}^{\infty} e^{-k^2 t}\hat{g}(k)dk + \int_{L_1} e^{-k^2 t}\hat{g}(-k)dk - \int_{L_1} 2ike^{-k^2 t}\hat{h}_0(k)dk \Big).$$

This is precisely the expression that appears in the analysis of the heat equation with $x \in [0, \infty)$. Since $\hat{g}(k)$ is analytic for $0 \leq \arg k \leq \pi$, the combination of the first two terms in equation (4.11) gives zero contribution [15], and equation (4.11) reduces to $q(0, t) = h_0(t)$. A similar analysis yields $q(l, t) = h_l(t)$.   **QED**

Theorem 4.1 is a particular case of the following more general result.

Proposition 4.1 *Let $q(x, t)$ satisfy equation (4.1) and the initial condition $q(x, 0) = g(x) \in C^2([0, l])$. Assume that there exists a solution of this problem such that $q(x, t) \to 0$ as $t \to \infty$. This solution is given by equations (4.3) and (4.4) where $\hat{g}(k)$ is defined by (4.5) and the functions $\nu(k), \lambda(k)$ are defined by*

$$(4.12) \qquad \nu(k) = \int_0^\infty e^{k^2 t} \left[q_x(0, t) - ikq(0, t) \right] dt, \qquad \frac{5\pi}{4} \leq \arg k \leq \frac{7\pi}{4},$$

$$(4.13) \qquad \lambda(k) = \int_0^\infty e^{k^2 t} \left[q_x(l, t) - ikq(l, t) \right] dt, \qquad \frac{\pi}{4} \leq \arg k \leq \frac{3\pi}{4}.$$

The boundary values of $q(x, t)$ satisfy the constraints

$$\int_0^\infty e^{k^2 t} \left[q_x(0, t) - ikq(0, t) \right] dt$$

$$(4.14) \qquad = e^{ikl} \int_0^\infty e^{k^2 t} \left[q_x(l, t) - ikq(l, t) \right] dt + \int_0^l e^{ikx} q(x, 0) dx,$$

for $\frac{\pi}{4} \leq \arg k \leq \frac{3\pi}{4}$ and

$$e^{-ikl} \int_0^\infty e^{k^2 t} \left[q_x(0, t) - ikq(0, t) \right] dt$$

$$(4.15) \qquad = \int_0^\infty e^{k^2 t} \left[q_x(l, t) - ikq(l, t) \right] dt + e^{-ikl} \int_0^l e^{ikx} q(x, 0) dx,$$

for $\frac{5\pi}{4} \leq \arg k \leq \frac{7\pi}{4}$.

Proof   It is possible to verify directly that if the boundary values of q satisfy equations (4.14) and (4.15), then the function $q(x, t)$ defined by equation (4.3) satisfies the heat equation together with the boundary values $q(x, 0)$, $q(0, t)$, $q_x(0, t)$, $q(l, t)$, $q_x(l, t)$. Rather than give this proof, we use the method introduced in [15] to derive equations (4.3), (4.4) and (4.12)–(4.15). We emphasise that this derivation assumes *a priori* that $q(x, t)$ exists. However, this assumption can be eliminated, since *a posteriori*, one can verify directly that $q(x, t)$ defined by equation (4.3) solves the given IBV problem. An alternative derivation based on Green's functions is presented in [26].

Equation (4.1) admits the Lax pair [15]

$$(4.16a) \qquad\qquad\qquad\qquad \mu_x + ik\mu = q$$

$$(4.16b) \qquad\qquad\qquad\qquad \mu_t + k^2 \mu = q_x - ikq,$$

where k is a complex-valued parameter. Indeed, it is straightforward to verify that equations (4.16) are compatible if and only if q satisfies the heat equation. Let the sectors $I, ..., VI$ of the complex k-plane be defined by $\arg k$ in the closed intervals

$$\left[0, \frac{\pi}{4} \right], \quad \left[\frac{\pi}{4}, \frac{3\pi}{4} \right], \quad \left[\frac{3\pi}{4}, \pi \right], \quad \left[\pi, \frac{5\pi}{4} \right], \quad \left[\frac{5\pi}{4}, \frac{7\pi}{4} \right], \quad \left[\frac{7\pi}{4}, 2\pi \right],$$

respectively (see Figure 4).

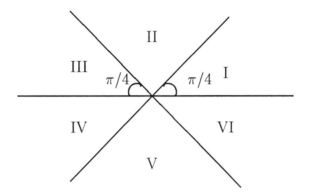

Figure 4
The definitions of the sectors I,...,VI

Let $\widetilde{q}(x,t,k)$ be defined by

(4.17) $$\widetilde{q}(x,t,k) = q_x(x,t) - ikq(x,t).$$

The general solution of equation (4.16a) is

$$\mu(x,t,k) = e^{-ikx}\mu(0,t,k) + \int_0^x e^{-ik(x-x')}q(x',t)dx'.$$

Equations (4.16a) and (4.16b) are compatible, thus the solution $\mu(x,t,k)$ defined above is also a solution of (4.16b) if and only if $\mu(0,t,k)$ solves

$$\mu_t(0,t,k) + k^2\mu(0,t,k) = \widetilde{q}(0,t,k).$$

A particular solution of this equation is

$$\mu(0,t,k) = \int_0^t e^{-k^2(t-t')}\widetilde{q}(0,t',k)dt'.$$

Thus, the function $\mu^{(46)}$ defined by

(4.18a)
$$\mu^{(46)} = e^{-ikx}\int_0^t e^{-k^2(t-t')}\widetilde{q}(0,t',k)dt' + \int_0^x e^{-ik(x-x')}q(x',t)dx', \quad k \in IV \cup VI$$

solves both equations (4.16). The superscript indicates the region where the above function is analytic and bounded. The boundedness of $\mu^{(46)}$ is a consequence of the following elementary facts. The function $e^{ik\xi}$ for $\xi \in \mathbb{R}^+$ is bounded in $I \cup II \cup III$, while the function $e^{k^2\tau}$ for $\tau \in R^+$ is bounded in $II \cup V$. Thus $\mu^{(46)}$ is bounded in the intersection of $I \cup III \cup IV \cup VI$ with the lower half plane, which is $IV \cup VI$.

Similarly, the functions $\mu^{(13)}$, $\mu^{(2)}$ and $\mu^{(5)}$ defined below also solve both equations (4.16):

$$\mu^{(13)} = e^{-ik(x-l)} \int_0^t e^{-k^2(t-t')} \widetilde{q}(l, t', k) dt'$$

(4.18b)
$$- \int_x^l e^{-ik(x-x')} q(x', t) dx', \quad k \in I \cup III$$

$$\mu^{(2)} = -e^{-ik(x-l)} \int_t^\infty e^{-k^2(t-t')} \widetilde{q}(l, t', k) dt'$$

(4.18c)
$$- \int_x^l e^{-ik(x-x')} q(x', t) dx', \quad k \in II$$

$$\mu^{(5)} = -e^{-ikx} \int_t^\infty e^{-k^2(t-t')} \widetilde{q}(0, t', k) dt'$$

(4.18d)
$$+ \int_0^x e^{-ik(x-x')} q(x', t) dx', \quad k \in V.$$

In the domain of their overlap, the above functions are related by

(4.19a)
$$\mu^{(13)} - \mu^{(46)} = -e^{-ikx-k^2 t} s(k), \quad k \in \mathbb{R}$$

(4.19b)
$$\mu^{(13)} - \mu^{(2)} = e^{-ikx-k^2 t} e^{ikl} \lambda(k), \quad \arg k = \frac{\pi}{4}, \frac{3\pi}{4},$$

(4.19c)
$$\mu^{(5)} - \mu^{(46)} = -e^{-ikx-k^2 t} \nu(k), \quad \arg k = \frac{5\pi}{4}, \frac{7\pi}{4},$$

where the functions $\hat{g}(k)$, $\nu(k)$, $\lambda(k)$ are defined by equations (4.5), (4.12)–(4.13). Indeed, let $\Delta\mu$ denote the difference of any two solutions of equations (4.16). Since these solutions satisfy equation (4.16a), it follows that $\Delta\mu = e^{-ikx} f(t, k)$. Similarly, since these solutions satisfy equation (2.16b), it follows that $\Delta\mu = e^{-k^2 t} g(x, k)$. Thus $\Delta\mu = e^{-ikx-k^2 t} \rho(k)$, and the function $\rho(k)$ can be determined by evaluating this equation at any suitable x, t. For example, evaluating the equation $\mu^{(13)} - \mu^{(46)} = e^{-ikx-k^2 t} \rho(k)$ at $x = t = 0$, it follows that $\rho(k) = -s(k)$. Similarly for equations (4.19b,c).

Integration by parts of equations (4.18) implies that if μ denotes $\mu^{(46)}$ or $\mu^{(13)}$ or $\mu^{(2)}$ or $\mu^{(5)}$, then for μ off the contour depicted in Figure 2,

(4.20)
$$\mu(x, t, k) = \frac{q(x, t)}{ik} + O\left(\frac{1}{k^2}\right), \quad k \to \infty.$$

Equations (4.19) and (4.20) define a Riemann–Hilbert (RH) problem with respect to the directed contour depicted in Figure 5. The unique solution of the RH problem is given by [9]

(4.21)
$$\mu = -\frac{1}{2\pi i} \int_L e^{-ilx-l^2 t} \rho(l) \frac{dl}{l - k}$$

(The contour L differs from the contour of figure 1 in the direction of the rays defined by $\arg k = \pi/4$, $\arg k = 3\pi/4$, since $\rho(k)$ is defined by $e^{ikl} \lambda(k)$ and not by $-e^{ikl} \lambda(k)$). Equation (4.21) together with the equation obtained by considering the large k behaviour of equation (4.21) yields equations (4.3) and (4.4).

It will now be shown that if $q(x,t) \to 0$ as $t \to \infty$ then the boundary values of $q(x,t)$ satisfy equations (4.14) and (4.15). To derive equation (4.14) we note that $\mu^{(2)}$, in addition to satisfying equation (4.18c), also satisfies the equation

$$(4.22) \qquad \mu^{(2)} = -\int_t^\infty e^{-k^2(t-t')}\widetilde{q}(x,t',k)dt', \quad k \in II.$$

This equation follows from equation (4.16b) and the assumption that $q(x,t) \to 0$ as $t \to \infty$. Equation (4.14) follows by evaluating equations (4.18c) and (4.22) at $x = 0$ and $t = 0$. Similarly, $\mu^{(5)}$, in addition to satisfying equation (4.18d), also satisfies the equation

$$(4.23) \qquad \mu^{(5)} = -\int_t^\infty e^{-k^2(t-t')}\widetilde{q}(x,t',k)dt', \quad k \in V.$$

Equation (4.15) follows by evaluating equations (4.18d) and (4.23) at $x = l$ and $t = 0$. **QED**

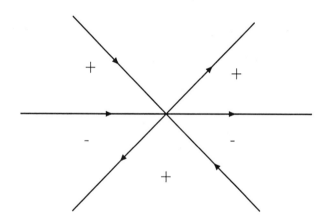

Figure 5
The directed contour L of the Riemann–Hilbert problem
for the heat equation

Remark 4.1: Equations (4.14) and (4.15) involve the five functions $q(x,0)$, $q(0,t)$, $q_x(0,t)$, $q(l,t)$, $q_x(l,t)$. Consider the following IBV problem for the heat equation: $q(x,0) = g(x)$,

$$\alpha_1 q_x(0,t) + \alpha_2 q(0,t) = h_0(t), \ \ \beta_1 q_x(l,t) + \beta_2 q(l,t) = h_l(t).$$

where $\alpha_1, \alpha_2, \beta_1, \beta_2$ are given constants and $g(x), h_0(t), h_l(t)$ are given functions. Equations (4.14), (4.15) and the above equations are four algebraic equations for the following four unknown functions:

$$\int_0^\infty e^{k^2 t} q(0,t)dt, \quad \int_0^\infty e^{k^2 t} q_x(0,t)dt, \quad \int_0^\infty e^{k^2 t} q(l,t)dt, \quad \int_0^\infty e^{k^2 t} q_x(l,t)dt.$$

The above functions uniquely specify $\hat{g}(k), \nu(k), \lambda(k)$ (see equations (4.5), (4.12)–(4.13)), and hence $q(x,t)$ (see equations (4.3) and (4.4)).

As a particular example of the general case, consider the case that $\alpha_1 = \beta_1 = 0$ and $\alpha_2 = \beta_2 = 1$. Then equation (4.14) and the equation resulting from equation (4.15) by replacing k with $-k$ yield

$$\left[1 - e^{2ikl}\right] \int_0^\infty e^{k^2 t} q_x(l,t)dt$$

$$= e^{ikl}[\hat{g}(k) - \hat{g}(-k)] + 2ike^{ikl}\hat{h}_0(k) - ik(1 + e^{2ikl})\hat{h}_l(k)$$

for $\frac{\pi}{4} \le \arg k \le \frac{3\pi}{4}$ and

$$\left[1 - e^{-2ikl}\right] \int_0^\infty e^{k^2 t} q_x(0,t)dt$$

$$= \hat{g}(k) - e^{2ikl}\hat{g}(-k) - 2ike^{ikl}\hat{h}_l(k) + ik(1 + e^{-2ikl})\hat{h}_0(k),$$

for $\frac{\pi}{4} \le \arg k \le \frac{3\pi}{4}$. Using the above equations in (4.12) and (4.13) to replace $\int_0^\infty e^{k^2 t} q_x(0,t)dt$ and $\int_0^\infty e^{k^2 t} q_x(l,t)dt$, equations (4.12) and (4.13) become equations (4.6) and (4.7).

5. IBV Problems for Inhomogeneous Linear PDE's in 2 Variables

It is straightforward to construct a Lax pair for any inhomogeneous linear PDE in 2 variables with x-independent coefficients.

Proposition 5.1 Let $q(x,y)$ satisfy the linear inhomogeneous PDE

$$(5.1) \qquad\qquad M(\partial_x, \partial_y)q = f(x,y),$$

where $M(\partial_x, \partial_y)$ is a linear operator of ∂_x and ∂_y with x-independent coefficients. This equation possesses the Lax pair

$$(5.2a) \qquad\qquad \mu_x + ik\mu = q, \qquad k \in \mathbb{C},$$

$$(5.2b) \qquad\qquad M(\partial_x, \partial_y)\mu = g(x,y,k)$$

where $g(x,y,k)$ is any solution of

$$(5.2c) \qquad\qquad g_x + ikg = f.$$

Proof Applying the operator M on equation (5.2a) and using the fact that M and $\partial_x + ik$ commute it follows that

$$(\partial_x + ik)g = Mq,$$

which is equation (5.1).   **QED**

Example A Lax pair of $iq_t + q_{xx} = f$ is given by $\mu_x + ik\mu = q$ together with

(5.3) $$\mu_t + ik^2\mu = iq_x + kq - ig.$$

Indeed, using (5.2a) to eliminate μ_{xx}, the equation $i\mu_t + \mu_{xx} = g$ becomes equation (5.3).

We will use this example to illustrate how to solve IBV problems for inhomogeneous equations.

Theorem 5.1 *Let $q(x,t) \in \mathbb{C}$ satisfy the equation*

(5.4) $$iq_t + q_{xx} = f, \qquad x,t \in \mathbb{R}^+,$$

where $f(x,t)$ is a given Schwartz function. Let q satisfy the initial condition

(5.5) $$q(x,0) = q_1(x),$$

and any one of the following boundary conditions
  (a)

(5.6a) $$q(0,t) = q_2(t),$$

  (b)

(5.6b) $$q_x(0,t) = q_2(t),$$

  (c)

(5.6c) $$q_x(0,t) + i\alpha q(0,t) = q_2(t),$$

where $q_1(x)$, $q_2(t) \in S(\mathbb{R}^+)$, and α is constant with $\arg \alpha \neq \frac{3\pi}{2}$.

The IBV problem satisfying (a), or (b), or (c) with $\arg \alpha$ outside the interval $[\pi, \frac{3\pi}{2}]$ has a unique solution such that $q(x,t) \to 0$ as $x \to \infty$. The IBV problem satisfying (c) with $\pi \leq \arg \alpha < \frac{3\pi}{2}$ has a unique solution such that $q(x,t) \to 0$ as $x \to \infty$ provided that $q_1(x)$, $q_2(t)$ and $f(x,t)$ satisfy the global constraint

(5.7) $$\int_0^\infty e^{-i\alpha x} q_1(x)dx = i \int_0^\infty \int_0^\infty e^{i\alpha^2 t - i\alpha x} f(x,t)dxdt + i \int_0^\infty e^{i\alpha^2 t} q_2(t)dt.$$

The solution of any of the above problems is given by

$$q(x,t) = \frac{1}{2\pi} \int_{-\infty}^\infty e^{-ikx - ik^2 t} \left[\hat{q}_1(k) - i \int_0^t \int_0^\infty f(x,\tau)e^{ik^2\tau + ikx}dxd\tau \right] dk +$$

(5.8) $$\frac{1}{2\pi} \int_{\hat{L}} e^{-ikx - ik^2 t} \nu(k)dk,$$

where $\hat{L}$ consists of the negative real axis and of the negative imaginary axis (see Figure 1) and the functions $\hat{q}_1(k)$ and $\nu(k)$ are defined as follows:

(5.9) $$\hat{q}_1(k) = \int_0^\infty e^{ikx} q_1(x)dx, 0 \leq \arg k \leq \pi;$$

$\nu(k)$ for (a), or (b), or (c) is given by

(5.10a) $$\nu(k) = 2k\hat{q}_2(k) + \hat{q}_1(-k) - i\hat{f}(k),$$

(5.10b) $\nu(k) = 2i\hat{q}_2(k) - \hat{q}_1(-k) + i\hat{f}(k)$

(5.10c) $\nu(k) = \dfrac{\alpha + k}{\alpha - k}[\hat{q}_1(-k) - i\hat{f}(k)] - \dfrac{2ik\hat{q}_2(k)}{\alpha - k},$

where $\pi \le \arg k \le \frac{3\pi}{2}$, and

(5.11) $\hat{q}_2(k) = \displaystyle\int_0^\infty e^{ik^2 t} q_2(t)dt, \ \hat{f}(k) = \int_0^\infty \int_0^\infty e^{ik^2 t - ikx} f(x,t)dxdt.$

If $\arg \alpha = \frac{3\pi}{2}$ the solution $q(x,t)$ in addition to the term given by equation (5.8), it also contains the term $2i\alpha\sigma \exp[-i\alpha x - i\alpha^2 t]$, σ constant. This term decays to zero as $x \to \infty$ but it oscillates in t. If $\pi \le \arg \alpha < \frac{3\pi}{2}$ and the global constraint is not satisfied then the solution blows up exponentially (like $\exp[-i\alpha x - i\alpha^2 t]$) as $t \to \infty$.

The proof of this theorem is similar to the proof of the corresponding theorem for the homogeneous case (see [15]). The above result is a particular case of a more general result:

Proposition 5.2. *Let $q(x,t) \in \mathbb{C}$ satisfy the equation (5.4) and the initial condition $q(x,0) = q_1(x) \in S(\mathbb{R}^+)$. Assume that there exists a solution of this problem, such that $q(x,t) \to 0$ as $x \to \infty$ and $q(x,t) \to 0$ as $t \to \infty$. This solution is given by equation (5.8) where $\hat{q}_1(k)$ is the Fourier transform of the initial data (see equation (5.9)) and $\nu(k)$ is defined by*

(5.12) $\nu(k) = \displaystyle\int_0^\infty e^{ik^2 t}(iq_x(0,t) + kq(0,t))dt, \pi \le \arg k \le \dfrac{3\pi}{2}.$

The boundary values of $q(x,t)$ and the function $f(x,t)$ satisfy the global constraint

$$\int_0^\infty e^{-ikx} q(x,0)dx =$$

(5.13) $\displaystyle\int_0^\infty e^{ik^2 t}\left[iq_x(0,t) - kq(0,t) + i \int_0^\infty e^{-ikx} f(x,t)dx \right] dt, \ \ \pi \le \arg k \le \dfrac{3\pi}{2}.$

We first note that theorem 5.1 is a consequence of the above proposition. Indeed suppose that either $q_x(0,t)$ or $q(0,t)$ are given. Then adding or subtracting (5.12) and (5.13), we find equations (5.10a) and (5.10b). Similarly, if $q_x(0,t) + i\alpha q(0,t)$ is given, multiplying equation (5.12) by $i(\alpha - k)/2k$, equation (5.13) by $-i(\alpha + k)/2k$ and adding the resulting equations, we find equation (5.10c). The constraint (5.7) follows from the evaluation of equation (5.13) at $k = \alpha$.

Proof. It is possible to verify directly that if the boundary values of q satisfy the constraint (5.13), then $q(x,t)$ defined by (5.8) solves equation (5.4) and the boundary values of this $q(x,t)$ are precisely the functions $q(0,t)$, $q_x(0,t)$, $q(x,0)$. Instead of giving this proof we will show how our method can be used to *derive* the relevant formulae. This derivation assumes a priori that $q(x,t)$ exists. However, this assumption can be eliminated, since a posteriori it can be verified directly that this $q(x,t)$ solves the given IBV problem.

Let

(5.14) $$\tilde{q}(x,t,k) = iq_x(x,t) + kq(x,t).$$

The general solution of equation (5.2a) is

$$\mu(x,t,k) = e^{-ikx}\mu(0,t,k) + \int_0^x e^{-ik(x-x')}q(x',t)dx'.$$

Equations (5.2a) and (5.3) are compatible, thus the solution $\mu(x,t,k)$ defined above also solves equation (5.3) iff $\mu(0,t,k)$ solves

$$\mu_t(0,t,k) + ik^2\mu(0,t,k) = \tilde{q}(0,t,k) - ig(0,t,k).$$

The function $g(x,t,k)$ solves equation (5.2c). We choose the following particular solutions of this equation:

(5.15) $$g^+(x,t,k) = -\int_x^\infty e^{-ik(x-x')}f(x',t)dx', \ \ k_I \geq 0,$$

(5.16) $$g^-(x,t,k) = \int_0^x e^{-ik(x-x')}f(x',t)dx', \ \ k_I \leq 0.$$

A particular solution $\mu(0,t,k)$ for $k_I \leq 0$ is

$$\mu(0,t,k) = \int_0^t e^{-ik^2(t-t')}\tilde{q}(0,t',k)dt'.$$

Thus, the function $\mu^{(4)}(x,t,k)$ defined by

(5.17) $$\mu^{(4)}(x,t,k) = e^{-ikx}\int_0^t e^{-ik^2(t-t')}\tilde{q}(0,t',k)dt' + \int_0^x e^{-ik(x-x')}q(x',t)dx'$$

solves both equations (5.2a) and (5.3). Similarly, the functions $\mu^{(3)}(x,t,k)$ and $\mu^{(12)}(x,t,k)$ defined by

(5.18) $$\mu^{(3)}(x,t,k) = -e^{-ikx}\int_t^\infty e^{-ik^2(t-t')}\tilde{q}(0,t',k)dt' + \int_0^x e^{-ik(x-x')}q(x',t)dx',$$

and by

(5.19) $$\mu^{(12)}(x,t,k) = -\int_x^\infty e^{-ik(x-x')}q(x',t)dx',$$

also solve both equations (5.2a) and (5.3).

The superscripts indicate that the functions $\mu^{(4)}$, $\mu^{(3)}$, $\mu^{(12)}$ are analytic and bounded in the fourth quadrant, the third quadrant, and the upper half of the complex k-plane. The boundness of these functions follows from the following elementary facts. The function $e^{ik\xi}$ for $\xi \in \mathbb{R}^+$ is bounded in the upper half of the complex k-plane, while the function $e^{ik^2\tau}$ for $\tau \in \mathbb{R}^+$ is bounded in the first and the third quadrants of the complex k-plane.

Integration by parts implies that if μ denotes $\mu^{(4)}$, or $\mu^{(3)}$, or $\mu^{(12)}$ then

$$(5.20) \qquad \mu(x,t,k) = \frac{q(x,t)}{ik} + O\left(\frac{1}{k^2}\right), \qquad k \to \infty, \ \ \arg \neq 0, \pi, \frac{3\pi}{2}.$$

In the domain of their overlap, the functions $\mu^{(4)}$, $\mu^{(3)}$, $\mu^{(12)}$ are simply related:

$(5.21a)$
$$\mu^{(12)}(x,t,k) - \mu^{(4)}(x,t,k) = -e^{-ikx-ik^2t}\left[s(k) - \hat{F}(t,k)\right], \qquad k \in \mathbb{R}^+,$$

$(5.21b)$
$$\mu^{(3)}(x,t,k) - \mu^{(12)}(x,t,k) = -e^{-ikx-ik^2t}[-s(k) + \nu(k) + \hat{F}(t,k)], \qquad k \in \mathbb{R}^-,$$

$$(5.21c) \qquad \mu^{(3)}(x,t,k) - \mu^{(4)}(x,t,k) = -e^{-ikx-ik^2t}\nu(k), \qquad k \in i\mathbb{R}^-,$$

where

$$(5.22) \qquad s(k) = \int_0^\infty e^{ikx}q(x,0)dx, \qquad 0 \leq argk \leq \pi,$$

$$(5.23) \qquad \nu(k) = \int_0^\infty e^{ik^2t}(iq_x(0,t) + kq(0,t))dt, \qquad \pi \leq argk \leq \frac{3\pi}{2},$$

$$(5.24) \qquad \hat{F}(t,k) = i\int_0^t \int_0^\infty e^{ikx+ik^2\tau}f(x,\tau)dxd\tau, k \in \mathbb{R}.$$

In order to derive equations (5.21) we note that if $\Delta = \mu^{(3)} - \mu^{(4)}$, then $\Delta_t + ik^2\Delta = 0$. Also if $\Delta = \mu^{(12)} - \mu^{(4)}$, then $\Delta_t + ik^2\Delta = ie^{-ikx}\int_0^\infty e^{ikx'}f(x',t)dx'$. These equations together with $\Delta_x + ik\Delta = 0$ imply equations (5.21).

Equations (5.20)-(5.21) define a Riemann-Hilbert problem.

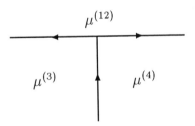

Figure 6.
The Riemann-Hilbert problem for the equation
$iq_t + q_{xx} = f(x,t)$ **in the quarter plane.**

The unique solution of this Riemann-Hilbert problem is given by

$$(5.25) \qquad \mu = -\frac{1}{2\pi i}\int_L \frac{e^{-ilx-il^2t}\rho(l,t)}{l-k}dl,$$

where μ is $\mu^{(12)}$ if $k \in I \cup II$, μ is $\mu^{(3)}$ if $k \in III$, μ is $\mu^{(4)}$ if $k \in IV$, the directed contour L is depicted in Fig. 7, and $\rho(k,t) = \nu(k) - s(k) + \hat{F}(t,k)$ for $k \in \mathbb{R}^-$, $\rho(k,t) = s(k) - \hat{F}(t,k)$ for $k \in \mathbb{R}^+$, $\rho(k,t) = \nu(k)$ for $k \in i\mathbb{R}^-$.

Equation (5.20) and the large k behavior of equation (5.25) yield equation (5.8). We now derive equation (5.13). This equation is a consequence of the fact that each solution of equations (5.2a) and (5.3) possesses two different representations. For example, $\mu^{(12)}$ in addition to equation (5.19) also satisfies

$$\mu^{(12)}(x,t,k) = e^{-ik^2t}\mu^{(12)}(x,0,k)$$

$$+ \int_0^t e^{-ik^2(t-t')} \left[\tilde{q}(x,t',k) - ig^+(x,t',k)\right] dt',$$

or

$$\mu^{(12)}(x,t,k) = -e^{-ik^2t} \int_x^\infty e^{-ik(x-x')}q(x',0)dx'$$

(5.26) $$+ \int_0^t e^{-ik^2(t-t')} \left[\tilde{q}(x,t',k) - ig^+(x,t',k)\right] dt',$$

where equation (5.19) has been used to evaluate $\mu^{(12)}(x,0,k)$. The exponentials appearing in (5.26) are bounded in II. But equation (5.19) implies that $\mu^{(12)}$ is well defined in both I and II. Rewriting the $\int_0^t$ in (5.26) as $\int_0^\infty - \int_t^\infty$, it follows that $\mu^{(12)}$ is analytic and bounded in $(I \cup II)$ iff

(5.27) $$\int_0^\infty e^{ik^2t} \left[\tilde{q}(x,t,k) - ig^+(x,t,k)\right] dt - \int_x^\infty e^{-ik(x-x')}q(x',0)dx' = 0, \quad k \in I.$$

Letting $x = 0$ in this equation and replacing k by $-k$ we find equation (5.13).

6. IBV Problems for Nonlinear Integrable PDE's in 2 Variables

The main advantage of our method is that *it yields Riemann-Hilbert problems whose x and t dependence is explicit. Furthermore, this dependence can be deduced from the dispersion relation of the linearized version of the given nonlinear equation.* For example for the nonlinear Schrödinger equation in the quarter plane this method yields the Riemann-Hilbert problem

(6.1a) $$\mu^+(x,t,k) = \mu^-(x,t,k)e^{(-ikx-2ik^2t)\sigma_3}S(k)e^{(ikx+2ik^2t)\sigma_3}, \qquad k \in \mathbb{R} \cup i\mathbb{R},$$

(6.1b) $$\mu = I + O\left(\frac{1}{k}\right), \quad k \to \infty, \quad \arg k \notin \mathbb{R} \cup i\mathbb{R},$$

where the 2×2 matrix valued function μ is sectionally holomorphic in k, $\sigma_3 = diag(1,-1)$ and the matrix $S(k)$ which contains the spectral data is a 2×2 matrix defined for k real and for k purely imaginary.

Recall that the RH problem associated with the solution of $iq_t + q_{xx} = 0$ in the quarter plane is formulated on $k \in \mathbb{R} \cup i\mathbb{R}^-$. Since the Lax pair of the NLS equation is constructed from the combination of the Lax pairs of $iq_t + q_{xx} = 0$ and of $-i\bar{q}_t + \bar{q}_{xx} = 0$, it is not surprising that the RH problem of NLS is formulated on $k \in \mathbb{R} \cup i\mathbb{R}$.

This type of formulation can be thought of as providing the *extension of the Ehrenpreis principle to nonlinear integrable equations*. For certain nonlinear equations, such as the N-wave interactions, the sine-Gordon equation and the equations for the stimulated Raman scattering, $S(k)$ can be explicitly computed in terms of the given initial and boundary data. For other equations, such as the NLS equation and the Korteweg-deVries equation, one needs to solve an additional inverse problem in order to compute some of the elements of $S(k)$. Unfortunately, this inverse problem appears to be nonlinear.

We note that for initial value problems the inverse spectral method can be used to show that the solution exists and is unique, and that for certain problems it decomposes into a number of solitons for large time and for $x/t = O(1)$. Although existence and uniqueness can be obtained by the usual PDE techniques, these techniques cannot be used to obtain the asymptotic behavior of the solution. For IBV problems, establishing existence by PDE techniques is much harder. Using the method presented here it is straightforward to establish existence and uniqueness of the problems for which $S(k)$ can be computed explicitly, and to derive the large time behavior of the solution for all IBV problems. In particular, since the associated Riemann-Hilbert problems have explicit x and t dependence, it is possible to apply the rigorous technique of [23] for the investigation of the large t asymptotics.

The IBV problem of the nonlinear Schrödinger equation in the quarter plane as well as similar problems for the Korteweg-deVries and for the sine-Gordon equation in laboratory coordinates have been investigated in [24]. In these papers it was realized that in addition to the spectral analysis of the x-part of the Lax pair, one also needs to perform a spectral analysis of the t-part. This novel idea led to the determination of the explicit time dependence of the spectral data. It turns out that the *simultaneous* analysis of the two parts of the Lax pair proposed here, simplifies substantially the derivation of the relevant results. Furthermore, it can be used to obtain an inverse problem for the "missing" spectral data. This is based on the investigation of the nonlinear analogue of equation (5.13).

An example of the IBV problem which can be solved as effectively as the usual Cauchy problems is given below [20].

Theorem 6.1. *Let* $b(\chi, \tau) \in \mathbb{R}$, $Y(\chi, \tau) \in \mathbb{C}$, $X(\chi, \tau) \in \mathbb{C}$ *satisfy the equations*

$$(6.2) \qquad \frac{\partial b}{\partial \chi} = i(\bar{X}Y - X\bar{Y}), \ \frac{\partial Y}{\partial \chi} = 2ibX, \ \frac{\partial X}{\partial \tau} = -\frac{i}{2}Y, \ \chi \in [0, l], \ \tau \in [0, 1],$$

where $l > 0$. *Let*

$$(6.3a) \qquad b(0, \tau) = b_0(\tau), \qquad Y(0, \tau) = Y_0(\tau), \qquad X(\chi, 0) = X_0(\chi),$$

where $b_0(\tau)$, $Y_0(\tau)$ *are differentiable for* $\tau \in [0, 1]$, *and* $X_0(\chi)$ *is differentiable for* $\chi \in [0, l]$. *Assume that*

$$(6.3b) \qquad\qquad\qquad b_0(\tau)^2 + |Y_0(\tau)|^2 = 1.$$

The unique solution of this IBV problem is given by

$$X(\chi,\tau) = 2i \lim_{k\to\infty} (k\Psi_1^+(\chi,\tau,k)),$$

(6.4) $$b(\chi,\tau) = -1 - 4i\frac{\partial}{\partial\tau} \lim_{k\to\infty} \left(k\overline{\Psi_2^+(\chi,\tau,\bar{k})}\right), k \in \mathbb{C}, k_I \neq 0,$$

where the scalar functions $\Psi_1^+(\chi,\tau,k)$ and $\Psi_2^+(\chi,\tau,k)$, $k \in \mathbb{C}$, can be obtained by solving the following Riemann-Hilbert problem,

$$\begin{pmatrix} \Psi_1^+(\chi,\tau,k) & \frac{\Phi_1^+(\chi,\tau,k)}{\rho_1(k)} \\[2mm] \Psi_2^+(\chi,\tau,k) & \frac{\Phi_2^+(\chi,\tau,k)}{\rho_1(k)} \end{pmatrix} =$$

(6.5a)

$$= \begin{pmatrix} -\frac{\overline{\Phi_2^+(\chi,\tau,k)}}{\overline{\rho_1(k)}} & \overline{\Psi_2^+(\chi,\tau,k)} \\[2mm] \frac{\overline{\Phi_1^+(\chi,\tau,k)}}{\overline{\rho_1(k)}} & -\overline{\Psi_1^+(\chi,\tau,k)} \end{pmatrix} \begin{pmatrix} 1 & \frac{\rho_2(k)}{\rho_1(k)}e^{2ik\chi+\frac{i\tau}{2k}} \\[2mm] -\frac{\overline{\rho_2(k)}}{\overline{\rho_1(k)}}e^{-2ik\chi-\frac{i\tau}{2k}} & \frac{1}{|\rho(k)|^2} \end{pmatrix}, k \in \mathbb{R},$$

$$\Phi_1^+(\chi,\tau,k) = 1 + O(\frac{1}{k}), \Phi_2^+(\chi,\tau,k) = O(\frac{1}{k}),$$

(6.5b) $$\Psi_1^+(\chi,\tau,k) = O(\frac{1}{k}), \ \Psi_2^+(\chi,\tau,k) = 1 + O(\frac{1}{k}), \ k \to \infty, \ k_I \neq 0.$$

This Riemann-Hilbert problem, which is specified through the scalar functions $\rho_1(k)$ and $\rho_2(k)$, $k \in \mathbb{R}$, has a unique solution. The functions $\rho_1(k)$ and $\rho_2(k)$ are constructed as follows: Let $(\mu_1(\tau,k),\mu_2(\tau,k))^T$ be the unique solution of

(6.6a) $$\frac{\partial}{\partial\tau}\begin{pmatrix} \mu_1(\tau,k) \\ \mu_2(\tau,k) \end{pmatrix} = \frac{1}{4k}\begin{pmatrix} ib_0(\tau) & -Y_0(\tau) \\ \bar{Y}_0(\tau) & -ib_0(\tau) \end{pmatrix}\begin{pmatrix} \mu_1(\tau,k) \\ \mu_2(\tau,k) \end{pmatrix},$$

(6.6b) $$\mu_1(1,k) = 1, \qquad \mu_2(1,k) = 0.$$

Let $(\nu_1(\chi,k),\nu_2(\chi,k))^T$ be the unique solution of

(6.7a) $$\frac{\partial}{\partial\chi}\begin{pmatrix} \nu_1(\chi,k) \\ \nu_2(\chi,k) \end{pmatrix} = \begin{pmatrix} -ik & X_0(\chi) \\ -\bar{X}_0(\chi) & ik \end{pmatrix}\begin{pmatrix} \nu_1(\chi,k) \\ \nu_2(\chi,k) \end{pmatrix},$$

(6.7b) $$\nu_1(0,k) = \mu_1(0,k)e^{-\frac{i}{4k}}, \qquad \nu_2(0,k) = \mu_2(0,k)e^{-\frac{i}{4k}}.$$

The functions $\rho_1(k)$ and $\rho_2(k)$ are defined by

(6.8) $$\rho_1(k) = \nu_1(l,k)e^{ikl}, \qquad \rho_2(k) = \nu_2(l,k)e^{-ikl}.$$

If $\rho_1(k) \neq 0$ for $\text{Im}\,k \geq 0$, the above Riemann-Hilbert problem reduces to solving a system of linear integral equations. In this case $X(\chi,\tau)$ and $b(\chi,\tau)$ are given by

$$X(\chi,\tau) = \frac{1}{\pi} \int_{-\infty}^{\infty} \frac{\bar{\rho}_2(k)}{\bar{\rho}_1(k)} e^{-2ik\chi - \frac{i\tau}{2k}} M_1(\chi,\tau,k)dk,$$

(6.9) $$b(\chi,\tau) = -1 + \frac{2}{\pi} \frac{\partial}{\partial \tau} \int_{-\infty}^{\infty} \frac{\rho_2(k)}{\rho_1(k)} e^{2ik\chi + \frac{i\tau}{2k}} \bar{M}_2(\chi,\tau,k)dk,$$

where the functions M_1 and M_2 are defined as the unique solution of the following system of linear integral equations

$$\begin{pmatrix} -M_2(\chi,\tau,k) \\ M_1(\chi,\tau,k) \end{pmatrix} =$$

(6.10) $$\begin{pmatrix} 0 \\ 1 \end{pmatrix} + \frac{1}{2i\pi} \int_{-\infty}^{\infty} \frac{\rho_2(l)}{\rho_1(l)} e^{2il\chi + \frac{i\tau}{2l}} \begin{pmatrix} \bar{M}_1(\chi,\tau,l) \\ \bar{M}_2(\chi,\tau,l) \end{pmatrix} \frac{dl}{l - (k - i0)}.$$

If $\rho_1(k_j) = 0$, $j = 1, 2, ...,$ $Im k_j \geq 0$, then the above Riemann-Hilbert problem reduces to solving a system of linear integral equations similar to (6.10) supplemented by a system of algebraic equations.

7. Acknowledgements

This work was partially supported by the EPSRC under grant number GR/J71885, by the Air Force Office of Scientific Research under grant number F49620-93-1-0088, and by the National Science Foundation under grant number DMS-9500311.

8. References

1. G.S. Gardner, J.M. Greene, M.D. Kruskal, R.M. Miura, Phys. Rev. Lett. **19**, 1095 (1967); Comm. Pure Appl. Math. **27**, 97 (1974).

2. P.D. Lax, Comm. Pure Appl. Math. **21**, 467 (1968).

3. F. Calogero, In *What is Integrability*, V.E. Zakharov, ed., Springer Verlag, 1992.

4. A.S. Fokas, A.R. Its, A. Kapaev, V. Novoksenov, *Painlevé Transcendents-Nonlinear Special Functions*, Kluver.

5. A.S. Fokas and S. Tanveer, A Hele-Shaw Problem and the Second Painlevé Equation, to appear in Math. Proc. Cambridge Phil. Soc.

6. A.S. Fokas, Inverse Problems, **10** L19 (1994).

7. A.S. Fokas and P.M. Santini, Physica D **44**, 99-130 (1990).

8. L-Y. Sung, J. Nonlinear Sci. **5**, 433-452 (1995).

9. M.J. Ablowitz and A.S. Fokas, *Complex Variables and Applications*, Cambridge University Press, 1997.

10. V.E. Zakharov and A. Shabat, Soviet Phys. JETP, **34**, 62 (1972).

11. A.S. Fokas and I.M. Gel'fand, Lett. Math. Phys. **32**, 189 (1994).

12. A.S. Fokas and V.E. Zakharov, J. Nonlinear Science, **2**, 109-134 (1992).

13. B. Noble, *Methods Based on the Wiener-Hopf Technique*, Pergamon Press, NY 1958.

14. J.J. Stoker, *Water Waves*, Interscience Publishers, NY 1957.

15. A.S. Fokas, A Unified Transform Method for Solving Linear and Certain Nonlinear PDE's, Proc. R. Soc. London A, 453, 1411–1443 (1997).

16. A.S. Fokas, Lax Pairs and a New Spectral Method for Solving Linear PDE's in Two Variables (preprint) 1996.

17. L. Ehrenpreis, *Fourier Analysis in Several Complex Variables*, Wiley-Interscience, 1970.

18. Ya.B. Lopatinski, Ukrain. Mat. Zh. **5**, 123-151 (1953); Z.Ja. Shapiro, Izv. Akad. Nauk Ser. Mat. **17**, 539-562 (1953).

19. A.S. Fokas and B.Pelloni, Several IBV Problems for the Linearized KdV equation, Proc. R. Soc. London A (in press).

20. A.S. Fokas and C.R. Menyuk, Integrability and Self-Similarity in Transient Stimulated Raman Scattering (preprint) 1996.

21. A.S. Fokas, D. Tsoubelis and L.Y. Sung, The Inverse Spectral Method for Colliding Gravitational Waves (preprint) 1997.

22. P. Bressloff and A.S. Fokas, A New Spectral Transform for Solving the Continuous and Spatially Discrete Heat Equation on Simple Trees (preprint) 1997.

23. P. Deift, X. Zhou, Announcement in Bull. Amer. Math. Soc. **26**, 119 (1992).

24. A.S. Fokas and A.R. Its, SIAM J. Math. Anal. **27** (1996); Mathematics and Computers in Simulation **37**, 293 (1994); Theoretical and Mathematical Phys. **92**, 2 (1992).

25. A.S. Fokas and V.E. Zakharov, eds., *Important Developments in Soliton Theory*, Springer Verlag, 1993.

26. P.C. Bressloff, A new Green's function method for solving linear PDE's in two variables, JMAA (in press).

DEPARTMENT OF MATHEMATICS, IMPERIAL COLLEGE OF SCIENCE, TECHNOLOGY & MEDICINE, LONDON, SW7 2BZ, U.K. AND INSTITUTE OF NONLINEAR STUDIES, CLARKSON UNIVERSITY, POTSDAM, NY 13699-5815

Proceedings of Symposia in Applied Mathematics
Volume **54**, 1998

Methods in the theory of quasi periodic motions

Giovanni Gallavotti

Abstract. Recent results on the theory of the Hamilton–Jacobi equation and the regularity of their solutions, in spite of the non regularity of the data, are reviewed, and discussed with attention to the cancellation mechanisms that make regularity possible.

§1 The Hamilton Jacobi equation.

This review deals on the work developed in collaboration with F.Bonetto, G. Gentile, V. Mastropietro.

The Hamilton Jacobi equation for an invariant torus in a ℓ dimensional hamiltonian system close to an integrable system takes one of the forms:

$$(\boldsymbol{\omega}_0 \cdot \partial)^2 \mathbf{h}(\boldsymbol{\psi}) = \varepsilon \, (\partial f)(\boldsymbol{\psi} + \mathbf{h}(\boldsymbol{\psi}))$$
$$(\boldsymbol{\omega}_0 \cdot \partial)\mathbf{h}(\boldsymbol{\psi}) = \varepsilon \, \mathbf{f}(\boldsymbol{\psi} + \mathbf{h}(\boldsymbol{\psi})) + \mathbf{N} \tag{1.1}$$

where $\boldsymbol{\psi}$ is a point on the ℓ–dimensional torus, $\mathbf{h}(\boldsymbol{\psi})$ a R^ℓ–valued function, $f, \mathbf{f}$ are functions on T^ℓ, $\mathbf{N}$ is a vector in R^ℓ and $\boldsymbol{\omega}_0$ is a diophantine vector in R^ℓ such that, considering the integer components lattice Z^ℓ, there are constants $C, \tau > 0$ with:

$$C|\boldsymbol{\omega}_0 \cdot \boldsymbol{\nu}| \geq |\boldsymbol{\nu}|^{-\tau} \qquad \forall \boldsymbol{\nu} \in Z^\ell, \boldsymbol{\nu} \neq \mathbf{0} \tag{1.2}$$

where $|\boldsymbol{\nu}| = \sqrt{\nu_1^2 + \ldots}$. The gradient ∂f is evaluated first and then computed at the point $\boldsymbol{\psi} + \mathbf{h}(\boldsymbol{\psi})$ (*i.e.* it is *not* the gradient of $f(\boldsymbol{\psi} + \mathbf{h}(\boldsymbol{\psi}))$).

The first of (1.1) corresponds to the construction of KAM tori in the case of the Thirring models (a general class of hamiltonian systems, [1]), and the second corresponds to the KAM tori for perturbations of non resonant harmonic oscillators, [2], [3]: in the latter case not only $\mathbf{h}$ is unknown but also $\mathbf{N}$, called a *counterterm* because one can think of it as a quantity that must be fixed in order to make the equations soluble.

We shall not be concerned here with the physical interpretations of (1.1).

Partially sponsored by the research program of the European Network on: "Stability and Universality in Classical Mechanics", # ERBCHRXCT940460.

Mathematics Subject Classification numbers: AMS 58F27, 70H20.

It was pointed out explicitly by Moser that the second equation is in some sense the fundamental one. Its theory is essentially equivalent to the KAM theorem theory, [2].

Here we shall consider a wider class of equations:

$$
\begin{array}{lll}
(1) & D\mathbf{h}(\boldsymbol{\psi}) = \varepsilon\,(\partial f)(\boldsymbol{\psi} + \mathbf{h}(\boldsymbol{\psi})) & \\
(2) & D\mathbf{h}(\boldsymbol{\psi}) = \varepsilon\,\mathbf{f}(\boldsymbol{\psi} + \mathbf{h}(\boldsymbol{\psi})) + \mathbf{N} &
\end{array} \tag{1.3}
$$

where D is a pseudodifferential operator like, respectively:

$$
\begin{array}{lll}
(1) & D = (\boldsymbol{\omega}_0 \cdot \partial)^2, & \text{or}\quad D = (-\Delta)^r \quad r \geq 0 \\
(2) & D = (\boldsymbol{\omega}_0 \cdot \partial), & \text{or}\quad D = (-\Delta)^r L \quad r \geq 0
\end{array} \tag{1.4}
$$

where Δ is the Laplace operator and L is a linear elliptic operator ("Dirac operator": for instance $\gamma \cdot \partial$ where γ are the Dirac's *gamma matrices*). The cases $r = 0$ in the first equation in (1.3) are remarkable as they *are not* differential equations.

It is easy to show by using the contraction method that (1.3), (1.4) with $r \geq 0$ do have a solution $\mathbf{h}$ or, respectively, $\mathbf{h}, \mathbf{N}$ for small ε if $\mathbf{f}$ is regular enough. This remains true, but it is far less easy, in the two remaining cases, in which D is a power of $(\boldsymbol{\omega}_0 \cdot \partial)$:

• The classical results that hold when f is assumed analytic on T^ℓ give analyticity in ε, ψ of the solution $\mathbf{h}$ and $\mathbf{N}$, for ε small, [2].
• In the differentiable case the first equation with $r = 0$ admits a solution in class $C^{(p-1)}(T^\ell)$ if $f \in C^{(p)}(T^\ell)$ and $p > 1$: this is an elementary result (contraction principle) for ε small. The solution is also $p - 1$ times differentiable in ε.
• In the case $D = (\boldsymbol{\omega}_0 \cdot \partial)^2$ and $f \in C^{(p)}(T^\ell)$ it admits a solution in class $C^{(p-4\tau-2)}$ if $p > 4\tau + 2$, a consequence of [4], where the contraction method is "replaced" by Nash's implicit functions theorem. The solution is also smooth in ε.
• In the case $D = \boldsymbol{\omega}_0 \cdot \partial$ and $f \in C^{(p)}(T^\ell)$ it admits a solution in class $C^{(p-2\tau-2)}$ if $p > 2\tau + 2$, as a consequence of Parasiuk's work, [5], see [6], based on Moser's technique; the solution is smooth in ε.

We address the question of the regularity in ε of the latter solutions. From the quoted results only differentiability in ε, to an order related to p, can be deduced.

We consider functions $f, \mathbf{f}$ which are finitely differentiable but that have the special form: $\mathbf{f} = (f_1, \ldots, f_\ell)$, with each f_j of the class $\hat{C}^{(p)}(T^\ell)$ introduced in [7], for some p.

Namely the class $\hat{C}^{(p)}$ consists of the R^ℓ valued functions such that $f(\boldsymbol{\psi}) = \sum_{\boldsymbol{\nu} \in Z^\ell} f_{\boldsymbol{\nu}}\, e^{i\boldsymbol{\nu}\cdot\boldsymbol{\psi}}$, $f_{\boldsymbol{\nu}} = f_{-\boldsymbol{\nu}}$, with $f_0 = 0$ and for $\boldsymbol{\nu} \neq \mathbf{0}$,

$$
f_{\boldsymbol{\nu}} = \sum_{n \geq p+\ell}^{N} \frac{P_{2k}^{(n)}(\boldsymbol{\nu})}{|\boldsymbol{\nu}|^{n+2k}} \tag{1.5}
$$

for some $N \geq p + \ell$ and some even polynomials $P_{2k}^{(n)}(\boldsymbol{\nu})$ of degree $2k$.

For instance we can choose $f_{j\boldsymbol{\nu}} = a_j|\boldsymbol{\nu}|^{-b}$, with $b = p + \ell$ and $\mathbf{a} = (a_1, \ldots, a_\ell)$ a vector in R^ℓ. In the following we shall deal explicitly with such a function: the analysis can be trivially extended to the class of functions (1.5).

Hence we discuss (1.3) only in cases in which f or, respectively $\mathbf{f}$ are special even functions. The results that follow do not hold in general for non even functions, see [7] for some simple counterexamples.

A *solution* of the second of (1.3) will be any pair $\mathbf{h} \in C^{(0)}(T^\ell)$ and $\mathbf{N} \in R^\ell$ that satisfies the identity:

$$- \int_{T^\ell} D\mathbf{F}(\psi) \cdot \mathbf{h}(\psi)d\psi = \int \mathbf{F}(\psi) \cdot (\varepsilon \mathbf{f}(\psi + \mathbf{h}(\psi)) + \mathbf{N}) \, d\psi \qquad (1.6)$$

for all $\mathbf{F} \in C^\infty(T^\ell)$. Similarly we define solutions of the first of (1.3).

In [7],[8] the following theorem is proved.

Theorem: *Equations (1.3) admit solutions that are analytic in ε for ε small if f or, respectively, $\mathbf{f}$ are in $\hat{C}^{(p)}(T^\ell)$ and p is large enough. For instance:*
(i) if $D = (\omega_0 \cdot \partial)^2$ the first of (1.3) admits a $C^{(<p-6\tau-3)}(T^\ell)$ solution if $f \in \hat{C}^{(p)}(T^\ell)$ provided $p - 6\tau - 3 > 0$.
(ii) if $D = \omega_0 \cdot \partial$ the second of (1.3) admits a $C^{(<p-3\tau-2)}(T^\ell)$ solution if $f \in \hat{C}^{(p)}(T^\ell)$ provided $p - 3\tau - 2 > 0$.
(iii) if $D = 1$ the first of (1.3) admits a $C^{(<p-2)}(T^\ell)$ solution if $f \in \hat{C}^{(p)}(T^\ell)$ provided $p - 2 > 0$.

The analyticity in ε is perhaps unexpected even in the case $D = 1$ since the solutions $\mathbf{h}(\psi)$ are *not* analytic in ψ. Furthermore it is not a general property valid whenever the equations admit a solution, because of the mentioned counterexamples.

One can prove the result because there are remarkable cancellations that we call *ultraviolet cancellations*, for reasons that become clear if one considers the analogy between the KAM problem and field theory problems, see [9]. The cancellations *are not* the ones pointed out by Eliasson, [10], that are specific to the KAM problems and that permit to derive the KAM theorem by a majorant series method: the latter cancellations are called *infrared cancellations* and have been recently revisited in several papers. The papers reproduce Eliasson's work with minor changes or additions but treat the simplest forms of the problem thus providing a simple understanding of the new methods (see for instance [1] which has the advantage of being short but uncompromising with the key difficulties; further developments can be found, *along the same lines*, in [11], [12], [13], [9] where also the relevant references to other papers can be found).

The names used for the two types of cancellations are taken from Quantum Fields theory which leads to problems, and techniques for their solutions, that are very close to the ones used in the theory of the Lindstedt series, see [9].

In §2 we discuss the Lindstedt formal series solution of (1.3): we shall deal with the second of (1.3) with $D = \omega_0 \cdot \partial$ as it is simpler than the first of (1.3) with $D = (\omega_0 \cdot \partial)^2$. Actually the simplest problem is the first equation in (1.3) with $D = 1$ or $D = (-\Delta)^r$: the case $D = 1$ is *non trivial* and it is not chosen to illustrate the analysis only because we also want to make clear that the small divisors can be treated. But the ideas and methods apply to all cases. In §3 we discuss the cancellation mechanism through some simple examples.

Note that the above quoted results of Moser give existence for a wider range of p's in the case of (1.1) while the results of [5], see [6], give existence in a wider range of p's if $\ell > 2$, but in a different range if $\ell = 2$ (the results cannot be really compared, even for $\ell = 2$, because $\hat{C}^{(p)} \subset C^{(p)}$).

§2 *Lindstedt series.*

Consider the second of Eq. (1.3) with $f_{j\boldsymbol{\nu}} = a_j|\boldsymbol{\nu}|^{-b}$ replaced by $f_{j\boldsymbol{\nu}}\,e^{-\kappa|\boldsymbol{\nu}|}$ and we write $\mathbf{h}(\boldsymbol{\psi}) = \sum_{k=1}^{\infty} \varepsilon^k \mathbf{h}^{(k)}(\boldsymbol{\psi})$. The parameter κ is taken $\kappa > 0$ and, after computing the coefficients $\mathbf{h}_{\boldsymbol{\nu}}^{(k)}$ of the Forurier series of the function $\mathbf{h}^{(k)}(\boldsymbol{\psi})$ which will depend on κ, we shall perform the limit $\kappa \to 0$ (*Abel's summation*). This is done because, contrary to what one could fear, the limit is well defined, if b is large enough, *for all k*.

It is easy to find the coefficients via a set of rules that associate with $\mathbf{h}^{(k)}$ a family of "Feynman graphs" ϑ and assign to each graph a *value* which is a function of $\boldsymbol{\nu} \in Z^\ell$ so that summing all values, at fixed $\boldsymbol{\nu}$, of the graphs that are associated with $\mathbf{h}^{(k)}$ one gets the Fourier component of order $\boldsymbol{\nu}$ of $\mathbf{h}^{(k)}$.

The graphs have the form of a tree graph (a very unusual situation for the Feynman graphs that arise in quantum field theory) like:

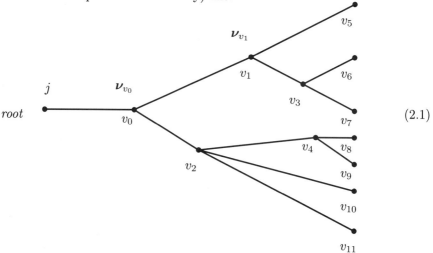

$$(2.1)$$

Fig.1. A graph ϑ with $p_{v_0} = 2, p_{v_1} = 2, p_{v_2} = 3, p_{v_3} = 2, p_{v_4} = 2$ and $k = 12$, $\prod p_v! = 2^4 \cdot 6$, and some decorations. The line numbers, distinguishing the lines, and the arrows, pointing at the root, are not shown. The lines length should be 1 but it is drawn of arbitrary size.

We lay down one after the other, on a plane, k pairwise distinct unit segments oriented from one extreme to the other: respectively the *initial point* and the *endpoint* of the oriented segment. The oriented segment will also be called *arrow*, *branch* or *line*. The segments are supposed to be numbered from 1 to k.

The rule is that after laying down the first segment, the *root branch*, with the endpoint at the origin and otherwise arbitrarily, the others are laid down one after the other by attaching an endpoint of a new branch to an initial point of an old one and by leaving free the new branch initial point. The set of initial points of the object thus constructed will be called the set of the graph *nodes* or *vertices*. A graph of *order k* is therefore a partially ordered set of k nodes with top point the endpoint of the root branch, also called the *root* (which is not a node); in general there will be several "bottom nodes" (at most $k-1$ and 7 in Fig.1).

We denote by $\leq$ the ordering relation, and say that two nodes v, w are "comparable" if $v < w$ or $w < v$.

With each graph node v we associate an *external momentum* or *mode* which is simply an integer component vector $\boldsymbol{\nu}_v \neq \mathbf{0}$; with the root of the graph (which is not regarded as a node) we associate a label $j = 1, \ldots, \ell$.

For each node v, we denote by v' the node immediately following v and by $\lambda_v \equiv v'v$ the branch connecting v to v', (v will be the initial point and v' the endpoint of λ_v). If v is the node immediately preceding the root r (*highest node*) then we shall write $v' = r$, for uniformity of notation (recall that r is not a node).

We consider "comparable" two lines λ_v, λ_w, if v, w are comparable.

If p_v is the number of branches entering the node v, then each of the p_v branches can be thought as the root branch of a *subgraph* having root at v: the subgraph is uniquely determined by v and one of the p_v nodes w immediately preceding v. Hence if $w' = v$ it can be denoted ϑ_{vw}.

We shall call *equivalent* graphs which can be overlapped by
(1) changing the angles between branches emerging from the same node, or
(2) permuting the subgraphs entering into a node v
in such a way that all the labels match.

The number of (non equivalent numbered) graphs with k branches is bounded by $4^k k!$, [14].

The "value" of a graph ϑ is now given by

$$\mathrm{Val}(\vartheta) = \prod_{v < r} \frac{(i\boldsymbol{\nu}_{v'} \cdot \mathbf{f}_{\boldsymbol{\nu}_v})}{i\boldsymbol{\omega}_0 \cdot \boldsymbol{\nu}_{\lambda_v}} , \qquad (2.2)$$

where
(1) v' is the node immediately following v in ϑ, and $\lambda_v = v'v$ is the line emerging from v and entering v';
(2) r is the *root* of the graph, and $i\boldsymbol{\nu}_r$ denotes the unit vector in the jth direction, $i\boldsymbol{\nu}_r = \mathbf{e}_j$, $j = 1, \ldots, \ell$;
(3) $\boldsymbol{\nu}_v$ is the *external momentum* associated with the *node* v, $\boldsymbol{\nu}_{\lambda_v} = \sum_{w \leq v} \boldsymbol{\nu}_w$ is the *momentum* flowing through the line λ_v, and $\boldsymbol{\nu}(\vartheta)$ is the momentum flowing through the line entering the root. *Only external momenta such that $\boldsymbol{\nu}_\lambda \neq \mathbf{0}$ for all λ will be allowed*: the generality of this property was the contribution of Poincaré to the analysis of the Lindstedt–Newcomb series, [15], Ch. IX (T. II).

It can be convenient to introduce the notations

$$g_\lambda \equiv \frac{1}{\boldsymbol{\omega}_0 \cdot \boldsymbol{\nu}_\lambda} , \qquad D(\vartheta) = \prod_{\lambda \in \vartheta} g_\lambda , \qquad (2.3)$$

so that (2.2) becomes

$$\mathrm{Val}(\vartheta) = D(\vartheta) \prod_{v < r} (\boldsymbol{\nu}_{v'} \cdot \mathbf{f}_{\boldsymbol{\nu}_v}) \qquad (2.4)$$

g_λ is called the *propagator* of the line λ. Let us denote by $\mathcal{T}(k, \boldsymbol{\nu})$ the set of non equivalent labeled graphs of order k with $\boldsymbol{\nu}(\vartheta) = \boldsymbol{\nu}$ and $i\boldsymbol{\nu}_r = \mathbf{e}_j$; then

$$h_{j\boldsymbol{\nu}}^{(k)} = \frac{1}{k!} \sum_{\vartheta \in \mathcal{T}(k, \boldsymbol{\nu})} \mathrm{Val}(\vartheta) = \frac{1}{k!} \sum_{\vartheta^0 \in \mathcal{T}^0(k)} W(\vartheta^0, \boldsymbol{\nu}) , \qquad (2.5)$$

where $\vartheta = (\vartheta^0, \{\boldsymbol{\nu}_\lambda\})$, if ϑ is a labeled graph, while ϑ^0 is a graph bearing no external momentum labels, $\mathcal{T}^0(k)$ is the set of such graphs of order k, and

$$W(\vartheta^0, \boldsymbol{\nu}) \equiv \sum_{\{\boldsymbol{\nu}_x\} : \boldsymbol{\nu}(\vartheta) = \boldsymbol{\nu}} \mathrm{Val}(\vartheta) . \qquad (2.6)$$

§3 *The basic ultraviolet cancellation.*

We consider the simplest graph that cannot be naively estimated uniformly in the ultraviolet cut off κ: it is the "brusch" graph:

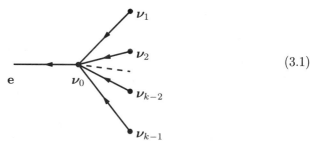

$$\text{(3.1)}$$

Recalling that $\mathbf{f}_{\boldsymbol{\nu}} = \mathbf{a}|\boldsymbol{\nu}|^{-b}$, its value is given by:

$$\frac{\mathbf{e} \cdot \mathbf{a}}{(\boldsymbol{\omega}_0 \cdot \boldsymbol{\nu})|\boldsymbol{\nu}_0|^b} \prod_{j=1}^{k-1} \frac{\boldsymbol{\nu}_0 \cdot \mathbf{a}}{(\boldsymbol{\omega}_0 \cdot \boldsymbol{\nu}_j)|\boldsymbol{\nu}_j|^b} \qquad \boldsymbol{\nu} = \sum_{j=0}^{k-1} \boldsymbol{\nu}_j \qquad \text{(3.2)}$$

Then we see that the sum of all the contributions of the above type, with fixed k and various $\boldsymbol{\nu}_j$ such that $\sum_j \boldsymbol{\nu}_i = \boldsymbol{\nu}$, to $\mathbf{h}_{\boldsymbol{\nu}}^{(k)}$ can be bounded by:

$$C^k |\boldsymbol{\nu}_0|^{k-1} |\mathbf{a}|^k |\boldsymbol{\nu}|^\tau |\boldsymbol{\nu}_0|^{-b} \prod_{j=1}^{k-1} |\boldsymbol{\nu}_j|^{\tau-b} \qquad \text{(3.3)}$$

which, to be summable over $\boldsymbol{\nu}_j$ requires that $\mathbf{f}$ is in $\hat{C}^{(>k-1+2\tau)}(T^\ell)$, so that the contributions from the graphs just considered may not make sense for large k *no matter how regular f is* (*i.e.* for $k \geq b+1-2\tau-\ell$).

But the problem really arises only if $|\boldsymbol{\nu}_0| \gg |\boldsymbol{\nu}_j|$: in the cases $|\boldsymbol{\nu}_0| \ll |\boldsymbol{\nu}_j|$ we can replace $|\boldsymbol{\nu}_0|$ by $|\boldsymbol{\nu}_j|$ and $|\boldsymbol{\nu}|$ by $k \prod_v |\boldsymbol{\nu}_v|$ in the bounds and, by (3.3), this would require only $f \in \hat{C}^{(>2\tau+1)}$.

Therefore we add, for each $\boldsymbol{\nu}_j$ with $|\boldsymbol{\nu}_j| \ll |\boldsymbol{\nu}_0|$, the graph with $\boldsymbol{\nu}_0, \boldsymbol{\nu}_j$ replaced by $\boldsymbol{\nu}_0 + 2\boldsymbol{\nu}_j, -\boldsymbol{\nu}_j$, and noting that both graphs contribute to the same Fourier component $\boldsymbol{\nu}$ of $\mathbf{h}$, we see that this replaces:

$$\frac{(\boldsymbol{\nu}_0 \cdot \mathbf{a})}{(\boldsymbol{\omega}_0 \cdot \boldsymbol{\nu})|\boldsymbol{\nu}_0|^b (\boldsymbol{\omega}_0 \cdot \boldsymbol{\nu}_j)} \rightarrow \frac{1}{(\boldsymbol{\omega}_0 \cdot \boldsymbol{\nu})} \left(\frac{(\boldsymbol{\nu}_0 \cdot \mathbf{a})}{|\boldsymbol{\nu}_0|^b (\boldsymbol{\omega}_0 \cdot \boldsymbol{\nu}_j)} - \frac{(\boldsymbol{\nu}_0 + 2\boldsymbol{\nu}_j) \cdot \mathbf{a}}{|\boldsymbol{\nu}_0 + 2\boldsymbol{\nu}_j|^b (\boldsymbol{\omega}_0 \cdot \boldsymbol{\nu}_j)} \right) \quad \text{(3.4)}$$

the other factors in (3.2) remain unchanged by the parity properties of $\mathbf{f}$.

More generally if $|\boldsymbol{\nu}_0| \gg |\boldsymbol{\nu}_j|$, $j = 1, \ldots, s$ and $|\boldsymbol{\nu}_0| \ll |\boldsymbol{\nu}_j|$, $j > s$ we combine all the graphs with $\boldsymbol{\nu}_0$ replaced by $\boldsymbol{\nu}_0 + \sum_{j=1}^s (1 - \sigma_j)\boldsymbol{\nu}_j$ and with $\boldsymbol{\nu}_j$ replaced by $\sigma_j \boldsymbol{\nu}_j$ for $j = 1, \ldots, s$ and $\sigma_j = \pm 1$. The sum of the considered graph values can be expressed by an interpolation formula; let us define $\boldsymbol{\nu}(\mathbf{t}_{v_0}) = \boldsymbol{\nu}_0 + 2\sum_{j=1}^s t_j \boldsymbol{\nu}_j$ fot $\mathbf{t}_{v_0} = (t_1, \ldots, t_s)$, then the sum is:

$$\int_1^0 \left(\prod_{j=1}^s dt_j \frac{\partial}{\partial t_j} \right) \left\{ \frac{\mathbf{e} \cdot \mathbf{a}}{\boldsymbol{\omega}_0 \cdot \boldsymbol{\nu}} \frac{1}{|\boldsymbol{\nu}_{v_0}(\mathbf{t}_{v_0})|^b} \prod_{j=1}^{k-1} \frac{\boldsymbol{\nu}_{v_0}(\mathbf{t}_{v_0}) \cdot \mathbf{a}}{\boldsymbol{\omega}_0 \cdot \boldsymbol{\nu}_j} \frac{1}{|\boldsymbol{\nu}_j|^b} \right\} \quad \text{(3.5)}$$

and the *cancellation* is expressed by the fact that in the above interpolation formula there are no contributions from the boundary values $t_j = 1$. It is important to note, also, that the factors containing $\boldsymbol{\omega}_0$ are $\mathbf{t}$–*independent*, hence they (*i.e.* the "small divisors") are *not affected* by the interpolation.

The (3.5) can be bounded as before *but replacing each of the s vectors $\boldsymbol{\nu}_{v_0}$ which are large with the corresponding* $|\boldsymbol{\nu}_j|$, and (trivially) by replacing the other $k - 1 - s$ and $\boldsymbol{\nu}_0$'s by the correponding $|\boldsymbol{\nu}_j|$; *i.e.* for a suitable $\tilde{C}$ the bound (3.3) is replaced by:

$$\tilde{C}^k |\boldsymbol{\nu}|^\tau |\boldsymbol{\nu}_0|^{-b} \prod_{j=1}^{k-1} |\boldsymbol{\nu}_j|^{\tau+1-b} \tag{3.6}$$

and one needs (to sum over $\boldsymbol{\nu}_j$ at fixed $\boldsymbol{\nu}$) only $b > \ell + \tau + 1$, *i.e.* $f \in \hat{C}^{(>\tau+1)}(T^\ell)$, or $b > \ell + 2\tau + 1$ to sum without fixing $\boldsymbol{\nu}$, (*i.e.* $f \in \hat{C}^{(2\tau+1)}$).

This is the key "ultraviolet" cancellation introduced and used in [7] and [8]. Clearly there remains quite a lot of work to do:

(1) how does one treat the cases in which the $\boldsymbol{\nu}_0$ and the $\boldsymbol{\nu}_j$ are "close"? In such cases $\boldsymbol{\nu}_{v_0}(\mathbf{t}_{v_0}) = \boldsymbol{\nu}_0 + 2\sum_{j=1}^{k-1} t_j \boldsymbol{\nu}_j$ may vanish for some choices of the interpolation variables t_j. This type of situation is a manifestation of rather general class of difficulties known in field theory as *overlapping divergences*: it must clearly be avoided.

(2) how does one treat the cases in which the graph is more elaborated, with many nodes with bifurcations, and therefore many of the above problems arise *simultaneously?*

(3) how does one treat the *infrared* cancellations of [10]? are they obvious even considering the necessity of collecting graphs so that no ultraviolet divergence arises? In principle exhibiting the ultraviolet cancellations may require different *non commuting* collections of terms and this would ruin the analysis (again a "overlapping divergences problem").

In the next section we discuss the basic ideas about (1): this will give an idea also for the solution of the problem (2). The problem in (3) is real and it arises, in fact. However the incompatibility is somewhat trivial and one can eliminate it by giving up performing some ultraviolet cancellations. This leads eventually, see [8], only to replace the minimum value for p that one would naively expect from Eliasson method and his extension of Siegel's small divisor bound ($p > 3\tau + 1$) with $p > 3\tau + 2$.

§4 Overlapping divergences?

Consider the case of fig.(3.1): we set $p_{v_1} = k - 1$ and $w_j = v_j$ and rename v_1 the graph node previously called v_0 and $w_1, \ldots, w_{p_{v_1}}$ the $k - 1$ endnodes.

The change in notation is made to adhere to the notation in [8] which is apt to treat also the more structered graphs.

Then we introduce the notion of *scale* of a vector $\boldsymbol{\nu} \in Z^\ell$ ($\boldsymbol{\nu}$ will also be called a "mode"). This is done by simply declaring it of *scale* h if $2^{h-1} \le |\boldsymbol{\nu}| < 2^h$.

We shall set $\boldsymbol{\nu}_{w_j} \equiv \boldsymbol{\nu}_{\lambda_{w_j}}$, again for ease of reference to [8]. Note that in the graph (3.1) $\boldsymbol{\nu}_{w_j}$ and $\boldsymbol{\nu}_{w_j}$ accidentally coincide due to the graph simplicity.

Fixed $\boldsymbol{\nu}$ and $\{\boldsymbol{\nu}_{\lambda_w}\}_{w \in \mathcal{B}_{v_1}}$ let h_{v_1} be *the scale of* $\boldsymbol{\nu}_{v_1}$ (it depends on $\boldsymbol{\nu}$ and on the $\boldsymbol{\nu}_{\lambda_{w_j}} \equiv \boldsymbol{\nu}_{w_j}$ because $\boldsymbol{\nu}_{v_1} = \boldsymbol{\nu} - \sum \boldsymbol{\nu}_{w_j}$). This means that $\boldsymbol{\nu}_{v_1}$ is such that $2^{h_{v_1}-1} \le |\boldsymbol{\nu}_{v_1}| < 2^{h_{v_1}}$ and we refer to this relation by saying that h_{v_1} is *compatible* with $\boldsymbol{\nu}_1$.

With the new notations the value of the graph is:

$$\frac{\mathbf{e} \cdot \mathbf{a}}{|\boldsymbol{\nu}_{v_1}|^b \boldsymbol{\omega}_0 \cdot \boldsymbol{\nu}} \prod_{j=1}^{p_{v_1}} \frac{\boldsymbol{\nu}_1 \cdot \mathbf{a}}{|\boldsymbol{\nu}_j|^b \boldsymbol{\omega}_0 \cdot \boldsymbol{\nu}_j} \qquad \boldsymbol{\nu} = \sum_{j=0}^{k-1} \boldsymbol{\nu}_j \qquad (4.1)$$

We say that w_j is *out of order* with respect to v_1 if:

$$2^{h_{v_1}} > 2^o p_{v_1} |\boldsymbol{\nu}_{\lambda_w}| \qquad o = 5 \qquad (4.2)$$

where p_v is the number of branches entering v.

We denote $\mathcal{B}_{1v_1}$ the set of nodes $w \in \mathcal{B}_{v_1}$ which are out of order with respect to v_1: this set can be denoted also $\mathcal{B}_{1v_1}(\boldsymbol{\nu}, \{\boldsymbol{\nu}_{\lambda_w}\}_{w \in \mathcal{B}_{v_1}})$. The number of elements in $\mathcal{B}_{1v_1}$ will be denoted $q_v = |\mathcal{B}_{1v_1}|$. The notion of w being out of order with respect to v_1 depends on $\{\boldsymbol{\nu}_{\lambda_w}\}_{w \in \mathcal{B}_{v_1}}$ and $\boldsymbol{\nu}$.

Given a set $\{\boldsymbol{\nu}_{\lambda_w}\}_{w \in \mathcal{B}_{v_1}}$ for all choices of $\sigma_w = \pm 1$ we define the map:

$$U(\{\boldsymbol{\nu}_{\lambda_w}\}_{w \in \mathcal{B}_{v_1}}) \equiv (\{\sigma_w \boldsymbol{\nu}_{\lambda_w}\}_{w \in \mathcal{B}_{v_1}}) \qquad (4.3)$$

and given a set $C \subseteq \mathcal{B}_{v_1}$ we call $\mathcal{U}(C)$ the set of all transformations U such that $\sigma_w = 1$ for $w \notin C$; we also set $\sigma_w = 1$ if $w \notin \mathcal{B}_{1v_1}$

If $[2^{h-1}, 2^h)$ is a scale interval I_h, $h = 1, 2, \ldots$ we call

- the first quarter of I_h the *lower part* $I_h^- = [2^{h-1}, \frac{5}{4}2^{h-1})$ of I_h,
- the fourth quarter of I_h the *upper part* $I_h^+ = [\frac{7}{8}2^h, 2^h)$ of I_h, and
- the remaining part the *central part* I_h^c.

We shall group branch momenta $\{\boldsymbol{\nu}_{\lambda_w}\}_{w \in \mathcal{B}_{v_1}}$ into collections by proceeding iteratively. *The collections will be built so that in each collection the cancellation discussed in Remark 4.2 above can be exhibited.*

Fixed $\boldsymbol{\nu}$ and h choose $\{\boldsymbol{\nu}_{\lambda_w}^1\}_{w \in \mathcal{B}_{v_1}}$ such that $|\boldsymbol{\nu}_{v_1}^1| \in I_h^c$: such $\{\boldsymbol{\nu}_{\lambda_w}^1\}_{w \in \mathcal{B}_{v_1}}$ is called a *representative*. Given the representative we define:

- the *branch momenta collection* to be the set of the $\{\boldsymbol{\nu}_{\lambda_w}\}_{w \in \mathcal{B}_{v_1}}$ of the form

$$U(\{\boldsymbol{\nu}_{\lambda_w}^1\}_{w \in \mathcal{B}_{v_1}}), \qquad U \in \mathcal{U}(\mathcal{B}_{1v_1}(\boldsymbol{\nu}, \{\boldsymbol{\nu}_{\lambda_w}^1\}_{w \in \mathcal{B}_{v_1}})) ; \qquad (4.4)$$

- the *external momenta collection*, for $U \in \mathcal{U}(\mathcal{B}_{1v_1}(\boldsymbol{\nu}, \{\boldsymbol{\nu}_{\lambda_w}^1\}_{w \in \mathcal{B}_{v_1}}))$, to be the set of momenta

$$\boldsymbol{\nu}_w^{1U} = \sigma_w \boldsymbol{\nu}_w, \qquad \boldsymbol{\nu}_{v_1}^{1U} = \boldsymbol{\nu} - \sum_{w \in \mathcal{B}_{v_1}} \sigma_w \boldsymbol{\nu}_{\lambda_w}^1, \qquad (4.5)$$

Note that the elements of the above constructed external momenta collection need not be necessarily contained in I_h^c (see below).

We consider then another representative $\{\boldsymbol{\nu}_{\lambda_w}^2\}_{w \in \mathcal{B}_{v_1}}$ such that $|\boldsymbol{\nu}_{v_1}^2| \in I_h^c$ *and* not belonging to the branch momenta collection associated with $\{\boldsymbol{\nu}_{\lambda_w}^1\}_{w \in \mathcal{B}_{v_1}}$, if there are any left; and we consider the corresponding branch momenta and external momenta collections as above. We proceed in this way until all the representatives such that $\boldsymbol{\nu}_1$ is in I_h^c, for the given h, have been put into some collection of branch momenta.

We then repeat the above construction with the interval I_h^- replacing the I_h^c, always being careful not to consider representatives $\{\boldsymbol{\nu}_{\lambda_w}\}_{w \in \mathcal{B}_{v_1}}$ that appeared as members of previously constructed collections. It is worth pointing out that not all the external momenta $\boldsymbol{\nu}_{v_1}^U$, $U \in \mathcal{U}(\mathcal{B}_{1v_1}(\boldsymbol{\nu}, \{\boldsymbol{\nu}_{\lambda_w}^1\}_{w \in \mathcal{B}_{v_1}}))$, are in I_h^-, but they are all in the corridor $I_{h-1}^+ \cup I_h^-$, by (4.2).

Finally we consider the interval I_{h-1}^+, (if $h = 1$ we simply skip this step). The construction is repeated for such intervals.

Proceeding iteratively in this way starting from $h = 1$ and, after exhausting all the $h = 1$ cases, continuing with the $h = 2, 3 \ldots$ cases, we shall have grouped the sets of branch momenta into collections obtainable from a representative $\{\boldsymbol{\nu}_{\lambda_w}\}_{w \in \mathcal{B}_{v_1}}$ by applying the operations $U \in \mathcal{U}(\mathcal{B}_{1v_1}(\boldsymbol{\nu}, \{\boldsymbol{\nu}_{\lambda_w}^1\}_{w \in \mathcal{B}_{v_1}}))$ to it. *Note that, in this way, when the interval I_{h-1}^+ is considered, all the remaining representatives are such that* $|\boldsymbol{\nu}_{v_1}^U| \in I_{h-1}^+$ *for all* $U \in \mathcal{U}(\mathcal{B}_{1v_1}(\boldsymbol{\nu}, \{\boldsymbol{\nu}_{\lambda_w}^1\}_{w \in \mathcal{B}_{v_1}}))$.

Note that the graphs with momenta in each collection are just the graphs involved in the parity cancellation described in the previous section. In fact if U is generated by the signs $\{\sigma_w\}_{w \in \mathcal{B}_v}$, we have:

$$
\begin{aligned}
\boldsymbol{\nu}_{v_1}^U &= \Big(\prod_{w \in \mathcal{B}_{1v_1}} U_{v_1 w}^{\sigma_w} \{\boldsymbol{\nu}_x\} \Big)_{v_1} " = " \boldsymbol{\nu} + \sum_w (1 - \sigma_w) \boldsymbol{\nu}_w \\
(U(\{\boldsymbol{\nu}_{\lambda_{\tilde{w}}}\}_{\tilde{w} \in B_{v_1}}))_w &= \sum_{z \leq w} \Big(\prod_{\tilde{w} \in \mathcal{B}_{1v_1}} U_{v_1 \tilde{w}}^{\sigma_{\tilde{w}}} \{\boldsymbol{\nu}_x\} \Big)_z " = " \sigma_w \boldsymbol{\nu}_w
\end{aligned}
\tag{4.6}
$$

where, given the sets of momenta $\{\boldsymbol{\nu}_x\}$ and $\{\boldsymbol{\nu}_{\lambda_{\tilde{w}}}\}$, $(\{\boldsymbol{\nu}_x\})_v$ denotes the external momentum in $\{\boldsymbol{\nu}_x\}$ corresponding to the node v and $(\{\boldsymbol{\nu}_{\lambda_{\tilde{w}}}\})_w$ denotes the branch momentum in $\{\boldsymbol{\nu}_{\lambda_{\tilde{w}}}\}$ corresponding to the branch λ_w. The " $=$ " mean that, in the special case of (3.1) that we treat, the equality between $\boldsymbol{\nu}_w$ and $\boldsymbol{\nu}_{\lambda_w}$ implies that the r.h.s. of (4.6) takes the simple form indicated. The general case of more structured graphs is more involved, [8].

Remark. The complexity of the above construction is due to the necessity of avoiding overcountings: *i.e.* overlapping of cancellations whereby the same graph value is used to "compensate" two or more other graph values. In fact it is possible that, for some $U \in \mathcal{U}(\mathcal{B}_{1v_1}(\boldsymbol{\nu}, \{\boldsymbol{\nu}_{\lambda_w}\}_{w \in \mathcal{B}_{v_1}})$, one has:

$$
\mathcal{B}_{1v_1}(\boldsymbol{\nu}, U(\{\boldsymbol{\nu}_{\lambda_w}\}_{w \in \mathcal{B}_{v_1}})) \neq \mathcal{B}_{1v_1}(\boldsymbol{\nu}, \{\boldsymbol{\nu}_{\lambda_w}\}_{w \in \mathcal{B}_{v_1}}),
\tag{4.7}
$$

because the scale of $\boldsymbol{\nu}_{v_1}^U$ may be $h - 1$, while that of $\boldsymbol{\nu}_{v_1}$ may be h; so that if one considered, for instance, I_{h-1}^+ before I_h^- overcountings would be possible, and in fact they would occur.

It is convenient rewriting the sum that we want to estimate, *i.e.* the contribution to $\mathbf{h}^{(k)}$ from the sum of values of graphs like (3.1), as:

$$
\sum_{\boldsymbol{\nu}} |\boldsymbol{\nu}|^s \Big| \sum_{h_{v_1}} \sum_{\{\boldsymbol{\nu}_{\lambda_w}\}_{w \in \mathcal{B}_{v_1}}}^* \sum_{U \in \mathcal{U}(B_{1v_1})} \frac{\mathbf{e} \cdot \mathbf{a}}{\boldsymbol{\omega}_0 \cdot \boldsymbol{\nu} |\boldsymbol{\nu}_{v_1}^U|^b} \Big(\prod_{w \in \mathcal{B}_{1v_1}} \sigma_w \Big) \prod_{j=1}^{p_{v_1}} \frac{\boldsymbol{\nu}_{v_1} \cdot \mathbf{a}}{\boldsymbol{\omega}_0 \cdot \boldsymbol{\nu}_{w_j} |\boldsymbol{\nu}_{w_j}|^b}
\tag{4.8}
$$

where $\boldsymbol{\nu}_{v_1}(\mathbf{t}_{v_1}) = \boldsymbol{\nu}_{v_1} + 2 \sum_{j=1}^{k-1} t_j \boldsymbol{\nu}_j$, $\boldsymbol{\nu}_w \equiv \boldsymbol{\nu}_{\lambda_w}$, and the sum $\sum_{\{\boldsymbol{\nu}_{\lambda_w}\}_{w \in \mathcal{B}_{v_1}}}^*$ means sum over the above defined representatives such that $\boldsymbol{\nu}_{v_1}$ is "compatible" with h_{v_1}, *i.e.* it has scale h_{v_1} (see lines preceding (4.1)); and we abridge $\mathcal{B}_{1v_1}(\boldsymbol{\nu}, \{\boldsymbol{\nu}_{\lambda_w}\}_{w \in \mathcal{B}_{v_1}})$ by $\mathcal{B}_{1v_1}$ in conformity with the notations introduced after (4.2). The summation over $\boldsymbol{\nu}$ is weighted with the weight $|\boldsymbol{\nu}|^s$ in order to obtain also estimates of the s–th derivatives of $\mathbf{h}^{(k)}(\boldsymbol{\psi})$.

The explicit sum over the scales h_{v_1} is introduced to simplify the bounds analysis that we perform now. Note that $\boldsymbol{\nu}^U_{v_1}$ is, in general, not compatible with h_{v_1}, $i.e.$ we are grouping together also terms with different scale label (but the difference in scale is at most one, see (4.11) below).

We can now apply interpolation in (3.5) to the node v and rewrite (4.8) as:

$$
\sum_{\boldsymbol{\nu}} |\boldsymbol{\nu}|^s \Big| \sum_{h_{v_1}} \sum_{\{\boldsymbol{\nu}_{\lambda_w}\}_{w \in \mathcal{B}_{v_1}}}^{*} \Big[\int_1^0 \Big(\prod_{w \in \mathcal{B}_{1v_1}} dt_w \Big) \cdot
$$
$$
\cdot \Big(\prod_{w \in \mathcal{B}_{1v_1}} \frac{\partial}{\partial t_w} \Big) \Big(\frac{1}{(\boldsymbol{\omega}_0 \cdot \boldsymbol{\nu})|\boldsymbol{\nu}_{v_1}(\mathbf{t}_{v_1})|^b} \prod_{j=1}^{k-1} \frac{\boldsymbol{\nu}_{v_1}(\mathbf{t}_{v_1}) \cdot \mathbf{a}}{(\boldsymbol{\omega}_0 \cdot \boldsymbol{\nu}_{w_j})|\boldsymbol{\nu}_{w_j}|^b} \Big)
$$

(4.10)

where if $\mathcal{B}_{1v_1} = \emptyset$ no interpolation is made; by the definition of nodes out of order, see (4.2), and by the iterative grouping of the representatives it is:

$$
2^{h_{v_1} - 2} \leq |\boldsymbol{\nu}_{v_1}(\mathbf{t}_{v_1})| < 2^{h_{v_1}} ,
$$

(4.11)

so that the interpolation formulae discussed in §3 *can be used* because no singularity ($i.e.$ no division by 0) arises in performing the $\mathbf{t}_{v_1}$–integrations.

It is now easy to check, by computing the derivatives, that each of the terms in (4.10) can be conveniently bounded by its (easily estimated) maximum value (over τ) and the resulting bound has the form obtained in the simple case of §3, (3.6), in which some of the $\boldsymbol{\nu}_{w_j}$ were $\ll |\boldsymbol{\nu}_{v_1}|$ and some were $\gg |\boldsymbol{\nu}_{v_1}|$. The $\boldsymbol{\nu}_w$ of the nodes out of order play the role of the small $\boldsymbol{\nu}_{w_j}$'s and the others can be used as bounds on $\boldsymbol{\nu}_{v_1}$. *No further care needs to be exercized in the bounds.* The details are not very interesting, see [8].

The extension to more structured graphs is not really more difficult. One has to be careful in interpreting $\boldsymbol{\nu}_{w_j}$ as the momentum flowing on the brach $v_1 w_j$, $i.e.$ as the sum of the node momenta of all the nodes above v_1. See [7] for the detailed algebra that is necessary. Therefore the above analysis gives a rather complete idea of the ultraviolet cancellations mechanism. The infrared cancellations, $i.e.$ the cancellations discussed by Eliasson does not interfere with the above ultraviolet cancellation and the analysis is done in [8]. Note that in the case of the first of (1.3) with $D = (-\Delta)^r$, or the second with $D = (-\Delta)^r L$, there is no need of infrared cancellations ($i.e.$ no small divisors are present).

The above analysis seems to imply that the validity condition for the theorem should be $p > 2\tau + 1$ instead of $3\tau + 2$, §1. The "extra" τ is due to the fact that the product of the small divisors, in general graphs "without resonances" (see [8]) is essentially estimated, following [10] (see also [7]) by $\prod_v |\boldsymbol{\nu}_v|^{3\tau}$, instead of the $\prod_v |\boldsymbol{\nu}_v|^{2\tau}$ that occurs in the simple eaxmple chosen here. The "extra" 1 is due to the fact that general graphs may contain "resonances" (see [10], and §3 in [8]) which interfere with the ultraviolet cancellations causing an extra loss of one unit in the differentiability condition.

Acknowledgements. : it includes the text of a conference delivered at the meeting "Recent Advances in partial differenzial equations and applications", Venezia, june 10–14, 1996 and describes the results in [7], [8].

References

[1] G. Gallavotti: *Twistless KAM tori*, Communications in Mathematical Physics, **164**, 145–156 (1994).

[2] J. Moser: *Convergent series expansions for quasi periodic motions*, Mathematische Annalen, **169**, 136-176 (1967).

[3] G. Gallavotti: Classical Mechanics and Renormalization Group, Lectures notes, 1983 Erice School, Ed. A. Wightman, G. Velo, Plenum, N.Y. (1985).

[4] J. Moser: *A rapidly convergent iteration method and nonlinear differential equations II*, Annali della Scuola Normale Sup. di Pisa, **20**, 499–535, (1966).

[5] I.S. Parasyuk: *On instability zones of the Schrödinger equation with a quasi periodic potential*, Ukr. Mat. Zh. **30**, 70-78 (1985).

[6] L.A. Pastur, A. Figotin: *Spectra of random and almost-periodic operators*, Grundlehren der Math. Wissenschaften **297**, Springer, Berlin (1992).

[7] F. Bonetto, G. Gallavotti, G. Gentile, V. Mastropietro: *Lindstedt series, ultraviolet divergences and Moser's theorem*, Preprint IHES/P/95/102 (1995).

[8] F. Bonetto, G. Gallavotti, G. Gentile, V. Mastropietro: *Quasi linear flows on tori: regularity of their linearization.* in mp_arc@ math. utexas. edu, #96-251 and chao-dyn@ xyz. lanl. gov #9605017.

[9] G. Gallavotti, G. Gentile, V. Mastropietro: *Field theory and KAM tori*, Mathematical Physics Electronic Journal, **1**, (5): 1–13 (1995), (http://mpej@ math. utexas. edu).

[10] H. Eliasson: *Absolutely convergent series expansions for quasi-periodic motions*, Mathematical Physics Electronic Journal, **2**, 33 pp, 1996, http://mpej @math. utexas.edu.

[11] G. Gallavotti, G. Gentile: *Majorant series method for KAM tori*, Ergodic theory and applications, **15**, p. 857–869, 1995.

[12] G. Gentile, V. Mastropietro: *KAM theorem revisited*, Physica D, **90**, 225–234 (1996).

[13] G. Gentile, V. Mastropietro: *Methods for the analysis of the Lindstedt series for KAM tori and renormalizability in classical mechanics. A review with some applications*, Reviews of Mathematical Physics, **8**, 393–444 (1996).

[14] F. Harary, E. Palmer: *Graphical enumeration*, Academic Press, N.Y., 1973.

[15] Poincaré, H.: *Les Méthodes Nouvelles de la Mécanique Céleste*, Gauthiers Villars, Paris, 1893.

Fisica, Università di Roma, P.le Moro 2, 00185 Roma, Italy and
I.H.E.S., 91440 Bures s/Yvette, France

Internet: Authors' preprints also freely downloadable (latest version) at
                    http://chimera.roma1.infn.it
                http://www.math.rutgers.edu/~giovanni
Mathematical Physics Preprints (mirror) pages.
e-mail: *giovanni@ipparco.roma1.infn.it*

Proceedings of Symposia in Applied Mathematics
Volume **54**, 1998

FOURIER ANALYSIS AND
NONLINEAR WAVE EQUATIONS

Sergiu Klainerman

Department Mathematics, Princeton University, Princeton, NJ 08544-1000

While Fourier Analysis has always played a major role in the linear theory of hyperbolic equations it is only quite recently that it started to affect in a fundamental way the nonlinear theory. The nonlinear theory seems in fact to require some of the deeper results and methods of Harmonic Analysis developed in the last thirty years. A well known example is the global regularity result for the critical semilinear wave equation in $\mathbb{R}^{3+1}$,

$$\Box\phi + \phi^5 = 0.$$

The proof, as developed by Grillakis and Struwe rests on two main ideas. The first, due to Struwe[Str], is a non-concentration argument the second, due to Grillakis[Gr1], [Gr2], is to use a particular form of the Strichartz–Pecher estimates. The Strichartz–Pecher type inequalities have in fact been used for subcritical semilinear wave equations, before Grillakis. Strichartz[S1] himself was motivated by a question, concerning non-linear wave equations, raised by I. Segal[Se], an important part of his achievement[1] was to recognize that Segal's question can be answered with the help of a freshly proved result in Fourier Analysis, the Restriction Theorem of Stein and Thomas[St]. Motivated by questions of regularity and asymptotic behavior for semi–linear wave equations Brenner[Br], Pecher[Pe], Ginibre–Velo[G-V], Kapitanski[Ka], Lindblad–Sogge[L-So] have in fact found significant generalizations of the original Strichartz estimates. Recently Georgiev–Lindblad–Sogge[G-L-So] found some very interesting weighted Strichartz estimates and applied them to settle a conjecture of Strauss which generalizes an older result of F. John.

---

[1] The Strichartz–Pecher inequalities are however intimately connected with a previous work of Strichartz[2] concerning $L^p - L^q$ estimates at fixed t.

The usefulness of the Strichartz–Pecher estimates seems, at least at first glance, limited to non-linear wave equations of the type, $\Box \phi = F(\phi)$. However, virtually all interesting wave equations involve derivatives in the nonlinear terms. One needs thus apriori estimates for the first derivatives of ϕ which cannot be derived by Strichartz type inequalities. In fact, the only known estimate of derivatives of ϕ in terms of F alone is the classical energy estimate,

$$\|\partial \phi(t, \cdot)\|_{L^2(\mathbb{R}^n)} \leq \|\partial \phi(0, \cdot)\|_{L^2(\mathbb{R}^n)} + \int_0^t \|F(\tau, \cdot)\|_{L^2(\mathbb{R}^n)} d\tau.$$

It is well known that for space dimensions $n > 1$ one cannot replace, in this estimate, the $L^2(\mathbb{R}^n)$ norm by any other L^p norm without losing derivatives. Taking space-time estimates may help, yet one can show[2], using a construction introduced originally by Fefferman in his famous disc multiplier counterexample, that even if F is bounded there are no estimates of the first derivatives of ϕ in any local space-time $L_t^1 L_x^p$ norm, for $p > 2$. If one allows a small loss of derivatives such estimates are conjectured to exist (see [So]) but they are known to be extremely hard to prove. In fact some of these conjectures imply, and seem significantly harder, than the well-known Bochner–Riesz conjectures of Harmonic Analysis.

It seems thus that, in dimensions $n > 1$, one is reduced to work with energy estimates. They have, of course, been used extensively for local existence and stability results. They have also been used successfully to prove the existence of global, small, solutions, for important physical equations such as the Einstein field equations in General Relativity[Ch–Kl]. It is known however that they are too crude[3] for a global regularity theory.

To circumvent the difficulty of lack of good estimates for $\Box \phi = F$ one is forced to,

1) Give up working with traditional L^p - based Sobolev norms. Use instead norms which take advantage of the specific symbol of the wave equation, in the frequency space .

2) Use methods which are sensitive to the specific nonlinear "null" structure of the equations. Many important equations have such special features due to the symmetries of the original Lagrangean.

The need to develop such methods can be easiest understood from the point of view of optimal regularity of the initial conditions which assures local existence together with uniqueness and continuous dependence of the data or, in short, "well posedness". The classical local existence results, based only on energy estimates and Sobolev inequalities, require data in the

---

[2]The proof of this result was provided to me by T. Wolff in a private communication.

[3]As we shall see below they lose a little more than one derivative over a scale invariant estimate.

Sobolev space H^s with $s > 1 + s_c$. The critical exponent s_c corresponds to the scaling properties of the equations, one does not expect uniqueness and any reasonable continuous dependence on the data for $s < s_c$. On the other hand one might expect "well posedness" for $s \geq s_c$. The proof of such a statement would imply, in the case of interesting examples of conservative P.D.E's, global existence for subcritical nonlinearities and a crucial step in the proof of global existence for critical ones such as Wave Maps in $2 + 1$ space-times or Yang–Mills equations in $4 + 1$ space-times.

In the case of systems of equations of the form $\Box \phi = F(\phi)$, where the nonlinear terms F does not depend on the derivatives of ϕ, the problem of optimal " well posedness" can be completely understood with the help of the Strichartz–Pecher estimates. This work was carried out in recent years by Lindblad and Sogge[L-So] and also Kapitanski[Ka]. As explained above, the case of equations nonlinear in the derivatives of ϕ is far more difficult. Consider the following examples of such equations:

(I) $$\Box \phi = Q(\phi, \partial \phi) + C(\phi),$$

with $Q(\phi, \partial \phi)$ a quadratic form, linear in the first derivatives of ϕ which we denote by $\partial \phi$, and C cubic in ϕ. The Yang–Mills equations take, roughly, such a form in the Lorentz gauge.

(II) $$\Box \phi = \Gamma(\phi) Q(\partial \phi, \partial \phi)$$

with Q quadratic in $\partial \phi$ and Γ an arbitrary smooth function of ϕ. This is the case of Wave Maps from Minkowski space to Riemannian manifold, expressed in local coordinates.

(III) $$\Box_{g(\phi)} \phi = \Gamma(\phi) Q(\partial \phi, \partial \phi)$$

where $\Box_{g(\phi)}$ denotes the D'Alembertian with respect to a metric depending on ϕ. The Einstein-Vacuum equations can be, roughly, written in this form relative to the so called wave coordinates.

In a series of papers, written mainly in collaboration with M. Machedon, we have been able to make progress on this problem, in the case of equations of type (I) and (II). Here is a short summary of our main results:

i) In both cases (I) and (II) we have developed methods which go a long way to show that the initial value problem is well posed for $s > s_c$, without any additional assumptions on the nonlinear terms, provided that the space dimension $n \geq 5$. My student S. Selberg is now in the process of completing the program.

ii) For lower dimensions one needs an additional assumption[4] on the quadratic terms $Q(\phi, \partial \phi)$ and respectively $Q(\partial \phi, \partial \phi)$. These assumptions are

---

[4] The quadratic term Q should be expressed in terms of null forms. As shown by Lindblad [L] the result may fail if this "null condition" is not verified

verified in the interesting cases such as Yang–Mills and Wave–Maps equations.

iii) In the general case of Wave–Maps defined from Minkowski space $\mathbb{R}^{n+1}$ to an arbitrary Riemannian manifold, expressed relative to a local chart, we have verified well-posedness for $s > s_c$ in all dimensions[5] $n \geq 2$. The case of dimensions $n \geq 3$ has been discussed in [Kl–Ma4] while the case of dimension $n = 2$ was settled in [Kl–Se]. In [Kl–Ma5] we have considered the Wave–Maps problem defined from $\mathbb{R}^{n+1}$ to a Lie group, without any local chart assumption on the target manifold. Instead of local coordinates we have expressed the equations relative to a frame of invariant vector fields, similar to that used by Helein[He], see also [Ch–Za]. We were able to prove, for a slightly simplified model problem, well-posedness for $s > s_c$ in dimension $n = 3$. The case of dimension 2 remains open.

iv) In the general case of equations of type II verifying the null condition, Machedon and I were recently able to prove, see [Kl–Ma6], well posedness for $s > s_c$ in all dimensions $n \geq 3$. The proof of the result introduces an important technical innovation, which I will discuss below, responsible for all our recent advances.

v) In the case of Yang–Mills equations the critical exponents are $s_c = \frac{1}{2}$ for $n = 3$ and $s_c = 1$ for $n = 4$. In four space dimensions, $n = 4$, and in the simpler case of abelian gauge theories such as Maxwell–Klein–Gordon, Machedon and I were recently able to prove, see [Kl–Ma8] well posedness for all $s > s_c$. The nonabelian case is somewhat harder and requires both the technique of [Kl–Ma6] as well as an extension of the new type of "Strichartz inequalities" derived in [Kl–Ma7]. Recently such estimates were proved by Tataru. The general result is going to appear in [Kl–Ta].

In the case $n = 3$ Machedon and I have proved, see [Kl–Ma3], global well posedness for $s = 1$, corresponding to the energy norm. This extends the previous global regularity result in [E–Mo]. Using an adaptation of our techniques S. Cuccagna was recently able to prove well posedness for $s > \frac{3}{4}$ in the simpler case of Abelian gauge theories. He makes use in an essential way of the Strichartz type" estimates of [Kl–Ma7]. I am confident that the new methods introduced in [Kl–Ma6] will extend the result to the non-abelian case. The range $\frac{1}{2} < s < \frac{3}{4}$ remains wide open. It is clear the methods we have used so far break-down.

Finally I will try to discuss below some of the main ingredients in the proof of the results mentioned above. For simplicity I will restrict myself to equations of the type,

$$(1) \qquad\qquad \Box \phi^I = \sum_{J,K} \Gamma^I_{JK} B^I_{JK}(\phi^J, \phi^K)$$

---

[5]The critical exponent for Wave -Maps is $s_c = \frac{n}{2}$.

in $\mathbb{R}^{n+1}$. Here $\square = -\partial_t^2 + \Delta$ denotes the standard D'Alembertian, Γ_{JK}^I are constants and the B_{JK}^I any of the null forms,

$$(2a) \qquad Q_0(\phi, \psi) = \partial_\alpha \phi \cdot \partial^\alpha \psi = -\partial_t \phi \partial_t \psi + \sum_{i=1}^n \partial_i \phi \partial_i \psi$$

$$(2b) \qquad Q_{\alpha\beta}(\phi, \psi) = \partial_\alpha \phi \partial_\beta \psi - \partial_\beta \phi \partial_\alpha \psi \quad 0 \le \alpha < \beta \le n.$$

Consider the space-time norms $N_{s,\delta}(\phi) = \|\Lambda_+^s \Lambda_-^\delta \phi\|_{L^2(\mathbb{R}^{n+1})}$ where $\tilde{\phi}$ denotes the space-time Fourier transform of ϕ, $w_\pm(\tau, \xi) = 1 + \left| |\tau| \pm |\xi| \right|$ and $\widetilde{\Lambda_\pm \phi} = w_\pm \tilde{\phi}$. I will denote by by $H_{s,\delta}$ the space of tempered distributions in $\mathbb{R}^{n+1}$ for which the norm $N_{s,\delta}(\phi)$ is finite. Estimates involving the homogeneous[6] version of these norms, for the optimal exponents $\delta = \pm\frac{1}{2}$ have first appeared in [Kl–Ma1]. These type of estimates were used in the proof of global well–posedness of the Yang–Mills equations in $\mathbb{R}^{3+1}$, see[Kl–Ma3]. The inhomogeneous version, which was first used in [Kl–Ma4], was inspired from Bourgain's work [B] on the KdV equations[7]. The inhomogeneous norms appear naturally in connection with Bourgain's time cut–off idea which allows one, essentially, to replace the symbol $\tau^2 - |\xi|^2$ of $\square$ by $w_+ \cdot w_-(\tau, \xi)$.

To show well posedness in H^s it suffices, roughly, to prove the following bilinear estimate,

$$(3) \qquad N_{s,\delta}\left(\Lambda_+^{-1}\Lambda_-^{-1} Q(\phi, \psi)\right) \le C N_{s,\delta}(\phi) N_{s,\delta}(\psi).$$

This estimate holds true, see [Kl–Ma4] and [Kl–Se], for all $s > s_c = \frac{n}{2}$ and any n in the case of the null form $Q = Q_0$. It is however false[8] for $n = 3$, and the null forms $Q_{\alpha\beta}$. The key new idea in [Kl–Ma6], which overcomes this difficulty, is to consider the auxiliary norms

$$(4) \qquad M_\delta(\phi) = \sup_{\||b\|| \le 1, \, N_{0,\delta}(u) \le 1} \int \int |w_+^{\frac{1}{2}} w_-^{\frac{1}{2}} w_-^{2\delta-1} \tilde{\phi}(\tau, \xi) \cdot \widetilde{ub}(\tau, \xi)| d\tau d\xi$$

where,

$$(5) \qquad \||b\|| = \| \int |\tilde{b}(\tau, \cdot)| d\tau \|_{L^2(\mathbb{R}^3)}.$$

---

[6]i.e. with $w_+(\tau, \xi)$ replaced by $\left| |\tau| \pm |\xi| \right|$.

[7]See also [K–P–V].

[8]For s close to the critical exponent s_c. This is proved in [Kl–Ma].

We denote by B the space of tempered distributions b for which $|||b|||$ is finite. Observe that if $b \in B$ then,

$$(6) \qquad\qquad \sup_{t \in \mathbb{R}} \|b(t, \cdot)\|_{L^2(\mathbb{R}^3)} \leq |||b||| < \infty.$$

The crucial properties of the space B are:

(B_1) For any $\delta > \frac{1}{2}$, $H_{0,\delta} \subset B$

(B_2) If $u \in H_{0,\delta}$, $\delta > \frac{1}{2}$ and $b \in B$ then, for all $\delta \geq \frac{1}{2}$ and $s > 1$, we have[9],

$$\Lambda_-^{-\delta} \Lambda_+^{-s}(ub) \in B.$$

This allows one to replace the false estimate (3) with the following result, in $\mathbb{R}^{3+1}$,

THEOREM. *Assume that,*

$$(7a) \qquad\qquad N_{s,\delta}(\phi),\ N_{s,\delta}(\psi) \leq 1$$

$$(7b) \qquad\qquad M_\delta(\phi),\ M_\delta(\psi) \leq 1$$

Then, for all $\delta > \frac{1}{2}$ and $s > \frac{1}{2} + 2\delta$,

$$(8a) \qquad\qquad N_{s,\delta}\left(\Lambda_-^{-1}\Lambda_+^{-1}Q(\phi,\psi)\right) \leq C$$

$$(8b) \qquad\qquad M_\delta\left(\Lambda_+^{-1}\Lambda_-^{-1}Q(\phi,\psi)\right) \leq C$$

The difference between the null form Q_0 and $Q_{\alpha\beta}$ can be easily seen from the cancellation properties of their symbols. Indeed writing,

$$(9) \qquad \widetilde{Q(\phi,\psi)} = \int q(\tau - \lambda, \xi - \eta; \lambda, \eta)\tilde{\phi}(\tau - \lambda, \xi - \eta)\tilde{\psi}(\lambda, \eta)\,d\lambda\,d\eta$$

we have, in the case of the null form Q_0, $q_0(\tau, \xi; \lambda, \eta) = \tau\lambda - \xi \cdot \eta$ and, in the case of the null forms Q_{ij}, $q_{ij}(\tau, \xi; \lambda, \eta) = \xi_i \eta_j - \xi_j \eta_i$. Also, for Q_{0i}, $q_{0i}(\tau, \xi; \lambda, \eta) = \tau\eta_i - \lambda\xi_i$. Observe that $2q_0(\tau, \xi; \lambda, \eta) = (|\tau + \lambda|^2 - |\xi + \eta|^2) - (|\tau|^2 - |\xi|^2) - (|\lambda|^2 - |\eta|^2)$. Hence,

$$(10a)\ \ |q_0(\tau, \xi; \lambda, \eta)| \leq C\left(w_+ w_-(\tau + \lambda, \xi + \eta) + w_+ w_-(\tau, \xi) + w_+ w_-(\lambda, \eta)\right).$$

---

[9]The property B_2 breaks-down if we replace the space B by $H_{0,\delta}$.

On the other hand for the null forms $q = q_{\alpha\beta}$ we only have,

$$(10b) \qquad |q(\tau,\xi;\lambda,\eta)| \leq c w_+^{\frac{1}{2}}(\tau,\xi) w_+^{\frac{1}{2}}(\lambda,\eta) w_+^{\frac{1}{2}}(\tau+\lambda,\xi+\eta) W^{\frac{1}{2}}(\tau,\xi;\lambda,\eta)$$

where $W(\tau,\xi;\lambda,\eta)$ is the maximum of the weights $w_-(\tau,\xi)$, $w_-(\lambda,\eta)$, $w_-(\tau+\lambda,\xi+\eta)$.

The formulas 10a, 10b allow us to split the corresponding integrals in three parts, each of which exhibits a cancellation relative to the degenerate weights w_-. Thus, in the case of the null form $Q = Q_0$ the estimate 3 reduces to the following product estimates:

PROPOSITION 1. *Let $s > \frac{n}{2}$, $\delta > \frac{1}{2}$. Then,*

$$(11a) \qquad \|\phi \cdot \psi\|_{s,\delta} \leq c\|\phi\|_{s,\delta}\|\psi\|_{s,\delta}$$
$$(11b) \qquad \|\phi \cdot \psi\|_{s-1,\delta-1} \leq c\|\phi\|_{s-1,\delta-1}\|\psi\|_{s,\delta}$$

Proposition 1 can be reduced further to the following sharp[10] bilinear estimates for solutions of linear wave equations:

PROPOSITION 2. *Let ϕ, ψ be solutions of the homogeneous wave equations $\Box\phi = \Box\psi = 0$ in $\mathbb{R}^{n+1}$ subject to the standard I.V.P $\phi = f$, $\psi = g$ $\partial\phi = \partial\psi = 0$ at $t = 0$. Then,*

$$(12a) \qquad \left\| \||\tau| - |\xi|\|^{\frac{1}{2}} \widetilde{\phi\psi} \right\|_{L^2(\mathbb{R}^{n+1})} \leq c\|f\|_{L^2(\mathbb{R}^n)}\|g\|_{\dot{H}^{\frac{n}{2}}(\mathbb{R}^n)}$$

$$(12b) \qquad \left\| \frac{\||\tau| - |\xi|\|^{\frac{1}{2}}}{(|\tau| + |\xi|)^{\frac{n-2}{2}}} \widetilde{\phi\psi} \right\|_{L^2(\mathbb{R}^{n+1})} \leq c\|f\|_{L^2(\mathbb{R}^n)}\|g\|_{\dot{H}^{\frac{n}{2}}(\mathbb{R}^n)}$$

Estimates of the type of Proposition 2 were first proved for dimension $n \geq 3$ in [Kl–Ma1] and for dimension $n = 2$ in [Kl–Se] with the help of the estimates in [K-l-Ma7].

As explained above the estimate 3 is false in the case of the null forms $Q = Q_{\alpha\beta}$ and we have to use the auxiliary norms M_δ, see [Kl–Ma6].

Another important ingredient in our approach consists in a generalized version of the Strichartz–Pecher inequalities[Kl–Ta]. Consider the wave equation,

$$\Box\phi = 0$$

subject to the initial conditions $\phi(0,x) = f(x)$, $\partial_t\phi(0,x) = 0$. The standard Strichartz–Pecher inequalities can be written in the form,

---

[10]They hold for the optimal exponent $s = \frac{n}{2}$.

(13)
$$\|\phi^2\|_{L_t^q L_x^r} \le \|\Lambda^s f\|_{L^2}^2$$

where $\Lambda = (-\Delta)^{\frac{1}{2}}$ and

(14a)
$$s = \frac{n}{2}(1 - \frac{1}{r}) - \frac{1}{2q}$$

(14b)
$$0 \le \frac{1}{q} \le \min(\gamma(r), 1)$$

(14c)
$$(\frac{1}{q}, \gamma(r)) \ne (1, 1)$$

with $\gamma(r) = \frac{n-1}{2}\alpha(r)$ and $\alpha(r) = 1 - \frac{1}{r}$. The generalized version is,

THEOREM. *Under the same conditions (14a–14c) and for*

(14d)
$$0 \le \sigma < n\alpha(r) - \frac{2}{q},$$

we have,

(15)
$$\|\Lambda^{-\sigma}\phi^2\|_{L_t^q L_x^r} \le \|\Lambda^{s-\frac{\sigma}{2}} f\|_{L^2}^2$$

The theorem was first proved in the special case $q = r = 2$ in [Kl–Ma7]. The general case will appear in [Kl–Ta].

Finally in the end of this lecture I will mention some of the main open problems.

1. Clarify the remaining issues concerning the case s arbitrarily close to s_c. The most conspicuous unsolved problem is that of the Yang–Mills equations in $\mathbb{R}^{3+1}$.

2. With all the progress we have made on the question of optimal well posedness our methods fall short of treating the really interesting case $s = s_c$. This must require a truly new idea. One can distinguish two serious difficulties. The first has to do with the nature of the estimates and the corresponding spaces of functions needed. Roughly speaking we need to define norms involving the space–time L^2 norm of fractional powers of the D' Alembertian $\Box$. To treat the critical case $s = s_c$ it seems necessary to work with the quantity $\|\Box^{\frac{1}{2}}\phi\|_{L^2}$ and its dual $\|\Box^{-\frac{1}{2}}\phi\|_{L^2}$. Unfortunately, however, these cannot be defined in any straightforward way[11]. Is there a substitute? The second difficulty has to do with the fact that in the interesting examples one may not even expect to get "well posedness" for $s = s_c$, even for data small in the $\| \ \|_{s_c}$ norm. What we expect however to be true is the following weaker version. "The solutions corresponding to any smooth initial data, small in the $\| \ \|_{s_c}$ norm, stay smooth for all time". Here are two outstanding conjectures related to this:

---

[11] For $s > s_c$ this difficulty can be avoided by using an inhomogeneous analogue of these norms such as that introduced by Bourgain on his work on the KdV equations.

CONJECTURE 1. *All solutions of Wave–Maps equations, corresponding to smooth data, defined from the $2 + 1$ Minkowski space–time to a negatively curved Riemannian space are globally smooth.*

CONJECTURE 2. *All solutions, corresponding to smooth data, of the Yang–Mills equations in the $4 + 1$ Minkowski space-time are globally smooth.*

3. In the supercritical case $s_c > 1$ we expect singularities to develop. In the special case of Wave Maps defined from $\mathbb{R}^{3+1}$ to S^2 Shatah[Sh] has constructed a special example of break-down solutions. We don't have however any general theory. One fundamental challenge, connected with the issues discussed above, is to construct global, generalized, solutions consistent with the development of singularities. One would like to prove a well–posedness result for data in a dimension-less space of functions consistent with singularities. More precisely the Sobolev space H^{s_c}, which was discussed above, is dimension-less but does not allow singularities such as those exhibited by Shatah. In elliptic theory one works with Morrey spaces, what are the correct spaces for wave equations? In a first approximation one would like to find such a space and its corresponding norm and prove that small data[12] relative to that norm lead to global in time solutions.

4. Another outstanding problem is to extend the results we have derived so far for equations of type I and II to those of type III. The case of quasilinear equations is far more difficult and will necessarily require new ideas. The most interesting examples, such as Einstein equations and the equations of Elasticity are quasilinear. As I have mentioned above the classical well posedness results require $s > 1 + s_c$. At first glance it seems that even a small improvement of this result requires an analogue of the Strichartz inequalities for linear wave equations with very rough coefficients, significantly rougher than $C^{1,1}$. It is known however that these inequalities fail if the coefficients are that rough, see [Sm–So]. I believe however that a careful analysis of the specific nonlinear problem will allow one to overcome this difficulty. In other words the corresponding linear problem has coefficients which, though rough in the sense discussed above, satisfy additional assumptions. In particular, the fact that the coefficients themselves are related to the original nonlinear problem, significantly enhances the regularity properties of the corresponding solutions of the Eikonal equation[13]. This fact has played a crucial role in the proof of the nonlinear stability of the Minkowski metric, see [Ch–Kl] chapter 13, and I believe that it will prove to be equally important in this context.

---

[12]This result would be analogous to the well–known global result of Glimm for 1–D conservation laws which requires small data in BV.

[13]The Eikonal equation plays a fundamental role in the construction of a parametrix to the linearized problem.

References

(1) [B] J. Bourgain, "Fourier transform restriction phenomena for certain lattice subsets and applications to non–linear evolution equations," I, II, *Geom. Funct. Analysis* **3** (1993), 107–156, 202–262.

(2) [Br] P. Brenner, "On $L^p - L^{p'}$ estimates for the wave equations," Math. Z. **145**(1975), 251–254.

(3) [Ch–Kl] D. Christodoulou and S. Klainerman, "The global nonlinear stability of the Minkowski Space," Princeton Mathematical Series, 41.

(4) [Ch–Za] Christodoulou D. and A. Shadi–Tahvildar–Zadeh, "On the regularity of spherically symmetric wave–maps," *Comm. P. Appl. Math.* **46**(1993), 1041–1091.

(5) [E–M] D. Eardley, V. Moncrief, "The global existence for Yang–Mills–Higgs fields in M^{3+1}," C-M-P 83 (1982), 171–212.

(6) [He] F. Helein, "Regularity of weakly harmonic maps from a surface into a manifold with symmetries," *Manuscr. Math.* **70**(1991), 203–218.

(7) [G–V] J. Ginibre, G. Velo, "Generalized Strichartz inequality for the wave equation," to appear in J.F.A.

(8) [Gr1] M. Grillakism "Regularity and asymptotic behavior of the wave equation with a critical nonlinearity," Ann. of Math. **132**(1990), 485-509.

(9) [Gr2] M. Grillakis, "Regularity of the wave equation with a critical nonlinearity," Comm. Pure Appl. Math. **45**(1992), 749-774.

(10) [Ka] L. V. Kapitanski, "Weak and yet weaker solutions of semilinear wave equations," Comm. P.D.E. **19**(1994), 1629-1676.

(11) [K–P–V] C. Kenig, G. Ponce, and L. Vega, "The Cauchy problem for the Korteweg–De Vries equation in Sobolev spaces of negative indices," *Duke Math. J.* **71** No. 1 (1994), 1–21.

(12) *[Kl–Ma 1] S. Klainerman and M. Machedon, "Space-time estimates for null forms and the local existence theorem," Comm. Pure Appl. Math. **46**(1993), 1221-1268.*

(13) *[Kl–Ma 2] S. Klainerman and M. Machedon, "On the Maxwell–Klein–Gordon equation with finite energy," Duke Math. J. **74** No. 1 (1994), 19–44.*

(14) *[Kl–Ma 3] S. Klainerman and M. Machedon, "Finite energy solutions for the Yang-Mills solutions in $\mathbb{R}^{3+1}$," Ann. of Math. **142** (1995), 39–119.*

(15) *[Kl–Ma 4] S. Klainerman and M. Machedon, "Smoothing estimates for null forms and applications," Duke Math. J. **81** (1995), 99-103.*

(16) *[Kl–Ma 5] S. Klainerman and M. Machedon, "On the regularity properties of a model problem related to wave maps," to appear in Duke*

Math. J.

(17) [Kl–Ma 6] S. Klainerman and M. Machedon, "Estimates for null forms and the spaces," $H_{s,\delta}$, International Math Research Notices 17(1996), 853-866.

(18) [Kl–Ma 7] S. Klainerman and M. Machedon, with appendices by J. Bourgain and D. Tataru, "Remark on Strichartz type inequalities," International Math Research Notices (1996), No. 5.

(19) [Kl–Ma 8] S. Klainerman and M. Machedon, "On the optimal local regularity for gauge field theories," preprint.

(20) [Kl–Se] S. Klainerman and S. Selberg, "Remark on the optimal regularity for equations of Wave Maps type," to appear in Comm. P.D.E.

(21) [Kl–Ta] "On the optimal local regularity for the Yang–Mills equations in $\mathbb{R}^{4+1}$," preprint.

(22) [L–So] H. Lindblad and C. Sogge, "On existence and scattering with minimal regularity for semilinear wave equations," J. Funct. Anal. (1995), 357–426.

(23) [Pe] "Non–linear small data scattering for the wave and Klein–Gordon equations," Math. Z. 185(1984), 261-270.

(24) [Se] I. Segal "Space–time decay for solutions of wave equations," Adv. in Math. 22(1976), 305–311.

(25) [Sh] Shatah, "Weak solutions and development of singularities in the SU(2) -σ model," C.P.A.M. 41(1988), 459-469.

(26) [Sm–So] H. Smith and C. Sogge, "On Strichartz and eigenfunction estimates for low regularity metrics," preprint.

(27) [So] C. Sogge, "Propagation of singularities and maximal functions in the plane," Invet. Math. 104(1991), 349–376.

(28) [St] Stein, E. M., Harmonic Analysis, Princ. Univ. Press (1993).

(29) [Str] M. Struwe, "Global regular solutions to the u^5 Klein–Gordon equations," Ann. Sc. Norm. Pisa 15(1988), 495–513.

(30) [S1] R. S. Strichartz, "Restrictions of Fourier transform to quadratic surfaces and decay of solutions of wave equations," Duke Math. J. 44(1977), 705-714.

(31) [S2] R. S. Strichartz, "Convolutions with kernels having singularities on the sphere," Trans. A.M.S. 148(1970), 461-471.

Proceedings of Symposia in Applied Mathematics
Volume **54**, 1998

THE KdV ZERO-DISPERSION LIMIT AND DENSITIES OF DIRICHLET SPECTRA

C. David Levermore

Dedicated to Peter Lax and Louis Nirenberg
on the fete of their 70th birthdays
by a grateful student

ABSTRACT. This paper serves two purposes. First, it reviews the history of the zero-dispersion limit problem for the Korteweg-de Vries (KdV) equation, with emphases placed on the influence of Peter Lax. The theory developed by Lax and the author, which characterizes the limit of the conserved densities and fluxes in terms of the solution of a maximization problem, will be explained and expanded in the light insights gained over the last fifteen years. Second, it summarizes and clarifies recent results obtained by Ercolani, Zhang and the author that characterize the maximizer as the limiting density of half-line Dirichlet spectra of the associated Schrödinger operator. This enables one both to strengthen the limits asserted for the conserved densities and fluxes, and to establish the limit of the associated Weyl functions.

1. INTRODUCTION

Just over twenty years ago, Peter Lax called me into his office at the Courant Institute to propose a thesis topic. His pitch went something like this. The zero-dispersion limit problem for the Korteweg-de Vries equation might shed light on the generation of oscillations in numerical schemes. Moreover, it would be a great chance to exercise the (still relatively new) inverse scattering transform methods, in which he knew I had interest. While I was not convinced by his first line at the time, I was by his second. I was hooked. The problem proved far richer than even he suspected. I hope that some of this richness is conveyed below.

1991 *Mathematics Subject Classification.* Primary 35-01, 35Q53; Secondary 58F19, 34E20.

Key words and phrases. Korteweg-de Vries equation, zero-dispersion limit, Weyl functions, Dirichlet spectra, Herglotz representation.

Presented 11 June 1996 at *Recent Advances in Partial Differential Equations and Applications*, a conference celebrating the 70th birthdays of Peter D. Lax and Louis Nirenberg, held in Venice, Italy during the week of 10–14 June 1996.

This work was supported in part by the NSF under grant DMS-9404570.

1.1 Origins. Historically, the zero-dispersion limit problem was to determine

$$(1.1) \qquad\qquad \lim_{\epsilon \to 0} u^\epsilon(x,t)\,,$$

where u^ϵ solves the Korteweg-de Vries (KdV) initial-value problem

$$(1.2a) \qquad\qquad \partial_t u^\epsilon - 6u^\epsilon \partial_x u^\epsilon + \epsilon^2 \partial_{xxx} u^\epsilon = 0\,,$$
$$(1.2b) \qquad\qquad u^\epsilon(x,0) = v(x)\,,$$

for initial data v that is independent of ϵ. The modern history of this problem traces its origin back to the early 1960's. Motivated by shallow water theory, Whitham [**83**] studied this problem using his theory of (strongly) nonlinear "geometrical optics" [**84**], a modulation theory. About the same time Kruskal, motivated by (1.2) as an approximation to the Fermi-Pasta-Ulam (FPU) lattice, began looking into the problem [**50**]. It is important to understand that, while the parallel of this problem to the zero-viscosity limit problem was obvious to all, it was not the main motivation to study the KdV zero-dispersion limit.

Of course, just by looking at (1.2) one would guess that

$$(1.3) \qquad\qquad \lim_{\epsilon \to 0} u^\epsilon(x,t) = u(x,t)\,,$$

where u solves the Hopf initial-value problem

$$(1.4) \qquad\qquad \partial_t u - 6u\partial_x u = 0\,, \qquad u(x,0) = v(x)\,,$$

at least so long as the solution of this problem remains classical. This is indeed the case [**55**]. However, the solution of (1.4) generally develops an infinite derivative in a finite time, the so-called break time. As it does so, its third derivative becomes larger even faster, which is inconsistent with our neglect of the third derivative term in (1.2a) when deriving (1.4). The problem is therefore to understand how the limit (1.1) behaves after this break time.

Numerical experiments with small values of ϵ had shown that a train of oscillations would emerge and fan out from the impending singularity. Moreover, as ϵ was decreased it was seen that the wavelength of these oscillations scaled like $O(\epsilon)$, while their envelope had a clear $O(1)$ limiting behavior. The limit (1.1) could at best be a weak limit after such oscillations emerge. To leading order, Whitham [**83**] approximated the oscillations as a modulation of the family of spatially periodic traveling wave solutions of (1.2a). He showed that the resulting hyperbolic system of three modulation equations could be put in Riemann invariant form, thereby giving one of the first clues of the integrability of the KdV equation. There were however some problems with his picture. First, it was local in both space and time. This is to say that it did not connect regions with oscillations, which are described by his modulation equations, to regions without oscillations, which are described by (1.4). In particular, there was no connection back to the initial data (1.2b). Second, his modulation equations are genuinely nonlinear, so their solutions also generally develop an infinite derivative in a finite time. He suggested that jump discontinuities (shocks) might form that separate regions described by different solutions of the modulation equations or regions described the solutions of the modulation and Hopf equations. Following up this idea, Gurevich and Pitaevski [**37**] computed the self-similar onset of oscillations described by matching solutions of the Whitham and Hopf equations. However, the overall idea was not correct. Rather, much richer behavior would be discovered.

Numerical experiments again provided a clue. It should be pointed out that at the time such experiments were extremely difficult to carry out because: (1) the grid size Δx needs to satisfy $\Delta x \ll \epsilon$ to resolve the oscillations; (2) one needs $\epsilon \ll 1$ to produce enough oscillations to suggest a limiting behavior. Still, numerical work suggested that, after the Whitham theory breaks down, the solution behaves as a modulated family of quasiperiodic solutions. Such a picture was beyond the grasp of classical mathematical tools.

Fortunately, a golden age was dawning that would revolutionize mathematics. Peter Lax was one of the leaders of this revolution. We will examine this revolution from the narrow perspective of the zero-dispersion limit problem of the KdV equation.

1.2 The Golden Age and Its Impact. Motivated by numerical discovery of the KdV soliton by Zabusky and Kruskal [87], the KdV initial-value problem (1.2) was solved by Gardner, Greene, Kruskal and Miura [34] for initial data $v(x)$ that decays sufficiently rapidly as $|x| \to \infty$. As abstracted by Lax [51], their critical observation was that the KdV equation (1.2) is the solvability condition for the linear system

$$(1.5) \qquad \mathcal{L}(t)f = \lambda f, \qquad \partial_t f = \mathcal{B}(t)f,$$

where the operators $\mathcal{L}$ and $\mathcal{B}$ are given by

$$(1.6a) \qquad \mathcal{L}(t)f \equiv -\epsilon^2 \partial_{xx}f + uf,$$
$$(1.6b) \qquad \mathcal{B}(t)f \equiv -4\epsilon^2 \partial_{xxx}f + 3\big(u\partial_x + \partial_x u\big)f;$$

namely, it can be recast as

$$(1.7) \qquad \partial_t \mathcal{L} = [\mathcal{B}, \mathcal{L}] \equiv \mathcal{B}\mathcal{L} - \mathcal{L}\mathcal{B}.$$

Here λ in (1.5) is an eigenvalue of the Schrödinger operator $\mathcal{L}(t)$ given by (1.6a). The operators $\mathcal{L}$ and $\mathcal{B}$ are called a Lax pair for the KdV equation. Even before the seminal discovery of Gardner, Greene, Kruskal and Miura, when faced with mounting evidence that the KdV equation possessed an infinity of conserved quantities, Lax postulated the KdV equation could be recast in the abstract setting (1.7) because it evolves $\mathcal{L}$ as an infinitesimal similarity transform, thereby conserving its spectrum.

The method to solve the KdV initial-value problem [34,35] is an example of what has become known as an inverse spectral or inverse scattering transform (IST) method. Loosely speaking, it goes as follows. Given $u(x,0) = v(x)$, the potential of $\mathcal{L}(0)$:
   1) one computes the asymptotics of the eigenfunctions f as $|x| \to \infty$ (the scattering data);
   2) the evolution of the scattering data is determined explicitly;
   3) the potential $u(x,t)$ of $\mathcal{L}(t)$ is then obtained through inverse scattering theory [16,26].

Guided by this framework the KdV initial-value problem was solved for periodic initial data first for a number of special cases [21,43,52,63,71] and then completely by McKean and Trubowitz [62]. The case of initial data with differing limits as $x \to \pm\infty$ was similarly solved [11].

The framework was also applied to solve the initial-value problem for many other systems. There were the nonlinear Schrödinger (NLS) equation [72,89,90], the sine-Gordon equation [2], the AKNS system [3], the Kadomtsev-Petviashvili (KP) equation [61], and many other partial differential equations. There were also systems of ordinary differential equations of arbitrary dimension like the Toda lattice [28,29,40,47], the Ablowitz-Ladik lattice [4,6,66,77], and more. A good review of the early history of these discoveries can be found in [70]. The influence of this body of work on mathematics has been profound! For example, in geometry it has provided new tools for the construction and classification of such things as Riemann matrices [1], constant mean curvature tori [7,86], harmonic maps on spheres and tori [20,41,74,76], and evolving knotted filaments [10]. It also plays a important role in string theory [49].

These new tools gave rise to two new strategies for approaching the KdV zero-dispersion limit problem:

A)   Flaschka, Forest and McLaughlin [31] developed a multiphase modulation theory for quasiperiodic KdV solutions, thereby generalizing the Whitham theory. Like that theory, this theory is local in space and time.

B)   Lax and Levermore [54,55] analyzed the scattering transform as ϵ tends to zero, obtaining a global characterization of the limit in the case of $v(x) \leq 0$. Venakides extended this to $v(x) \geq 0$ [78] and to periodic $v(x)$ [80].

The linear analogs of these strategies are (A) geometrical optics and (B) the stationary phase method applied to a Fourier integral. This paper will focus on strategy B. Section 2 gives needed background about the scattering transform. Section 3 lays out the basic results obtained by strategy B. Section 4 presents recent results obtained with Ercolani and Zhang that illuminate and strengthen the basic results. Finally, some open problems and perspectives will be given.

2. INTEGRABILITY OF THE KdV EQUATION

In this section we review the relevant facts regarding the complete integrability of the KdV initial-value problem by the scattering transform method. Throughout this section $\epsilon > 0$ will be considered to be a *fixed* positive constant, and hence, no implicit ϵ dependence will be indicated. The KdV initial-value problem (1.2) is therefore written simply as

$$(2.1) \qquad \partial_t u - 6u\partial_x u + \epsilon^2 \partial_{xxx} u = 0, \qquad u(x,0) = v(x),$$

while the linear system (1.5) becomes

$$(2.2a) \qquad \mathcal{L}(t)f \equiv -\epsilon^2 \partial_{xx} f + uf = \lambda f,$$

$$(2.2b) \qquad \partial_t f = \mathcal{B}(t)f \equiv -4\epsilon^2 \partial_{xxx} f + 3(u\partial_x + \partial_x u)f,$$

where λ is an eigenvalue of the Schrödinger operator $\mathcal{L}(t)$.

2.1. The Scattering Transform. For every u that decays sufficiently rapidly as $|x| \to \infty$, the L^2-spectrum of the Schrödinger operator (2.2a) consists of the nonnegative semi-axis $\lambda \geq 0$ along with a finite set (possibly empty) of negative simple eigenvalues $\lambda_1, \cdots, \lambda_N$. The asymptotic behavior of an eigenfunction $f = f(k,x)$ corresponding to a $\lambda = k^2 > 0$ in the continuous spectrum is given by

$$(2.3) \quad f(k,x) \sim \begin{cases} \exp\left(\dfrac{-ikx}{\epsilon}\right), & \text{as } x \to -\infty, \\[2mm] \dfrac{1}{T(k)}\exp\left(\dfrac{-ikx}{\epsilon}\right) + \dfrac{R(k)}{T(k)}\exp\left(\dfrac{ikx}{\epsilon}\right), & \text{as } x \to +\infty, \end{cases}$$

where $T(k)$ and $R(k)$ are the so-called transmission and reflection coefficients. The asymptotic behavior of a real unit normalized eigenfunction $f = f_j(x)$ corresponding to a discrete eigenvalue $\lambda_j = -\eta_j^2 < 0$ is given by

$$(2.4) \qquad f_j(x) \sim \exp\left(\frac{-\eta_j x + \chi_j}{\epsilon}\right), \qquad \text{as } x \to +\infty.$$

The inverse theory prescribes that the fundamental scattering data consist of the reflection coefficient $R(k)$, the eigenvalues λ_j, and the norming exponents χ_j. The transmission coefficient $T(k)$, all other asymptotic information, and even u itself can be determined from this fundamental set [16,26]. The full generality of this determination will not be needed here.

As u evolves according to the KdV equation (2.1a) then for every positive $\lambda = k^2$ in the continuous spectrum there is a solution $f(k,x,t)$ of the linear system (2.2) that evolves as
(2.5)

$$f(k,x,t) \sim \begin{cases} \exp\left(\dfrac{-ikx - 4ik^3 t}{\epsilon}\right), & \text{as } x \to -\infty, \\[2mm] \dfrac{1}{T(k)}\exp\left(\dfrac{-ikx - 4ik^3 t}{\epsilon}\right) + \dfrac{R(k)}{T(k)}\exp\left(\dfrac{ikx + 4ik^3 t}{\epsilon}\right), & \text{as } x \to +\infty, \end{cases}$$

while the eigenvalues $\lambda_j = -\eta_j^2$ remain fixed in time and there is a solution $f_j(x,t)$ of (2.2) that evolves as

$$(2.6) \qquad f_j(x,t) \sim \exp\left(\frac{-\eta_j x + 4\eta_j^3 t + \chi_j}{\epsilon}\right), \qquad \text{as } x \to +\infty,$$

where the $T(k)$, $R(k)$, and χ_j are the initial scattering data. Comparing (2.5) and (2.6) with (2.3) and (2.4), the time evolution of the scattering data for $u(x,t)$ can be read off as

$$(2.7) \qquad \chi_j(t) = \chi_j + 4\eta_j^3 t, \qquad R(k,t) = R(k)\exp\left(i\frac{8k^3 t}{\epsilon}\right).$$

Hence, given $R(k)$, η_j, and χ_j computed from the initial data $u(x,0)$, the solution $u(x,t)$ of the KdV equation (2.1) is then determined by inverse scattering from the $R(k,t)$, η_j, and $\chi_j(t)$ given by (2.7).

2.2. The Weyl Function and Conservation Laws. Given the solution $f(k,x,t)$ of system (2.2) that satisfies the asymptotics (2.5), we introduce the Weyl function, $m(k,x,t)$, by

$$(2.8) \qquad m \equiv \epsilon \partial_x \log f.$$

A direct calculation starting from (2.2a) shows that m satisfies the Ricatti equation

$$(2.9) \qquad -\epsilon \partial_x m = k^2 - u + m^2.$$

Proceeding from (2.2b) while using (2.9) to eliminate all explicit occurrences of u yields

$$(2.10) \qquad -\epsilon \partial_t \log f = -6k^2 m - 2m^3 + \epsilon^2 \partial_{xx} m.$$

Cross-differentiating (2.8) and (2.10) shows that m satisfies the modified KdV equation [68]

$$\text{(2.11)} \qquad \partial_t m + \partial_x \big(-6k^2 m - 2m^3 + \epsilon^2 \partial_{xx} m \big) = 0 \,.$$

This single local conservation law depends on the parameter k. The analytic dependence of the Jost function on k [16] gives the expansions

$$\text{(2.12a)} \qquad m = \epsilon \, \partial_x \log f = -ik + \sum_{\ell=0}^{\infty} \frac{\rho_\ell}{(2ik)^{\ell+1}} \,,$$

(2.12b)

$$-6k^2 m - 2m^3 + \epsilon^2 \partial_{xx} m = -\epsilon \, \partial_t \log f = i4k^3 + \sum_{\ell=0}^{\infty} \frac{\nu_\ell}{(2ik)^{\ell+1}} \,.$$

Placing these expansions into the modified KdV equation (2.11) then yields

$$\text{(2.13)} \qquad \partial_t \rho_\ell + \partial_x \nu_\ell = 0 \,, \qquad \text{for } \ell = 0, 1, \cdots,$$

thereby indicating that the KdV equation (2.1a) may posses an infinite family of nontrivial local conservation laws with densities ρ_ℓ and fluxes ν_ℓ [69].

The densities ρ_ℓ are computed in terms of u by formally substituting the expansion (2.12a) into the Ricatti equation (2.9). The first five densities are found to be

$$\text{(2.14)} \qquad \begin{aligned} \rho_0 &= -u \,, \\ \rho_1 &= -\epsilon \, \partial_x u \,, \\ \rho_2 &= u^2 - \epsilon^2 \partial_{xx} u \,, \\ \rho_3 &= 4\epsilon \, u \partial_x u - \epsilon^3 \partial_{xxx} u \,, \\ \rho_4 &= -2u^3 + 6\epsilon^2 u \partial_{xx} u + 5\epsilon^2 (\partial_x u)^2 - \epsilon^4 \partial_{xxxx} u \,. \end{aligned}$$

In general, the coefficients ρ_ℓ satisfy the recursion relation

$$\text{(2.15)} \qquad \rho_{\ell+1} = \sum_{j=1}^{\ell} \rho_{\ell-j} \rho_{j-1} + \epsilon \, \partial_x \rho_\ell \,, \qquad \text{for } \ell = 1, 2, \cdots.$$

Notice that ρ_1 and ρ_3 are perfect derivatives, a property shared by all the densities with odd indices. This can be seen by decomposing m into m_e and m_o, its even and odd components as a function of k, and observing that the odd component of the Ricatti equation (2.9) gives the relation

$$\text{(2.16)} \qquad m_e = -\tfrac{1}{2}\epsilon \, \partial_x \log m_o \,.$$

Hence, only the densities ρ_{2n}, for $n = 0, 1, \cdots$, can give nontrivial local conservation laws, which in fact they do [69]. Of course, any of these densities can be modified by a multiplicative constant or an additive perfect derivative without changing any of the essential mathematics, but the normalization adopted here in terms of the expansions (2.12) is the most natural for our purposes.

2.3. Hamiltonian Structure and the KdV Hierarchy. Associated with each of the nontrivial locally conserved densities ρ_{2n} is a conserved functional H_n that can be expressed as

$$(2.17) \qquad H_n = \int_{-\infty}^{\infty} \tfrac{1}{2} \rho_{2n} \, dx \,, \qquad \text{for } n = 0, 1, \cdots .$$

The first three conserved functionals H_n so obtained from (2.14) are simply
$$(2.18)$$
$$H_0 = -\int_{-\infty}^{\infty} \tfrac{1}{2} u \, dx \,, \qquad H_1 = \int_{-\infty}^{\infty} \tfrac{1}{2} u^2 \, dx \,, \qquad H_2 = -\int_{-\infty}^{\infty} u^3 + \tfrac{1}{2} \epsilon^2 (\partial_x u)^2 \, dx \,.$$

The KdV equation (2.1a) can be recast in the Hamiltonian form

$$(2.19) \qquad \partial_t u + \partial_x \frac{\delta H_2}{\delta u} = 0 \,,$$

where the Hamiltonian is H_2 given in (2.18). The infinite family of conserved functionals H_n are independent and satisfy the Poisson commutation relation

$$(2.20) \qquad 0 = \{H_n, H_m\} \equiv \int_{-\infty}^{\infty} \frac{\delta H_n}{\delta u} \partial_x \frac{\delta H_m}{\delta u} \, dx \,, \qquad \text{for } n, m = 0, 1, \cdots ,$$

and are therefore in involution [33,88]. Each H_m except H_0 is a Hamiltonian which generates a member of the so-called KdV hierarchy commuting flows [27]. Let t_m denote the time variable associated with m^{th} KdV flow as generated by H_{m+1} through the equation

$$(2.21) \qquad \partial_{t_m} u + \partial_x \frac{\delta H_{m+1}}{\delta u} = 0 \,, \qquad \text{for } m = 0, 1, \cdots .$$

Recalling H_1 and H_2 from (2.18), for $m = 0$ equation (2.21) is the flow of positive x-translation by t_0 while for $m = 1$ it is the KdV flow (2.19) with t_1 being identified with t. This observation motivated the introduction of the $\tfrac{1}{2}$ into (2.17). By the Poisson commutation (2.20), every H_n is conserved by each of these flows.

Because these flows commute, they may be solved simultaneously for $u = u(x, \mathbf{t})$ where, for the moment, the time vector $\mathbf{t} = (t_0, t_1, \cdots)$ is understood to have all but finitely many t_m zero. Associated with each such $\mathbf{t}$ is the odd polynomial $p(\,\cdot\,, \mathbf{t})$ defined by

$$(2.22) \qquad p(\eta, \mathbf{t}) = \sum_{m=0}^{\infty} t_m 4^m \eta^{2m+1} \,.$$

The simultaneous evolution of the scattering data is then given by

$$(2.23) \qquad \chi_j(\mathbf{t}) = \chi_j + p(\eta_j, \mathbf{t}) \,, \qquad R(k, \mathbf{t}) = R(k) \exp\left(\frac{-2p(ik, \mathbf{t})}{\epsilon} \right) ,$$

and $u(x, \mathbf{t})$ is determined by inverse scattering. This solution may be extended to all those time vectors $\mathbf{t}$ in the class

$$(2.24) \qquad \mathcal{E} \equiv \left\{ \mathbf{t} \, : \, \lim_{m \to \infty} |t_m|^{\frac{1}{m}} = 0 \right\},$$

which corresponds to extending the class of mappings $\eta \mapsto p(\eta, \mathbf{t})$ to those that are entire, odd, and have real symmetry.

Under the mth flow the density ρ_{2n} satisfies the local conservation law

$$(2.25) \qquad \partial_{t_m}\rho_{2n} + \partial_x \nu_{2m,2n} = 0, \qquad \text{for } m,n = 0,1,\cdots.$$

The flux $\nu_{2m,2n}$ for the density ρ_{2n} under the mth flow is determined by the expansion

$$(2.26) \qquad -\epsilon\,\partial_{t_m}\log f = -4^m (ik)^{2m+1} + \sum_{\ell=0}^{\infty} \frac{\nu_{2m,\ell}}{(2ik)^{\ell+1}}.$$

Note that for $m = 0$, corresponding to the x-translational flow, one has $\nu_{0,\ell} = \rho_\ell$, while for $m = 1$, corresponding to the KdV flow, one has $\nu_{2,\ell} = \nu_\ell$ as determined by (2.12b). By setting $n = 0$ in (2.25) one recovers the mth flow (2.21) in the KdV hierarchy. There is a deep relationship [30] between the coefficients in the expansion (2.12a) and some of those in the expansions (2.26). One finds that

$$(2.27) \qquad (2n+1)\,\rho_{2n} - \nu_{2n,0} = \text{a perfect } x\text{-derivative}.$$

As stated earlier, an additive perfect derivative does not alter the essential mathematics in that $(2n+1)\,\rho_{2n}$ and $\nu_{2n,0}$ are densities for the same conserved quantity.

3. THE ZERO-DISPERSION LIMIT

The problem of the zero-dispersion limit for the KdV considers the solution $u^\epsilon(x,\mathbf{t})$ of the whole KdV hierarchy (2.9) as a function of ϵ for some initial data $v(x)$ that is independent of ϵ, and tries to determine the limiting behavior of the conserved densities ρ_n^ϵ and fluxes $\nu_{m,n}^\epsilon$ as ϵ tends to zero. Lax and Levermore [54,55] analyzed the limiting behavior of the scattering and inverse scattering transform using a WKB analysis of (2.2) and a kind of steepest descent argument to obtain a characterization of the (weak) limits in terms of the solution of a variational problem. Below we review some aspects of this theory.

3.1. WKB Analysis. For small ϵ approximate scattering data of $v(x)$ may be computed using the WKB method. For simplicity, we present only the case when the initial data $v(x)$ is a single nonpositive well with a minimum value of $-\eta_{max}^2$ as depicted in Figure 3.1. This was the case treated in [54,55]. Other classes of initial data have been analyzed and will be discussed later.

The WKB analysis yields eigenvalues $\lambda_j^\epsilon = -\eta_j^{\epsilon\,2}$ that are distributed within the open interval $(-\eta_{max}^2, 0)$ so as to be consistent with the so-called Weyl asymptotic density $\varphi(\eta)$ with respect to the spectral variable η, which is given by

$$(3.1) \qquad \varphi(\eta) = \int_{x_-(\eta)}^{x_+(\eta)} \frac{\eta}{\sqrt{-v(x)-\eta^2}}\,dx,$$

where $x_-(\eta) < x_+(\eta)$ are determined by $v(x_\pm) = -\eta^2$ as shown in Figure 3.1. Specifically, $\eta = \eta_j^\epsilon$ is the unique positive solution of

$$(3.2) \qquad \frac{1}{\pi\epsilon}\Phi(\eta) = j - \tfrac{1}{2}, \qquad \text{for } j = 1,\cdots,N^\epsilon,$$

where $\Phi(\eta)$ is defined by

$$(3.3) \qquad \Phi(\eta) \equiv \int_\eta^{\eta_{max}} \varphi(\xi)\,d\xi = \int_{x_-(\eta)}^{x_+(\eta)} \sqrt{-v(x)-\eta^2}\,dx,$$

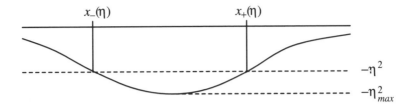

FIGURE 3.1. A typical initial data $v = v(x)$ considered here. The defining relations for the functions $x = x_\pm(\eta)$ are also indicated.

and N^ϵ is defined by

$$(3.4) \qquad N^\epsilon \equiv \mathrm{Int}\left(\frac{1}{\pi\epsilon}\Phi(0)\right) = \mathrm{Int}\left(\frac{1}{\pi\epsilon}\int_{-\infty}^{\infty}\sqrt{-v(x)}\,dx\right).$$

Here $\mathrm{Int}(\,\cdot\,)$ rounds its argument to the closest integer value, with half-integers rounded down. So determined, the WKB eigenvalues are strictly ordered as

$$(3.5) \qquad -\eta_{max}^2 < \lambda_1^\epsilon < \cdots < \lambda_j^\epsilon < \cdots < \lambda_{N^\epsilon}^\epsilon < 0.$$

The corresponding norming exponents obtained from the WKB analysis are

$$(3.6) \qquad \chi_j^\epsilon = \chi(\eta_j^\epsilon), \qquad \text{for } j = 1, \cdots, N^\epsilon,$$

where the so-called asymptotic norming exponent $\chi(\eta)$ is defined by

$$(3.7) \qquad \chi(\eta) = \eta\,x_+(\eta) + \int_{x_+(\eta)}^{\infty}\left(\eta - \sqrt{\eta^2 + v(x)}\right)dx.$$

Finally, the reflection coefficient is found to be zero to all orders for $k \neq 0$.

3.2. Reflectionless Approximation. Motivated by this calculation, we choose to replace the exact initial data $v(x)$ by the reflectionless potential $v^\epsilon(x)$ corresponding to the above WKB scattering data. While this device is not the best one might hope for, it is partially justified a posteriori by the result [**55**] that v^ϵ converges strongly to v. The solution $u^\epsilon(x, \mathbf{t})$ of the whole KdV hierarchy corresponding to this reflectionless initial data can be constructed from the approximate scattering data by the Kay-Moses formula [**48**],

$$(3.8) \qquad u^\epsilon(x, \mathbf{t}) = -2\epsilon^2\partial_{xx}\log\tau^\epsilon(x, \mathbf{t}),$$

where the so-called tau-function $\tau^\epsilon(x, \mathbf{t})$ is the $N^\epsilon \times N^\epsilon$ determinant

$$(3.9) \qquad \tau^\epsilon(x, \mathbf{t}) = \det\left(I + G^\epsilon(x, \mathbf{t})\right).$$

Here the matrix G^ϵ has the form

$$(3.10) \qquad G^\epsilon(x, \mathbf{t}) = \left(\frac{1}{\eta_j^\epsilon + \eta_l^\epsilon}\exp\left(\frac{a(\eta_j^\epsilon, x, \mathbf{t}) + a(\eta_l^\epsilon, x, \mathbf{t})}{\epsilon}\right)\right)_{j,l},$$

where

$$(3.11) \qquad a(\eta, x, \mathbf{t}) \equiv -\eta\,x + p(\eta, \mathbf{t}) + \chi(\eta).$$

Associated conserved densities and fluxes are given by [**30,70,85**]

(3.12a) $\frac{2n+1}{2}\rho^\epsilon_{2n}(x,\mathbf{t}) = -\epsilon^2 \partial_{xt_n} \log \tau^\epsilon(x,\mathbf{t})$,

(3.12b) $\frac{2n+1}{2}\nu^\epsilon_{2m,2n}(x,\mathbf{t}) = \epsilon^2 \partial_{t_m t_n} \log \tau^\epsilon(x,\mathbf{t})$.

It is important to understand that these are *not* the same densities and fluxes as were defined in (2.12a) and (2.26) respectively. They do, however, agree with those up to additive derivatives so as to give the same integrated conservation laws.

The Kay-Moses formula (3.8) is the special case of (3.12a) obtained by setting $n = 0$ while using (2.14) and the fact that $\partial_{t_0}\tau^\epsilon = -\partial_x\tau^\epsilon$. Moreover, a comparison of (3.12a) with (3.12b) when $m = 0$ yields

$$(2n+1)\,\rho^\epsilon_{2n} = -2\epsilon^2 \partial_{xt_n} \log\tau^\epsilon(x,\mathbf{t}) = 2\epsilon^2 \partial_{t_0 t_n} \log\tau^\epsilon(x,\mathbf{t}) = \nu^\epsilon_{2n,0}\,.$$

This is equivalent to (2.27) modulo the freedom to add a perfect x-dervative to the conserved densities which does not alter the essential content of the conservation laws.

The class of $\mathbf{t}$ for which the above constructions are valid may be extended from the class $\mathcal{E}$ of (2.24) to the real-valued weighted ℓ^1 space

(3.13) $$\mathcal{T} \equiv \left\{ \mathbf{t} \; : \; \sum_{m=0}^{\infty} |t_m| 4^m \eta_{max}^{2m+1} < \infty \right\}.$$

The space $\mathcal{T}$ shall be considered as a representation of the dual space of c_0, the space of sequences that converge to zero, and, as such, to be endowed with its weak-$*$ topology. With this topology for $\mathcal{T}$ the matrix G^ϵ and the function τ^ϵ are clearly continuous over $\mathbb{R} \times \mathcal{T}$.

At this point we make the important remark that for each $\epsilon > 0$ the function

(3.14) $(x,\mathbf{t}) \mapsto \log \tau^\epsilon(x,\mathbf{t})$ is smooth, positive and convex.

These properties are all consequences of the fact that the matrix $G^\epsilon(x,\mathbf{t})$ of (3.10) is Hermitian positive [**55**], which by (3.9) immediately gives that $\tau^\epsilon(x,\mathbf{t}) > 1$. The regularity and positivity are then obvious. The convexity was first noticed in the study of the semiclassical limit of the nonlinear Schrödinger equation [**44,45,46**], but the proof applies to the KdV case without change. Indeed, this convexity will hold for any tau-function, $\tau^\epsilon(x,\mathbf{t})$, that is a linear combination of real exponentials with positive coefficients.

3.3. Lax-Levermore Theory. One can establish [**54,55**] the existence of the limit

(3.15) $\lim_{\epsilon \to 0} \epsilon^2 \log \tau^\epsilon = q$ in $C^1(\mathbb{R} \times \mathcal{T})$.

The limit $q(x,\mathbf{t})$ is characterized by the maximization problem

(3.16) $q(x,\mathbf{t}) = \dfrac{2}{\pi} \max \left\{ Q(\psi; x,\mathbf{t}) \; : \; \psi \in \mathcal{A} \right\},$

where the admissible set $\mathcal{A}$ is defined by

(3.17) $\mathcal{A} \equiv \left\{ \psi \in L^1([0, \eta_{max}]) \; : \; 0 \le \psi \le \varphi \right\},$

and the quadratic functional $Q(\psi; x, \mathbf{t})$ is defined by

(3.18)
$$Q(\psi; x, \mathbf{t}) = \frac{2}{\pi} \int_0^{\eta_{max}} a(\eta, x, \mathbf{t})\, \psi(\eta)\, d\eta$$
$$+ \frac{1}{\pi^2} \int_0^{\eta_{max}} \int_0^{\eta_{max}} \log\left|\frac{\eta - \xi}{\eta + \xi}\right| \psi(\eta)\psi(\xi)\, d\eta\, d\xi\,,$$

with $a(\eta, x, \mathbf{t})$ given by (3.11). The initial data $v(x)$ enters this problem through the Weyl asymptotic density $\varphi(\eta)$ defined by (3.1), which determines the admissible set $\mathcal{A}$ in (3.17), and through the asymptotic norming exponent $\chi(\eta)$ defined by (3.7), which arises in the $a(\eta, x, \mathbf{t})$ appearing in (3.18).

Upon analyzing the maximization problem (3.16–3.18), one finds [55] that for every $(x, \mathbf{t}) \in \mathbb{R} \times \mathcal{T}$ the functional $Q(\psi; x, \mathbf{t})$ is bounded above and strictly concave over $\mathcal{A}$. Hence, there exists a unique $\psi = \psi^*(x, \mathbf{t})$ in the admissible set $\mathcal{A}$ at which the maximum in (3.16) is attained:

(3.19)
$$q(x, \mathbf{t}) = Q(\psi^*(x, \mathbf{t}); x, \mathbf{t})\,.$$

Moreover, $\psi^*(x, \mathbf{t})$ depends continuously on $(x, \mathbf{t})$ when $\mathcal{A}$ is identified as a set of densities endowed with the weak topology of measures. The uniqueness and continuity of ψ^* imply [55] that $q(x, \mathbf{t})$ is a continuously differentiable function of $(x, \mathbf{t})$ with

(3.20a)
$$\partial_x q(x, \mathbf{t}) = -\frac{2}{\pi} \int_0^{\eta_{max}} \eta\, \psi^*(\eta, x, \mathbf{t})\, d\eta\,,$$

(3.20b)
$$\partial_{t_n} q(x, \mathbf{t}) = \frac{2}{\pi} \int_0^{\eta_{max}} 4^n \eta^{2n+1}\, \psi^*(\eta, x, \mathbf{t})\, d\eta\,,$$

where we have used $\partial_{t_n} p(\eta, \mathbf{t}) = 4^n \eta^{2n+1}$ by (2.22). Because $0 \in \mathcal{A}$, we see that $q(x, \mathbf{t})$ is nonnegative. Because the map $(x, \mathbf{t}) \mapsto q(x, \mathbf{t})$ is realized as a supremum of linear functions of $(x, \mathbf{t})$, it is intrinsically convex. The regularity and convexity of the function q also imply that its Hessian matrix of distributional second derivatives is measurable function that is given by a Hessian matrix of classical derivatives almost everywhere. Its values in terms of ψ^* are given by

$$\begin{pmatrix} \partial_{xx} q(x, \mathbf{t}) & \partial_{xt_n} q(x, \mathbf{t}) \\ \partial_{t_m x} q(x, \mathbf{t}) & \partial_{t_m t_n} q(x, \mathbf{t}) \end{pmatrix}_{m,n}$$

(3.21)
$$= \begin{pmatrix} -\dfrac{2}{\pi} \displaystyle\int_0^{\eta_{max}} \eta\, \partial_x \psi^*(\eta, x, \mathbf{t})\, d\eta & \dfrac{2}{\pi} \displaystyle\int_0^{\eta_{max}} 4^n \eta^{2n+1}\, \partial_x \psi^*(\eta, x, \mathbf{t})\, d\eta \\ -\dfrac{2}{\pi} \displaystyle\int_0^{\eta_{max}} \eta\, \partial_{t_m} \psi^*(\eta, x, \mathbf{t})\, d\eta & \dfrac{2}{\pi} \displaystyle\int_0^{\eta_{max}} 4^n \eta^{2n+1}\, \partial_{t_m} \psi^*(\eta, x, \mathbf{t})\, d\eta \end{pmatrix}_{m,n},$$

where the derivatives on ψ^* are understood in the sense of distributions.

Combining (3.15) and (3.21) with (3.12) leads to the weak limits

(3.22)
$$\frac{2n+1}{2} \lim_{\epsilon \to 0} \rho_{2n}^\epsilon(x, \mathbf{t}) = -\partial_{xt_n} q(x, \mathbf{t}) = -\frac{2}{\pi} \int_0^{\eta_{max}} 4^n \eta^{2n+1}\, \partial_x \psi^*(\eta, x, \mathbf{t})\, d\eta\,,$$

$$\frac{2n+1}{2} \lim_{\epsilon \to 0} \nu_{2m,2n}^\epsilon(x, \mathbf{t}) = \partial_{t_m t_n} q(x, \mathbf{t}) = \frac{2}{\pi} \int_0^{\eta_{max}} 4^n \eta^{2n+1}\, \partial_{t_m} \psi^*(\eta, x, \mathbf{t})\, d\eta\,,$$

for the conserved densities and fluxes associated with the solution $u^\epsilon(x, \mathbf{t})$. The above limits for ρ^ϵ_{2n} are weak with respect to the x variable, while those for $\nu^\epsilon_{2m,2n}$ are weak with respect to the t_m variable. In particular, (3.22) gives

$$(3.23) \qquad \lim_{\epsilon \to 0} u^\epsilon(x, \mathbf{t}) = -2\,\partial_{xx} q(x, \mathbf{t}) = \frac{4}{\pi} \int_0^{\eta_{max}} \eta\,\partial_x \psi^*(\eta, x, \mathbf{t})\,d\eta\,,$$

in the sense of x-distributions, which resolves the question raised in (1.1).

Remark 3.1: The general limit formulas for all conserved densities and fluxes of the KdV hierarchy (3.22) were not derived in [54,55], but first appeared in the context of the semiclassical limit of the defocusing nonlinear Schrödinger hierarchy [44,45,46]. However, these generalizations are straightforward given (3.12) once the strong limit of $\epsilon^2 \log \tau^\epsilon$ is established as was done in [55].

Remark 3.2: The significance of this result is that it provides a *global* characterization of the zero-dispersion limit for all the conserved densities and fluxes of the KdV hierarchy as a function of the initial data through the maximization problem (3.16). This is not just of theoretical interest, as it has proven computationally effective [65].

Remark 3.3: Of course, the story of the zero-dispersion limit goes far beyond this characterization. The variational problem associated with (3.16) was transformed into a Riemann-Hilbert problem and systems of hyperbolic equations were found that locally describe the evolution of the zero-dispersion limit [54,55,57]. These so-called Whitham modulation equations were found independently by Flaschka, Forest and McLaughlin [31] by the averaging of families of quasi-periodic KdV solutions, thereby generalizing the early work of Whitham [83,84]. While this modulation approach provides a far better picture of what the KdV solution might look like for ϵ small, it is intrinsically local in nature and cannot connect regions in $(x, \mathbf{t})$-space that are governed by the averaging different families of KdV solutions. In particular, it does not relate the dynamics back to the initial data. On the other hand, the global nature of the characterization (3.15) provides a prescription of how the modulation equations arising from averaging different families partition $(x, \mathbf{t})$ space [55]. Venakides tried to bridge the gap between these two approaches by developing a picture of the dynamics of related spectra [79,80] and a heuristic theory of the microstructure of solutions based on quantizing the Lax-Levermore variational problem [81]. Recently, a powerful asymptotic method for analyzing integrable systems has been introduced by Deift and Zhou [12,17,19]; it is currently being applied to extend the Lax-Levermore theory so as to establish the limiting microstructure of KdV solutions [18,64]. All but the most recent of these results and more are surveyed in [22,56].

Remark 3.4: Here we have treated initial data $v(x)$ of the simple single-well form shown in Figure 3.1. With varying degrees of rigor, Venakides analyzed the cases when v was positive [78] and periodic [80]. As the limit of the inverse spectral transform takes quite a different character in each of these settings, these works were far from direct generalizations of the earlier work. However, once the appropriate maximization problem is identified, the associated variational problem is again transformed into a Riemann-Hilbert problem and the *same* hyperbolic systems of Whitham modulation equations are found to locally describe the evolution of the zero-dispersion limit.

Remark 3.5: This kind of analysis has also been carried out on other integrable systems. The semiclassical limit of the defocusing cubic nonlinear Schrödinger (NLS) equation was the first to be treated after the KdV equation [45,46]. Because that work introduced the hierarchical approach shown above, it therefore included the defocusing modified KdV (mKdV) equation, which is the next equation in the NLS hierarchy. Both the KdV and defocusing NLS hierarchies have self-adjoint Lax operators $\mathcal{L}$, so it was a surprise to some when a zero-dispersion limit for the focusing mKdV equation was the next to be treated [23]. In that case $\mathcal{L}$ is the *nonself-adjoint* Zakharov and Shabat operator [89]. The continuum limit of the Toda lattice was next [15,64]. It was the first discrete system to so analyzed. Most recently, a continuum limit for Ablowitz-Ladik lattice has been carried out [75]. Each of these limits has presented new difficulties to be overcome. The one common thread through each of these cases has been the convexity of a log-tau function like (3.14). For no system has all integrable classes of initial data and boundary conditions been treated.

4. DENSITIES OF DIRICHLET SPECTRA

The moment formulas (3.22) that globally characterize the KdV zero-dispersion limit are expressed in terms of the maximizer $\psi^*(\eta, x, \mathbf{t})$. In [24] it was shown that this maximizer is the limit of densities of Dirichlet spectra associated with the Schrödinger operator $\mathcal{L}(\mathbf{t})$ considered over half-lines. Moreover, a direct relationship was established in [24] between this maximizer and the zero-dispersion limit of the logarithm of the Jost functions associated with the inverse spectral transform. All the KdV conserved densities are encoded in the spatial derivative of these functions, known as Weyl functions. The Weyl functions are densities of measures that converge in the weak sense to a limiting measure. This limiting measure encodes all of the weak limits of the KdV conserved densities. Similar results have been established for the fluxes [25]. These results are surveyed and clarified below.

4.1. Jost Functions and Dirichlet Spectra. Given any solution of the KdV hierarchy $u^\epsilon(x, \mathbf{t})$, there is a unique eigenfunction $f^\epsilon(k, x, \mathbf{t})$ of the associated Schrödinger operator $\mathcal{L}^\epsilon(\mathbf{t})$ that is analytic in the upper-half complex k-plane and satisfies

$$(4.1) \qquad \mathcal{L}^\epsilon(\mathbf{t})f^\epsilon \equiv -\epsilon^2 \partial_{xx} f^\epsilon + u^\epsilon(x, \mathbf{t})f^\epsilon = k^2 f^\epsilon ,$$

and has the large k expansion of

$$(4.2) \quad f^\epsilon(k, x, \mathbf{t}) = \exp\left(\frac{-ikx + p(ik, \mathbf{t})}{\epsilon}\right)\left(1 + \frac{b_1^\epsilon(x, \mathbf{t})}{ik} + \cdots + \frac{b_n^\epsilon(x, \mathbf{t})}{(ik)^n} + \cdots\right).$$

This function is known as the Jost function of $\mathcal{L}^\epsilon(\mathbf{t})$. Rather than work with $f^\epsilon(k, x, \mathbf{t})$, it is natural to introduce [16] the normalized Jost function $g^\epsilon(k, x, \mathbf{t})$ by

$$(4.3) \qquad f^\epsilon(k, x, \mathbf{t}) = \exp\left(\frac{-ikx + p(ik, \mathbf{t})}{\epsilon}\right) g^\epsilon(k, x, \mathbf{t}).$$

As a function of k, the function g^ϵ is not only analytic wherever f^ϵ is analytic, but it is analytic at infinity. For this reason we will work with g^ϵ here, rather than work with f^ϵ as was done in [24]. This new presentation is, I believe, cleaner.

One can express $\epsilon \log g^\epsilon(k, x, \mathbf{t})$ in terms of Dirichlet spectra associated with the Schrödinger operator $\mathcal{L}^\epsilon(\mathbf{t})$, defined in (2.2), considered over half-lines. Specifically, the normalized Jost function g^ϵ can be written in the form

$$(4.4) \qquad g^\epsilon(k, x, \mathbf{t}) = \prod_{j=1}^{N^\epsilon} \frac{ik + \xi_j^\epsilon(x, \mathbf{t})}{ik - \eta_j^\epsilon},$$

where each ξ_j^ϵ lies in $[-\eta_{max}, \eta_{max}]$ and each nonzero $\mu_j^\epsilon(y, \mathbf{t}) = -\xi_j^{\epsilon\,2}(y, \mathbf{t})$ is an eigenvalue for the Schrödinger operator $\mathcal{L}^\epsilon(\mathbf{t})$ considered over either $L^2([y, \infty))$ or $L^2((-\infty, y])$ with a Dirichlet boundary condition at $x = y$ depending on whether $\xi_j^\epsilon(y, \mathbf{t})$ is negative or positive, respectively. The set of whole-line eigenvalues $\{\lambda_j^\epsilon\}$ is strictly ordered as in (3.5) while classical spectral theory [16] states that those $\{\mu_j^\epsilon(x, \mathbf{t})\}$ that are half-line Dirichlet eigenvalues are simple and can be ordered so as to interlace with the $\{\lambda_j^\epsilon\}$:

$$(4.5) \quad -\eta_{max}^2 < \lambda_1^\epsilon \leq \mu_1^\epsilon(x, \mathbf{t}) \leq \cdots \leq \lambda_j^\epsilon \leq \mu_j^\epsilon(x, \mathbf{t}) \leq \cdots \leq \lambda_{N^\epsilon}^\epsilon \leq \mu_{N^\epsilon}^\epsilon(x, \mathbf{t}) \leq 0,$$

where equalities may occur at a given $(x, \mathbf{t})$ only as triple coincidences of the form

$$(4.6) \qquad \cdots < \mu_{j-1}^\epsilon(x, \mathbf{t}) = \lambda_j^\epsilon = \mu_j^\epsilon(x, \mathbf{t}) < \cdots.$$

This convention determines the labeling of the $\{\xi_j^\epsilon(x, \mathbf{t})\}$; however, it should be noted that the ordering of the labels does not correspond to the numerical ordering of the $\{\xi_j^\epsilon(x, \mathbf{t})\}$.

The normalized Jost function can be related directly to the conserved densities and fluxes of the KdV hierarchy through the large k asymptotics of its logarithmic derivatives. Specifically,

$$(4.7) \quad \begin{aligned} \epsilon \partial_x \log g^\epsilon(k, x, \mathbf{t}) &= \sum_{\ell=0}^{\infty} \rho_\ell^\epsilon(x, \mathbf{t}) \left(\frac{1}{2ik}\right)^{\ell+1}, \\ -\epsilon \partial_{t_m} \log g^\epsilon(k, x, \mathbf{t}) &= \sum_{\ell=0}^{\infty} \nu_{2m,\ell}^\epsilon(x, \mathbf{t}) \left(\frac{1}{2ik}\right)^{\ell+1}. \end{aligned}$$

Therefore, the limiting form for all these quantities will be encoded in the limiting form of $\epsilon \log g^\epsilon(k, x, \mathbf{t})$.

The function $\epsilon \log g^\epsilon(k, x, \mathbf{t})$ is analytic over all complex k on the Riemann sphere $\overline{\mathbb{C}} \equiv \mathbb{C} \cup \{\infty\}$ outside the imaginary interval $[-i\eta_{max}, i\eta_{max}]$, that is, in

$$(4.8) \qquad \Omega \equiv \left\{ k \in \overline{\mathbb{C}} : k \notin [-i\eta_{max}, i\eta_{max}] \right\}.$$

Moreover, it has the Herglotz representation

$$(4.9) \qquad \epsilon \log g^\epsilon(k, x, \mathbf{t}) = \frac{1}{\pi} \int_{-\eta_{max}}^{\eta_{max}} \frac{d\sigma^\epsilon(\eta; x, \mathbf{t})}{ik - \eta},$$

where $d\sigma^\epsilon(\eta; x, \mathbf{t}) = w^\epsilon(\eta; x, \mathbf{t})\, d\eta$ with

$$(4.10) \qquad w^\epsilon(\eta; x, \mathbf{t}) = \int_\eta^{\eta_{max}} \epsilon\pi \sum_{j=1}^{N^\epsilon} \left[\delta\big(\xi - \eta_j^\epsilon\big) - \delta\big(\xi + \xi_j^\epsilon(x, \mathbf{t})\big) \right] d\xi.$$

The ordering (4.5) ensures that the density $w^\epsilon(\,\cdot\,; x, \mathbf{t})$ is nonnegative and supported within $[-\eta_{max}, \eta_{max}]$.

Remark 4.1: The dependence of the $\{\xi_j^\epsilon(x, \mathbf{t})\}$ on x is as follows. As $x \to -\infty$ one has $\xi_j^\epsilon(x, \mathbf{t}) \to -\eta_j^\epsilon$. As x increases, but away from triple coincidences, $\xi_j^\epsilon(x, \mathbf{t})$ strictly increases in either $(-\eta_j^\epsilon, -\eta_{j+1}^\epsilon)$ or $(\eta_{j+1}^\epsilon, \eta_j^\epsilon)$. A triple coincidence of the type (4.6) can only occur when $\xi_{j-1}^\epsilon(x, \mathbf{t})$ jumps from $(-\eta_{j-1}^\epsilon, -\eta_j^\epsilon]$ to $(\eta_j^\epsilon, \eta_{j-1}^\epsilon)$ while $\xi_j^\epsilon(x, \mathbf{t})$ jumps from $(\eta_{j+1}^\epsilon, \eta_j^\epsilon]$ to $(-\eta_j^\epsilon, -\eta_{j+1}^\epsilon)$ as x increases. This jumping mechanism is repeated $N^\epsilon(N^\epsilon - 1)/2$ times until finally $\xi_j^\epsilon(x, \mathbf{t}) \to \eta_j^\epsilon$ as $x \to \infty$. The dependence on $\mathbf{t}$ is much more complicated (cf. [**79**]).

4.2. Limits for Jost Functions and Dirichlet Spectra. The Herglotz representation (4.9) is exploited in [**24**] to show that the limit of $\epsilon \log g^\epsilon(k, x, \mathbf{t})$ as ϵ tends to zero can be evaluated in terms of the maximizer $\psi^*(\eta, x, \mathbf{t})$. To state the result, we first extend the functions $\psi^*(\,\cdot\,, x, \mathbf{t})$ and $\phi(\,\cdot\,)$, which are defined over $[0, \eta_{max}]$, to the whole real line as odd functions with their support contained within $[-\eta_{max}, \eta_{max}]$. One then has:

THEOREM 4.1. *The normalized Jost functions defined by (4.3) satisfy*

$$(4.11) \qquad \lim_{\epsilon \to 0} \epsilon \log g^\epsilon(k, x, \mathbf{t}) = \frac{1}{\pi} \int_{-\eta_{max}}^{\eta_{max}} \frac{d\sigma(\eta; x, \mathbf{t})}{ik - \eta},$$

where $d\sigma(\eta; x, \mathbf{t}) = w(\eta; x, \mathbf{t}) \, d\eta$ with

$$(4.12) \qquad w(\eta; x, \mathbf{t}) = \int_{\eta}^{\eta_{max}} \left[\phi(\xi) - \psi^*(\xi, x, \mathbf{t}) \right] d\xi.$$

The limit (4.11) is uniform over compact subsets of $\Omega \times \mathbb{R} \times \mathcal{T}$. In particular, the right side of (4.11) is analytic over the domain Ω given by (4.8).

Because, by (3.17), $\psi^*(\eta, x, \mathbf{t}) \leq \phi(\eta)$ over $[0, \eta_{max}]$, for every $(x, \mathbf{t})$ the w given by (4.12) is nonnegative and vanishes identically for those η outside the interval $[-\eta_{max}, \eta_{max}]$. Moreover, it also implies that $w(\eta; x, \mathbf{t})$ is bounded above uniformly in $(x, \mathbf{t})$ by the even extension of $\Phi(\eta)$ over $[-\eta_{max}, \eta_{max}]$, where $\Phi(\eta)$ was defined in (3.3) and is bounded above by $\Phi(0)$. The associated measure $d\sigma(\eta; x, \mathbf{t})$ is therefore a nonnegative, finite, regular Borel measure that is supported within the interval $[-\eta_{max}, \eta_{max}]$, so the right side of (4.11) is a Herglotz representation over Ω.

It follows from Theorem 4.1 that for every $\gamma \in C([-\eta_{max}, \eta_{max}], \mathbb{C})$ one has the limit

$$(4.13) \qquad \lim_{\epsilon \to 0} \int_{-\eta_{max}}^{\eta_{max}} \gamma(\eta) \, d\sigma^\epsilon(\eta; x, \mathbf{t}) = \int_{-\eta_{max}}^{\eta_{max}} \gamma(\eta) \, d\sigma(\eta; x, \mathbf{t}),$$

where the limit is uniform over compact subsets of $(\gamma, x, \mathbf{t})$ in $C([-\eta_{max}, \eta_{max}], \mathbb{C}) \times \mathbb{R} \times \mathcal{T}$. The convergence of $d\sigma^\epsilon$ to $d\sigma$ given in (4.13) is that of the weak topology of Borel measures. In [**24**] this convergence is strengthened as the following result concerning the convergence of their densities, w^ϵ and w, defined by (4.10) and (4.12).

THEOREM 4.2. *The measures dw^ϵ and dw defined by*

$$(4.14) \qquad dw^\epsilon(\eta; x, \mathbf{t}) = -\epsilon\pi \sum_{j=1}^{N^\epsilon} \left[\delta(\eta - \eta_j^\epsilon) - \delta(\eta + \xi_j^\epsilon(x, \mathbf{t})) \right] d\eta,$$

$$dw(\eta; x, \mathbf{t}) = -\left[\phi(\eta) - \psi^*(\eta, x, \mathbf{t}) \right] d\eta,$$

satisfy

(4.15) $$\lim_{\epsilon \to 0} dw^\epsilon(\eta; x, \mathbf{t}) = dw(\eta; x, \mathbf{t}),$$

where the convergence is in the weak topology of Borel measures over $[-\eta_{max}, \eta_{max}]$ and is uniform over compact subsets of $(x, \mathbf{t})$ in $\mathbb{R} \times \mathcal{T}$.

Upon restricting the measures in (4.15) to $[0, \eta_{max}]$ and using the consequence of definition (3.2) that $\phi(\eta)$ is the limiting density of the η_j^ϵ, one immediately obtains [24] the following characterization of the maximizer $\psi^*(\eta, x, \mathbf{t})$ in terms of the limiting density of the $\{\xi_j^\epsilon(x, \mathbf{t})\}$:

THEOREM 4.3. *In the weak topology of Borel measures over $[0, \eta_{max}]$ one has*

(4.16) $$\lim_{\epsilon \to 0} \epsilon \pi \sum_{j=1}^{N^\epsilon} \delta\big(\eta + \xi_j^\epsilon(x, \mathbf{t})\big) \, d\eta = \psi^*(\eta, x, \mathbf{t}) \, d\eta,$$

where the limit is uniform over subsets of $(x, \mathbf{t})$ that are compact in $\mathbb{R} \times \mathcal{T}$.

Only terms that have support in $[0, \eta_{max}]$ will contribute to the sum on the left side of (4.16). More specifically, only those terms for which $\xi_j^\epsilon(x, \mathbf{t})$ is nonpositive will contribute.

Remark 4.2: Under stronger (virtually unverifiable) hypotheses, and with weaker conclusions, a formal version of Theorem 4.3 is given in [**79**].

4.3. Limits for Weyl Functions and Conserved Densities. Formula (3.22) recovers the limiting form of the conserved densities and fluxes in terms of *distributional* derivatives of ψ^*. For the conserved densities, understood as densities with respect to dx, the convergence of the limit can be strengthened to that of the *weak topology of (finite) Borel measures in x*.

This analysis is mediated by a normalized form of the Weyl function introduced in (2.8), which, when g^ϵ is given by (4.3), is

(4.17) $$n^\epsilon(k, x, \mathbf{t}) \equiv \epsilon \, \partial_x \log g^\epsilon(k, x, \mathbf{t}) = \epsilon \sum_{j=1}^{N^\epsilon} \frac{\partial_x \xi_j^\epsilon(x, \mathbf{t})}{ik + \xi_j^\epsilon(x, \mathbf{t})}.$$

Basic properties of n^ϵ can be read off from the following.

LEMMA 4.4. *Each $n^\epsilon(k, x, \mathbf{t})$ is Herglotz over Ω with the representation*

(4.18) $$n^\epsilon(k, x, \mathbf{t}) = \frac{1}{\pi} \int_{-\eta_{max}}^{\eta_{max}} \frac{d\sigma_x^\epsilon(\eta; x, \mathbf{t})}{ik - \eta}.$$

where $d\sigma_x^\epsilon(\eta; x, \mathbf{t}) = w_x^\epsilon(\eta, x, \mathbf{t}) \, d\eta$ with

(4.19) $$w_x^\epsilon(\eta, x, \mathbf{t}) = \epsilon \pi \sum_{j=1}^{N^\epsilon} \partial_x \xi_j^\epsilon(x, \mathbf{t}) \, \delta\big(\eta + \xi_j^\epsilon(x, \mathbf{t})\big).$$

Note that formulas (4.18) and (4.19) follow directly from (4.17), so all that all one must check is that $d\sigma_x^\epsilon(\eta; x, \mathbf{t})$ is a nonnegative, finite Borel measure. By (4.19), these properties will follow from the fact (cf. [**24**]) that $\partial_x \xi_j^\epsilon(x, \mathbf{t}) \geq 0$.

The measures $d\sigma_x^\epsilon$ are called the associated Weyl spectral measures. Upon expanding (4.18) about $k = \infty$ and comparing the result with (4.7), one finds that

$$(4.20) \qquad \rho_\ell^\epsilon(x,\mathbf{t}) = \frac{2^{\ell+1}}{\pi} \int_{-\eta_{max}}^{\eta_{max}} \eta^\ell \, d\sigma_x^\epsilon(\eta; x, \mathbf{t}) = \frac{2^{\ell+1}}{\pi} \int_{-\eta_{max}}^{\eta_{max}} \eta^\ell \, w_x^\epsilon(\eta, x, \mathbf{t}) \, d\eta \, .$$

The zero-dispersion limit of the measures $d\sigma_x^\epsilon$ will therefore contain the limits of all the conserved densities.

The trick is to consider the family of product measures over $[-\eta_{max}, \eta_{max}] \times \mathbb{R}$ given by

$$(4.21) \qquad d\sigma_x^\epsilon(\eta; x, \mathbf{t}) \, dx = w_x^\epsilon(\eta, x, \mathbf{t}) \, d\eta \, dx \, .$$

Two facts now come into play. First, the total variation of these measures is bounded as

$$(4.22) \qquad \begin{aligned} \int_{-\infty}^\infty \int_{-\eta_{max}}^{\eta_{max}} w_x^\epsilon(\eta, x, \mathbf{t}) \, d\eta \, dx &= \int_{-\infty}^\infty \partial_x \left(\epsilon\pi \sum_{j=1}^{N^\epsilon} \xi_j^\epsilon(x, \mathbf{t}) \right) dx \\ &= 2\epsilon\pi \sum_{j=1}^{N^\epsilon} \eta_j^\epsilon \leq 2\epsilon\pi \eta_{max} N^\epsilon \, , \end{aligned}$$

which is uniformly bounded in both ϵ and $\mathbf{t}$. Hence, the $d\sigma_x^\epsilon \, dx$ lie in a set that is compact in the weak topology of Borel measures over $[-\eta_{max}, \eta_{max}] \times \mathbb{R}$. Second, for any compactly supported differentiable function $h = h(x)$ and $(k, \mathbf{t}) \in \Omega \times \mathcal{T}$ one has by Theorem 4.1 that

$$(4.23) \qquad \begin{aligned} \lim_{\epsilon \to 0} \int_{-\infty}^\infty h(x) \, n^\epsilon(k, x, \mathbf{t}) \, dx &= -\lim_{\epsilon \to 0} \int_{-\infty}^\infty \partial_x h(x) \, \epsilon \log g^\epsilon(k, x, \mathbf{t}) \, dx \\ &= \int_{-\infty}^\infty \partial_x h(x) \, \frac{1}{\pi} \int_{-\eta_{max}}^{\eta_{max}} \frac{d\sigma(\eta; x, \mathbf{t})}{ik - \eta} \, dx \, , \end{aligned}$$

where $d\sigma(\eta; x, \mathbf{t})$ is given as in (4.11) and the limit is uniform over $(k, \mathbf{t})$ in compact subsets of $\Omega \times \mathcal{T}$. From these facts we can infer the following.

THEOREM 4.5. *For each* $\mathbf{t} \in \mathcal{T}$ *there exist a Borel measure* $w_x(\eta, x, \mathbf{t}) \, d\eta \, dx$ *over* $[-\eta_{max}, \eta_{max}] \times \mathbb{R}$ *such that in the weak topology of Borel measures one has*

$$(4.24) \qquad \lim_{\epsilon \to 0} w_x^\epsilon(\eta, x, \mathbf{t}) \, d\eta \, dx = w_x(\eta, x, \mathbf{t}) \, d\eta \, dx \, ,$$

where the density $w_x(\eta, x, \mathbf{t})$ *is given by*

$$(4.25) \qquad w_x(\eta, x, \mathbf{t}) = -\partial_x \int_\eta^{\eta_{max}} \psi^*(\xi, x, \mathbf{t}) \, d\xi \, .$$

Moreover, the limit (4.24) *is uniform over compact subsets of* $\mathcal{T}$. *In particular, in the weak topology of Borel measures over* $\mathbb{R}$ *one has*

$$(4.26) \qquad \lim_{\epsilon \to 0} n^\epsilon(k, x, \mathbf{t}) \, dx = \frac{1}{\pi} \int_{-\eta_{max}}^{\eta_{max}} \frac{w_x(\eta; x, \mathbf{t})}{ik - \eta} \, d\eta \, dx \, ,$$

where the limit is uniform over $(k, \mathbf{t})$ *in compact subsets of* $\Omega \times \mathcal{T}$.

When Theorem 4.5 is combined with (4.20), it is seen that the conserved densities $\rho_\ell^\epsilon(x, \mathbf{t})$ will converge as densities of the Lebesgue measure dx. Similar things can be proved [25] about the $\nu_{2m,\ell}^\epsilon(x, \mathbf{t})$ appearing in (4.7). Specifically, we have the following:

THEOREM 4.6. *In the weak topology of (finite) Borel measures over* $\mathbb{R}$ *one has*

$$
\lim_{\epsilon \to 0} \rho_\ell^\epsilon(x, \mathbf{t}) \, dx = -\frac{2^{\ell+1}}{\pi} \partial_x \int_{-\eta_{max}}^{\eta_{max}} \eta^\ell \int_\eta^{\eta_{max}} \psi^*(\xi, x, \mathbf{t}) \, d\xi \, d\eta \, dx \,,
$$

(4.27)

$$
\lim_{\epsilon \to 0} \nu_{2m,\ell}^\epsilon(x, \mathbf{t}) \, dt_m = \frac{2^{\ell+1}}{\pi} \partial_{t_m} \int_{-\eta_{max}}^{\eta_{max}} \eta^\ell \int_\eta^{\eta_{max}} \psi^*(\xi, x, \mathbf{t}) \, d\xi \, d\eta \, dt_m \,.
$$

Here the first limit is uniform over compact subsets of $\mathcal{T}$ *while the second is uniform over compact subsets of* $\mathbb{R} \times \mathcal{T}_m$, *where* $\mathcal{T}_m \equiv \mathcal{T}/\{t_m \in \mathbb{R}\}$.

If one uses the fact that $\psi^*(\eta, x, \mathbf{t})$ is an odd function of η to evaluate the right sides of (4.27), one finds that the weak limits of the conserved densities and fluxes are given by

$$
\lim_{\epsilon \to 0} \rho_{2n}^\epsilon(x, \mathbf{t}) = -\frac{4}{\pi} \frac{4^n}{2n+1} \partial_x \int_0^{\eta_{max}} \eta^{2n+1} \psi^*(\eta, x, \mathbf{t}) \, d\eta \,,
$$

(4.28)

$$
\lim_{\epsilon \to 0} \nu_{2m,2n}^\epsilon(x, \mathbf{t}) = \frac{4}{\pi} \frac{4^n}{2n+1} \partial_{t_m} \int_0^{\eta_{max}} \eta^{2n+1} \psi^*(\eta, x, \mathbf{t}) \, d\eta \,,
$$

for $\ell = 2n$, $n = 0, 1, \cdots$ and

$$
(4.29) \qquad \lim_{\epsilon \to 0} \rho_{2n+1}^\epsilon(x, \mathbf{t}) = 0 \,, \qquad \lim_{\epsilon \to 0} \nu_{2m,2n+1}^\epsilon(x, \mathbf{t}) = 0 \,,
$$

for $\ell = 2n+1$, $n = 0, 1, \cdots$. Formulas (4.28) agree with those given in (3.22).

Remark 4.3: In the context of the semiclassical limit of the defocusing nonlinear Schrödinger hierarchy, it is known [46] that some conserved densities converge in the weak-L_{loc}^1 topology, not just in the weak topology of measures. I conjecture that *all* the conserved densities converge in the weak-L_{loc}^1 topology, both for the semiclassical limit of the defocusing nonlinear Schrödinger hierarchy and for the zero-dispersion limit of the KdV hierarchy, thereby strengthening Theorem 4.6.

Remark 4.4: We expect similar results to hold for any class of initial data for which the KdV equation is now known to be completely integrable. The same story should be true for the semiclassical limit of the defocusing nonlinear Schrödinger equation, where structures similar to those for KdV exist [44,45,46]. It is less obvious, but likely, that similar results hold for the semiclassical limit of the odd flows in the focusing NLS hierarchy with real-valued data, a case that includes the modified KdV equation, while it excludes the NLS equation itself. This limit was established in [23] even though the associated spectral problem is not self-adjoint, a structural difference that would make the extension of the above results to this case interesting.

5. CONCLUDING REMARKS

There are still many unresolved questions regarding the KdV zero dispersion limit. To begin with, can the replacement of the initial data by its WKB approximation (3.8) be avoided? Such a replacement has been used in all extensions of this theory so far, and most researchers (including me) believe it can be removed, but it has been resistant to our efforts to do so. One can also ask how much the analysis of the different cases of initial data can be unified? The fact that the limiting dynamics in all cases are governed by the same set of *hyperbolic* systems certainly points in that direction, as it shows the limit to have a local nature. Finally, there are the question of microstructure addressed in Remark 3.3 and the conjecture of Remark 4.3.

There are also many unresolved questions regarding the existence of analogous limits for other integrable equations. Remarks 3.3 and 4.4 may have given the impression that such limits should be easy to obtain for any integrable system. If so, this impression is wrong. Indeed, early numerical evidence does not yet indicate that such a dynamical limit even exists for either the focusing NLS equation [9,45] or the KP equations [60]. One thing the numerics does make clear is that these limits, if they exist, are not described by modulations of quasiperiodic motion. Rather, the microstructure that is seen appears to be chaotic in both time and space, forming intermittent concentrations. Furthermore, on the theoretical level, one sees that properties like (3.14) fail. Recently however, Bronski [8] has given striking numerical evidence that the nonself-adjoint Zakharov-Shabat eigenvalue problem associated with the focusing NLS equation does have remarkable limiting semiclassical behavior. This gives hope that progress can be made, perhaps by describing the motion by a modulation of geodesic flow on manifolds of negative curvature. The challenge to us here is simple: here is an integrable system, complete with the latest mathematical technology; does it have semiclassical limit dynamics?

Peter Lax initiated [53] the modern study of such limits in the discrete setting. (You may recall from the beginning of this article that understanding numerical schemes was one of his motivations to study the KdV zero-dispersion limit.) The simplest such integrable system is the Toda lattice. Following early work on its modulation theory [5,36] and self-similar solutions [13,82], its continuum limit was recently established [15,64]. Richer behavior is seen in the solutions of the Ablowitz-Ladik (AL) lattice. The AL lattice is to the Toda lattice in the discrete setting, what the NLS equations are to the KdV equation in the continuous setting. Following work on its modulation theory [66,67], its continuum limit was recently established in a regime similar to the defocusing NLS semiclassical limit [75]. The AL lattice also has regimes in which its continuum limit is similar to the focusing NLS semiclassical limit. In studying such regimes the AL lattice offers an advantage over NLS in that numerical experiments will be relatively easier to carry out on the AL lattice. This is because the AL small parameter *is* the lattice spacing, while the NLS small parameter must be resolved by the computational grid spacing.

Of course, the question of how to analyze nonintegrable nonlinear systems in such limits is wide open. Simple modulation theories have been worked out [84], and there have been some very preliminary analytical results [38], but the gaps in our knowledge are enormous. Because of the difficulty in resolving the small parameter mentioned above, numerical experiments with continuum systems have not yet given any breakthrough insights. For this reason, many numerical studies

have been done with discrete nonintegrable systems [**14,36,39,42,58,59,73**]. These
show a wealth of phenomena, not all of which have appeared in the integrable
context. In some cases, very preliminary analytical results have been obtained
[**14,36,59**], even regarding transitions [**59**]. At this stage, this component of the
investigation is truly an effort in experimental mathematics. I now think Peter Lax
understood this would be the case over twenty years ago.

What are the scientific and mathematical pay-offs of these investigations? My
own view is that nonlinear zero-dispersion (or semiclassical) phenomena are so
poorly understood mathematically, that should they arise in a physical setting,
they may not be recognized. For example, the "two fluid" theory of superfluids is
most likely a heuristic attempt to capture such phenomena. One pay-off may be
therefore a deeper understanding of superfluids, super conductivity, semiclassical
limits of nonlinear quantum mean-field theories generally, and the quantum Hall
effect in particular. Another example would be the nonlinear propagation of laser
pulses in optical fibers, where the semiclassical limit of defocusing NLS provides
an idealized description of optical shocks and wave breaking [**9,32**]. A pay-off here
may be in telecommunications. On the theoretical side, these limits are examples
of macroscopic limits for nonequilibrium statistical mechanical systems. Loosely
speaking, for the integrable examples that have been done, Lax-Levermore gives
the macroscopic dynamics, while microstructure theories mentioned in Remark 3.3
identify microstates associated with macroscopic phases [**56**]. That systems like
KdV or Toda, which are microscopically integrable, exhibit an infinity of macroscopic phases should not be surprising. On the other hand, it may be that the
integrable focusing NLS has only one macroscopic phase; at this point we don't
know. I believe the resolution of this question will lead to insights about macroscopic limits for nonintegrable systems.

Acknowledgment. I am grateful to N.M. Ercolani for his helpful comments, and
to him and T. Zhang for allowing me to draw so heavily on the material in [**24**].

References

1. E. Arbarello and C. De Concini, *On Sets of Equations Characterizing Riemann Matrices*, Ann. of Math. **120** (1984), 119–140.

2. M.J. Ablowitz, D.J. Kaup, A.C. Newell, and H. Segur, *Method for Solving the Sine-Gordon Equation*, Phys. Rev. Lett. **30** (1973), 1262–1264.

3. M.J. Ablowitz, D.J. Kaup, A.C. Newell, and H. Segur, *The Inverse Scattering Transform – Fourier Analysis for Nonlinear Problems*, Stud. Appl. Math. **53** (1974), 249–315.

4. M.J. Ablowitz and J.F. Ladik, *Nonlinear Differential-Difference Equations and Fourier Analysis*, J. Math. Phys. **17** (1976), 1011–1018.

5. A.M. Bloch and Y. Kodama, *Dispersive Regularization of the Whitham Equation for the Toda Lattice*, SIAM J. Appl. Math. **52** (1992), 909–928.

6. N.N. Bogolyubov Jr. and A.K. Prikarpat-skii, *The Inverse Periodic Problem for a Discrete Approximation of a Nonlinear Schrödinger Equation*, Sov. Phys. Dokl. **27** (1982), 113–116.

7. A. Bobenko, *All Constant Mean Curvature Tori in $\mathbb{R}^3$, $\mathbb{S}^3$, $\mathbb{H}^3$ in Terms of Theta Functions*, Math. Ann. **290** (1991), 209–245.

8. J.C. Bronski, *Semiclassical Eigenvalue Distribution of the Zakharov-Shabat Eigenvalue Problem*, Physica D **97** (1996), 376–397.

9. J.C. Bronski and D.W. McLaughlin, *Semiclassical Behavior in the NLS Equation: Optical Shocks-Focusing Instabilities*, in *Singular Limits of Dispersive Waves*, N.M. Ercolani, I.R. Gabitov, C.D. Levermore and D. Serre eds., NATO ASI Series B, vol. 320, Plenum, New York, 1994, pp. 21–38.

10. A. Calini, *A Note on a Bäcklund Transformation for the Continuous Heisenberg Model*, Phys. Lett. A **203** (1995), 333–344.

11. A. Cohen and T. Kappeler, *Scattering and Inverse Scattering for Steplike Potentials in the Schrödinger Equation*, Indiana U. Math. J. **34** (1985), 127–180.

12. P. Deift, A. Its and Z. Zhou, *Long-Time Asymptotics for Integrable Nonlinear Wave Equations*, in *Important Developments in Soliton Theory*, T. Fokas and V.E. Zakharov eds., Springer-Verlag, New York, 1993, pp. 181–204.

13. P. Deift, S. Kamvissis, T. Kriecherbauer, and X. Zhou, *The Toda Rarefaction Problem*, Comm. Pure & Appl. Math. **49** (1996), 35–83.

14. P. Deift, T. Kriecherbauer and S. Venakides, *Forced Lattice Vibrations I, II*, Comm. Pure & Appl. Math. **48** (1995), 1187–1250, 1251–1298.

15. P. Deift and K. T-R McLaughlin, *A Continuum Limit of the Toda Lattice*, in *Memoirs of the AMS*, 1997 (to appear).

16. P. Deift and E. Trubowitz, *Inverse Scattering on the Line*, Comm. Pure & Appl. Math. **32** (1979), 121–251.

17. P. Deift, S. Venakides, and X. Zhou, *The Collisionless Shock Region for the Long-Time Behavior of Solutions of the KdV Equation*, Comm. Pure & Appl. Math. **47** (1994), 199–206.

18. P. Deift, S. Venakides and X. Zhou, *New Results in Small Dispersion KdV by an Extension of the Steepest Descent Method for Riemann Hilbert Problems*, Proc. Nat. Acad. Sci. USA (submitted 1996).

19. P. Deift and X. Zhou, *A Steepest Descent Method for Oscillatory Riemann-Hilbert Problems: Asymptotics for the mKdV Equation*, Annals of Math. **137** (1993), 295–368.

20. J. Dorfmeister, F. Pedit and H. Wu, *Weierstrass Type Representation of Harmonic Maps into Symmetric Spaces*, Comm. Anal. Geom. (1997) (to appear).

21. B.A. Dubrovin and S.P. Novikov, *A Periodicity Problem for the Korteweg-de Vries and Sturm-Liouville Equations*, Soviet Math. Dokl. **15** (1974), 1597–1601.

22. N. Ercolani, I. Gabitov, C.D. Levermore and D. Serre eds., *Singular Limits of Dispersive Waves*, (Proceedings of the NATO Advanced Research Workshop, 8–12 July 1991, Lyon, France) NATO ASI Series B, vol. 320, Plenum, New York, 1994.

23. N. Ercolani, S. Jin, C.D. Levermore and W. MacEvoy, *The Zero-Dispersion Limit of the NLS/mKdV Hierarchy for the Nonselfadjoint ZS Operator*, preprint (1993).

24. N. Ercolani, C.D. Levermore and T. Zhang, *The Behavior of the Weyl Function in the Zero-Dispersion KdV Limit*, Comm. Math. Phys. **183** (1997), 119–143.

25. N. Ercolani, C.D. Levermore and T. Zhang, *The KdV Zero-Dispersion Limit via Dirichlet Spectra and Weyl Functions*, preprint (1997).

26. L. Faddeev, *Inverse Problem in Quantum Theory of Scattering*, J. Math. Phys. **4** (1963), 72–104.

27. L.D. Faddeev and L.A. Takhtajan, *Hamiltonian Methods in the Theory of Solitons*, Springer-Verlag, Berlin, 1987.

28. H. Flaschka, *The Toda Lattice I: Existence of Integrals*, Phys. Rev. B **9** (1974), 1924–1925.

29. H. Flaschka, *On the Toda Lattice II: Inverse Scattering Solution*, Prog. Theo. Phys. **51** (1974), 703–716.

30. H. Flaschka, *Construction of Conservation Laws for Lax Equations: Comments on a Paper by G. Wilson*, Quart. J. Math. Oxford **34** (1983), 61–65.

31. H. Flaschka, M.G. Forest and D.W. McLaughlin, *Multiphase Averaging and the Inverse Spectral Solutions of the Korteweg-de Vries Equation*, Comm. Pure & Appl. Math. **33** (1980), 739–784.

32. M.G. Forest and K.T-R. McLaughlin, *Onset of Oscillations in Nonsoliton Pulses in Nonlinear Dispersive Fibers*, J. Nonlinear Sci., (accepted 1997).

33. C.S. Gardner, *The Korteweg-de Vries Equation and Generalizations IV: The Korteweg-de Vries Equation as a Hamiltonian System*, J. Math. Phys. **12** (1971), 1548–1551.

34. C.S. Gardner, J.M. Greene, M.D. Kruskal and R.M. Miura, *Method for Solving the Korteweg-de Vries Equation*, Phys. Rev. Lett. **19** (1967), 1095–1097.

35. C.S. Gardner, J.M. Greene, M.D. Kruskal and R.M. Miura, *The Korteweg-de Vries Equation and Generalizations VI: Methods for Exact Solutions*, Comm. Pure & Appl. Math. **27** (1974), 97–133.

36. J. Goodman and P. Lax, *On Dispersive Difference Schemes I*, Comm. Pure & Appl. Math. **41** (1988), 591–613.

37. A.V. Gurevich and L.P. Pitaevski, *Nonstationary Structure of a Collisionless Shock Wave*, Sov. Phys. JETP **38** (1974), 291–297.

38. E. Grenier, *Limite semi-classique de l'équation de Schrödinger non linéaire en temps petit*, CRAS **320** (1995), 691–694.

39. M.H. Hays, C.D. Levermore, and P.D. Miller, *Macroscopic Lattice Dynamics*, Physica D **79** (1994), 1–15.

40. M. Hénon, *Integrals of the Toda Lattice*, Phys. Rev. B **9** (1974), 1921–1923.

41. N.J. Hitchin, *Harmonic Maps from a 2-Torus to the 3-Sphere*, J. Diff. Geom. **31** (1990), 627–710.

42. T. Hou and P.D. Lax, *Dispersive Approximations in Fluid Dynamics*, Comm. Pure & Appl. Math. **44** (1991), 1–40.

43. A.R. Its and V.B. Matveev, *The Periodic Korteweg-de Vries Equation*, Funct. Anal. Appl. **9** (1975), 67–??.

44. Shan Jin, *The Semiclassical Limit of the Defocusing Nonlinear Schrödinger Flows*, Ph.D. Dissertation, University of Arizona, 1991.

45. S. Jin, C.D. Levermore and D.W. McLaughlin, *The Behavior of Solutions of the NLS Equation in the Semiclassical Limit*, in *Singular Limits of Dispersive Waves*, N.M. Ercolani, I.R. Gabitov, C.D. Levermore and D. Serre eds., NATO ASI Series B, vol. 320, Plenum, New York, 1994, pp. 235–255.

46. S. Jin, C.D. Levermore and D.W. McLaughlin, *The Semiclassical Limit of the Defocusing NLS Hierarchy*, Comm. Pure & Appl. Math. (submitted 1997).

47. M. Kac and P. van Moerbeke, *On an Explicitly Solvable System of Nonlinear Differential Equations Related to the Toda Lattice*, Adv. in Math. **16** (1975), 160–169.

48. I. Kay and H.E. Moses, *Reflectionless Transmission through Dielectrics and Scattering Potentials*, J. Appl. Phys. **27** (1956), 1503–1508.

49. I.M. Krichever, *Tau-Function of the Universal Whitham Hierarchy*, Comm. Pure & Appl. Math. **47** (1994), 437–475.

50. M.D. Kruskal, *Asymptotology in Numerical Computation: Progress and Plans on the Fermi-Pasta-Ulam Problem*, in *IBM Symposium on Large-Scale Problems in Physics*, IBM Data Processing Division, White Plains, 1965, 43–62.

51. P.D. Lax, *Integrals of Nonlinear Equations of Evolution and Solitary Waves*, Comm. Pure & Appl. Math. **21** (1968), 467–490.

52. P.D. Lax, *Periodic Solutions of the KdV Equation*, Comm. Pure & Appl. Math. **28** (1975), 141–188.

53. P.D. Lax, *On Dispersive Difference Schemes*, Physica D **18** (1986), 250–254.

54. P.D. Lax and C.D. Levermore, *The Zero Dispersion Limit of the Korteweg-de Vries Equation*, Proc. Nat. Acad. Sci. USA **76** (1979), 3602–3606.

55. P.D. Lax and C.D. Levermore, *The Small Dispersion Limit of the Korteweg-de Vries Equation I, II, III*, Comm. Pure & Appl. Math. **36** (1983), 253–290, 571–593, 809–829.

56. P.D. Lax, C.D. Levermore and S. Venakides, *The Generation and Propagation of Oscillations in Dispersive IVPs and Their Limiting Behavior*, in *Important Developments in Soliton Theory*, T. Fokas and V.E. Zakharov eds., Springer-Verlag, New York, 1993, pp. 205–241.

57. C.D. Levermore, *The Hyperbolic Nature of the Zero-Dispersion KdV Limit*, Comm. P.D.E. **13** (1988), 495–514.

58. C.D. Levermore and J.G. Liu, *Oscillations Arising in Numerical Experiments*, in *Singular Limits of Dispersive Waves*, N.M. Ercolani, I.R. Gabitov, C.D. Levermore, and D. Serre, eds., NATO ASI Series B, vol. 320, Plenum, New York, 1993, pp. 329–346.

59. C.D. Levermore and J.G. Liu, *Large Oscillations Arising in a Dispersive Numerical Scheme*, Physica D **99** (1996), 191–216.

60. W. MacEvoy, private communication, 1993.

61. S.V. Manikov, *The Inverse Scattering Transform for the Time Dependent Schrödinger Equation and Kadomtsev-Petviashvili Equation*, Physica D **3** (1981), 420–427.

62. H.P. McKean and E. Trubowitz, *Hill's Operator and Hyperelliptic Function Theory in the Presence of Infinitely Many Branch Points*, Comm. Pure & Appl. Math. **29** (1976), 143–226.

63. H.P. McKean and P. Van Moerbeke, *The Spectrum of Hill's Equation*, Invent. Math. **30** (1975), 217–274.

64. K.T-R. McLaughlin, *A Continuum Limit of the Toda Lattice*, Ph. D. Thesis, New York University, 1994.

65. D.W. McLaughlin and J. Strain, *Calculating the Weak Limit of KdV*, Comm. Pure & Appl. Math. **47** (1994), 1–46.

66. P.D. Miller, N.M. Ercolani, I.M. Krichever, and C.D. Levermore, *Finite Genus Solutions to the Ablowitz-Ladik Equations*, Comm. Pure & Appl. Math. **48** (1995), 1369–1440.

67. P.D. Miller, N.M. Ercolani, and C.D. Levermore, *Modulation of Multiphase Waves in the Presence of Resonance*, Physica D **92** (1996), 1–27.

68. R.M. Miura, *Korteweg-de Vries Equation and Generalizations I: A Remarkable Explicit Nonlinear Transformation*, J. Math. Phys. **9** (1968), 1202–1204.

69. R.M. Miura, C.S. Gardner and M.D. Kruskal, *Korteweg-de Vries Equation and Generalizations II: Existence of Conservation Laws and Constants of Motion*, J. Math. Phys. **9** (1968), 1204–1209.

70. A.C. Newell, *Solitons in Mathematics and Physics*, CBMS-NSF Regional Conference Series in Applied Matematics, vol. 48, SIAM, Philadelphia, 1985.

71. S.P. Novikov, *The periodic Problem for the Korteweg-de Vries Equation*, Funct. Anal. Appl. **8** (1974), 236–246.

72. E. Previato, *Hyperelliptic Quasi-periodic and Soliton Solutions of the Nonlinear Schrödinger Equation*, Duke Math. J. **52** (1985), 329–377.

73. R. Rosales and C. Turner, *The Small Dispersion Limit for a Nonlinear Semi-Discrete System of Equations: Part I*, Stud. Appl. Math. (1996) (to appear).

74. G.B. Segal, *Loop Groups and Harmonic Maps*, in *Advances in Homotopy Theory*, London Math. Soc. Lecture Notes **139**, 153–164, Cambridge Univ. Press, 1989.

75. S. Shipman, *A Continuum Limit of a Finite Discrete Nonlinear Schrödinger System*, Ph.D. Dissertation, University of Arizona, 1997.

76. K.K. Uhlenbeck, *Harmonic Maps into Lie Groups*, J. Diff. Geom. **30** (1989), 1–50.

77. V.E. Vekslerchik, *Finite Nonlinear Schrödinger Chain*, Phys. Lett. A **174** (1993), 285–288.

78. S. Venakides, *The Zero Dispersion Limit of the Korteweg-de Vries Equation with Nontrivial Reflection Coefficient*, Comm. Pure & Appl. Math. **38** (1985), 125–155.

79. S. Venakides, *The Generation of Modulated Wavetrains in the Solution of the Korteweg-de Vries Equation*, Comm. Pure & Appl. Math. **38** (1985), 883–909.

80. S. Venakides, *The Zero Dispersion Limit of the Korteweg-de Vries Equation with Periodic Initial Data*, AMS Trans. **301** (1987), 189–225.

81. S. Venakides, *Higher Order Lax-Levermore Theory*, Comm. Pure & Appl. Math. **43** (1990), 335–362.

82. S. Venakides, P. Deift and R. Oba, *The Toda Shock Problem*, Comm. Pure & Appl. Math. **44** (1991), 1171–1242.

83. G.B. Whitham, *Non-Linear Dispersive Waves*, Proc. Roy. Soc. London Ser. A **283** (1965), 238–261.

84. G.B. Whitham, *Linear and Nonlinear Waves*, J. Wiley, New York, 1974.

85. G. Wilson, *On Two Constructions of Conservation Laws for Lax Equations*, Quart. J. Math. Oxford **32** (1981), 491–512.

86. H. Wente, *Counterexample to a Conjecture of Hopf*, Pacific J. Math. **121** (1986), 193–246.

87. N.J. Zabusky and M.D. Kruskal, *Interaction of "Solitons" in a Collisionless Plasma and the Recurrence of Initial States*, Phys. Rev. Lett. **15** (1965), 240–243.

88. V.E. Zakharov and L. Faddeev, *The Korteweg-de Vries Equation: A Completely Integrable Hamiltonian System*, Funct. Anal. Appl. **5** (1971), 280–287.

89. V.E. Zakharov and A.B. Shabat, *Exact Theory of Two-dimensional Self-focusing and One-dimensional Self-modulation of Waves in Nonlinear Media*, Sov. Phys. JETP **34** (1972), 62–69.

90. V.E. Zakharov and A.B. Shabat, *Interaction Between Solitons in a Stable Medium*, Sov. Phys. JETP **37** (1973), 823–828.

DEPARTMENT OF MATHEMATICS, UNIVERSITY OF ARIZONA, TUCSON, AZ 85721, USA.

E-mail address: lvrmr@math.arizona.edu

Proceedings of Symposia in Applied Mathematics
Volume **54**, 1998

On Boltzmann equation
and its applications

Pierre-Louis Lions

CEREMADE
Université Paris-Dauphine
Place de Lattre de Tassigny
75775 Paris Cedex 16

dedicated to the memory of R.J. DiPerna

Abstract.

Our goal here is to present some of the recent mathematical results on
Boltzmann equations concerning the global existence of weak solutions
and their compactness properties. But before we do so we recall briefly
some general facts on this famous problem (history, modern applications
to science and industry, mathematical difficulties...). And we conclude
with a discussion of some important open problems.

Summary

I. Introduction : the scientific context.

We want to begin this report by a few general and well-known facts on the Boltzmann's equation before even writing it down : the precise formulation of the equation will be given in section II below. We explain in section III some of the mathematical difficulties associated with this equation and we briefly present there a survey of existing results on the subject. Sections IV-VI are devoted to a general overview of some of the recent results. And we conclude in section VII with a list of open problems and with some general perspectives (wishful thinking ?).

First of all, the equation was introduced by L. Boltzmann [5] and J.C. Maxwell [33],[34] more than a century ago (in 1872) and it seems fair to say that there are still many mathematical issues associated with it that are not yet understood.

From a Physics viewpoint, it is an important model since it is one of the few models that attempt to describe physical situations which are not at a thermodynamical equilibrium and their trend towards equilibrium.

A bit more specifically, the Boltzmann's equation provides a mathematical model for the statistical evolution (in time) of a large number of particles interacting through various physical phenomena that can be thought of as collisions. Of course, other interactions such as electromagnetic forces can be present - an example which is important for charged particles-, in which case additional terms have to be inserted in the model. But, in order to keep this presentation both short and elementary, such possibilities will not be considered here. Let us also point out that the word particle is used above in a rather loose sense since, depending on the context, these "particles" can be gas molecules or charged particles such as electrons or ions... Typical physical contexts in which Boltzmann's equation can be used include gas dynamics, lasers and plasmas. More precisely, it is used in gas dynamics for the description of a moderately rarefied gas like the upper layers of the atmosphere (at an altitude of say 70-120 kms) : in such circumstances, classical hydrodynamics or gas dynamics models, which postulate that at every point of the fluid and at every time thermodynamical equilibrium has been achieved, are no more valid and one has to use a more detailed description such as the one given by Boltzmann's equation.

In fact, one can say that the recent mathematical progress on Boltzmann's equation has been somehow triggered by industrial and

engineering applications involving flights in a rarefied atmosphere like
the reentry problem for the space shuttles. Indeed, these practical problems have led various groups of applied mathematicians (in France in
particular) to "go back" to the study of Boltzmann's equation. Also
the renewed interest in the problem is very much related to the progress
in power and speed of computers which makes this model accessible
to numerical simulations : and, when we present the precise mathematical formulation, we shall see that the equation is set on $\mathbb{R}^6$ (plus
time) and thus requires large-scale computations. We will come back
on this particular point later on. Finally, let us also observe that the
increasing scientific activity on this model is probably a good example of a trend in the physical modelling of complex systems namely the
growing use of mesoscopic models. Mesoscopic model refers to the mesoscopic scale which could be thought of as being intermediate between the
macroscopic scale and a "true microscopic scale" like the atomic scale.
Developing such intermediate models allows to consider situations where
classical macroscopic models fail because of microscopic effects while a
full treatment at the microscopic scale is prohibitive : for instance, it is
obviously impossible to determine the evolution of even a rarefied gas
by a study of all its atoms (just think of the size of Avogadro number!).

Finally, let us warn the reader that, as usual, the model we are going
to investigate mathematically namely the classical Boltzmann's equation is just the basic one. It admits many important variants (known as
kinetic models) which have to be used in realistic applications (which
often involve the effect of chemistry, of electromagnetic forces...). However, from a mathematical viewpoint, the classical model concentrates
all the essential mathematical difficulties.

II. Presentation of the model.

The Boltzmann's equation is given by the following equation

$$(1) \qquad \frac{\partial f}{\partial t} + v \cdot \nabla_x f \; = \; Q(f, f) \qquad \text{for} \quad x \in \mathbb{R}^N \, , \; v \in \mathbb{R}^N \, , \; t \geq 0$$

where $N \geq 2$, $f = f(x, v, t)$ is a nonnegative function on $\mathbb{R}^{2N} \times [0, \infty)$
and $Q(f, f)$ is a non-local (in the v variable), quadratic operator de-
tailed below. Here and everywhere below $\nabla_x f$ stands for the spatial
gradient and we denote indifferently by $z_1 \cdot z_2$ or (z_1, z_2) the usual scalar

product between z_1 and z_2 in $\mathbb{R}^N$ and by $|z|$ or $\|z\|$ the usual norm on $\mathbb{R}^N$. The unknown function f corresponds at each time t to the density of particules at the point x with velocity v. If the operator Q were 0, (1) would simply mean that the particles do not interact and f would constant along particle paths ($\dot{x} = v$, $\dot{v} = 0$) i.e. straight lines $(x+vt, v)$ in the phase space $\mathbb{R}^{2N}$. This conservation no longer holds if collisions occur which modify the velocities of the particles, in which case the rate of changes of f due to collisions has to be specified. Such a description was obtained by L. Boltzmann [5] and J.C. Maxwell [33],[34] and involves an integral operator described below. This model is derived under the assumption of stochastic independence of pairs of particles at (x, t) with different velocities (molecular class assumption). For further detail on the derivation of this model (Boltzmann collision operator), we refer the reader to S. Chapman and T.G. Cowling [10], H. Grad [21], C. Cercignani [8], C. Truesdell and R.G. Muncaster [41] and the references therein.

Before discussing the crucial point of the structure of Q, let us immediately mention that most of this survey is concerned with the case when the equation is set in the whole space ($x \in \mathbb{R}^N$) and when the density goes to 0 at infinity (as $|(x, v)| \to +\infty$). Most applications of the Boltzmann's equations require a setting where x lies in a region of $\mathbb{R}^N$ with appropriate boundary conditions : this fact creates additional difficulties that are mentioned (and "solved") in section VI.

We now detail the structure of Boltzmann collision operator B. If φ is a smooth function of v (say $\varphi \in C_0^\infty(\mathbb{R}^N)$ - C^∞ functions with compact support) then $Q(f, f)$ is a function of v given by

$$(2) \qquad Q(\varphi, \varphi) = Q^+(\varphi, \varphi) - Q^-(\varphi, \varphi)$$

$$(3) \qquad Q^+(\varphi, \varphi) = \int_{\mathbb{R}^N} dv_* \int_{S^{N-1}} d\omega\, \varphi(v')\, \varphi(v'_*)\, B(v - v_*, \omega)$$

$$(4) \qquad Q^-(\varphi, \varphi) = \int_{\mathbb{R}^N} dv_* \int_{S^{N-1}} d\omega\, \varphi(v')\, \varphi(v'_*)\, B(v - v_*, \omega)$$
$$= \varphi(v)(\varphi * A)(v)$$

where $A(v) = \int_{S^{N-1}} d\omega\, B(v, \omega)$. The velocities v', v'_* are given by

$$(5) \qquad v' = v - (v - v_*, \omega)\omega \quad , \quad v'_* = v_* + (v - v_*, \omega)\omega \ .$$

The function B appearing in (3),(4) is given - it is called the collision kernel or the scattering cross section - : it depends on the nature of the interaction between particles and always satisfies at least

(6)
$$\begin{cases} B(z,\omega) \geq 0 \quad \text{on} \quad \mathbb{R}^N \times S^{N-1}, \\ B(z,\omega) \quad \text{is a function of} \quad |z|, |(z,\omega)| \quad \text{only}. \end{cases}$$

A particular example - the so-called hard spheres model - is given by :
$B = |(z,\omega)|$.

Before discussing a bit more the nature of B, let us first recall the significance of the velocities v, v_*, v', v'_* : one way to think of these velocities - which is not entirely accurate from a physical viewpoint - is to say that v, v_* are the velocities of two "colliding" point-particules before an elastic collision that will bring them to have velocities v', v'_*. Elastic collisions must obey the conservation of momentum and kinetic energy that is

$$v' + v'_* = v + v_* \quad , \quad |v'|^2 + |v'_*|^2 = |v|^2 + |v_*|^2 .$$

And (5) is just a description of the set of solutions of these equations, parametrized by the angle ω : write indeed $v' = v - w$, $v'_* = v_* + w$ and thus $|w|^2 = (v - v_*, w)$ i.e. $w = (v - v_*, \omega)\omega$ for some $\omega \in S^{N-1}$.

Of course, $Q(f, f)$ in (1) means $Q(f(x, \cdot, t), f(x, \cdot, t))$ provided such a quantity makes sense or in other words provided the integrals in (2) (or (3)-(4)) make sense first for a smooth φ and next for a solution of (1). The second part of this question is of course related to a priori estimates and regularity informations on the solutions. But, even the first part is a serious mathematical difficulty since realistic collision kernels can be rather singular as we explain now. Indeed, if the hard spheres model indicated above does not present real singularities, for inverse power intermolecular potentials, B takes the following form

$$B(z, \omega) = b(\theta)|z|^\gamma \quad \text{with} \quad \gamma = 1 - 2(N-1)(s-1)^{-1}$$

where $s > 1$ is the exponent of the potential ("short-range" potentials, the Coulomb potential is excluded...), θ is the angle between $v - v_*$ and ω so that $\cos \theta = (v - v_*, \omega)|v - v_*|^{-1}$. In addition, b is smooth except at $\theta = \pm\frac{\pi}{2}$ where it has a singularity of the form $|\cos \theta|^{-\alpha}$ with $\alpha = \frac{s+1}{s-1}$ if $N = 3$. In other words, B presents singularities of an arbitrary high order (as $s \to 1_+$) when $(v - v_*, \omega)$ goes to 0, condition that corresponds to the so-called grazing collisions (notice that when $(v - v_*, \omega) = 0$, $v' = v$, $v'_* = v_*$). In particular, one sees that Q^+ and Q^- do not make sense for all smooth φ ! On the other hand, when writing Q, one may take advantage of the cancellation namely

$$Q(\varphi, \varphi) = \int_{\mathbb{R}^N} dv_* \int_{S^{N-1}} d\omega \, B(v - v_*, \omega) \{ \varphi(v'_*) \varphi(v') - \varphi(v_*) \varphi(v) \}$$

and one can check that the integrals make sense if $s > 2$ ($N = 3$): write indeed $\varphi(v'_*) \varphi(v') - \varphi(v_*) \varphi(v) = (v - v_*, \omega) \{ \varphi(v)(\nabla \varphi(v_*), \omega) - \varphi(v_*)(\nabla \varphi(v), \omega) \}$ + terms of order $|(v - v_*, \omega)|^2$ which give a finite contribution to the integrals if $s > 2$, then, since B depends only on $|z|$ and θ, integrating in ω the quantity $|v - u_*| \cos \theta \{ \varphi(v)(\nabla \varphi(v_*), \omega) - \varphi(v_*)(\nabla \varphi(v), \omega) \}$ amounts to integrate with respect to θ only the quantity $[(v - v_*, \varphi(v) \nabla \varphi(v_*) - \varphi(v_*) \nabla \varphi(v))] \cos^2 \theta$ which is bounded by $|v - v_*|^2 \cos^2 \theta$. And the integrals make sense if $s > 2$.

However, these elementary computations require quite a lot of smoothness on φ (two bounded derivatives) and such smoothness is not available for solutions of (1). This is the reason why almost nothing is known mathematically on (1) when B contains the singularities described above - we shall come back on this striking fact in section VII. A classical approach consists in avoiding this severe difficulty : one neglects grazing collisions, truncating b near $\theta = \pm \frac{\pi}{2}$, assuming that $b \sin \theta \in L^1(0, 2\pi)$. Then, Q^+ and Q^- make sense if $1 - 2 \frac{(N-1)}{s-1} > -N$ i.e. $s > \frac{3N-1}{N+1}$ or $s > 2$ if $N = 3$. In all that follows, we shall use this truncation procedure (called angular cut-off assumption, see H. Grad [21], C. Cercignani [8], C. Truesdell and R.G. Muncaster [41] for instance) and we shall assume in addition to (6)

$$(7) \quad \begin{cases} B \in L^1_{\mathrm{loc}}(\mathbb{R}^N \times S^{N-1}) \,, \quad \dfrac{1}{1 + |\xi|^2} \displaystyle\int_{|z - \xi| \le R} dz \int_{S^{N-1}} d\omega \, B(z, \omega) \to 0 \\[2mm] \qquad \text{as} \quad |\xi| \to +\infty \qquad (\forall \, R \in (0, \infty)) \,. \end{cases}$$

Let us emphasize the fact that, to our best knowledge, all numerical simulations, of Boltzmann models use collision kernels B which satisfy (7) (and usually much more...). And the numerical approximation of models without angular cut-off seems to be an interesting open question.

We wish to conclude this presentation with a different collision model where, on the contrary to what we ended up doing above, one neglects "all collisions but the grazing ones". Then, phenomenological arguments introduced by Landau (see E.M. Lifschitz and L.P. Pitaevskii [27]) and by S. Chapman and T.G. Cowling [10]) lead to another collision operator

$$(8) \quad Q(f, f) = \frac{\partial}{\partial v_i} \left\{ \left(\int_{\mathbb{R}^N} dv_* \, a_{ij}(v - v_*) \left[f(v_*) \frac{\partial f}{\partial v_j}(v) - f(v) \frac{\partial f(v_*)}{\partial v_j} \right] \right\}$$

in which case the equation (1) is called the Landau equation (it is also called sometimes the Fokker-Planck equation). The matrix $(a_{ij}(z))$ is symmetric, nonnegative, even in z and satisfies

(9)
$$\text{a.e. } z \in \mathbb{R}^N , \quad a_{ij}(z)\eta_i\eta_j > 0 \quad \text{if } \eta \cdot z = 0 , \ \eta \neq 0 , \ a_{ij}(z)z_iz_j = 0 ,$$

(10)
$$a_{ij} \in L^1(\mathbb{R}^N) + L^\infty(\mathbb{R}^N) \quad ; \quad \frac{\partial a_{ij}}{\partial z_j} , \ \frac{\partial^2 a_{ij}}{\partial z_i \, \partial z_j} \in \mathcal{M}(\mathbb{R}^N) + L^\infty(\mathbb{R}^N) ,$$

where we use the usual convention of implicit summation over repeated indices and where $\mathcal{M}(\mathbb{R}^N)$ denotes the space of bounded measures on $\mathbb{R}^N$. A typical example is given by : $N \geq 3$, $a_{ij}(z) = \frac{a(z)}{|z|^{N-2}}\left\{\delta_{ij} - \frac{z_iz_j}{|z|^2}\right\}$ where $a \in W^{2,\infty}(\mathbb{R}^N)$ (a bounded, Lipschitz and ∇a Lipschitz), $a > 0$ on $\mathbb{R}^N$, a is even in z. Justifications of the collision operator given in (8) can be found in L. Desvillettes [12] (through an asymptotic expansion of Boltzmann collision operators with small parameters) and in P. Degond and B. Lucquin-Desreux [11] (via an expansion of a physically realistic Boltzmann collision operator around grazing collisions). These works strongly suggest that, in addition to the intrinsic interest in Landau equation, some insight on the Boltzmann equation when one does not make the angular cut-off might be gained by an analysis of the Landau model.

III. Mathematical difficulties.

We investigate in this report the Cauchy problem associated with (1) and we thus prescribe initial conditions on f :

(11)
$$f|_{t=0} \ = \ f_0 \quad \text{on} \quad \mathbb{R}^{2N}$$

where $f_0 \geq 0$ is a given function on $\mathbb{R}^{2N}$.

III.1 A priori estimates.

Even if (1) seems at first sight rather simple, tremendous mathematical difficulties arise when we investigate it. To explain what they are, let us first recall all the known a priori estimates. As it is often the case with physically realistic (at scales which are not microscopic) nonlinear equations, there are very few a priori estimates and they are all deduced from conservation laws or identities that involve physically natural quantities. Because of (5),(6), it is not difficult to check that we have at least formally

$$(12) \quad \int_{\mathbb{R}^N} dv \, Q(f,f)\psi \; = \; 0 \qquad \text{for} \quad \psi(v) = 1 \,, \; v_j \, (1 \le j \le N) \,, \; |v|^2$$

(conservation of mass or of the number of particles, of momentum and kinetic energy). Indeed, (5)-(6) yield the following formal fact valid for any function φ of (v, v_*, v', v'_*)

$$(13) \quad \begin{cases} \displaystyle \iint_{\mathbb{R}^{2N}} dv \, dv_* \int_{S^{N-1}} d\omega \, B(v-v_*, \omega)\varphi \; = \\ \displaystyle = \iint_{\mathbb{R}^{2N}} dv \, dv_* \int_{S^{N-1}} d\omega \, B(v-v_*, \omega)\tilde{\varphi} \end{cases}$$

denoting by $\tilde{\varphi}$ any function obtained from φ by permuting the variables (v, v_*, v', v'_*) in one of the following manners : $v \to v_*, \; v_* \to v, \; v' \to v'_*,$ $v'_* \to v' \,; \; v \to v', \; v_* \to v'_*, \; v' \to v, \; v'_* \to v_*$ or $v \to v'_*, \; v_* \to v', \; v' \to v_*,$ $v'_* \to v.$

Then, (12) combined with (1) yields the following formal conserved quantities

$$(14) \quad \begin{cases} \displaystyle \iint_{\mathbb{R}^{2N}} dx \, dv \, f(x,v,t)\psi \quad \text{is independent of} \quad t \\ \text{for} \quad \psi \equiv 1 \,, \; v_j \, (1 \le j \le N) \,, \; |v|^2 \,, \; |x-vt|^2 \,. \end{cases}$$

Notice that this only gives weighted (with weights $1, |v|^2, |x-vt|^2$) bounds in $L^1(\mathbb{R}^{2N})$ uniformly in $t \ge 0$ - recall that $f \ge 0$.

The only other known bound is deduced from the entropy identity based upon the following observation (deduced from (13)) where we denote by $f_* = f(v_*), \; f' = f(v'), \; f'_* = f(v'_*)$:

$$(15) \quad \begin{cases} \displaystyle \int_{\mathbb{R}^N} Q(f,f) \log f \, dv \; = \\ \displaystyle = -\frac{1}{4} \iint_{\mathbb{R}^{2N}} dv \, dv_* \int_{S^{N-1}} d\omega \, B(f'f'_* - ff_*) \log \frac{f'f'_*}{ff_*} \; \le \; 0 \end{cases}$$

and if $B > 0$ (say), the inequality is strict unless f is a Maxwellian :
$f = M(\rho, u, T)$ where we set for $\rho \geq 0$, $u \in \mathbb{R}^N$, $T > 0$ (if $T = 0$,
$M = \rho \delta_u(v)$)

$$(16) \qquad M(\rho, u, T)(v) = (4\pi T)^{-N/2} \rho \exp -\left(\frac{|v-u|^2}{4T}\right) .$$

This is nothing but the famous H-theorem due to L. Boltzmann [5].
From (15) we deduce multiplying (1) by $\log f$, integrating over $\mathbb{R}^{2N}_{x,v}$
and using (14) with $\psi \equiv 1$ the following formal identity :

$$(17) \qquad E[f(t)] + \frac{1}{4} \int_0^t ds \int_{\mathbb{R}^N} dx\, D = E[f_0]$$

where $E[f] = \iint_{\mathbb{R}^{2N}} f \log f\, dx\, dv$, $D[f](x,t) = \iint_{\mathbb{R}^{2N}} dv\, dv_* \int_{S^{N-1}} d\omega \cdot$
$(f'f'_* - ff_*) \log \frac{f'f'_*}{ff_*} \geq 0$, $E[f_0] = E[f(0)]$ in view of (11).

In particular, $E[f(t)] \leq E[f_0]$ for all $t \geq 0$. Using then the inequality

$$a|\log a| \leq a \log a + 2a \log^- b + 2(b-a)^+ \quad , \quad \forall\, a, b \geq 0\,,$$

with $a = f(t)$, $b = \exp -\frac{1}{2}\left(|v|^2 + |x-vt|^2\right)$, we find

$$\begin{cases} \iint_{\mathbb{R}^{2N}} f(t)|\log f(t)|\, dx\, dv \leq \\[2mm] \leq E(t) + 2 \iint_{\mathbb{R}^{2N}} f(t)(|v|^2 + |x-vt|^2) + 2(2\pi)^N \,. \end{cases}$$

In conclusion, we have derived the following a priori estimates : for
each $A \in (0, \infty)$, there exists $C \in (0, \infty)$ depending only on A such that
if f_0 satisfies

$$(18) \qquad \iint_{\mathbb{R}^{2N}} dx\, dv\, f_0\left(1 + |v|^2 + |x|^2 + |\log f_0|\right) \leq A$$

then we have the following a priori bounds

$$(19) \qquad \sup_{t \geq 0} \iint_{\mathbb{R}^{2N}} dx\, dv\, f(t)\left(1 + |v|^2 + |x-vt|^2 + |\log f(t)|\right) \leq C\,,$$

$$(20) \qquad \int_0^\infty dt \int_{\mathbb{R}^N} dx\, D[f] \leq C\,.$$

Remark III.1. The same a priori estimates hold for the Landau equation with the same proof.   □

Remark III.2. (Towards Fluid Mechanics).

When the frequency of collisions increases, one expects from a physical viewpoint that the gas will thermalize (go to a local in (x, t) thermodynamical equilibrium) and thus should become amenable to classical Fluid and Gas Dynamics models. Formally, this is the case since if we insert in front of Q a large parameter $\frac{1}{\varepsilon}$ ($\varepsilon \to 0_+$, ε is called the mean free path), the entropy identity forces solutions to "go to" local Maxwellians $M(\rho(x, t), u(x, t), T(x, t))$ and thus $\int_{\mathbb{R}^N} f \, dv \underset{\varepsilon}{\to} \rho(x, t)$, $\int_{\mathbb{R}^N} f v_i \, dv \underset{\varepsilon}{\to} (\rho u_i)(x, t)$, $\int_{\mathbb{R}^N} f v_i v_j \, dx \underset{\varepsilon}{\to} (\rho u_i u_j + \rho T \delta_{ij})$, $\int_{\mathbb{R}^N} f v_i |v|^2 \, dv \underset{\varepsilon}{\to} \rho u_i [|u|^2 + (N+2)T]$.

Next, using (12) and averaging (1) with respect to dv, $v \, dv$, $|v|^2 \, dv$, we recover formally the compressible Euler equations with the law of perfect gases (and $\gamma = \frac{N+2}{N}$ or $\gamma = \frac{5}{3}$ if $N = 3$, i.e. the monoatomic case).

Other formal limits as ε goes to 0 - called hydrodynamical limits like the preceding one - yield other models of Fluid Mechanics like the incompressible, homogeneous Navier-Stokes equations. We do not want to discuss more this subtle question and we refer instead to C. Cercignani [8] for the classical facts and to the recent works by C. Bardos, F. Golse and D. Levermore [2],[3],[4] for an account of the state of the art on this important matter and for partial results that rely upon what is discussed below. Let us note for future purposes that this limit only involves the so-called macroscopic quantities that is velocity averages with respect to various weights $\psi(v)$.   □

III.2 Difficulties and partial results.

The most obvious difficulty is the lack of bounds (a priori estimates) on solutions of (1) : in particular the bounds collected in (19)-(20) are not enough even to make sense of $Q(f, f)$ in L^1_{loc} !

This formidable difficulty has led mathematicians to analyse (1) in the following directions - leaving aside D. Hilbert's attempt to solve (1) by asymptotic series based, in some sense, on the hydrodynamical limit described in Remark III.2 above [23] :

1) The problem is well-posed locally in t : for appropriate classes of "smooth" initial conditions - and finding the largest one is a delicate problem -, there exists a unique "smooth" solution on a maximal time interval $[0, T_{\max}[$ and if $T_{\max} < \infty$, the norm (in the class...) of the solution blows up as t goes to $T_{\max}$. This type of infinite-dimensional version of Cauchy-Lipschitz theorems (for ordinary differential equations) was initiated by T. Carleman [6] in 1932 (see also C. Cercignani [8] and the bibliography in R.J. DiPerna and P.L. Lions [15]).

2) One can build particular solutions of (1) : the most celebrated class of solutions is probably the space-homogeneous one namely solutions of (1) that do not depend on x. Again, T. Carleman [7] studied this question, see also L. Arkeryd [1] and the bibliography in [8],[15] and in L. Desvillettes [13].

3) One can construct global (in t) unique and "smooth" solutions which initially are either small or close to an equilibrium (a Maxwellian $M(\rho, u, T)$ for $\rho > 0$, $T > 0$, $u \in \mathbb{R}^N$) and remain for all $t \geq 0$ small or close to that equilibrium. Of course, "small" and "close" are to be defined precisely and this involves various functional spaces (wich vary in the numerous works on that subject). The key to understand these results is to observe that, because of 1) above, solutions exist on arbitrarily long time intervals if we choose the initial conditions small enough or close enough to an equilibrium. Then, the global existence follows from the properties of the linearized equation around 0 or the equilibrium : in the second case, the entropy identity translates into a "force attracting solutions towards the equilibrium" while, in the first case, the linearized equation around 0 is simply $\left(\frac{\partial f}{\partial t} + v \cdot \nabla_x f = 0\right)$ which has dispersive effects and "pushes back" solutions to 0. There has been many important works in that direction and we simply advise the interested reader to look for the bibliography in [8],[15].

Quite clearly, all the regimes described above in which solutions of (1) are known to exist are rather particular and thus leave the global structure of (1) quite in the dark. Then, one might (or even should ?) try a different approach to (1) and one such attempt, directly inspired from J. Leray's pioneering work on Navier-Stokes equations [24],[25],[26], is described in the rest of these notes. One looks for *global* but *weak solutions*, the price that we may expect to pay being that uniqueness is not known even if some "substitutes" are often (and will be here) available such as compactness of the set of solutions, uniqueness between one weak and one strong solution... Let us also mention that there are now several instances in Mathematical Physics where weak solutions are the only solutions which are available and that in each case, in the process of building up global weak solutions, much insight was gained on various issues like the qualitative properties of weak and smooth solutions, what is the crucial feature of the model that may be kept in more elaborate and realistic models, or a better understanding in numerical computations and an increased confidence in them... In fact, numerical approximation is by itself a complete justification of the search for weak solutions : indeed, any numerical procedure will create approximated solutions which often enjoy the same a priori bounds than those expected from solutions of the original problem. Then, the question is : to which object(s) converge these approximated solutions ? The study of (global) weak solutions provides an answer to that question. Also, this theme indicates what is going to be the crucial part of the analysis since we have to build a weak solution out of approximated solutions which have all the expected bounds.

Then, it is quite clear that the basic problem is to be able to pass to the limit in the equation for sequences of solutions (even smooth ones !) or approximated solutions. This form of "stability" or "compactness" properties for solutions is then made difficult by the lack of a priori bounds and by the nonlinearity of the problem. Typically, in the case of (1), we do not have other bounds than (19)-(20) and thus we can not expect better than weak L^1 compactness and convergence. And, since Q is quadratic, we might not be able to pass to the limit and we might lose the equation. An academic but elementary example of such a phenomenon is the following : $\varphi_n(x,t) = e^{in(x-t)}$ is a solution for all $n \geq 1$ of the equation $\left(\frac{\partial \varphi}{\partial t} + \frac{\partial \varphi}{\partial x} = 1 - |\varphi|^2\right)$, φ_n converges weakly to 0 and 0 is not a solution. Such issues in conjunction with nonlinear partial differential equations are at the basis of L. Tartar's compensated-compactness theory [38],[39],[40] - see also F. Murat [35],[36],[37]...

In the case of the Boltzmann's equation (1), we shall see in section V why these passages to the limit are possible allowing to obtain the global existence results of weak solutions stated in section IV. And the arguments will be a good illustration of some rather general trends in nonlinear partial differential equations namely 1) the study of specific nonlinearities, 2) the study of the behavior (or possible behaviors) of solutions the interplay between "compactness" and these specific nonlinearities, 3) the arguments are related to many branches of Analysis.

IV. Global weak solutions.

We present in this section a global existence result of weaksolutions due to R.J. DiPerna and P.L. Lions [15],[17] which we state immediately before even defining precisely what we mean by weak solutions since this is a bit technical.

Theorem IV.1. *Let $f_0 \geq 0$ on $\mathbb{R}^{2N}$ satisfy :* $\iint_{\mathbb{R}^{2N}} dx\, dv\, f_0(1 + |x|^2 + |v|^2 + |\log f_0|) < \infty$. *Then there exists a "weak solution" $f \in C([0,\infty); L^1(\mathbb{R}^{2N}))$ of (1) satisfying (11).* $\square$

Remark IV.1. Additional regularity properties to the bounds (19)-(20) are not known nor is the uniqueness of weak solutions that we define below. In fact, the original notion of weak solutions given in [17] was reinforced by the author in [29],[30],[31] where it is shown that any weak solution in that sense coincides with a strong solution (say bounded for instance) on $\mathbb{R}^{2N} \times [0,T]$ whenever the latter exists. However, the precised notion of weak solutions introduced in [29],[30],[31] is too complicated to be given here. $\square$

We now (finally !) explain what we mean by weak solution in the above result (let us mention that these solutions are called *renormalized solutions* in [17],[15]) : $f \in C([0,\infty); L^1(\mathbb{R}^{2N})) \geq 0$ on $\mathbb{R}^{2N} \times [0,\infty)$ is said to be a weak solution of (1), (11) if f satisfies (11), (19)-(20), if $\frac{Q^-(f,f)}{1+f} \in L^\infty(0,\infty; L^1(\mathbb{R}_x^N \times B_R))$, $\frac{Q^+(f,f)}{1+f} \in L^1(0,T; L^1(\mathbb{R}_x^N \times B_R))$ for all $R,T \in (0,\infty)$ and if we have for all $\beta \in C^1([0,\infty),\mathbb{R})$ with $\beta'(t)(1+t)$ bounded on $[0,\infty)$

(21)
$$\begin{cases} \left\{\frac{\partial}{\partial t} + v \cdot \nabla_x\right\}\beta(f) = \beta'(f)Q^+(f,f) - \beta'(f)Q^-(f,f) \\ \qquad\qquad \text{in } \mathcal{D}'(\mathbb{R}^{2N} \times (0,\infty))\,. \end{cases}$$

Observe that $\beta'(f)Q^\pm(f,f)$ make sense in L^1_{loc} since $\beta'(f)Q^\pm(f,f) = [\beta'(f)(1+f)]\frac{Q^\pm(f,f)}{1+f}$. In addition, we may require that f satisfies
(22)
$$\begin{cases} \frac{\partial}{\partial t}\left(\int_{\mathbb{R}^N} f\, dv\right) + \text{div}_x\left(\int_{\mathbb{R}^N} fv\, dv\right) = 0 \quad \text{in } \mathcal{D}'(\mathbb{R}^N \times (0,\infty)), \\[2mm] \iint_{\mathbb{R}^{2N}} fv\, dv \quad \text{is independent of } t \geq 0 \end{cases}$$

(23)
$$\begin{cases} \iint_{\mathbb{R}^{2N}} f(t)|v|^2\,dx\,dv \;\le\; \iint_{\mathbb{R}^{2N}} f_0|v|^2\,dx\,dv\,, \\[2mm] \iint_{\mathbb{R}^{2N}} f(t)|x-vt|^2\,dx\,dv \;\le\; \iint_{\mathbb{R}^{2N}} f_0|v|^2\,dx\,dv \end{cases}$$

(24)
$$E[f(t)] + \frac{1}{4}\int_0^t ds \int_{\mathbb{R}^N} dx\, D[f] \;\le\; E[f_0]\,.$$

Remark IV.2. Exactly as in J. Leray's construction of weak solutions for three-dimensional incompressible Navier-Stokes equations [24], it is not known if the equality holds in (24) - in fact the parallel is somewhat justified in view of the results of C. Bardos, F. Golse and D. Levermore [2],[3],[4] on hydrodynamical limits. □

Remark IV.3. It is not known if equalities hold in (23) : this is very much related to obtaining bounds on $f\psi(v)$ in $L^\infty(0,T;L^1(\mathbb{R}^{2N}))$ ($\forall\, T \in (0,\infty)$) for some ψ such that $\psi(v)|v|^{-2}$ goes to $+\infty$ as $|v|$ goes to $+\infty$. For the same reason, the local conservation of momentum namely

$$\frac{\partial}{\partial t}\left(\int_{\mathbb{R}^N} fv_i\,dv\right) + \frac{\partial}{\partial x_j}\left(\int_{\mathbb{R}^N} fv_iv_j\,dv\right) \;=\; 0$$

is not known to hold. □

Remark IV.4. Let us conclude this section by some comments on the formulation (21) of (1). First of all, (21) is to be expected from (1) by using the chain rule. And if $Q = 0$, then (cf. section II) (1) simply means that f is constant along particle paths ($\dot{x} = v\,,\ \dot{v} = 0$) while (21) means that $\beta(f)$ is constant along the same particle paths and the two informations are clearly equivalent (choose $\beta\,1$ to 1 on $[0,\infty)$!). Now, when the Boltzmann collision operator is present in (1), since we have no bounds in L^1_{loc} on Q^+ or Q^- - except in one very particular case when f depends only on one spatial coordinate say x_1 and when B satisfies various truncation assumptions that we do not want to detail, this bound is due to C. Cercignani [9] -, (1) does not make sense while (21) does since $\frac{Q^+(f,f)}{1+f}$, $\frac{Q^-(f,f)}{1+f} \in L^1_{\text{loc}}$. In the event that $Q^+(f,f)$, $Q^-(f,f)$ are known to be in L^1_{loc} then one can easily check that (1) holds in $\mathcal{D}'(\mathbb{R}^{2N} \times (0,\infty))$ (take a sequence of β in (21) going to the identity...). Finally, let us mention that the bound on $\frac{Q^+(f,f)}{1+f}$ follows from (21) and the bound on $\frac{Q^-(f,f)}{1+f}$. The latter is obtained easily since we have in view of (4)

$$\frac{Q^-(f,f)}{1+f} = \frac{f}{1+f} f * A \leq f * A$$

and (7) implies

$$\int_{\mathbb{R}^N} dx \int_{B_R} dv\, f * A = \int_{\mathbb{R}^N} dx \int_{\mathbb{R}^N} dv_* f(v_*) \int_{|v-v_*| \leq R} A(v)\, dv$$

$$\leq C \iint_{\mathbb{R}^{2N}} dx\, dv_* (1+|v_*|^2) f(v_*) . \quad \square$$

V. Compactness results.

We first state in section V.1 the main results concerning compactness and stability properties of weak solutions of (1) which play a crucial role in the proof of Theorem IV.1 (as explained in section III). Then, in sections V.2 and V.3 we briefly present the main mathematical tools that allow to prove these results.

V.1 Results.

We consider a sequence of (weak) solutions f^n of (1) with initial conditions f_0^n and we assume that f_0^n satisfies (18) uniformly in n and that f^n satisfies (19)-(20) uniformly in r. These bounds imply, extracting subsequences if necessary, that f_0^n, f^n converge weakly in $L^1(\mathbb{R}^{2N})$, $L^1(\mathbb{R}^{2N} \times (0,T))$ ($\forall\, T \in (0,\infty)$) to some f_0, f respectively.

The following result is taken from R.J. DiPerna and P.L. Lions [15].

Theorem V.1. *We have for all $\psi \in L^\infty(\mathbb{R}_v^N)$ with compact support*

$$(25) \quad \begin{cases} \displaystyle \int_{\mathbb{R}^N} f^n(x,v,t)\psi(v)\, dv \xrightarrow[n]{} \int_{\mathbb{R}^N} f(x,v,t)\psi(v)\, dv \\ \text{in}\quad L^q(0,T;L^1(\mathbb{R}^N))\ , \quad \forall\, 1 \leq q < \infty\ , \forall\, T \in (0,\infty) \end{cases}$$

$$(26) \quad \begin{cases} \displaystyle\int_{\mathbb{R}^N} Q^+(f^n, f^n)\psi(v)\,dv \underset{n}{\to} \int_{\mathbb{R}^N} Q^+(f, f)\psi(v)\,dv\,, \\[2ex] \displaystyle\int_{\mathbb{R}^N} Q^-(f^n, f^n)\psi(v)\,dv \underset{n}{\to} \int_{\mathbb{R}^N} Q^-(f, f)\psi(v)\,dv \end{cases}$$

in measure on $(|x| < R) \times (0, T)$, $\forall\, R, T \in (0, \infty)$, and $f \in C([0, T];$ $L^1(\mathbb{R}^{2N}))$ is a weak solution of (1), (11). □

Remark V.1. Let us recall that g^n converges in measure on Ω to g if meas $\{z \in \Omega \,/\, |g^n(z) - g(z)| \geq \alpha\} \to_n 0$ for all $\alpha > 0$. □

Remark V.2. Once (26) is obtained, the fact that we can pass to the limit in the equation seems natural but it turns out that some delicate argument is needed (even if the original argument of [15] has been some-what simplified in P.L. Lions [31] with the use of Theorem V.3 below). Up to some unpleasant technicalities, it is not difficult to understand that (25) yields (26) : indeed, take for instance the case of Q^-

$$\int_{\mathbb{R}^N} Q^-(f^n, f^n)\psi(v)\,dv =$$
$$= \iint_{\mathbb{R}^{2N}} f^n(x, v, t)\, f^n(x, v_*, t)\, A(v - v_*)\, \psi(v)\,dv\,dv_*$$

and thus (this double integral is essentially a product (or a sum of products) of integrals of the type (25) i.e. macroscopic quantities. This observation clearly shows that (25) is the key point of Theorem V.1. It means that macroscopic quantities have some compactness properties. The origin of this compactness is presented in section V.2 below. □

In view of (25). a natural question to ask is whether the compacti-fication of f^n takes place in all variables : is f^n relatively compact in L^1_{loc}? From an intuitive physical viewpoint, one might expect this think-ing that collisions "organize" and regularize the distribution function of particles. The answer turns out to depend heavily on the form of the collision operator since it is true for the Landau operator but false in general for the Boltzmann operator as shown in P.L. Lions [28].

Theorem V.2.

1) (Landau collision operator). f^n converges to f in $L^1(\mathbb{R}^{2N} \times (0, T))$ ($\forall\, T \in (0, \infty)$).

2) (Boltzmann collision operator). If f^n converges to f in $L^1((|x| < R) \times (|v| < R) \times (0, T))$ ($\forall\, R \in (0, \infty)$) for some $T > 0$ then f_0^n converges to f_0 in $L^1(\mathbb{R}^{2N})$. □

Remark V.3. It is also possible to give examples of sequences of smooth (and "small") solutions of (1) such that $f^n(1)$ converges to $f(1)$ in $L^1(\mathbb{R}^{2N})$ if an only if $f^n(0)$ converges to $f(0)$ in $L^1(\mathbb{R}^{2N})$. $\quad\square$

Remark V.4. It is very much tempting to believe that, if we do not make the assumption of angular cut-off and keep the nonintegrable singularity of B, we might be in case 1). $\quad\square$

Remark V.5. The proof of the above result is given in [28] : it is rather intricate and uses the velocity averaging results described in section V.2 below. It is very much related to the study of classical hypoelliptic operators - except that we are of course in a nonlinear situation and that the compactification that takes place is somewhat weaker than for standard hypoelliptic operators. $\quad\square$

This result leads to the next question : if we assume that f_0^n converges to f_0 in $L^1(\mathbb{R}^{2N})$, what can we say about f^n ? This is solved in P.L. Lions [29],[30] and relies on a new compactness result described in section V.3 below. The precise result is given by the

Theorem V.3.

1) $Q^+(f^n, f^n) \underset{n}{\to} Q^+(f, f)$ in measure on $(|x| < R) \times (|v| < R) \times (0, T)$ ($\forall\, R, T \in (0, \infty)$).

2) If $f_0^n \underset{n}{\to} f_0$ in $L^1(\mathbb{R}^{2N})$, then $f^n \underset{n}{\to} f$ in $C([0, T]; L^1(\mathbb{R}^{2N}))$ ($\forall\, T \in (0, \infty)$). $\quad\square$

Remark V.6. The above result has many consequences, some of which are given in P.L. Lions [29],[30],[31] : it allows to precise the notion of weak solutions, prove the strong convergence in L^1 towards equilibrium (to a Maxwellian) as t goes to $+\infty$, simplify the proof of Theorem V.1 namely the passage to the limit in (1) under weak-L^1 convergences... This last fact can also be modified to yield the convergence, up to subsequences, of various numerical procedures such as splitting methods (...), justifying thus our rather vague claims at the end of section III.2. $\square$

Remark V.7. Part 1) of Theorem V.3 can be deduced from the results presented in section V.3 below by a rather complicated argument that we do not wish to detail here (see [29] for more details). Part 2) follows from Part 1) with some extra work that can be found in [30],[31]. $\quad\square$

V.2 Velocity averaging.

A typical example of the so-called velocity averaging results is the following

Theorem V.4. *Let* $m \geq 0$, *let* $f, g \in L^p(\mathbb{R}^{2N} \times \mathbb{R})$ *with* $1 < p \leq 2$, *satisfy*

$$(27) \qquad \frac{\partial f}{\partial t} + v \cdot \nabla_x f = (-\Delta_v + 1)^{m/2} g \qquad in \quad \mathcal{D}'(\mathbb{R}^{2N} \times \mathbb{R}) .$$

Then, for all $\psi \in C_0^\infty(\mathbb{R}_v^N)$, $\int_{\mathbb{R}^N} f(x, v, t)\psi(v) \, dv$ *belongs to the (Besov) space* $B_2^{s,p}(\mathbb{R}^N \times \mathbb{R})$ - *and in particular to the Sobolev space* $H^{s',p}(\mathbb{R}^N \times \mathbb{R})$ *for* $0 < s' < s$ - *with* $s = \frac{p-1}{p}(1+m)^{-1}$.

Remark V.8. If $m = 0$, $\int_{\mathbb{R}^N} f\psi \, dv \in H^{s,p}$ with $s = \frac{p-1}{p}$. If $p \in (1,2)$, it is not known if the above s is optimal while if $p = 2$ it is optimal. Similar results are available if $2 < p < \infty$ or in more general settings : we refer the reader to R.J. DiPerna, P.L. Lions and Y. Meyer [18]. □

Remark V.9. The first results in this direction were obtained in F. Golse, B. Perthame and R. Sentis [20], F. Golse, P.L. Lions, B. Perthame and R. Sentis [19] (where the case $m = 0$ is considered). The case $m \geq 0$, $p = 2$ was obtained by R.J. DiPerna and P.L. Lions [16] while the general case is due to R.J. DiPerna, P.L. Lions and Y. Meyer [18] - two related strategies are proposed in [18] that both rely on Fourier analysis, one uses some harmonic analysis namely product-Hardy spaces and interpolation argument while the second one uses classical multipliers theory and Littlewood-Paley dyadic blocks. However, the main idea that explains such "partial" regularity phenomena is elementary and described below in extremely rough terms. □

As indicated in the preceding remark, we give a carricatural - but accurate ! - explanation of the phenomena illustrated by Theorem V.4. If the Fourier transform (27) in (x, t), we see that we gain decay (that is regularity) in (ξ, τ) - dual variables of (x, t) - provided $|\tau + v \cdot \xi| \geq \delta |(\tau, \xi)|$ for some $\delta > 0$. On the other hand the set of v on which we do not gain that regularity namely $\{v \in \text{Supp } \psi \, / \, |\tau + v \cdot \xi| < \delta |(\tau, \xi)|\}$ has a measure of order δ hence contribute little to the integral $\left(\int_{\mathbb{R}^N} \hat{f}(\xi, v, \tau)\psi(v) \, dv \right)$. Balancing the two contributions, we obtain some partial (fractional) regularity.

V.3 The gain term and Radon transforms.

The main new ingredient in the proof of Theorem V.3 is the next result taken from P.L. Lions [29]. Let us first introduce some notations : we set $Q^+(f,g) = \int_{\mathbb{R}^N} dv_* \int_{S^{N-1}} d\omega \, B(v-v_*,\omega) f' g'_*$ for $f, g \in C_0^\infty(\mathbb{R}^N_v)$. We assume (to simplify the presentation) that B satisfies

$$(28) \quad B(z,\omega) = \varphi\left(|z|, \frac{|(z,\omega)|}{|z|}\right) \qquad \text{where} \quad \varphi \in C_0^\infty((0,\infty) \times (0,1)) \,.$$

Such a condition means that B vanishes for $|z|$ small and large, nearby grazing collisions $((z,\omega) = 0)$ and nearby "exchange" collisions $(|(z,\omega)| = |z|)$ i.e. these collisions where $v' = v_*$, $v'_* = v$.

Theorem V.4. *The operator Q^+ is bounded from $\mathcal{M}(\mathbb{R}^N) \times H^s(\mathbb{R}^N)$, $H^s(\mathbb{R}^N) \times \mathcal{M}(\mathbb{R}^N)$ into $H^{s+\frac{N-1}{2}}(\mathbb{R}^N)$ for all $s \in \mathbb{R}^N$.* □

(The space $\mathcal{M}(\mathbb{R}^N)$ is the space of bounded measures on $\mathbb{R}^N$).

The above gain of regularity $\frac{N-1}{2}$ derivatives is due to the fact that Q^+ can be written as a "linear combination" of translates of some Radon-like transforms (a variant of this proof making direct connection with the Radon transform has been recently given by B. Wennberg [42])

$$(28) \quad L_0\varphi(v) = \int_{S^{N-1}} B(v,\omega)\,\varphi((v,\omega)\omega)\,d\omega \,, \quad \forall\, \varphi \in C_0^\infty(\mathbb{R}^N)$$

or

$$(29) \quad L_0\varphi(v) = \int_{S^{N-1}} B(v,\omega)\,\varphi(v-(v,\omega)\omega)\,dw \,, \quad \forall\, \varphi \in C_0^\infty(\mathbb{R}^N) \,.$$

In both cases, one integrates φ over the set $\{(v,\omega)\omega \,/\, \omega \in S^{N-1}\} = \{v - (v,\omega)\omega \,/\, \omega \in S^{N-1}\}$ which is the sphere centered at $\frac{v}{2}$ and of radius $\frac{|v|}{2}$. These operators are rather special Fourier integral operators often called generalized Radon transform. The crucial fact is that the set over which φ is integrated moves with v - except that all these spheres go through 0 but this does not create difficulties since B vanishes if $(v,\omega)\omega = 0$ (grazing collisions) or if $v-(v,\omega)\omega = 0$ ("exchange" collisions). This is the main reason why one can prove
(30)
L_0 is bounded from $H^s(\mathbb{R}^N)$ into $H^{s+\frac{N-1}{2}}(\mathbb{R}^N)$ for all $s \in \mathbb{R}^N$

($\frac{N-1}{2}$ comes from the stationary phase principle...).

VI. Realistic boundary conditions.

We now want to consider cases when (1) is set in $(x, v, t) \in \Omega \times \mathbb{R}^N \times (0, \infty)$. Of course, (11) is now imposed on $\Omega \times \mathbb{R}^N$. And, we have to complement (1), (11) with boundary conditions on $\partial\Omega$. There exist many classical possibilities which are all physically relevant and we refer the reader to C. Cercignani [8] for a general presentation of this subject. We mention here only one example namely the specular boundary conditions : we then assume that $\partial\Omega$ is smooth say of class C^1 and bounded (so that either Ω is bounded or Ω^c is bounded) and we denote by n the unit outward normal to $\partial\Omega$. We then impose
(30)
$$f(x, v, t) \; = \; f(x, \, v - 2(v, n(x))n(x), \, t) \quad \text{for } x \in \partial\Omega, \; v \in \mathbb{R}^N, \; t \leq 0.$$

In the case Ω is bounded or if $\Omega = \mathcal{O}^c$, $\mathcal{O}$ is bounded and if f goes to 0 as $|(x, v)|$ goes to ∞, then the proof of Theorem IV.1 (and of the other results above) can be "easily" modified and yields similar results when we impose boundary conditions like (30) - see K. Hamdache [22] for more details.

However, some realistic situations do not fall into these categories. Indeed, if we go back to the industrial example of a flight in a rarefied atmosphere, we are naturally led to the following situation : $\Omega \equiv \mathcal{O}^c$ where $\mathcal{O}$ is a C^1 bounded connected domain in $\mathbb{R}^N$, we may impose (30) - or more complicated boundary conditions - but we cannot impose that f goes to 0 as $|(x, v)|$ goes to $+\infty$: indeed, at "infinity" the flow of gas is a constant flow and the natural condition is then to impose

(31) $$f(x, v, t) \; \rightarrow \; M_0 \qquad \text{as} \quad |x|, |v| \rightarrow +\infty$$

where $M_0 = M(\rho_0, u_0, T_0)$ is a fixed Maxwellian at infinity ($\rho_0 > 0$, $T_0 > 0$, $u_0 \in \mathbb{R}^N$).

Then, the difficulty is obvious since we simply lose all bounds : recall from section III.1 that the bounds we obtained follow from the collisions invariants and are weighted L^1 estimates. All these L^1 quantities become infinite if we impose (31). This difficulty was solved in P.L. Lions [32] by the introduction of various quantities related to the entropy $E[f]$ that provide similar bounds to those obtained in section III.1. With these bounds, one easily adapts all the results given above - see [32] for more details.

The first quantity is the relative entropy with respect to an appropriate local Maxwellian : we set $M(x) = M(\rho_0, u(x), T_0)$ where u is smooth and vanishes in a neighborhood of $\overline{\mathcal{O}}$, $u = u_0$ for $|x|$ large, $|u(x)| \leq |u_0|$ on $\mathbb{R}^N$, $\|\nabla u(x)\| \leq \varepsilon_0 < \frac{1}{2}$. Next, we consider $f \log \frac{f}{M} + M - f$ and we recall the elementary (convexity) inequality

$$a \log \frac{a}{b} + b - a \geq 0 \qquad \text{for all} \quad a, b \geq 0 .$$

Exactly as in section III.1 we then deduce at least formally

$$\frac{\partial}{\partial t} \left(\int_{\mathbb{R}^N} f \log \frac{f}{M} + M - f \, dv \right) +$$

$$+ \operatorname{div}_x \left(\int_{\mathbb{R}^N} v \left\{ f \frac{f}{M} + M - f \right\} dv \right) + \frac{1}{4} D[f(t)]$$

$$= - \int_{\mathbb{R}^N} f v \cdot \nabla_x \log M \, dv + \int_{\mathbb{R}^N} v \cdot \nabla_x M_0 \, dv$$

$$\leq \frac{\varepsilon_0}{2T_0} \left(\int_{\mathbb{R}^N} f |v|^2 \, dv + |u_0| \int_{\mathbb{R}^N} f |v| \, dv + C \right) 1_K(x)$$

where $K = \operatorname{Support}(|\nabla u|)$ is a compact set of Ω and C denotes various positive constants independent of $x, t, f...$

Then, we integrate this inequality over Ω and we observe that $f \log \frac{f}{M} + M - f$ goes to 0 as $|x|, |v|$ go to $+\infty$ in view of (31) and of the choice of M while

$$\int_{\mathbb{R}^N} \left(f \log \frac{f}{M} + M - f \right) v \cdot n(x) \, dx = 0 \qquad \text{for} \quad x \in \partial\Omega$$

in view of (30) and the choice of M. We then find for all $t \geq 0$

$$(32) \quad \begin{cases} \displaystyle\iint_{\Omega \times \mathbb{R}^N} f(t) \log \frac{f(t)}{M} + M - f(t) \, dx \, dv + \int_0^t ds \int_\Omega dx \, D[f(s)] \leq \\[2ex] \displaystyle\leq \iint_{\Omega \times \mathbb{R}^N} f_0 \log \frac{f_0}{M} + M - f_0 \, dx \, dv + \\[2ex] \displaystyle+ \frac{\varepsilon_0}{2T_0} \int_0^t ds \iint_{K \times \mathbb{R}^N} f(|v|^2 + C) \, dx \, dv + Ct . \end{cases}$$

Using Lemma IV.1 below (that we state without proof), we deduce that if $f_0 \geq 0$ satisfies

$$(33) \qquad \iint_{\Omega \times \mathbb{R}^N} f_0 \log \frac{f_0}{M} + M - f_0 \, dx \, dv \leq A$$

for some $A \in (0, \infty)$, then, for all $T \in (0, \infty)$, there exists $C > 0$ $(= C(A, T))$ such that

$$(34) \qquad \iint_{\Omega \times \mathbb{R}^N} f(t) \log \frac{f(t)}{M} + M - f(t) \, dx \, dv \ \leq \ C$$

$$(35) \quad \sup_{t \in [0,T]} \ \sup_{x_0 \in \mathbb{R}^N} \int_{(x_0 + B_R) \cap \Omega} dx \int_{\mathbb{R}^N} dv \, f(t) \, (1 + |v|^2 + |\log f(t)|) \ \leq \ C$$

$$(36) \qquad \int_0^T dt \int_\Omega dx \, D[f(t)] \ \leq \ C$$

completing thus the proof of the desired a priori estimates.

Lemma VI.1. *Let* $M = M(\rho, u, T)$ $(\rho > 0, \ T > 0, \ u \in \mathbb{R}^N)$, *let* $\delta \in (0, \frac{1}{4T})$, $K \in (0, \infty)$ *then there exists* $C \left(= \rho(e^K [\pi(1 - 4\delta T)]^{N/2} - 1) \right) > 0$ *such that for all* $h \in L^1(\mathbb{R}_v^N)$, $h \geq 0$

$$(37) \quad \int_{\mathbb{R}^N} \left[h \log \frac{h}{M} + M - h \right] dv \ \geq \ \int_{\mathbb{R}^N} h(\delta |v - u|^2 + K) - C . \quad \square$$

The other quantity used in P.L. Lions [**32**] to derive a priori estimates in other realistic situations is the following truncated entropy $\left(\iint f \log^+ (\frac{f}{M}) \, dx \, dv \right)$ where M is a Maxwellian. It is shown in [**32**] that the following inequality holds for all smooth nonnegative φ

$$(38) \qquad \int_{\mathbb{R}^N} Q(\varphi, \varphi) \log^+ \left(\frac{\varphi}{M} \right) dv \ \leq \ 3 \int_{\mathbb{R}^N} A * M \varphi \log^+ \frac{\varphi}{M} \, dv$$

(in addition, it is possible to "control" a term like $D[f]$ from the difference between the right-hand side and the left-hand side). Let us also observe that since $\int_{\mathbb{R}^N} Q(\varphi, \varphi) \log M \, dv = 0$, we can recover the usual inequality $\left(\int_{\mathbb{R}^N} Q(\varphi, \varphi) \log \varphi \, dv \leq 0 \right)$ by letting M go to 0.

VII. Perspectives.

First of all, let us draw a list of relevant open questions

1. Uniqueness of weak solutions.

2. Entropy identity and conservation of kinetic energy for weak solutions.

3. Regularity of solutions (if f_0 is smooth) at least in the case when $B = \varphi\big(|z|, \frac{|(z,\omega)|}{|z|}\big)$, $\varphi \in C_0^\infty((0,\infty) \times (0,1))$.

4. Hydrodynamical limits (recovering weak solutions of incompressible Navier-Stokes equations).

5. Stationary problems (exterior problem, shock problem...).

6. Compactness results for models without angular cut-off.

All these questions are completely open (and some may even seem out of reach...) but the results of section V.3 (for problem 3.) and some preliminary results due to L. Desvillettes [14] together with Theorem V.2 (for problem 6.) give some hope of making progress (step by step) on the Boltzmann's equation.

References.

[1] L. Arkeryd, *On the Boltzmann equation. I and II*. Arch. Rat. Mech. Anal., **45** (1972), p. 1–34.

[2] C. Bardos, F. Golse and D. Levermore, *Sur les limites asymptotiques de la théorie cinétique conduisant à la dynamique des fluides incompressibles*. C.R. Acad. Sci. Paris, **309** (1989), p. 727–732.

[3] C. Bardos, F. Golse and D. Levermore, *Fluid dynamic limits of kinetic equations. I : Formal derivations*. J. Stat. Phys., **63** (1991), p. 323–344.

[4] C. Bardos, F. Golse and D. Levermore, *Fluid dynamic limits of kinetic equations. II : Convergence proofs for the Boltzmann equation*. Preprint.

[5] L. Boltzmann, *Weitere Studien über das Wärme gleichgenicht unfer Gasmolekülar*. Sitzungsberichte der Akademie der Wissenschaften, Wien, **66** (1872), p. 275–370. (Trans. *Further studies on the thermal equilibrium of gaz molecules*, In Kinetic Theory vol. 2, p. 88-174, ed. S.G. Brush, Oxford, Pergamon, 1966).

[6] T. Carleman, Acta Math., **60** (1933), 91.

[7] T. Carleman, Problèmes mathématiques dans la théorie cinétique des gaz. Notes written by Carleson and Frostman, Uppsala, Almqvist and Wikselles, 1957.

[8] C. Cercignani, The Boltzman equation and its applications. Springer Berlin, 1988.

[9] C. Cercignani, *A remarkable estimate for the solutions of the Boltzmann equations*. Appl. Math. Letters, **5** (1992), p. 59–62.

[10] S. Chapman and T.G. Cowling, The mathematical theory of non-uniform gases. 2nd edit., Cambridge Univ. Press, 1952.

[11] P. Degong and B. Lucquin-Desreux, *The Fokker-Planck asymptotics of the Boltzmann collision operator in the Coulomb case*. Preprint.

[12] L. Desvillettes, *On an asymptotics of the Boltzmann equation when the collisions become grazing*. Preprint.

[13] L. Desvillettes, *Some applications of the method of moments for the homogeneous Boltzmann and Kac equations*. Arch. Rat. Mech. Anal., **132** (1993), p. 387–404.

[14] L. Desvillettes, *About the regularizing properties of the non cut-off Kac equation.* Preprint.

[15] R.J. DiPerna and P.L. Lions, *On the Cauchy problem for Boltzmann equations : global existence and weak stability.* Ann. Math., **130** (1989), p.321–366.

[16] R.J. DiPerna and P.L. Lions, *Global weak solutions of Vlasov-Maxwell systems.* Comm. Pure Applied Math., **62** (1989), p. 729–757.

[17] R.J. DiPerna and P.L. Lions, *Global solutions of Boltzmann's equation and the entropy inequality.* Arch. Rat. Mech. Anal., **114** (1991), p. 47–55.

[18] R.J. DiPerna, P.L. Lions and Y. Meyer, L^p *regularity of velocity averages.* Ann. I.H.P. Anal. Nonlin., **8** (1991), p. 271–287.

[19] F. Golse, P.L. Lions, B. Perthame and R. Sentis, *Regularity of the moments of the solutions of a transport equation.* J. Funct. Anal., **76** (1988), p. 110–125.

[20] F. Golse, P.L. Lions and R. Sentis, *Un résultat pour les équations de transport et application au calcul de la limite de la valeur propre principale d'un opérateur de transport.* C.R. Acad. Sci. Paris, **301** (1985), p. 341–344.

[21] H. Grad, *Principles of the kinetic theory of gases,* in Flügge's Handbuch der Physik, XII, Springer, Berlin, (1958), p. 205–294.

[22] K. Hamdache, *Global existence for weak solutions for the initial boundary value problems of Boltzmann equation.* Arch. Rat. Mech. Anal., **119** (1992), p. 309–353.

[23] D. Hilbert, Math. Ann., **72** (1912), 562.

[24] J. Leray, *Etude de diverses équations intégrales nonlinéaires et de quelques problèmes que pose l'hydrodynamique.* J. Maths. Pures Appl., **12** (1933), p. 1–32.

[25] J. Leray, *Essai sur les mouvements plans d'un fluide visqueux que limitent des parois.* J. Maths. Pures Appl., **13** (1934), p. 331–418.

[26] J. Leray, *Essai sur le mouvement d'un liquide visqueux emplissant l'espace.* Acata Math., **63** (1934), p. 193–248.

[27] E.M. Lifschitz and L.P. Pitaevskii, Physical kinetics. Oxford, Pergamon, 1981.

[28] P.L. Lions, *On Boltzmann and Landau equations.* Phil. Trans. R. Soc. London, A**346** (1994), p. 191–204.

[29] P.L. Lions, *Compactness in Boltzmann's equation via Fourier integral operators; Part I.* J. Math. Kyoto Univ., 1994.

[30] P.L. Lions, *Compactness in Boltzmann's equation via Fourier integral operators and applications. II.* J. Math. Kyoto Univ., 1994.

[31] P.L. Lions, *Compactness in Boltzmann's equation via Fourier integral operators and applications. III.* J. Math. Kyoto Univ., 1994.

[32] P.L. Lions, *Conditions at infinity for Boltzmann's equation.* Comm. P.D.E., **19** (1994), p. 335–367.

[33] J.C. Maxwell, *On the dynamical theory of gases.* Phil. Trans. Roy. Soc. London, **157** (1866), p. 49–88.

[34] J.C. Maxwell, Scientific papers. Vol. 2, Cambridge, Cambridge Univ. Press, 1890 (Reprinted by Dover Publications, New-York, 1965).

[35] F. Murat, *Compacité par compensation.* Ann. Sc. Norm. Sup. Pisa, **V** (1978), p. 498–507.

[36] F. Murat, *Compacité par compensation, II.* In Proceedings of the International Meeting on Recent Methods in Nonlinear Analysis. E. De Giorgi, E. Magenes, U. Mosco eds., Bologna, Pitagora, 1979.

[37] F. Murat, *Compacité par compensation, III.* Ann. Sc. Norm. Sup. Pisa, **VIII** (1981), p. 69–102.

[38] L. Tartar, In Nonlinear Analsyis and Mechanics, Heriot-Watt symposium, IV. London, Pitman, 1979.

[39] L. Tartar, In Systems of nonlinear partial differential equations. Dordrecht, Reidel, 1983.

[40] L. Tartar, In Macroscopic modelling of turbulent flows. Lecture Notes in Physics ♯ 230, Berlin, Springer, 1985.

[41] C. Truesdell and R.G. Muncaster, Fundamentals of Maxwell's kinetic theory of a simple monoatomic gas. New-York, Academic Press, 1960.

[42] B. Wennberg, *Regularity estimates for the Boltzmann equation.* Preprint, 1994.

Proceedings of Symposia in Applied Mathematics
Volume **54**, 1998

Simplified Asymptotic Equations for Slender Vortex Filaments

Andrew J. Majda

Dedicated to Peter Lax and Louis Nirenberg with admiration and friendship

Below we show how formal but concise asymptotic expansions can be utilized to study the motion of slender tubes of vorticity at high Reynolds numbers in such a fashion so that qualitative insight into the folding, wrinkling and bending of vortex tubes can be analyzed in a simplified context. Beautiful simplified asymptotic equations emerge with remarkable properties which we describe here.

The key technical idea is to expand the Biot-Savart law determining the velocity from the vorticity.

$$
(1) \qquad v(\mathbf{x}, t) = \frac{1}{4\pi} \int \frac{\mathbf{x} \times \omega}{|\mathbf{x}|^3}(\mathbf{y}, t) dy
$$

in a suitable asymptotic expansion provided that the vorticity, $|\omega|$, is large and confined to a narrow tube around a curve, $\mathbf{X}(s, t)$, defining the centerline. Of course, it is crucial to develop such expansions in a fashion consistent with the Navier-Stokes equations at high Reynolds numbers. In this manner, simplified dynamical equations emerge for the motion of the curve $\mathbf{X}(s, t)$ in various asymptotic regimes as well as equations for the interaction of many such strong vortex filaments represented by a finite collection of such curves $\{X_j(s, t)\}_{j=1}^N$.

In section 1 we describe the simplest such theory involving the self-induction approximation for the motion of a single vortex filament. We also describe Hasimoto's remarkable transformation which reduces these simplified vortex dynamics to those for the cubic nonlinear Schrödinger equation, a remarkable completely integrable P.D.E. with soliton behavior and heteroclinic instabilities (Ablowitz and Segur, Lamb). We also demonstrate that despite the great beauty of the self-induction approximation, it fails to allow for any vortex stretching which is one of the most prominent features in incompressible fluid flow with small viscosity!! Some of these shortcomings are addressed by a recent asymptotic theory (Klein and Majda, 1991 a, b) which allows for self-stretch of a single filament. This theory is described in section 2.

1991 *Mathematics Subject Classification.* Primary 76C05; Secondary 35Q30, 35Q35.

Andrew J. Majda was supported in part by NSF Grant DMS-9596102-001, ARO Grant DAAH0-95-11-0345 and ONR Grant N00014-96-1-0043.

In sections 3 and 4 we discuss simplified equations for the dynamics of many interacting vortex filaments. Such situations frequently arise in many practical contexts such as the two trailing wake vortices shed from the wing tips of aircraft (Van Dyke). In section 3 we briefly discuss the simplified dynamics for the interaction of a finite number of exactly parallel vortex filaments which reduces to the familiar interaction of point vortices in the plane (Chorin and Marsden, Aref). In section 4 we describe a recent asymptotic theory (Klein, Majda, and Damodaran) for the interaction of *nearly parallel vortex filaments* with remarkable properties incorporating features both of self-induction from section 1 and mutual point vortex interaction as described in section 3. We end this paper in section 5 with a discussion as well as a list of interesting open mathematical problems regarding the material here.

1. The Self-Induction Approximation, Hasimoto's Transform, and the Nonlinear Schrödinger Equation

First we give some precise geometric assumptions on the nature and strength of the evolving vortex filament which lead to a concise self-consistent asymptotic development of the Biot-Savart law in (1) which is also consistent with the Navier-Stokes equations. Here and throughout this paper, we do not supply the lengthy details of the asymptotic derivation but instead refer the interested reader to the published literature. Consider a concentrated thin tube of vorticity so that the following assumptions are satisfied:

(2)

    A) The vorticity is essentially non-zero only inside a tube of cross-sectional radius δ with $\delta \ll 1$.

    B) The vorticity ω is large inside this tube so that the circulation $\Gamma = \int \omega \cdot \mathbf{n}\, d\mathbf{s}$ satisfies $\mathrm{Re} = \Gamma/\nu \to \infty$ where ν is the viscosity and Re is the Reynolds number.

    C) Despite A) and B), the radius of curvature of the filament centerline remains bounded strictly away from zero.

    D) The cross-sectional radius, δ, from A) and the Reynolds number, Re, from B) are balanced so that $\delta = (\mathrm{Re})^{-\frac{1}{2}}$.

The assumptions in (2) indicate that there are primarily two scales in the assumed vortex-filament motion, an outer scale which is order $O(1)$ defined by the radius of curvature of the filament and an inner scale, δ, $\delta \ll 1$ describing the cross sectional radius of the filament. The large value of the vorticity within the filament is expressed concisely in non-dimensional terms through the requirement in D) above where $\delta = (\mathrm{Re})^{-\frac{1}{2}}$.

For such a thin filament of vorticity, consider the centerline curve $\mathcal{L}(t)\colon\ s \to \mathbf{X}(s,t)$. To a high degree of approximation for $\delta \ll 1$ this centerline should move with the fluid so that

$$(3) \qquad\qquad \frac{d\mathbf{X}}{dt} = \mathbf{v}(\mathbf{X}(s,t),t).$$

To get the velocity $\mathbf{v}(\mathbf{X}(s,t),t)$, we need to simplify the Biot-Savart law in (1) near the centerline $\mathbf{X}(s,t)$ under the geometric assumptions in (2). Under these circumstances concise matched asymptotic expansions (Callegari and Ting, Ting and Klein) establish that the right hand side of (3) is given by

$$(4) \qquad \frac{\partial X}{\partial t}(s,t) = \kappa \, \mathbf{b}(s,t) + \left(\ln \left(\frac{1}{\delta} \right) \right)^{-1} (\widetilde{C}(t)\,\mathbf{b} + \mathbf{Q}_f) + o(1)$$

for $\delta \ll 1$. Here $\mathbf{b} = \mathbf{t} \times \mathbf{n}$ is the binormal to the curve with $\mathbf{t}$ the tangent vector and $\mathbf{n}$ the normal while κ is the curvature of the curve. We change the time variable in (4) by $\bar{t} = \ln(1/\delta)t$ and retain only the leading order asymptotic term in (4) to obtain the

Self-Induction Equation for an Isolated Vortex filament:

$$\frac{\partial \mathbf{X}}{\partial \bar{t}} = \kappa \mathbf{b}.$$

This equation was first derived through a different procedure by Arms and Hama.

1.1. Hasimoto's Transform and the Cubic Nonlinear Schrödinger Equation.
For further developments we consider a generalization of the self-induction equation involving a perturbed binormal law.

$$(5) \qquad \frac{\partial \mathbf{X}(\tilde{s}, \bar{t})}{\partial \bar{t}} = (\kappa \mathbf{b})(\tilde{s}, \bar{t}) + \tilde{\delta}\mathbf{v}(\tilde{s}, \bar{t}),$$

where $\tilde{s}$ is the arclength along the curve and $\tilde{\delta}\mathbf{v}(\tilde{s}, \bar{t})$ is a small general perturbation velocity. Closely following the procedure of Hasimoto for the case of $\tilde{\delta} \equiv 0$ we derive a connection between (5) and a perturbed Schrödinger equation for the related filament function.

The Hasimoto Transform for a Perturbed Binormal Law

First we recall some basic notations and identities. The Serret-Frenet formulae

$$(6) \qquad \mathbf{X}_{\tilde{s}} = \mathbf{t}, \quad \mathbf{t}_{\tilde{s}} = \kappa \mathbf{n}, \quad \mathbf{n}_{\tilde{s}} = T\mathbf{b} - \kappa \mathbf{t}, \quad \mathbf{b}_{\tilde{s}} = -T\mathbf{n}$$

describe the variation of the intrinsic basis $\mathbf{t}, \mathbf{n}, \mathbf{b}$ along $\mathcal{L}$ in terms of the curvature κ and torsion T of the curve. Hasimoto considers the *filament function*

$$(7) \qquad \psi(\bar{s}, \bar{t}) = \kappa(\bar{s}, \bar{t})e^{i\Phi} \text{ with } \Phi = \int_0^{\bar{s}} T(s', \bar{t})ds'.$$

He replaces the principal and binormal unit vectors $\mathbf{n}$ and $\mathbf{b}$ by the complex vector function

$$(8) \qquad \mathbf{N}(\tilde{s}, \bar{t}) = (\mathbf{n} + i\mathbf{b})(\tilde{s}, \bar{t}) \exp[i\Phi(\tilde{s}, \bar{t})],$$

with Φ from (7) and by its complex conjugate $\overline{\mathbf{N}}$ to span the planes normal to $\mathbf{t}$. The new basis satisfies the orthogonality relations

$$(9) \qquad \mathbf{t} \cdot \mathbf{t} = 1, \quad \mathbf{N} \cdot \mathbf{N} = 0, \quad \mathbf{N} \cdot \overline{\mathbf{N}} = 2,$$

and the Serret-Frenet equations yield its variations along $\mathcal{L}$

$$(10) \qquad \mathbf{N}_{\tilde{s}} = -\psi \mathbf{t}, \quad \mathbf{t}_{\tilde{s}} = \frac{1}{2}(\overline{\psi}\mathbf{N} + \psi\overline{\mathbf{N}}).$$

Differentiating the perturbed binormal law (5) with respect to $\tilde{s}$ we obtain the dynamic behavior of the tangent vector

$$(11) \qquad \mathbf{t}_{\bar{t}} = \kappa_{\tilde{s}}\mathbf{b} - \kappa T\mathbf{n} + \tilde{\delta}\mathbf{v}_{\tilde{s}},$$

which may be expressed in terms of the new basis as

$$(12) \qquad \mathbf{t}_{\bar{t}} = \frac{1}{2}i(\psi_{\tilde{s}}\overline{\mathbf{N}} + \overline{\psi}_{\tilde{s}}\mathbf{N}) + \tilde{\delta}\mathbf{v}_{\tilde{s}}.$$

An evolution equation for the filament function is next derived by computing two independent representations of the cross-derivative $\mathbf{N}_{\tilde{s}\tilde{t}}$ of N and by comparing their respective components along $\mathbf{N}$ and $\mathbf{t}$. We first decompose $\mathbf{N}_{\tilde{t}}$ as

(13) $$\mathbf{N}_{\tilde{t}} = \alpha\mathbf{N} + \beta\overline{\mathbf{N}} + \gamma\mathbf{t}.$$

Then the orthogonality relations (9) and (12) yield the identities

$$\alpha + \bar{\alpha} = \frac{1}{2}(\mathbf{N} \cdot \overline{\mathbf{N}}_{\mathbf{t}} + \mathbf{N}_{\tilde{t}} \cdot \overline{\mathbf{N}}) = \frac{1}{2}(\mathbf{N} \cdot \overline{\mathbf{N}})_{\tilde{t}} \equiv 0,$$

(14) $$\beta = \frac{1}{4}(\mathbf{N} \cdot \mathbf{N})_{\tilde{t}} \equiv 0,$$

$$\gamma = \mathbf{t} \cdot \mathbf{N}_{\tilde{t}} = -\mathbf{t}_{\tilde{t}} \cdot \mathbf{N} = -i\psi_{\tilde{s}} - \tilde{\delta}\mathbf{N} \cdot \mathbf{v}_{\tilde{s}}.$$

Thus we find

(15) $$\mathbf{N}_{\tilde{t}} = i[R\mathbf{N} - (\psi_{\tilde{s}} - i\tilde{\delta}\mathbf{N} \cdot \mathbf{v}_{\tilde{s}})\mathbf{t}],$$

where $R = R(\tilde{s}, \tilde{t})$ is a yet unknown real valued function. The two representations of $\mathbf{N}_{\tilde{s}\tilde{t}}$ now follow from partial differentiation of (10) and (15) with respect to $\tilde{t}$ and $\tilde{s}$, respectively,

$$N_{\tilde{s}\tilde{t}} = -\psi_{\tilde{t}}\mathbf{t} - \frac{1}{1}2i(\psi\psi_{\tilde{s}}\overline{\mathbf{N}} - \psi\overline{\psi}_{\tilde{s}}\mathbf{N}) - \tilde{\delta}\psi\mathbf{v}_{\tilde{s}},$$

(16) $$N_{\tilde{t}\tilde{s}} = i[R_{\tilde{s}}\mathbf{N} - R\psi\mathbf{t} - [\psi_{\tilde{s}\tilde{s}} - i\tilde{\delta}(\mathbf{N} \cdot \mathbf{v}_{\tilde{s}})_{\tilde{s}}]\mathbf{t}$$
$$-\frac{1}{2}(\psi_{\tilde{s}} - i\tilde{\delta}\mathbf{N} \cdot \mathbf{v}_{\tilde{s}})(\overline{\psi}\mathbf{N} + \psi\overline{\mathbf{N}})].$$

Equating the coefficients of $\mathbf{t}$ and $\mathbf{N}$ in these relations we obtain

(17) $$\psi_{\tilde{t}} + \tilde{\delta}\psi\mathbf{v}_{\tilde{s}} \cdot \mathbf{t} = i[R\psi + \psi_{\tilde{s}\tilde{s}} - i\tilde{\delta}(\mathbf{N} \cdot \mathbf{v}_{\tilde{s}})_{\tilde{s}}]$$

and

(18) $$\mathbf{R}_{\tilde{s}} = \frac{1}{2}(\psi_{\tilde{s}}\overline{\psi} + \psi\overline{\psi}_{\tilde{s}}) + \frac{1}{2}i\tilde{\delta}(\psi\overline{\mathbf{N}} - \overline{\psi}\mathbf{N}) \cdot \mathbf{v}_{\tilde{s}}.$$

At this stage we recover Hasimoto's result

(19) $$R = \frac{1}{2}|\psi|^2 \text{ for } \tilde{\delta} = 0$$

by removing the perturbation term from (18). This yields the cubic nonlinear Schrödinger equation when inserted in (17). Hasimoto points out that an arbitrary integration function $A(\tilde{t})$ that might appear upon integration of (18) may be eliminated without loss of generality through a shift

$$\tilde{\Phi}(\tilde{s}, \tilde{t}) = \Phi(\tilde{s}, \tilde{t}) - \int_0^{\tilde{t}} A(t)dt,$$

of the phase function Φ in (7). In fact all the geometric information contained in the filament function is

(20) $$|\psi| = \kappa \quad \text{and} \quad T = [\arg(\psi)]_{\tilde{s}} = \Phi_{\tilde{s}} = \tilde{\Phi}_{\tilde{s}},$$

so that the phase shift does not influence the geometry of $\mathcal{L}$. In general, we can integrate (18) and insert their result into (17) to obtain the *exact perturbed Schrödinger equation*:

(21) $$\left(\frac{1}{i}\right)\psi_{\tilde{t}} = \psi_{\tilde{s}\tilde{s}} + \frac{1}{2}|\psi|^2\psi - \tilde{\delta}\big(i[(\mathbf{N} \cdot \mathbf{v}_{\tilde{s}})_{\tilde{s}} - \psi\mathbf{v}_{\tilde{s}} \cdot \mathbf{t}] + \psi\int_{\tilde{s}_0}^{\tilde{s}} \text{Im}(\psi\overline{\mathbf{N}}) \cdot \mathbf{v}_{\tilde{s}}d\tilde{s}\big).$$

In particular, for the special case with $\tilde{\delta} = 0$ in (21), we see that through the filament function

(22) $$\psi(\bar{s}, \bar{t}) = \kappa(\bar{s}, \bar{t}) \exp\left(i \int_0^{\bar{s}} T(s', \bar{t}) ds'\right)$$

the self induction equation reduces to the cubic nonlinear Schrödinger equation,

(23) $$\frac{1}{i}\psi_t = \psi_{ss} + \frac{1}{2}|\psi|^2\psi.$$

The equation in (23) is a famous exactly solvable equation by the inverse scattering method (Ablowitz and Segur, Lamb) so that one can utilize exact solutions of these equations in (23) to infer properties of vortex filament motion under the self-induction approximation. Next, we point out that despite the beauty of the self-induction equation for a vortex filament, the length of the curve $\mathbf{X}(s, t)$, representing the vortex tube cannot increase in time, i.e. the *vortex filament cannot have any self-stretching within the self-induction approximation*. Thus, if one wants to include other effects of bending and folding of a single vortex filament including local self-stretching, different asymptotic models are needed which incorporate new assumptions beyond those in (2) which allow for the new physics in simplified self-stretching mechanisms. This is the topic in section 2.

To establish this result we consider the general evolution of a curve, $\mathbf{X}(s, t)$, according to the law

(24) $$\frac{\partial \mathbf{X}}{\partial t} = \beta(s, t)\mathbf{n} + \gamma(s, t)\mathbf{b}$$

where $\mathbf{n}$ is the normal and $\mathbf{b}$ is the binormal. Here the parameter, s, is not necessarily the arc-length. We note that we have omitted a term $\alpha(s, t)\mathbf{t}$ on the right hand side of (24) without loss of generality since such terms yield only motion along the curve, $\mathbf{X}(s, t)$, itself which can always be eliminated by reparameterization. Consider the time-rate of change of the infinitesimal arc-length of the curve, $(\mathbf{X}_s \cdot \mathbf{X}_s)^{\frac{1}{2}}$. We compute from (24) that

(25) $$\frac{\partial}{\partial t}(\mathbf{X}_s \cdot \mathbf{X}_s)^{\frac{1}{2}} = \mathbf{t} \cdot \frac{\partial}{\partial s}(\beta(s, t)\mathbf{n} + \gamma(s, t)\mathbf{b}).$$

Utilizing the Frenet formulas in (6), we obtain the general identity

(26) $$\frac{\partial}{\partial t}(\mathbf{X}_s \cdot \mathbf{X}_s)^{\frac{1}{2}} = -\kappa\beta(\mathbf{X}_s \cdot \mathbf{X}_s)^{\frac{1}{2}}$$

where κ is the curvature. In particular, if the motion of the curve is always along the binormal, we have $\beta \equiv 0$ in (26) and the infinitesimal arc-length of a curve cannot increase in time. Thus, the self-induction approximation alone does not allow for vortex stretching!!

We end this section by computing the linearized self-induction equation about a straight line filament parallel to the x_3-axis. First, we rewrite the self-induction equation in the form

(27) $$\frac{\partial \mathbf{X}}{\partial t} = C_0 \mathbf{t} \times \frac{\partial^2 \mathbf{X}}{\partial \sigma^2}$$

where C_0 is a constant depending on the choice of reference time. We consider small perturbations of a straight line filament parallel to the x_3-axis which, without loss of generality, have the form

(28) $\mathbf{X} = (0, 0, \sigma) + \epsilon(x(\sigma, t), y(\sigma, t), 0)$

and $\epsilon \ll 1$. Inserting (28) into (27) and calculating the leading order behavior in ϵ
we obtain the
Linearized Self-Induction Equations Along a Straight Line Filament

(29) $$\frac{\partial \mathbf{X}}{\partial t} = J\left[C_0 \frac{\partial^2 \mathbf{X}}{\partial \sigma^2}\right]$$

for $\mathbf{X} = (x(\sigma, t), y(\sigma, t))$ where J is the standard skew-symmetric matrix

(30) $$J = \begin{pmatrix} 0 & -1 \\ 1 & 0 \end{pmatrix}.$$

If we introduce the complex notation

(31) $$\psi(\sigma, t) = x(\sigma, t) + iy(\sigma, t),$$

then the equation in (29) becomes the linear Schrödinger equation

(32) $$\frac{1}{i} \frac{\partial \psi}{\partial t} = C_0 \frac{\partial^2 \psi}{\partial \sigma^2}.$$

We hope that our slight abuse of notation in still denoting the perturbation quantity
$(x(\sigma, t), y(\sigma, t))$ by $\mathbf{X}$ in (29) has not confused the reader. We utilize (29) and
(32) in interpreting the asymptotic equations for nearly parallel interacting vortex
filaments described in section 4.

2. Simplified Asymptotic Equations with Self-Stretch for a Single Vortex Filament

Here we describe a simplified asymptotic equation which allows for some features of self-stretch of a single vortex filament including incipient formation of
kinks, folds, and hairpins (Klein and Majda 1991 a, b, 1993, and Klein, Majda,
McLaughlin 1992). This simplified asymptotic equation has the attractive features
that it incorporates some of the nonlocal features of vortex stretching represented
in the Biot-Savart law in (1) through a filament function ψ defined via the same
Hasimoto transform introduced in (7). In this fashion a singular perturbation of
the cubic nonlinear Schrödinger equation in (23) emerges where the nonlocal terms
represented by $I[\psi]$ compete directly with the cubic nonlinear terms in (23) which
as we have seen earlier arise from local self-induction.

The key idea in the derivation of these equations beyond those conditions
needed in the derivation of the self-induction equation in (2) is to allow the vortex
filament to have wavy perturbations on a scale ϵ that are short wavelength relative
to the $O(1)$ radius of curvature of the vortex filament but are long wavelength relative to the core thickness. Thus, the wavy perturbations of the vortex filament
satisfy

$$1 \gg \epsilon \gg \delta.$$

The choice of such scalings is motivated by numerical simulations (Chorin 1982,
1994) which demonstrate that such kinds of perturbations often develop folds and
hairpins in vortex filaments.

More precisely, the asymptotic filament equations with self-stretch to be described below arise from small amplitude short wavelength perturbations of a

straight line vortex filament aligned along the x_3-axis. These perturbations of the vortex center-line have the form

$$(33) \qquad \mathbf{X}(s, \bar{t}) = \bar{s}\mathbf{e}_3 + \epsilon^2 \mathbf{X}^{(2)}\left(\frac{\bar{s}}{\epsilon}, \frac{\bar{t}}{\epsilon^2}\right) + o(\epsilon^2).$$

where

$$(34) \qquad \mathbf{X}^{(2)} = \left(x^{(2)}\left(\frac{\bar{s}}{\epsilon}, \frac{\bar{t}}{\epsilon^2}\right), y^{(2)}\left(\frac{\bar{s}}{\epsilon}, \frac{\bar{t}}{\epsilon^2}\right), 0\right).$$

While we do not present the details here, the time scale $\bar{t}$ is the same one used for the local self-induction approximation so that the perturbations in (33) are not only short wavelength but also occur on rapid time scales relative to the local self-induction time scale, $\bar{t}$. We also note that while the vortex filaments in (33) vary rapidly in space, nevertheless they have only an $O(1)$ effect on the radius of curvature since

$$\frac{d^2\mathbf{X}(s,t)}{ds^2} = \left(\frac{d^2\mathbf{X}^{(2)}}{d\sigma^2}\right)\left(\sigma, \frac{\bar{t}}{\epsilon^2}\right)\Big|_{\sigma = \frac{\bar{s}}{\epsilon}}.$$

The key assumptions which emerge in the derivation (Klein and Majda 1991 a) require that the vortex filament thickness δ, the Reynolds number Re defined in (2), and the wavelength parameter satisfy the distinguished limit

$$(35) \qquad \epsilon^2[\ln\left(\frac{2\epsilon}{\delta}\right) + c] = 1$$

and

$$\delta = (\text{Re})^{-1/2}, \qquad \delta \ll 1.$$

The first condition in (35) guarantees that the requirements $1 \gg \epsilon \gg \delta$ are satisfied while the second condition in (35) is already familiar from (2) D) while the form of the perturbations in (33) automatically enforces (2) C).

2.1. Sketch of the Asymptotic Derivation. As in the derivation sketched in section 1 for the self-induction equation, one needs to develop a suitable asymptotic expansion of the Biot-Savart law in (1) with the ansatz in (33) for the filament equations under the assumptions in (35) and then insert this expansion into (3) to obtain a dynamical equation for the vortex filament center-line, $\mathbf{X}(s, \bar{t})$. Klein and Majda (1991 a, sections 3 and 4) show that the following *perturbed binormal law* emerges from this procedure under the assumptions in (35):

$$(36) \qquad \frac{\partial \mathbf{X}}{\partial \bar{t}} = \kappa\mathbf{b} + \epsilon^2 I[\mathbf{X}^{(2)}] \times \mathbf{e}_3.$$

Here $I[w]$ is the linear nonlocal operator

$$I[w] \overset{\text{def}}{=} \int_{-\infty}^{x} \frac{1}{|h^3|}[w(\sigma + h) - w(\sigma) - hw_\sigma(\sigma + h)$$

$$(37) \qquad \qquad + \frac{1}{2}h^2\mathcal{H}(1 - |h|)w_{\sigma\sigma}(\sigma)]dh,$$

where we use the notation $w_\sigma = \frac{\partial w}{\partial \sigma}$ and employ the Heaviside distribution $\mathcal{H}$. This operator $I[\mathbf{X}^{(2)}]$ incorporates the leading order nonlocal effects of the Biot-Savart law in (1) under the assumptions in (33) and (35).

We recall the Hasimoto transform from (7) which maps a curve $\mathbf{X}(\bar{s}, \bar{t})$ to a *filament function* $\psi(\bar{s}, \bar{t})$ via

$$(38) \qquad \psi(\bar{s}, \bar{t}) = \kappa(\bar{s}, \bar{t}) e^{i \int_0^{\bar{s}} T(s', \bar{t}) ds'}.$$

We claim that under the Hasimoto transform, the perturbed binormal law in (36) with the ansatz in (33) becomes the

Asymptotic Filament Equation with Self-Stretching

$$(39) \qquad \frac{1}{i} \psi_\tau = \psi_{\sigma\sigma} + \epsilon^2 \left(\frac{1}{2} |\psi|^2 \psi - I[\psi] \right)$$

in which the expansion parameter ϵ appears as a coupling constant. In (39) the arguments σ and τ are given by $\sigma = \frac{\bar{s}}{\epsilon^2}$ and $\tau = \frac{\bar{t}}{\epsilon^2}$ up to minor arc-length corrections (see Klein and Majda 1991 a) which we ignore here and in our discussion below. As advertised earlier in this section, the asymptotic equation in (39) has a linear nonlocal term, $I[\psi]$, which competes directly at the same order with the local self-induction represented by the cubic nonlinear terms, $\frac{1}{2} |\psi|^2 \psi$.

How can we compute the local rate of filament curve stretching from the solution ψ of (39)? With the ansatz in (33), we have the infinitesimal arc-length given by

$$\ell \left(\frac{\bar{s}}{\epsilon}, \frac{\bar{t}}{\epsilon^2} \right) \underset{\text{Def}}{=} (\mathbf{X}_{\bar{s}} \cdot \mathbf{X}_{\bar{s}})^{1/2}.$$

A lengthy calculation (see Klein and Majda 1991a, section 5) utilizing (25), (26) and (36) establishes that the time derivative $\dot{\ell}(\sigma, \tau)$ satisfies the *local curve stretching identity*

$$(40) \qquad \frac{\dot{\ell}(\sigma, \tau)}{\ell} = \frac{\epsilon^2}{4} i \int_{-\infty}^{\infty} \frac{1}{|h|} [\overline{\psi}(\sigma + h)\psi(\sigma) - \psi(\sigma + h)\overline{\psi}(\sigma)] dh.$$

We know that vortex stretching is a quadratically nonlinear process yet the effects of nonlocal self-stretch are incorporated in the filament equation in (39) through the linear operator $I[\psi]$. The formula in (40) resolves this apparent paradox since the relative changes in arc-length are a quadratically nonlinear functional of ψ.

The derivation of the filament equation in (39) from the perturbed binormal law in (36) proceeds by applying the general Hasimoto transform for a perturbed binormal law derived in (5)–(19) while respecting the rapid space-time scalings implicit in the special ansatz from (33). From (36) we have the correspondence

$$(41) \qquad \epsilon^2 I[\mathbf{X}^{(2)}] \times \mathbf{e}_3 = \tilde{\delta} \mathbf{v}(\bar{s}, \bar{t})$$

in the perturbed binormal law in (5). For small amplitude high frequency perturbations of the center-line with the form $\epsilon^2 \mathbf{X}^{(2)}(\frac{\bar{s}}{\epsilon}, \frac{\bar{t}}{\epsilon^2})$, we rescale the exact perturbed nonlinear Schrödinger equation derived in (19) via the variables $\sigma = \frac{s}{\epsilon}$, $\tau = \frac{t}{\epsilon^2}$ so the evolution equation for the filament function

$$(42) \qquad \frac{1}{i} \psi_\tau = \psi_{\sigma\sigma} + \left[\epsilon^2 \frac{1}{2} |\psi|^2 \psi - i\tilde{\delta}(\mathbf{N} \cdot \mathbf{v}_\sigma)_\sigma \right]$$

emerges at leading order with $\mathbf{N}$ defined earlier in (8). Next, identifying $\tilde{\delta} \mathbf{v}(\bar{s}, \bar{t})$ by utilizing (41), we obtain the evolution equation

$$(43) \qquad \frac{1}{i} \psi_\tau = \psi_{\sigma\sigma} + \epsilon^2 \frac{1}{2} |\psi|^2 \psi - i\epsilon^2 (\mathbf{N} \cdot I[\mathbf{X}_\sigma^{(2)}] \times \mathbf{e}_3)_\sigma.$$

The derivation of the asymptotic filament equation in (39) from the perturbed binormal in (36) is completed by applying the following identity in (43)

$$(44) \qquad\qquad -i(\mathbf{N} \cdot \mathbf{v}_\sigma)_\sigma = -I[\psi].$$

The proof of the identity in (44) is lengthy and given in section 5 of Klein and Majda (1991 a). The perceptive reader will note that our derivation of the perturbed Schrödinger equation from the perturbed binormal law in section 1 utilized arc-length coordinates while s in (33) is not exactly an arc-length coordinate, $\tilde{s}$; in fact, we have $s = \tilde{s}(1 + o(\epsilon^2))$. However, we have ignored these minor differences in our sketch presented here for pedagogical simplicity.

2.2. The Mathematical Structure of the Asymptotic Equation.

Here we discuss the mathematical properties of the filament equation with self-stretching from (39). We develop Fourier representations for the nonlocal operator $I[w]$, in (37). These formulas reveal the fact that (39) becomes a novel singular perturbation of the linear Schrödinger equation.

Explicit Fourier representation of the nonlocal operators. We discuss the operator $I[\cdot]$ in (37) acting on functions defined for all of $\mathbb{R}^1$ with rapid decay. The operator $I[\cdot]$ commutes with translation and therefore is given by convolution with a distribution kernel. For all of space and rapidly decreasing functions, the Fourier inversion formula yields

$$
\begin{aligned}
f(\sigma) &= (2\pi)^{-1} \int e^{i\sigma\xi} \hat{f}(\xi) d\xi, \\
\hat{f}(\xi) &= \int e^{-i\sigma\xi} f(\sigma) d\sigma.
\end{aligned}
$$

Thus, we have

$$(45) \qquad I[w](\sigma) = (2\pi)^{-1} \int e^{i\sigma\xi} \hat{I}(\xi)\hat{w}(\xi) d\xi$$

with the Fourier symbol $\hat{I}(\xi)$ of I defined by

$$(46) \qquad\qquad \hat{I}(\xi) = e^{-i\sigma\xi} I[e^{i\sigma\xi}].$$

From the definitions in (37) and (46), we compute explicitly that

$$(47) \qquad \hat{I}(\xi) = \int_0^\infty \frac{1}{h^3} \left[(e^{i\xi h} + e^{-i\xi h} - 2) - ih\xi(e^{i\xi h} - e^{-i\xi h}) - \theta(h)(\xi h)^2 \right] dh,$$

with $\theta(h) = \mathcal{H}(1 - |h|)$. Next, we determine the scaling properties of $\hat{I}(\xi)$. First we introduce real trigonometric functions in (47) and change variables with $\eta = \xi h$ to obtain

$$(48) \qquad \hat{I}(\xi) = \xi^2 \int_0^\infty \frac{1}{\eta^3} \left[2(\cos\eta - 1) + 2\eta\sin\eta - \theta\left(\frac{\eta}{|\xi|}\right) \eta^2 \right] d\eta.$$

Notice that only the function $\theta(\eta/|\xi|)$ has explicit dependence on ξ in the integrand in (48). We split the integral in (48) into a sum of convergent factors obtaining the decomposition

$$
\begin{aligned}
(49) \qquad \hat{I}(\xi) &= \xi^2 \left(\int_0^1 \frac{2(\cos\eta - 1) + 2\eta\sin\eta - \eta^2}{\eta^3} d\eta + \int_1^\infty \frac{2(\cos\eta - 1) + 2\eta\sin\eta}{\eta^3} d\eta \right) \\
&\quad - \xi^2 \int_0^\infty \left[\theta\left(\frac{\eta}{|\xi|}\right) - \theta(\eta) \right] \eta^{-1} d\eta.
\end{aligned}
$$

The term in braces in (49) contributes a finite constant while, by the definition of the function θ, we compute that

$$(50) \qquad \int_0^\infty \left[\theta\left(\frac{\eta}{|\xi|}\right) - \theta(\eta) \right] \eta^{-1} d\eta = \ln|\xi|.$$

With (49) and (50), we obtain the formula

$$\hat{I}(\xi) = -\xi^2 \ln|\xi| + C_0 \xi^2,$$

with the constant C_0 determined by the expression in braces in (49). The constant C_0 can be evaluated by a lengthy calculation utilizing the method of stationary phase. With that information, we have derived the *explicit formula for the symbol of* $I[\cdot]$,

$$\hat{I}(\xi) = -\xi^2 \ln|\xi| + C_0 \xi^2,$$

where

$$C_0 = \frac{1}{2} - \gamma = -0.0772\ldots$$

with γ given by Euler's constant.

Spectrum of the linearized operator and singular perturbation. The linearization of the asymptotic filament equation in (39) about the state $\psi \equiv 0$, is the linearized equation

$$(51) \qquad \frac{1}{i}\psi_t = \psi_{\sigma\sigma} - \epsilon^2 I[\psi].$$

The operator on the right hand side of (51) is symmetric and we calculate that the dispersion relation is given by

$$(52) \qquad \begin{aligned} \omega &= P_\epsilon(\xi), \\ P_\epsilon(\xi) &= -|\xi|^2 - \epsilon^2 C_0 |\xi|^2 + \epsilon^2 \xi^2 \ln|\xi|. \end{aligned}$$

The generator of the linearized equation (51) is the operator

$$(53) \qquad \psi_{\sigma\sigma} - \epsilon^2 I[\psi].$$

The spectrum of the linearized generator $\epsilon = 0$ is the familiar spectrum for the Schrödinger operator given by the interval $(-\infty, 0]$; the spectrum of the linearized generator in (53) is radically different for any $\epsilon > 0$ and reflects the singular character of this perturbation. We compute that $P_\epsilon(\xi)$ from (52) is a generalized eigenvalue of the linearized operator in (53) for any $\epsilon > 0$ and $\xi \in \mathbb{R}^1$. It follows immediately that, in contrast to the case with $\epsilon = 0$, the spectrum of the operator in (53) is $(a(\epsilon), +\infty)$ for any $\epsilon > 0$ where $a(\epsilon) \downarrow -\infty$ rapidly as $\epsilon \downarrow 0$; in fact, $|a(\epsilon)| = \mathcal{O}(\epsilon^2 \exp(2/\epsilon^2))$. Thus, a complete half-interval of new spectrum in $[0, \infty)$ is created by the effect of $I[\psi]$ for any $\epsilon > 0$. We conclude that $-\epsilon^2 I[\psi]$ is in fact a singular perturbation of the linear Schrödinger operator. We invite the reader to graph the dispersion relation $\omega = P_\epsilon(\xi)$ and to compare this graph with that for $\omega = -|\xi|^2$ from the linear Schrödinger equation.

Remark. The nature of the singular perturbation depends crucially on the sign of the coefficient preceding $I[\psi]$ in (39). For comparison we consider the equation

$$(54) \qquad \frac{1}{i}\psi_t^+ = \psi_{\sigma\sigma}^+ + \epsilon^2 \left(\frac{1}{2}|\psi^+|^2 \psi^+ + I[\psi^+] \right).$$

Then the dispersion relation for the linearized operator derived from the above equation is given by

$$\omega = P_\epsilon^+(\xi)$$
$$P_\epsilon^+(\xi) = -|\xi|^2 - \epsilon^2\xi^2 \ln|\xi| + \epsilon^2 C_0|\xi|^2$$

and the spectrum of the linearized operator is a $(-\infty, a_+(\epsilon)]$, where $a_+(\epsilon) \downarrow 0$ rapidly as $\epsilon \downarrow 0$; in this case, the spectrum for $\epsilon > 0$ is always well approximated by that of the linearized Schrödinger operator. Thus, at the linearized level, the perturbation in (54) is much milder than the one actually occurring in the asymptotic filament equation.

The asymptotic filament equations from (39) have global existence of solutions for all times and the elementary proof is sketched in Klein and Majda (1991 b). This proof reflects the character of the nonlocal operator $I[\psi]$ as a singular perturbation. However, solutions of these filament equations can lose their asymptotic validity as approximate solutions of the actual fluid motion at finite times (see Klein and Majda (1991 b)).

2.3. Asymptotic Equations for the Stretching of Vortex Filaments in a Background Flow Field.

Examples involving axisymmetric swirling vortices including the Burgers vortex provide simple analytic examples where a background strain flow can interact with a vortex tube and create vortex stretching. Localized vortex filaments with large strength and narrow cross section are also prominent fluid mechanical structures in applications such as secondary three-dimensional instability in mixing layers (Corcos and Lin) where larger scale vortices provide a slowly varying background flow field where strong vortex filaments are embedded. Motivated by these and other considerations, we develop here the asymptotic equations which generalize the filament equations with self-stretching from (39) when a background flow field is present.

We consider small amplitude wavy distortions of a straight line vortex filament as already described in (33) and (34) under the asymptotic conditions in (35). Besides the effects of local induction and self-stretch, we include the effect of a known background flow field which has the form

$$(55) \qquad\qquad \delta\mathbf{v}_b(\mathbf{x}, t) = \frac{1}{\epsilon^2} A\mathbf{x}$$

where the 3×3 matrix, A, representing the velocity gradients is given by

$$(56) \qquad\qquad A = \begin{pmatrix} s_{11} & -\frac{1}{2}\omega + s_{12} & 0 \\ \frac{1}{2}\omega + s_{12} & s_{22} & 0 \\ 0 & 0 & s_{33} \end{pmatrix}$$

with the incompressibility condition

$$(57) \qquad\qquad s_{11} + s_{22} + s_{33} = 0.$$

The background flow field described in (55) and (56) represents the local Taylor expansion of a general three-dimensional incompressible flow field which satisfies the additional *crucial requirement that the centerline along the x_3-axis for the unperturbed vortex filament remains invariant under the particle trajectory flow map associated with the background flow in (55).* This requirement is guaranteed by the block diagonal structure flow in (56). The coefficient, s_{33}, represents the rate of strain along the vortex filament due to the background flow while the 2×2 matrix

$$(58) \qquad S_2 = \begin{pmatrix} s_{11} & s_{12} \\ s_{12} & s_{22} \end{pmatrix}$$

represents the contribution of the deformation matrix perpendicular to the unperturbed vortex filament.

The same general procedure which we already outlined in section 2.1 can be utilized to pass from a perturbed binormal law as in (36) which also includes the background flow field in (55) to a modified filament equation including the effects of the background flow from (55) and (56) via the Hasimoto transform in (7) or (38). The reader interested in the details can consult section 2 of the paper by Klein, Majda, and McLaughlin. The result of this derivation is the
Asymptotic Filament Equation with Self-Stretch in a Background Flow Field,

$$(59) \qquad \frac{1}{i}\psi_\tau = \psi_{\sigma\sigma} + \epsilon^2 \left(\frac{1}{2}|\psi|^2\psi - I[\psi] \right) + \epsilon^2 \left[\left(\omega + \frac{5}{2}is_{33} \right)\psi \right.$$
$$\left. + is_{33}\sigma\psi_\sigma + \left(s_{12} - i\frac{s_{11} - s_{22}}{2} \right)\overline{\psi} \right]$$

The first two terms on the right hand side of (59) are already familiar to the reader from the filament equations with self-stretch without a background flow field in (39). The new terms due to the background flow field in (55) are in brackets on the right hand side of (59).

We consider each of these three terms separately in order to extract a physical interpretation of these effects. The separate effect of the first term yields the equation

$$(60) \qquad \frac{1}{i}\psi_\tau = \epsilon^2\omega\psi.$$

Thus, the effect of background rotation around the axis, st_0, on the perturbed filament corresponds to the rotation of ψ in the complex plane at a frequency, $\epsilon^2\omega$. The separate effect of the second background flow term in (59) is the advection operator

$$(61) \qquad \psi_\tau = -\epsilon^2 s_{33} \left(\frac{5}{2}\psi + \sigma\psi_\sigma \right)$$

and corresponds to growth or decay of amplitudes due to flow convergence or divergence in the normal plane and to advection along the x_3-axis. Finally, the last term in (59) involving the background flow yields the equation

$$(62) \qquad \left(\frac{1}{i} \right)\psi_\tau = \epsilon^2 \left(s_{12} - i\left[\frac{(s_{11} - s_{22})}{2} \right] \right)\overline{\psi}$$

and this represents the effect of the planar strain normal to the filament axis and given by S_2. We note that the complex conjugate of ψ, $\overline{\psi}$, appears on the right hand side of (59) and (62). The O.D.E. in (62) has both damping and growing solutions corresponding to the strain axes of the background flow field; this term both damps and drives the solutions of the asymptotic filament equation in (59).

2.3.1 Analytical Properties of the Filament Equation with a Background Flow.

Here, we develop a criterion for nonlinear stability of vortex filaments to appropriate short-wavelength perturbations through the equation in (59). We also describe several properties of the linearization of (59) in an instructive special case involving planar strain flows in directions orthogonal to the unperturbed vortex filament.

A. The nonlinear stability of vortex filaments in a background flow field.

Recent numerical simulations at moderately large Reynolds numbers (Ashurst et al., She et al., Majda) reveal that nearly columnar strong vortex filaments in a background flow field persist for some time despite the fact that the planar strains for the swirling flow in directions orthogonal to the filament axis are stronger than the strain component aligned with the vortex axis. Here, we give some precise nonlinear stability conditions utilizing (59) that are consistent with this scenario and provide additional analytic insight.

From the Hasimoto transform in (7), we see that $|\psi(\sigma, \tau)|$ has the physical significance of the curvature of the perturbed filament so that a natural quantity for measuring the stability of the columnar vortex is

$$\int |\psi(\sigma, \tau)|^2 d\sigma.$$

The unperturbed vortex filament is stable if there is a constant $\bar{\alpha} > 0$ so that

$$(63) \qquad \frac{\partial}{\partial \tau} \int |\psi(\sigma, \tau)|^2 d\sigma \leq -\bar{\alpha} \int |\psi(\sigma, \tau)|^2 d\sigma$$

for all solutions of (59) and for all $\tau \geq 0$. The condition in (63) guarantees that the mean square average of the curvature decays at least exponentially with time so that the columnar filament is stable for the specific class of short wavelength filament perturbations from (33) consistent with the asymptotic derivation.

Using (59), we calculate that

$$
\begin{aligned}
(64) \qquad \frac{\partial}{\partial \tau} \int |\psi|^2 d\sigma &= 2 \,\mathrm{Re} \int \psi_\tau \bar{\psi} d\sigma \\
&= 2\epsilon^2 \left\{ -s_{33}(\tau) \left[\int \mathrm{Re} \left(\sigma \psi_\sigma \bar{\psi} + \frac{5}{2} \psi \bar{\psi} \right) d\sigma \right\} \right. \\
&\quad \left. + \int \mathrm{Re} \left[\left(\frac{s_{11} - s_{22}}{2} + i s_{12} \right) \bar{\psi}^2 \right] d\sigma \right\},
\end{aligned}
$$

where Re in (64) denotes the real part. We compute further that

$$(65) \qquad \int \mathrm{Re} \left(\sigma \psi_\sigma \bar{\psi} + \frac{5}{2} \psi \bar{\psi} \right) d\sigma = 2 \int |\psi|^2 d\sigma$$

and

$$(66) \qquad \mathrm{Re} \left[\left(\frac{s_{11} - s_{22}}{2} + i s_{12} \right) \bar{\psi}^2 \right] = \psi \cdot M \psi,$$

with $\psi = {}^t(\psi_1, \psi_2)$, $\psi = \psi_1 + i\psi_2$, and M the symmetric 2×2 matrix,

$$(67) \qquad M = \begin{pmatrix} \frac{s_{11} - s_{22}}{2} & -s_{12} \\ -s_{12} & \frac{s_{22} - s_{11}}{2} \end{pmatrix}.$$

The matrix M has the two real eigenvalues $\lambda_{\pm} = \pm[\frac{(s_{11}-s_{22})^2}{4} + s_{12}^2]^{1/2}$ so that

(68) $$\psi \cdot M\psi \leq \lambda_+|\psi|^2.$$

Combining (64)–(68), we obtain that

(69) $$\frac{\partial}{\partial \tau} \int |\psi|^2 d\sigma \leq 2\epsilon^2[-2s_{33} + \lambda_+] \int |\psi|^2 d\sigma.$$

Looking back at the definition of stability in (63), we see that the *columnar vortex filament is nonlinearly stable to the short-wavelength perturbations in (33) provided that the strain matrix from the background flow satisfies*

(70) $$\left(\frac{(s_{11} - s_{22})^2}{4} + s_{12}^2\right)^{1/2} \leq 2s_{33}.$$

Of course, since the flow is incompressible, we have the identity

$$s_{33} = -(s_{11} + s_{22}),$$

which allows us to express the general criterion in (70) solely in terms of the strain matrix components orthogonal to the columnar filament. Obviously, a positive axial strain component is a necessary condition for the columnar filament to remain stable. We remark that the inequality in (70) does not involve the transverse rotation components for the background flow field ω from (59).

We interpret the criterion in (70) in the special case when the strain matrix is diagonal so that $s_{12} = 0$, $s_{11} = \gamma_1$, $s_{22} = \gamma_2$, $s_{33} = \gamma_{33}$ with $\gamma_1 + \gamma_2 + \gamma_3 = 0$. Without loss of generality, we assume that $\gamma_1 \geq \gamma_2$; then the stability criterion in (70) becomes

(71) $$\gamma_2 < 0, \quad \gamma_1 < \frac{3}{5}|\gamma_2|.$$

The regime of (71), where γ_1 and γ_2 satisfy

(72) $$\gamma_2 < 0, \quad \frac{1}{2}|\gamma_2| < \gamma_1 < \frac{3}{5}|\gamma_2|,$$

is very interesting for qualitative comparison with the computational results mentioned earlier. With the conditions in (72) we have the following.

(i) The axial component of the strain matrix is positive but γ_3 is the intermediate eigenvalue of the strain matrix.

(ii) The strongest strain effect is the flattening of the vortex core by the compressive strain γ_2 orthogonal to the filament.

(iii) Despite the structure in (i) and (ii) and also rapid planar rotation from ω, the columnar vortex is stable in time on the τ time scale according to the criterion in (71).

The conditions in (i) and (ii) are the ones typically observed for the nearly columnar vortices that emerge in the numerical experiments cited earlier conditioned on the large vorticity sets. Here, we have provided supporting analytic evidence for the stability of such columnar vortices to suitable short-wavelength perturbations on appropriate time scales. We have utilized the special case in (71) and (72) to elucidate this structure in a transparent fashion. Clearly, from the more general criterion in (70), we can even allow suitable temporal rotation of the planar strain axis in time and fulfill all three of the conditions from (i)–(iii).

B. Linearized theory for planar strain flows. Here, we develop the linearized theory for (59) in the important special case when the background flow field is a simple planar strain in the directions orthogonal to the unperturbed filament. This case has some physical importance since it occurs naturally in the development of secondary three-dimensional instabilities in mixing layers and other two-dimensional basic flows. Also the stability criteria just discussed are not satisfied; the unperturbed filament is linearly unstable with these background flows.

With constant planar strain flows as the background field, the filament equation in (59) is given by

$$(59) \quad \left(\frac{1}{i}\right)\psi_\tau = \psi_{\sigma\sigma} + \epsilon^2 \left(\frac{1}{2}|\psi|^2\psi - I[\psi] - i\gamma\overline{\psi}\right),$$

with γ the strain rate. To see the effect of the strain alone, we ignore all of the other terms and arrive at the ODE

$$(74) \quad \psi_\tau = \epsilon^2\gamma\overline{\psi}.$$

The simple ODE in (74) has both damping and growing solutions; thus, the background planar strain flow can be a source of instability for (73) at suitable wavelengths.

We linearize the equation in (73) at the unperturbed filament $\psi = 0$; thus, the linearized problem is given by

$$(75) \quad \left(\frac{1}{i}\right)\psi_\tau = \psi_{\sigma\sigma} + \epsilon^2(-I[\psi] - i\gamma\overline{\psi}).$$

We write the exact solution of (75) using the Fourier transform

$$(76) \quad \widehat{\psi}(k,\tau) = (2\pi)^{-1/2} \int_{-\infty}^{\infty} e^{ik\sigma}\psi(\sigma,\tau)d\sigma.$$

The effect of the straining term involving $\overline{\psi}$ in (75) is that the two modes k and $-k$ are coupled together so that the Fourier-transformed equation is no longer local. In fact, we calculate that the pair $\widehat{\psi}(k,\tau)$, $\overline{\widehat{\psi}}(-k,\tau)$ satisfy the coupled equations,

$$(77) \quad \begin{pmatrix} \widehat{\psi}(k,\tau) \\ \overline{\widehat{\psi}}(-k,\tau) \end{pmatrix}_\tau = i\left\{\widehat{\mathcal{L}}(k)\begin{bmatrix} 1 & 0 \\ 0 & -1 \end{bmatrix} - i\epsilon^2\gamma\begin{bmatrix} 0 & 1 \\ 1 & 0 \end{bmatrix}\right\} \times \begin{pmatrix} \widehat{\psi}(k,\tau) \\ \overline{\widehat{\psi}}(-k,\tau) \end{pmatrix}.$$

Here, $\widehat{\mathcal{L}}(k)$ is given by

$$(78) \quad \widehat{\mathcal{L}}(k) = -k^2 - \epsilon^2\widehat{I}(k),$$

with the Fourier symbol for the nonlocal operator I from (6.45) given by

$$\widehat{I}(k) = -k^2\ln|k| + C_0k^2,$$

with $C_0 = \frac{1}{2} - \gamma = -0.0772$ and γ Euler's constant.

Given the equation in (77) we explicitly calculate the solution,

$$(79) \quad \begin{pmatrix} \widehat{\psi}(k,\tau) \\ \overline{\widehat{\psi}}(-k,\tau) \end{pmatrix} = \{\cos[\tau p(k)]I + i\sin[\tau p(k)]P\} \times \begin{pmatrix} \widehat{\psi}(k,0) \\ \overline{\widehat{\psi}}(-k,0) \end{pmatrix},$$

with

$$(80) \quad p(k) = [\widehat{\mathcal{L}}^2(k) - \epsilon^4\gamma^2]^{1/2},$$

I the 2×2 identity matrix, and

$$(81) \qquad P = [p(k)]^{-1} \left\{ i\epsilon^2 \gamma \begin{bmatrix} 0 & 1 \\ 1 & 0 \end{bmatrix} + \widehat{\mathcal{L}}(k) \begin{bmatrix} 1 & 0 \\ 0 & -1 \end{bmatrix} \right\}.$$

The functions $\cos(z)$ and $\sin(z)$ are evaluated for complex variables z in (79). In particular, we have oscillatory behavior for the larger wave numbers

$$k \quad \text{with} \quad \widehat{\mathcal{L}}^2(k) > \epsilon^4 \gamma^2$$

and unstable growth for the sufficiently low wave numbers with

$$\widehat{\mathcal{L}}^2(k) < \epsilon^4 \gamma^2$$

in the behavior of the solutions of the linearized equation from (75). To summarize, we see that there are two effects of the planar strain flow as regards linearized theory: first, to introduce instability at low wave numbers; second to couple the modes k and $-k$ in a nonlocal fashion in Fourier space.

2.4. The Creation of Kinks, Folds, and Hairpins in Vortex Tubes Via the Filament Equations. Vortex tubes which develop kinks, folds, and hairpins are prominent structures in both numerical solutions (Chorin) and experiments in flows such as shear layers and trailing wakes (Corcos and Lin, van Dyke). The simplified asymptotic equations developed in (39) and (59) allow for suitable initial short wavelength perturbations as in (33) so that there is the real possibility that the nonlinearity in these equations together with the remarkable properties of solutions of the linear operators in these equations developed in 2.2 and 2.3.1 can interact and produce the incipient formations of kinks and hairpins in the vortex filaments. Here we present numerical solutions of the filament equations in (39) and (59) which confirm this possible behavior.

First we consider spatially periodic solutions of the filament equations with self-stretch in (39) without the effect of a background flow field. The equations in (39) have exact oscillatory solutions with the form

$$(82) \qquad \psi_H(\sigma, \tau) = A \exp(i[\xi\sigma + \Omega\tau])$$

provided that $\Omega(\xi, A, \epsilon)$ is given by

$$(83) \qquad \Omega(\xi, A, \epsilon) = -\xi^2 - \epsilon^2 \hat{I}(\xi) + \frac{1}{2}\epsilon^2 A^2$$

where the symbol, $\hat{I}(\xi)$, is given explicitly below (49). We claim that the exact solutions in (82) of the asymptotic equations correspond to helical vortex filaments. To see this, we recall the Hasimoto transform from (38); we have the general formulas

$$|\psi|(\sigma, \tau) \quad = \quad \text{curvature}$$
$$(84) \qquad\qquad \arg \psi \quad = \quad \int_0^\sigma T(\sigma, \tau) d\sigma,$$

with $T(\sigma, \tau)$ the torsion. From (84), the exact solutions in (82) have constant curvature and constant torsion and thus are necessarily helical vortex filaments.

The exact solutions in (82) are linearly unstable to suitable long wavelength perturbations (Klein and Majda 1991 b) and thus it is very interesting to see the nonlinear consequences of perturbations of such solutions. In Figure 1 we see the evolving curvature of the filament in plots of $|\psi|$ from a numerical solution with spatially periodic initial data involving a suitable small perturbation of the helical filament with the value of $\epsilon^2 = .25$. These snapshots indicate that a narrow highly

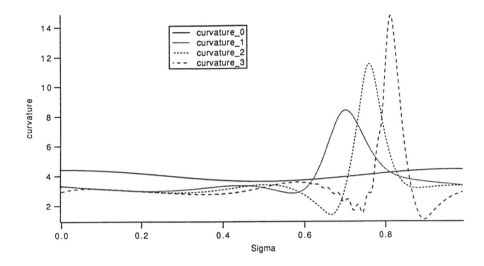

FIGURE 1. Time series of curvature distributions, $|\psi|$ vs. σ, for perturbation of the helix solution. Background helix: $|\psi H| = A = 4.0$, $\xi H = 2.0$; perturbation mode: $|\psi H|b = 0.4$, $\xi H + \beta = 3.0$; $\epsilon^2 = 0.25.$ — $t = 0$, ...$t = 1.6$, $---t = 1.75$, $------T = 1.9$.

localized spike in the curvature is developing in time. The corresponding spatial curves are given in Figure 2 and produce a periodic array of strong kinks in the vortex filament. Thus, solutions of the asymptotic equations for vortex filaments with self-stretch produce strong curvature spikes which correspond to the spontaneous generation of kinks. Such behavior is completely absent in corresponding solutions of the nonlinear Schrödinger equation in (23) which arises from the self-induction approximation without the self-stretch terms from (39) (Klein and Majda 1991 b).

While the solution depicted in Figures 1 and 2 continues to exist for all times, it loses its asymptotic validity as a physical solution of the vortex filament equations precisely near the final time depicted in Figure 1. The reason for this is that significant amplitudes at wavelengths on the scale of the vortex core thickness have been generated simultaneously as the kink in the vortex filament is being created so that the tacit assumptions in (35) regarding wavelength perturbations larger than the core thickness are violated as the solution evolves beyond these times. This makes physical sense since the subsequent evolution of the kink in the vortex filament as depicted in Figure 2 would clearly involve nontrivial flow in the vortex core which is ignored in the asymptotic equations. For a detailed quantitative explanation of these facts as well as more numerical solutions of the filament equations with self-stretch from (39), and a careful numerical validation study, we refer the interested reader to the paper of Klein and Majda (1992 b).

Next, we consider the birth of hairpin solutions from the asymptotic equations from (59) with self-stretch in a background flow field where we have a planar two dimensional strain flow perpendicular to the unperturbed vortex axis. Thus, we

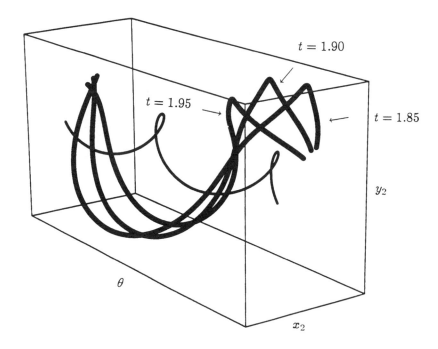

FIGURE 2. Time sequence of three dimensional plot of the filament
curve, showing the initial helix, followed by times just before, just
at, and just after the peak in maximum stretch rate. Viewed from
$\theta = 3$, $x_2 = 2$, $y_2 = 1$. Notice the formation of the kink structure
in the vortex filament.

describe numerical solutions of the nonlinear filament equation in (73). Two important facts emerge from the linearized theory developed in section 2.3.1 B): first,
the planar strain generates long wavelength instability; second, the equations non-
locally couple wavenumbers k and $-k$ through the strain flow terms and thus have
a propensity to creating growing standing modes. Next we will see that these two
facets of the dynamics from linear theory will combine with nonlinearity to produce
the birth of hairpin vortices.

Figure 3 depicts the initial curvature magnitude and the curvature magnitude
at a later time for a numerical solution of (73) with strain rate $\gamma = 2$ and $\epsilon^2 = .25$
for suitable long wavelength standing mode initial data. The plot at the later time
shows the emergence of two narrow localized curvature spikes. Since each narrow
curvature spike generates a kink in the vorticity and two kinks are required to create
a hairpin vortex filament, we might anticipate that this solution of (73) displays the
birth of a hairpin. The graph in Figure 4 of the actual filament curve at the later
time from Figure 3 confirms the formation of a spatially periodic array of localized
hairpins. The later time solution depicted in Figures 3 and 4 is near the time where
solution of the asymptotic equations in (73) loses its validity for the same reasons
mentioned earlier—the creation of the hairpin generates wavelengths on the order

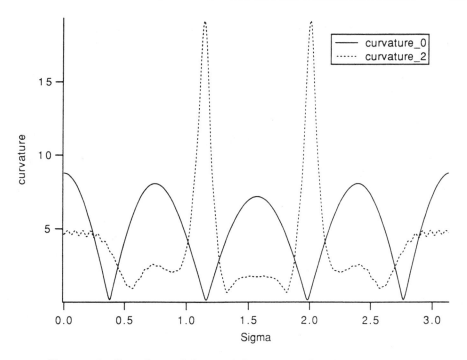

FIGURE 3. Snapshots of the spatial curvature distribution at times $t = 0.0$, $t = 1.66$ for perturbed standing mode initial data $\psi(\sigma, 0) = 4.0[\exp(4\pi i\sigma) + \exp(-4\pi i\sigma)] + .4[\exp(6\pi i\sigma) + \exp(-6\pi i\sigma)]$, and weak strain $\gamma = -2.0 + i0.0$.

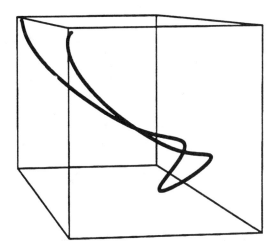

FIGURE 4. 3-dimentional representation of the filament geometry for the solution of Fig. 3 with viewpoint (3., -.4, .2).

of the core thickness. The interested reader can consult the paper by Klein, Majda, and McLaughlin for more details and further numerical simulations.

The Finite Effects of ϵ in the Asymptotic Theories. Despite the intuitive appeal and elegant mathematical underpinning related to physical effects in solutions of the asymptotic filament equations discussed throughout section 2, one can criticize this theory at the outset as being unrealistic because the requirements in (35) simultaneously require

$$(85) \qquad\qquad \epsilon \cong |\ln \mathrm{Re}|^{-\frac{1}{2}} \text{ and } \epsilon \ll 1.$$

Thus, the appearance of ϵ^2 in the simplified equations in (39) and (59) might imply that these equations study transcendentally small effects at ultrahigh Reynolds numbers. Furthermore the formation of kinks and hairpins in solutions of these asymptotic equations involved the moderate values of $\epsilon^2 = .25$ which seems unrealistically large.

Are such criticisms of the theory real or illusory? One way to address this is to develop detailed numerical simulations of the complete incompressible fluid equations for initial data like those from (33) which are compatible with the asymptotic theory. Recently, in an important paper, Klein and Knio have developed direct simulations of the incompressible fluid equations utilizing computational vortex methods in 3-D to address these issues. They emphatically confirm the quantitative predictions of the asymptotic filament equations described in this section by their direct numerical simulations even for the moderate values of $\epsilon^2 = .25$!! This provides dramatic evidence that the appealing physical asymptotic theory for vortex filaments developed in this section has quantitative and qualitative validity beyond the strict regime of asymptotic validity listed in (35) and (85).

On the other hand, numerical solutions of the asymptotic filament equations require only a few minutes on a desktop workstation while the direct simulations of Klein and Knio need to resolve stiff terms numerically which are filtered out by the asymptotics. In contrast, the direct numerical simulations of Klein and Knio require many hours on a very large supercomputer, the Cray C-90.

3. Interacting Parallel Vortex Filaments—Point Vortices in the Plane

The reader is probably already familiar with the motion of collections of point vortices in the plane (Chorin and Marsden). Here, for the purposes of motivating an important generalization presented in section 4, we regard a collection of point vortices in the plane as equivalently a collection of exactly parallel vortex filaments with no structural variation along the x_3-axis. For point vortices, the vorticity $\omega(x, t)$ has the postulated form as a superposition of δ-functions for all times,

$$(86) \qquad\qquad \omega(\mathbf{x}, t) = \sum_{j=1}^{N} \Gamma_j \delta(\mathbf{x} - \mathbf{X}_j(t))$$

$$\mathbf{X}_j(t)|_{t=0} = \mathbf{X}_j^0$$

where Γ_j is the circulation of the j-th vortex. From the vorticity stream form, the formal induced velocity associated with the vorticity in (86) is given by

$$(87) \qquad\qquad \mathbf{v}(\mathbf{x}, t) = \sum_{j=1}^{N} \Gamma_j \frac{(\mathbf{x} - \mathbf{X}_j)^{\perp}}{|\mathbf{x} - \mathbf{X}_j|^2}$$

with $\mathbf{x}^{\perp} = (-x_2, x_1)$. Ignoring the fact that the velocity of a point vortex is infinite at its center, a point vortex induces no motion at its center. Thus, using the particle trajectory equations

$$(88) \qquad \frac{d\mathbf{X}_j}{dt} = \mathbf{v}(\mathbf{X}_j, t), \quad \mathbf{X}_j|_{t=0} = \mathbf{X}_j^0$$

and the above formal fact we arrive at the
Equations for N-Interacting Exactly Parallel Vortex Filaments

$$(89) \qquad \frac{d\mathbf{X}_j}{dt} = \sum_{k \neq j} \Gamma_k \frac{(\mathbf{X}_j - \mathbf{X}_k)^\perp}{|\mathbf{X}_j - \mathbf{X}_k|^2}$$

$$\mathbf{X}_j|_{t=0} = \mathbf{X}_j^0.$$

It is worth noting here that such dynamic equations as in (89) have a self-consistent derivation through formal asymptotic expansions as the high Reynolds number limit of suitable solutions of the 2-D Navier-Stokes equations. The interested reader can consult the book by Ting and Klein for a detailed discussion.

It is well-known that the equations for point vortices are a Hamiltonian system, i.e. the dynamical equations in (89) can be rewritten in the generalized Hamiltonian form

$$(90) \qquad \Gamma_j \frac{d}{dt} X_j = J \nabla_{X_j} H_p$$

where J is the usual skew-symmetric matrix

$$(91) \qquad J = \begin{pmatrix} 0 & -1 \\ 1 & 0 \end{pmatrix}.$$

For the motion of point vortices in (89), the reader can readily verify that the Hamiltonian is given by

$$(92) \qquad H_p = 2 \sum_{j<k} \Gamma_j \Gamma_j \ln |\mathbf{X}_j - \mathbf{X}_k|.$$

Conserved Quantities. In general, according to Noether's Theorem (Arnold), symmetries in a Hamiltonian lead to conserved quantities. First, the translational symmetry $H(\mathbf{X}_1 + \mathbf{Y}, \mathbf{X}_2 + \mathbf{Y}, \dots, \mathbf{X}_N + \mathbf{Y}) = H(\mathbf{X}_1, \dots, \mathbf{X}_N)$ for all $\mathbf{Y}$ leads to the conserved quantity for the dynamics,

$$(93) \qquad M = \sum_{j=1}^{N} \Gamma_j \mathbf{X}_j(t).$$

Second the rotational symmetry of the Hamiltonian,

$$H(\mathcal{O}(\theta)\mathbf{X}_1, \mathcal{O}(\theta)\mathbf{X}_2, \dots, \mathcal{O}(\theta)\mathbf{X}_N) = H(\mathbf{X}_1, \dots, \mathbf{X}_N)$$

for any rotation matrix $\mathcal{O}(\theta)$ in the plane yields the conserved quantity for the dynamics,

$$(94) \qquad A = \sum_{j=1}^{N} \Gamma_j |\mathbf{X}_j(t)|^2.$$

We leave the verification that (93) and (94) are conserved quantities for the dynamics of point vortices as an exercise for the interested reader. However, we remark here that M and A are discrete versions of the continuous first two moments of vorticity $\int \mathbf{x} \omega \, dx$ and $\int |x|^2 \omega \, dx$ which are conserved quantities for the 2-D Euler equations.

Elementary Exact Solutions. It is well-known (Lamb, 1932) that the motion of two exactly parallel vortex filaments is exactly solvable; here we simply record these exact solutions in coordinates convenient for our discussion in section 4. For a pair of parallel vortex filaments, we can always rescale time so that one of the filaments has circulation 1 and the other filament has circulation Γ where Γ satisfies $-1 \leq \Gamma \leq 1$ with $\Gamma \neq 0$; thus, Γ represents the circulation ratio. We set $\mathbf{X}_j(t) = (x_j, y_j)$ and introduce the complex coordinate for each filament,

$$(95) \qquad\qquad \psi_j = x_j + iy_j, \quad \text{for} \quad j = 1, 2$$

with $i = (-1)^{1/2}$.

With these simplifications, the equations for a pair of point vortices have the form

$$(96) \qquad\qquad \begin{aligned} \frac{1}{i}\frac{\partial \psi_1}{\partial t} &= 2\Gamma \frac{\psi_1 - \psi_2}{|\psi_1 - \psi_2|^2} \\ \frac{1}{i}\frac{\partial \psi_2}{\partial t} &= -2 \frac{\psi_1 - \psi_2}{|\psi_1 - \psi_2|^2}. \end{aligned}$$

Since the nonlinear term in (96) is a function of only one variable, we introduce the coordinates $\psi = \psi_1 - \psi_2$ and $\phi = \psi_1 + \psi_2$ so that

$$(97) \qquad\qquad \psi_1 = \frac{1}{2}(\phi + \psi), \quad \psi_2 = \frac{1}{2}(\phi - \psi).$$

Then (96) becomes

$$(98) \qquad\qquad \begin{aligned} \frac{1}{i}\phi_t &= 2(\Gamma - 1)\frac{\psi}{|\psi|^2} \\ \frac{1}{i}\psi_t &= 2(\Gamma + 1)\frac{\psi}{|\psi|^2}. \end{aligned}$$

With the form of the equations in (98) it is easy to write down the exact solutions describing the motion of two point vortices. We choose the origin of coordinates so that at time $t = 0$, $\psi_1 = \left(\frac{1}{2}d, 0\right)$ and $\psi_2 = \left(-\frac{1}{2}d, 0\right)$ where d is the separation distance. Then, the equations in (98) have the explicit exact solutions for $\Gamma \neq -1$

$$\phi^0(t) = d\left(1 - \exp\left(\frac{2i(1 + \Gamma)t}{d^2}\right)\right)\frac{1 - \Gamma}{1 + \Gamma}$$

$$(99) \qquad\qquad \psi^0(t) = d\exp\left(\frac{2i(1 + \Gamma)t}{d^2}\right)$$

and for $\Gamma = -1$

$$\phi^0(t) = -\frac{4i}{d^2}t$$

$$(100) \qquad\qquad \psi^0(t) = (d, 0).$$

The case of $\Gamma = -1$ is called the *anti-parallel vortex pair* since the two vortices have equal and opposite circulations. For the anti-parallel pair the motion of the two vortices is uniform translation at a velocity related to the initial separation distance. In the general case, with $\Gamma \neq -1$, the two vortices simply rotate about their common center of mass at a fixed angular velocity. These *motions of exactly parallel vortex filaments are clearly stable within other exactly parallel vortex filament perturbations with the same fixed circulation ratio* although the anti-parallel pair with $\Gamma = -1$

exhibits weak instability involving mild $O(t)$ growth in time, as is clear from the formula in (100).

4. Asymptotic Equations for the Interaction of Nearly Parallel Vortex Filaments

Nearly parallel interacting vortex filaments with large strength and narrow cross-section are prominent fluid mechanical structures in mixing layers (Corcos and Lin) and trailing wakes (Van Dyke). A prominent and important example is the anti-parallel pair shed by the wing tips of an aircraft; the hazards of flying smaller aircraft into the turbulent wake generated by larger aircraft through the interaction of an anti-parallel vortex pair is a notable practical example of such an interaction.

How can we develop a simplified theory for the interaction of nearly parallel collections of vortex filaments? Here we will reverse our point of view when compared with our discussion in sections 1 and 2 on asymptotic models for individual vortex filaments. We will make an educated guess for an attractive plausible qualitative model for the interaction of nearly parallel vortex filaments and then indicate the fashion in which such simplified equations arise as a self-consistent asymptotic limit.

Consider N-vortex filaments which are all nearly parallel to the x_3-axis with their perturbed centerlines described by pairs of coordinates

$$(101) \qquad \mathbf{X}_j(\sigma, t) = (x_j(\sigma, t), y_j(\sigma, t)), \qquad 1 \le j \le N$$

where σ parameterizes the x_3-axis. The simplest theory of self-interaction for a nearly parallel vortex filament has been presented in equation (29) from section 1 and involves the linearized self-induction equation for each parallel vortex perturbation,

$$(102) \qquad \frac{\partial \mathbf{X}_j}{\partial t}(\sigma, t) = J\Gamma_j \frac{\partial^2 \mathbf{X}_j}{\partial \sigma^2}(\sigma, t), \qquad 1 \le j \le N.$$

On the other hand, the simplest theory for mutual interaction of exactly parallel vortex filaments is given by the equations for N-interacting point vortices described in section 3. Thus, by combining these two effects, we can propose
Simplified Equations for the Interaction of Nearly Parallel Vortex Filaments,

$$(103) \qquad \frac{\partial \mathbf{X}_j}{\partial t} = J\left[\Gamma_j \frac{\partial^2 \mathbf{X}_j}{\partial \sigma^2}\right] + J\left[\sum_{k \ne j}^{N} 2\Gamma_k \frac{(\mathbf{X}_j - \mathbf{X}_k)}{|\mathbf{X}_j - \mathbf{X}_k|^2}\right], \qquad 1 \le j \le N$$

where J is the skew-symmetric matrix in (91). In (103), the perturbed vortex filament curves have mutual interaction through the point vortex interaction in a layered fashion for each σ and simultaneously have self-interaction in the σ-variable through the linearized self-induction of each filament as described in (102). Solutions of these equations have remarkable mathematical and physical properties which we describe briefly below.

On the other hand, intuition suggests that these two physical effects in (103) might dominate for nearby interacting almost parallel vortex filaments where

1) the wavelength of perturbations is much longer than the separation
(104)   distance
2) the separation distance is much larger than the core thickness.

Recently, Klein, Majda, and Damodaran (1995) have developed a systematic asymptotic expansion of the Navier-Stokes equations in a suitable distinguished limit where 1) and 2) from (104) are satisfied and the simplified asymptotic equations for the interaction of nearly parallel vortex filaments from (103) emerge as the leading order limit equations. Next, we briefly describe the precise asymptotic conditions needed in the derivation.

First, it is assumed that the centerline of each nearly parallel vortex filament has the asymptotic form

$$(105) \qquad \mathbf{X}_j^\epsilon = s\mathbf{e}_3 + \epsilon^2 \mathbf{X}_j \left(\frac{s}{\epsilon}, \frac{t}{\epsilon^4} \right) + o(\epsilon^2)$$

for $1 \leq j \leq N$ and that each vortex filament has a cross-sectional core radius satisfying the relations

$$(106) \qquad \epsilon^2 = \ln\left(\frac{1}{\delta}\right), \quad \delta = \mathrm{Re}^{-1/2}, \quad \delta \ll 1$$

already familiar to the reader from (2) and (35). These conditions guarantee that condition 2) from (104) is automatically satisfied. From (105) we see that

the separation distance between vortices is $O(\epsilon^2)$ while
(107)   the wavelength of perturbations is $O(\epsilon)$.

Thus, in the very precise sense from (107) we see that for $\epsilon \ll 1$ we have guaranteed that the wavelengths of perturbations are much longer than the separation distance so that condition 1) in (104) is automatically satisfied.

We also note that, as in our discussion above (35) in section 2, the unit length scale for (105) is defined by the typical radius of curvature of each individual filament which is $O(1)$. This means that the vortices must be close within $O(\epsilon^2)$ on this length scale; in practical terms the interacting filaments can be fairly distant provided that the vortex perturbations have large radii of curvature. We also remark that the time scale in (105) is a more rapid time scale than the one utilized in (33) of section 2 for the self-stretch of an individual filament. In fact, the self-stretching mechanisms of an individual vortex filament studied extensively in section 2 are higher order asymptotic corrections with the ansatz in (105); the interested reader can find a detailed comparison and discussion of regimes for asymptotic vortex theories in section 6 of the paper of Klein, Majda and Damodaran. With the ansatz in (105) under the conditions in (106), the asymptotic equations in (1) emerge from a detailed asymptotic expansion of the Biot-Savart law from (1) under these assumptions in a similar fashion as described earlier in sections 1 and 2. The details are developed in section 2 of the paper by Klein, Majda and Damodaran.

4.1. Hamiltonian Structure and Conserved Quantities. We write the simplified equations for interacting nearly parallel vortex filaments from (103) in the form,

$$(108) \qquad \Gamma_j \frac{\partial \mathbf{X}_j}{\partial t} = J \left[\Gamma_j^2 \frac{\partial^2}{\partial \sigma^2} \mathbf{X}_j \right] + J \left[\sum_{k \neq j}^{N} 2\Gamma_j \Gamma_k \frac{(\mathbf{X}_j - \mathbf{X}_k)}{|\mathbf{X}_j - \mathbf{X}_k|^2} \right].$$

In order to find the Hamiltonian for this system of equations, we need to find a functional, $\mathcal{H}$, of the N-vortex filaments so that

$$(109) \qquad \Gamma_j \frac{\partial \mathbf{X}_j}{\partial t} = J \frac{\delta \mathcal{H}}{\delta \mathbf{X}_j}, \qquad 1 \leq j \leq N.$$

In (109), $\frac{\delta \mathcal{H}}{\delta \mathbf{X}_j}$, denotes the functional or variational derivative with respect to the curve $\mathbf{X}_j(\sigma)$ computed through the L^2 inner product for curves,

$$(110) \qquad (\mathbf{\Phi}, \mathbf{\Psi})_0 = \int \Phi(\sigma) \cdot \psi(\sigma) d\sigma.$$

In (110) we assume that the filament curves are either periodic in σ so that the integration range is over a period interval or that the filament curve perturbations vanish sufficiently rapidly together with their derivatives so that all the contributions from infinity vanish in the situation that the range of integration is the entire line.

To find the Hamiltonian satisfying (109) we write $\mathcal{H}$ in the form

$$(111) \qquad \mathcal{H} = \mathcal{H}_s + \mathcal{H}_p$$

where

$$(112) \qquad \frac{\delta \mathcal{H}_s}{\delta \mathbf{X}_j} = \Gamma_j^2 \frac{\partial^2}{\partial \sigma^2} \mathbf{X}_j$$

and

$$(113) \qquad \frac{\delta \mathcal{H}_p}{\delta \mathbf{X}_j} = \sum_{k \neq j}^{N} 2\Gamma_j \Gamma_k \frac{(\mathbf{X}_j - \mathbf{X}_k)}{|\mathbf{X}_j - \mathbf{X}_k|^2}$$

for $1 \leq j \leq N$. The functional, $\mathcal{H}_s$, satisfying (112) is given by

$$(114) \qquad \mathcal{H}_s = - \sum_{j=1}^{N} \frac{\Gamma_j^2}{2} \int \left| \frac{\partial \mathbf{X}_j}{\partial \sigma} \right|^2 d\sigma$$

while the functional, $\mathcal{H}_p$, satisfying (113) is merely the integral over σ of the familiar N-point vortex Hamiltonian discussed earlier in (92), i.e.,

$$(115) \qquad \mathcal{H}_p = 2 \sum_{j<k}^{N} \int \Gamma_j \Gamma_k \ln |\mathbf{X}_j(\sigma) - \mathbf{X}_k(\sigma)| d\sigma.$$

With (111)–(115), we conclude that the Hamiltonian satisfying (109) is given by

$$(116) \qquad \mathcal{H} = - \sum_{j=1}^{N} \frac{\alpha_j \Gamma_j^2}{2} \int \left| \frac{\partial \mathbf{X}_j}{\partial \sigma} \right|^2 d\sigma + 2 \sum_{j<k}^{N} \int \Gamma_j \Gamma_k \ln |\mathbf{X}_j(\sigma) - \mathbf{X}_k(\sigma)| d\sigma.$$

Of course, it follows immediately from the equations in (109) that

the Hamiltonian, $\mathcal{H}$, is conserved in time for solutions

of the asymptotic filament equations in (108).

Other Conserved Quantities. The equations in (108) have other symmetries which lead to integrated analogues of the conservation of the center of vorticity and angular momentum which are familiar to the reader for 2-D point vortices from (93) and (94).

The Hamiltonian remains unchanged under the transformations $\mathbf{X}_j \mapsto \mathbf{X}_j + \mathbf{Z}_0$ where $\mathbf{Z}_0$ is an arbitrary 2-vector. Thus, we define the *mean center of vorticity* by

$$(117) \qquad \mathbf{M} = \int \sum_{j=1}^{N} \Gamma_j \mathbf{X}_j(\sigma) d\sigma.$$

An elementary calculation establishes that $\frac{d\mathbf{M}}{dt} = 0$ for solutions of (108) so that

the mean center of vorticity, $\mathbf{M}$, is

conserved in time for solutions.

Similarly, the Hamiltonian remains invariant under the transformations $\mathbf{X}_j \mapsto \mathcal{O}(\theta)\mathbf{X}_j$ where $\mathcal{O}(\theta)$ is an arbitrary rotation matrix. We define the *mean angular momentum* by

$$(118) \qquad A = \int \sum_{j=1}^{N} \Gamma_j |\mathbf{X}_j(\sigma)|^2 d\sigma$$

and elementary calculations establish that

the mean angular momentum, A, defined in

(118) is conserved in time for solutions.

There is another conserved quantity which arises for solutions of the interacting filament equations which is not derived as a direct analogue in integrated form from the equations for 2-D point vortices. The Hamiltonian in (116) remains invariant under the translation, $\mathbf{X}_j(\sigma) \mapsto \mathbf{X}_j(\sigma + h)$ where h is arbitrary so that Noether's Theorem guarantees that there is another conserved quantity. By following the procedure for Noether's Theorem in the form stated in Arnold (1989), we claim that the quantity

$$(119) \qquad W = \int \sum_{j=1}^{N} \Gamma_j (J\mathbf{X}_j(\sigma)) \cdot \frac{\partial \mathbf{X}_j(\sigma)}{\partial \sigma} d\sigma$$

is conserved in time for solutions of the filament equations in (108). To verify this, we calculate that

$$
\begin{aligned}
(120) \qquad \frac{dW}{dt} &= 2\int \sum_{j=1}^{N} \Gamma_j J \frac{d\mathbf{X}_j}{dt} \cdot \frac{\partial \mathbf{X}_j}{d\sigma} d\sigma \\
&= \left\{ -2\int \left(\sum_{j=1}^{N} \Gamma_j^2 \frac{\partial^2 \mathbf{X}_j}{\partial \sigma^2} \cdot \frac{\partial \mathbf{X}_j}{\partial \sigma} \right) d\sigma \right\} \\
&\quad + \left\{ -4\int \left(\sum_{j<k} \Gamma_j \Gamma_k \frac{\mathbf{X}_j - \mathbf{X}_k}{|\mathbf{X}_j - \mathbf{X}_k|^2} \cdot \frac{\partial}{\partial \sigma}(\mathbf{X}_j - \mathbf{X}_k) \right) d\sigma \right\} \\
&= \{1\} + \{2\}.
\end{aligned}
$$

We claim that all of the integrands in both terms $\{1\}$ and $\{2\}$ are perfect derivatives so that all contributions on the right hand side of (120) vanish. Since

$$2\frac{\partial^2 \mathbf{X}_j}{\partial \sigma^2} \cdot \frac{\partial \mathbf{X}_j}{\partial \sigma} = \frac{\partial}{\partial \sigma}\left(\left|\frac{\partial \mathbf{X}_j}{\partial \sigma}\right|^2\right)$$

the integrand in $\{1\}$ is a perfect derivative while

$$\frac{\mathbf{X}_j - \mathbf{X}_k}{|\mathbf{X}_j - \mathbf{X}_k|^2} \cdot \frac{\partial}{\partial \sigma}(\mathbf{X}_j - \mathbf{X}_k) = \frac{\partial}{\partial \sigma}\ln|\mathbf{X}_j - \mathbf{X}_k|$$

so that the integrand in $\{2\}$ is also a perfect derivative. Thus,

the quantity, W, is conserved by solutions of the filament equations.

The Average Distance Functional. We briefly consider a generalized distance functional, I, defined by

$$(121) \qquad I = \frac{1}{2}\int \sum_{j,k}\Gamma_j\Gamma_k|\mathbf{X}_j - \mathbf{X}_k|^2 d\sigma.$$

We show below that I is conserved for special configurations consisting of identical vortex filaments. Through integration by parts, we compute in general that

$$\frac{dI}{dt} = -\int \sum_{j,k}^N\left[\Gamma_j\mathbf{X}_j \cdot \Gamma_k\frac{d\mathbf{X}_k}{dt} + \Gamma_k\mathbf{X}_k \cdot \Gamma_j\frac{d\mathbf{X}_j}{dt}\right] d\sigma$$

$$= -\sum_{j,k}^N\Gamma_j\Gamma_k(\Gamma_k - \Gamma_j)\int\left(\frac{\partial x_j}{\partial \sigma}\frac{\partial y_k}{\partial \sigma} - \frac{\partial x_k}{\partial \sigma}\frac{\partial y_j}{\partial \sigma}\right) d\sigma$$

where $\mathbf{X}_j(\sigma,t) = (x_j(\sigma,t), y_j(\sigma,t))$. From the above identity, we observe that

the average distance functional, I, is conserved in time for

solutions provided that for all the filaments, $\Gamma_j = \Gamma_k$.

In particular, for co-rotating filament pairs with the same structure in the vortex core, the distance functional, I, is conserved.

4.2. The Equations for Pairs of Interacting Filaments. Here we study solutions of the simplified equations for interacting filaments from (103) in the important special case involving two interacting nearly parallel vortex filaments. As in section 3, without loss of generality we assume that one of the filaments has circulation, $\Gamma_1 = 1$, while the other vortex filament has circulation $\Gamma_2 = \Gamma$ where Γ satisfies $-1 \leq \Gamma \leq 1$ with $\Gamma \neq 0$; thus, Γ represents the circulation ratio of the two interacting filaments. For the pair of interacting filaments with $\mathbf{X}_j(\sigma,t) = (x_j, y_j)$, it will be convenient to introduce the complex coordinates for each filament

$$(122) \qquad \psi_j = x_j(\sigma,t) + iy_j(\sigma,t)$$

for $j = 1,2$ with $i = \sqrt{-1}$.

The equations in (103) specialized to the case of two interacting filaments have the form

$$(123) \qquad \frac{1}{i}\frac{\partial \psi_1}{\partial t} = \frac{\partial^2\psi_1}{\partial\sigma^2} + 2\Gamma\frac{\psi_1 - \psi_2}{|\psi_1 - \psi_2|^2}$$

$$\frac{1}{i}\frac{\partial \psi_2}{\partial t} = \Gamma\frac{\partial^2\psi_2}{\partial\sigma^2} - 2\frac{\psi_1 - \psi_2}{|\psi_1 - \psi_2|^2}.$$

From Section 4.1 these equations have the following conserved quantities,

$$H = -\frac{1}{2}\int \left|\frac{\partial\psi_1}{\partial\sigma}\right|^2 d\sigma - \frac{\Gamma^2}{2}\int \left|\frac{\partial\psi_2}{\partial\sigma}\right|^2 d\sigma + 2\left[\int \Gamma \ln|\psi_1 - \psi_2|d\sigma\right],$$

$$(124)\quad M = \int \psi_1 d\sigma + \int \Gamma\psi_2 d\sigma,$$

$$A = \int |\psi_1|^2 d\sigma + \int \Gamma|\psi_2|^2 d\sigma,$$

$$W = i\int \psi_1 \frac{\partial\overline{\psi}_1}{\partial\sigma} + \Gamma\psi_2 \frac{\partial\overline{\psi}_2}{\partial\sigma} d\sigma.$$

Furthermore, for co-rotating pairs so that $\Gamma = 1$, we have the additional conserved quantity

$$(125)\quad I = \int |\psi_1 - \psi_2|^2 d\sigma$$

To make the nonlinear term in (123) a function of only one variable as in section 3 we introduce the coordinates $\psi = \psi_1 - \psi_2$, and $\phi = \psi_1 + \psi_2$ so that

$$(126)\qquad \psi_1 = \frac{(\phi + \psi)}{2},$$

$$\psi_2 = \frac{(\phi - \psi)}{2}.$$

The equations in (123) have the following form in the new variables,

The Equivalent Equation for a Nearly Parallel Filament Pair.

$$(127)\qquad \frac{1}{i}\phi_t = \frac{(1+\Gamma)}{2}\phi_{\sigma\sigma} + \frac{(1-\Gamma)}{2}\left[\psi_{\sigma\sigma} - 4\frac{\psi}{|\psi|^2}\right],$$

$$\frac{1}{i}\psi_t = \frac{(1-\Gamma)}{2}\phi_{\sigma\sigma} + \frac{(1+\Gamma)}{2}\left[\psi_{\sigma\sigma} + 4\frac{\psi}{|\psi|^2}\right].$$

4.2.1 Co-rotating Filament Pairs. For the special case of co-rotating filament pairs, we set $\Gamma = 1$ and the two equations in (127) completely decouple into two separate scalar equations given by

$$(128)\qquad \frac{1}{i}\phi_t = \phi_{\sigma\sigma}$$

$$\frac{1}{i}\psi_t = \psi_{\sigma\sigma} + \frac{4\psi}{|\psi|^2}.$$

The first equation in (128) is the linear Schrödinger equation while the second equation is a nonlinear Schrödinger equation with an unusual nonlinearity.

These equations have a dispersive wave-like behavior. We demonstrate this by writing down some elementary exact solutions. The complex function ψ will be a nonlinear plane wave solution of the form

$$\psi = Be^{i(k\sigma + \omega t)}$$

provided that

$$\omega = \frac{4}{B^2} - k^2.$$

With the exact solutions, we obtain general wave-like solutions in the original complex filament coordinates given by

$$
(129) \qquad
\begin{aligned}
\psi_1 &= \phi(\sigma, t) + \frac{B}{2} e^{i(k\sigma + \omega t)} \\
\psi_2 &= \phi(\sigma, t) - \frac{B}{2} e^{i(k\sigma + \omega t)}
\end{aligned}
$$

where ϕ is an arbitrary solution of the linear Schrödinger equation.

4.2.2 Linearized Stability for the Filament Pair.

It should be evident to the reader that any collection of exactly parallel vortex filaments which satisfies the N-point vortex equations in (89) is automatically an exact solution of the equation in (103) for N-interacting nearly parallel filaments. Thus, one can begin to understand the behavior of solutions of (103) by studying the linearized stability of the point vortex solutions within the class of general 3-D filament curve perturbations.

In (99) and (100) from section 3, we wrote down the well-known exact solutions for a pair of exactly parallel vortex filaments. As we remarked earlier in section 3, such solutions are essentially always stable to exactly parallel filament perturbations. Here, as the most important special case of the strategy outlined in the preceding paragraph, we study the linearized stability of these exact solutions within general filament pair perturbations satisfying the equation in (127). We will find the remarkable fact that *these exactly parallel configurations are always unstable at suitable long wavelengths for any negative circulation ratio, $-1 \leq \Gamma \leq 0$ and are always (neutrally) stable for any positive circulation ratio, $0 < \Gamma \leq 1$.*

We linearize the equations in (127) about the exact solutions ϕ^0, ψ^0 from (99) and (100) and obtain the linearized equations for perturbations, still denoted by ϕ and ψ, and given by

$$
(130) \qquad
\begin{aligned}
\frac{\partial \phi}{\partial t} &= i\frac{(1+\Gamma)}{2}\phi_{\sigma\sigma} + i\frac{(1-\Gamma)}{2}\left[\psi_{\sigma\sigma} + 4\frac{\overline{\psi}}{\overline{\psi}_0^2}\right], \\
\frac{\partial \psi}{\partial t} &= i\frac{(1-\Gamma)}{2}\phi_{\sigma\sigma} + i\frac{(1+\Gamma)}{2}\left[\psi_{\sigma\sigma} - 4\frac{\overline{\psi}}{\overline{\psi}_0^2}\right],
\end{aligned}
$$

where $\psi_0 = d e^{\frac{2i(1+\Gamma)t}{d^2}}$ and a bar indicates complex conjugation. To get rid of the time dependence that enters through $\overline{\psi}_0^2$, we go to a coordinate frame that rotates with the vortex filaments, through the variables χ and Θ defined by

$$
(131) \qquad
\begin{aligned}
\chi(\sigma, t) &= \psi(\sigma, t) e^{\frac{-2i(1+\Gamma)t}{d^2}}, \\
\Theta(\sigma, t) &= \phi(\sigma, t) e^{\frac{-2i(1+\Gamma)t}{d^2}}.
\end{aligned}
$$

With this transformation, the linearized equations of motion and the equations for
the corresponding complex conjugates are given by

$$
\begin{aligned}
\frac{1}{i}\chi_t &= \frac{(1+\Gamma)}{2}\left[\chi_{\sigma\sigma} - \frac{4}{d^2}\chi\right] + \frac{(1-\Gamma)}{2}\Theta_{\sigma\sigma} - \frac{2(1+\Gamma)}{d^2}\overline{\chi}, \\
(132)\qquad \frac{1}{i}\Theta_t &= \frac{(1-\Gamma)}{2}\chi_{\sigma\sigma} + \frac{(1+\Gamma)}{2}\left[\Theta_{\sigma\sigma} - \frac{4}{d^2}\Theta\right] + \frac{2(1-\Gamma)}{d^2}\overline{\chi}, \\
\frac{1}{i}\overline{\chi}_t &= -\frac{(1+\Gamma)}{2}\left[\overline{\chi}_{\sigma\sigma} - \frac{4}{d^2}\overline{\chi}\right] - \frac{(1-\Gamma)}{2}\overline{\Theta}_{\sigma\sigma} + \frac{2(1+\Gamma)}{d^2}\chi, \\
\frac{1}{i}\overline{\Theta}_t &= -\frac{(1-\Gamma)}{2}\overline{\chi}_{\sigma\sigma} - \frac{(1+\Gamma)}{2}\left[\overline{\Theta}_{\sigma\sigma} - \frac{4}{d^2}\overline{\Theta}\right] - \frac{2(1-\Gamma)}{d^2}\chi.
\end{aligned}
$$

We solve this 4×4 system with the Fourier transform. The coupling of χ and Θ
to their complex conjugates, $\overline{\chi}$ and $\overline{\Theta}$, in real space means that $\widehat{\chi}(\xi,t)$ is coupled
to $\overline{\widehat{\chi}}(-\xi,t)$ in wave number space and the same is true for Θ where ξ is the wave
number. Similar behavior also occurred earlier in section 2.3. In wave number
space the equations in (132) reduce to the 4×4 matrix-O.D.E.,

$$
(133)\qquad \frac{d}{dt}\begin{pmatrix} \widehat{\chi}(\xi,t) \\ \widehat{\Theta}(\xi,t) \\ \overline{\widehat{\chi}}(-\xi,t) \\ \overline{\widehat{\Theta}}(-\xi,t) \end{pmatrix} = iA(\xi)\begin{pmatrix} \widehat{\chi}(\xi,t) \\ \widehat{\Theta}(\xi,t) \\ \overline{\widehat{\chi}}(-\xi,t) \\ \overline{\widehat{\Theta}}(-\xi,t) \end{pmatrix},
$$

where

$$
(134)\qquad A(\xi) =
$$

$$
\begin{pmatrix}
-\frac{(1+\Gamma)}{2}\left(\xi^2 + \frac{4}{d^2}\right) & -\frac{(1-\Gamma)}{2}\xi^2 & -\frac{2(1+\Gamma)}{d^2} & 0 \\
-\frac{(1-\Gamma)}{2}\xi^2 & -\frac{(1+\Gamma)}{2}\left(\xi^2 + \frac{4}{d^2}\right) & \frac{2(1-\Gamma)}{d^2} & 0 \\
\frac{2(1+\Gamma)}{d^2} & 0 & \frac{(1+\Gamma)}{2}\left(\xi^2 + \frac{4}{d^2}\right) & \frac{(1-\Gamma)}{2}\xi^2 \\
-\frac{2(1-\Gamma)}{d^2} & 0 & \frac{(1-\Gamma)}{2}\xi^2 & \frac{(1+\Gamma)}{2}\left(\xi^2 + \frac{4}{d^2}\right)
\end{pmatrix}
$$

The matrix $A(\xi)$ has the eigenvalues $\lambda_\pm^1 = \pm\frac{1}{2}\sqrt{\mathcal{R} + 2\sqrt{\mathcal{P}}}$ and $\lambda_\pm^2 = \pm\frac{1}{2}\sqrt{\mathcal{R} - 2\sqrt{\mathcal{P}}}$,
where

$$
\begin{aligned}
(135)\qquad \mathcal{R} &= 8\frac{a^2}{d^4} + 8\frac{a^2\xi^4}{d^2} + a^2\xi^4 + b^2\xi^4 \\
\mathcal{P} &= 16\frac{a^4}{d^8} + 32\frac{a^2b^2\xi^2}{d^6} + 20\frac{a^2b^2\xi^4}{d^4} + 4\frac{b^4\xi^4}{d^4} + 8\frac{a^2b^2\xi^6}{d^2} + a^2b^2\xi^8
\end{aligned}
$$

with $a = 1 + \Gamma$ and $b = 1 - \Gamma$. We have growing solutions and linearized instability
whenever the $\lambda_\pm$ are imaginary. From the form of the $\lambda_\pm$ it is easy to see that since
both $\mathcal{R}$ and $\mathcal{P}$ are positive real numbers from (135), $\lambda_\pm^1$ is always a real number.
Thus, only $\lambda_\pm^2$, which has a minus sign under the radicand, can assume imaginary
values and yield instability. It is useful to define the growth factor $\mathcal{G}$, which depends
on the signs of the radicands in $\lambda_\pm^2$,

$$
(136)\qquad \mathcal{G} = \operatorname{sgn}\left(-(\mathcal{R} - 2\sqrt{\mathcal{P}})\right)\left|\sqrt{\mathcal{R} - 2\sqrt{\mathcal{P}}}\right|.
$$

Whenever the $\lambda_\pm^2$ are real, $\mathcal{G}$ is negative, and the perturbations are *neutrally
stable* with the oscillation frequency given by the absolute value of $\mathcal{G}$. If however,
any one of the $\lambda_\pm^2$ is imaginary, $\mathcal{G}$ is positive, and we have either exponentially
growing and damped solutions, with the growth/damping rate given by the absolute

value of $\mathcal{G}$. In general the growing solution dominates and we have an instability. From the definition of $\mathcal{G}$, we have $\mathcal{G} > 0$ and instability if and only if $\mathcal{R}^2 < 4\mathcal{P}$. Therefore we consider the quantity, $\mathcal{R}^2 - 4\mathcal{P}$, given explicitly from (135) by

$$
\begin{aligned}
\mathcal{Q} &= \mathcal{R}^2 - 4\mathcal{P} \\
(137) \quad &= [a^2 - b^2]\left[128\frac{a^2\xi^2}{d^6} + 80\frac{a^2\xi^4}{d^4} + 16\frac{b^2\xi^4}{d^2} + 16\frac{a^2\xi^6}{d^2} + (a^2 - b^2)\xi^8\right],
\end{aligned}
$$

where as before $a = 1 + \Gamma$ and $b = 1 - \Gamma$.

First we consider the situation with a positive circulation ratio so that $1 \geq \Gamma > 0$; in this case we have the inequality $a > b$ which guarantees by (137) that $\mathcal{R}^2 - 4\mathcal{P}$ is positive for all wave numbers ξ and there are no growing modes. Thus, we have linearized (neutral) stability for the vortex pair in this situation.

Next, we consider the situation with negative circulation ratios Γ with $-1 \leq \Gamma < 0$ which implies that we have $a < b$. From (137) in this case the only positive term in $\mathcal{Q} = \mathcal{R}^2 - 4\mathcal{P}$ is $(b^2 - a^2)b^2\xi^8$. At long wavelengths, $\xi^2 \ll 1$, this positive term is dominated in magnitude by the negative contributions to $\mathcal{Q}$ of order ξ^2, ξ^4, and ξ^6; thus, there is always long wavelength instability for the straight-line vortex pair for any negative circulation ratio. On the other hand, at short wavelengths, $\xi^2 \gg 1$, $\mathcal{R}^2 - 4\mathcal{P}$ is dominated by the positive factor $[a^2 - b^2]^2\xi^8$ and there is always short wavelength stability.

We summarize our analysis presented in the previous two paragraphs in the following fashion:

(138)   Within the simplified vortex filament equations in (123), straight line point vortex pairs have linearized long wavelength instability for arbitrary negative circulation ratios and linearized (neutral) stability for arbitrary positive circulation ratios.

For any fixed negative circulation ratio, the graph of the stability function $\mathcal{G}$ from (136) is positive over a finite interval of wave numbers extending from zero where $\mathcal{G}$ vanishes. For each negative circulation ratio there is a unique wave number with the largest growth rate.

4.3. Finite Time Collapse and Wave-like Behavior for Nearly Parallel Pairs of Vortex Filaments.

The results summarized in (138) on linearized stability suggest very different behavior of interacting pairs of filaments for negative circulation ratio with strong instability at long wavelengths as compared with the case for a positive circulation ratio where wave-like neutral stability without growth occurs. The special structure of the filament pair equations for the co-rotating case with $\Gamma = 1$ and presented in 4.2.1 also suggests that wave-like behavior dominates for positive circulation ratios.

Next we present numerical solutions of the filament pair equations in (123). For negative circulation ratios, we will see that the linearized instability and non-linearity conspire to produce a finite time local collapse of the two vortex filaments in a distinctive fashion for each different circulation ratio, Γ, with $-1 \leq \Gamma < 0$. On the other hand, the numerical solutions always indicate wave-like behavior without any finite time collapse for any positive circulation ratio.

4.3.1 Finite Time Collapse for Negative Circulation Ratios.
First, we consider the anti-parallel pair with the circulation ratio, $\Gamma = -1$. This configuration

is the one in the model equations created by the two trailing wake vortices shed
by the wing tips of an aircraft. For initial data we consider spatially periodic
perturbations of the filament pair with structure given by the most unstable mode
according to the linearized theory in 4.2.2; this mode yields symmetric perturbations
of the initial data as the reader can see from Figure 5A) which depicts the initial
data at $t = 0$; in Figure 5B) we display the projections of these filament curves
along the three coordinate planes. We display both types of data in the figures for
this section. In Figures 6 and 7 we display snapshots of the solution of the filament
pair equations in (123) at the times $t = .0800$ and $.0800$ as the filament pair
evolves toward local finite time collapse of the curves. The initial perturbations
are symmetric and the solutions retain this symmetry throughout the nonlinear
evolution. The nature of the filament collapse presented in these figures is very
intuitive when one considers the behavior locally of point vortices in two dimensions
with equal and opposite circulation summarized earlier in (100). As Figures 5-7
indicate, the points of closest separation move more rapidly and pinch off faster
than the rest of the vortex filament and the magnitude of the pinching necessarily
increases substantially as the interaction with local induction brings the filaments
together. The result of the interaction is finite time collapse.

Next we present evidence for universal behavior in the local nonlinear collapse
of solutions with $\Gamma = -.5$; however, the universal structure for this collapse depends
in an interesting fashion on Γ for these negative circulation ratios. We consider the
solution of the filament equations for $\Gamma = -.5$ with initial data involving perturbations along the most unstable linearized mode. In this case with $\Gamma = -.5$, the
perturbed filaments are no longer symmetric. The nonlinear solution exhibits finite
time collapse at $t = .1735$. Snapshots of the evolving solution at the times $t = 0$,
$.1500$, and $.1735$ are presented in Figures 8, 9 and 10, respectively. An important
qualitative feature is that there is much more local rotation without spatial translation in this situation with $\Gamma = -.5$ as compared to the earlier case with $\Gamma = -1$
where translation dominates; this is easy to understand with the rotating motion
of point vortices in (99) for $\Gamma \neq -1$. As depicted in Figures 9 and 10, this local
rotation gives a qualitatively different structure for the local collapse for $\Gamma = -.5$
which occurs when the filament with smaller vorticity loops around the filament
with larger circulation and is sucked in by it. This process is shown clearly by the
coordinate projections in Figures 9B) and 10B).

Both of the calculations just presented indicate that general three-dimensional
perturbations of a pair of vortex filaments with negative circulation ratio yield
a finite time collapse. Furthermore, the nature of the collapse process depends
strongly on the specific value of the circulation ratio, Γ, with $-1 \leq \Gamma < 0$. In fact,
this collapse is probably self-similar with a different structure for each negative circulation ratio. Much more numerical evidence for this behavior as well as a careful
validation study can be found in the paper by Klein, Majda, and Damodaran.

Of course, as the solutions approach closely near collapse, the solutions of (123)
are no longer valid as a physical approximation because the separation distance becomes comparable to the core size and 2) in (104) is violated. Nevertheless, it is very
interesting that the asymptotic solutions, having only linearized self-induction and
nonlinear potential vortex interaction, provide the only physical effects needed to
drive two filaments with negative circulation ratio very close together. At distances
where the core thickness and separation distance are comparable, many recent
numerical experiments with the full Navier-Stokes equations give strong evidence

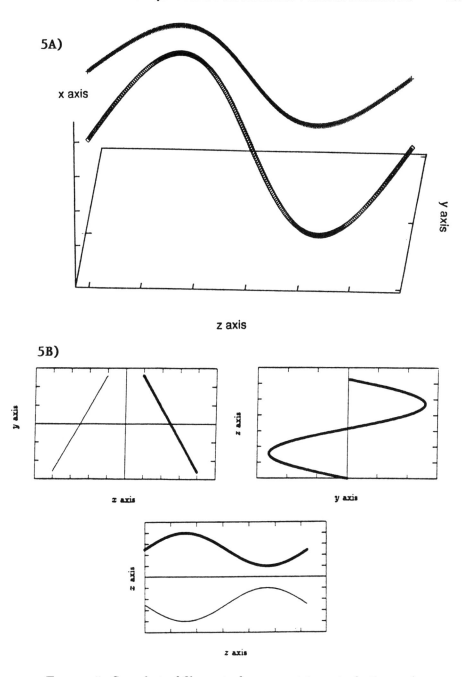

FIGURE 5. Snapshot of filaments for symmetric perturbation and $\Gamma = -1$ at time $t = 0$; part b) is the projection of the filaments on the coordinate axes.

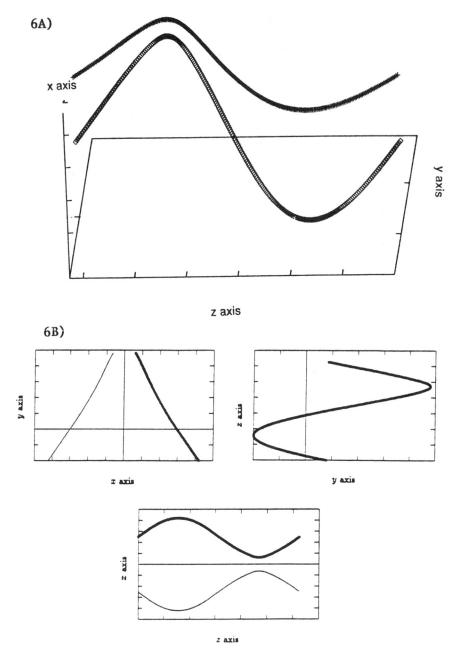

FIGURE 6. Snapshot of filaments for symmetric perturbation and
$\Gamma = -1$ at time $t = .0800$.

for substantial core flattening and vortex reconnection in the situation with negative circulation ratios (see Anderson and Greengard, Kerr and Hussain, Meiron et
al., Melander and Zabusky, Kida and Takaoka). Clearly the simplified asymptotic
equations cannot account for any of these effects. On the other hand, the numerical

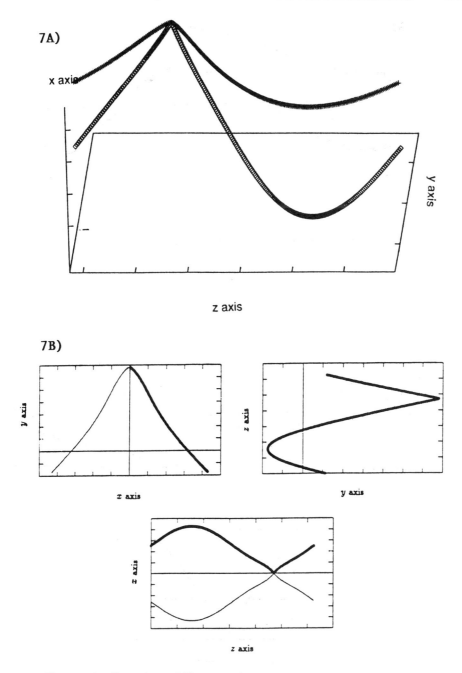

FIGURE 7. Snapshot of filaments for symmetric perturbation and $\Gamma = -1$ at time $t = .1146$.

solutions reported here take only a few minutes on a work station in contrast to the direct simulations just mentioned which typically require many hours on a large supercomputer.

8A)

x axis

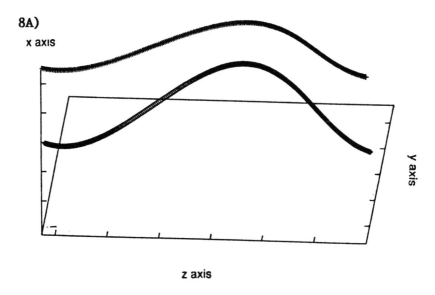

y axis

z axis

8B)

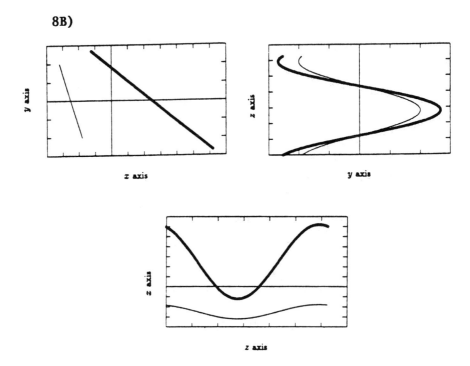

FIGURE 8. Snapshot of filaments for the large amplitude perturbation from (5.2) and $\Gamma = -.5$ at time $t = 0$.

**4.3.2 Wave-like Behavior Without Collapse for Positive Circulation
Ratios.** According to the predictions of linearized stability theory from 4.2.2, all
perturbations of an exactly parallel filament pair with any positive circulation ratio

9A)

x axis

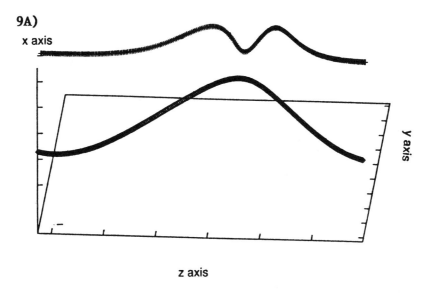

y axis

z axis

9B)

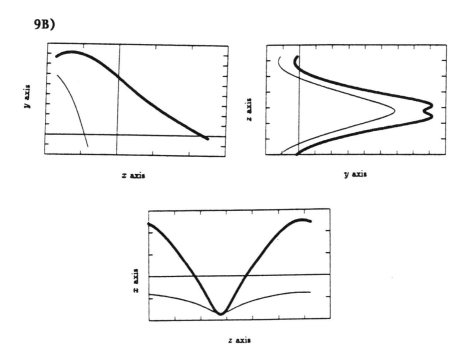

FIGURE 9. Snapshot of filaments for the large amplitude perturbation from (5.2) and $\Gamma = -.5$ at time $t = .1500$.

are neutrally stable and wave-like at all wave numbers. Numerical experiments for

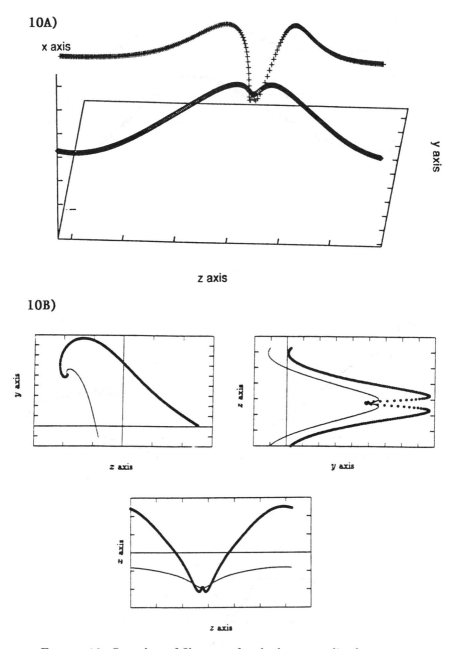

FIGURE 10. Snapshot of filaments for the large amplitude perturbation from (5.2) and $\Gamma = -.5$ at time $t = .1735$.

a fixed positive circulation ratio, Γ, with $0 < \Gamma \leq 1$ show that the fully nonlinear solutions of the filament pair equations in (123) continue to remain wave-like
throughout their evolution without any finite time collapse.

We illustrate this typical wave-like behavior through snapshots of a numerical solution depicted in Figure 11A), B), C) at the times $t = 0, .3$, and $.908$. The wave-like evolution of the solution is evident from these snapshots; we have singled out the time, $t = .908$, because this is the time of closest approach of the filament pair throughout the entire time history. Clearly, there is no evidence of collapse as occurred in the situation with negative circulation ratios described earlier.

5. Mathematical and Applied Mathematical Problems Regarding Asymptotic Vortex Filaments

There are several categories of interesting open problems related to the material described in this paper. We consider the following three areas below (which are not necessarily disjoint!!).

1: Applied Mathematics Issues
2: Mathematics Issues Regarding Properties of the Asymptotic Equations
3: Mathematics Issues Regarding Justification of the Asymptotic Approximation.

5.1. Applied Mathematics Issues.

#1: It is very interesting to extend the asymptotic theories developed in section 4 for interacting parallel vortices to model trailing wake flows such as the von Karman vortex street.

#2: Collections of nearly parallel interacting filamentary structures arise in many physical systems such as magneto-fluid dynamics, superfluids, high temperature superconductivity, etc. It is very interesting to develop the analogous asymptotic models as described here in sections 4 and 2 in these diverse physical contexts.

#3: As mentioned earlier in section 4, the self-stretch effects of a vortex filament studied in section 2 are higher order corrections in the theories described in section 4. It should be possible to devise asymptotic theories where both effects are included simultaneously for pairs of interacting filaments by considering the circulation ratio, Γ, as another small parameter with $\Gamma \ll 1$. Section 6 of the paper by Klein, Majda, and Damodaran should be a useful starting point as technical background. There the celebrated theory of Crow (1970) for linearized stability of the anti-parallel pair, another nonlinear asymptotic theory of Klein and Majda (1993) for the antiparallel pair which allows for self-stretch and reduces to Crow's theory after linearization, and the theory from section 4 are all compared and contrasted. We note here that the nonlinear asymptotic theory for the antiparallel pair due to Klein and Majda (1993) applies in a different asymptotic regime when compared with the theory from section 4. While (106) is satisfied for that theory, unlike the crucial requirement in (107), the wavelength $O(\epsilon)$ is comparable to the separation distance $O(\epsilon)$ but the amplitude of filament perturbations, $O(\epsilon^2)$, is much smaller than the $O(\epsilon)$ separation distance.

#4: The general equations from section 4 for interacting nearly parallel vortex filaments with their concise Hamiltonian structure and conserved quantities are an appealing candidate for simplified statistical theories of nearly parallel vortex filaments which are intermediate between two and three dimensional statistical theories for vortices (see Chorin (1988), (1994)). P.L. Lions and

11A)

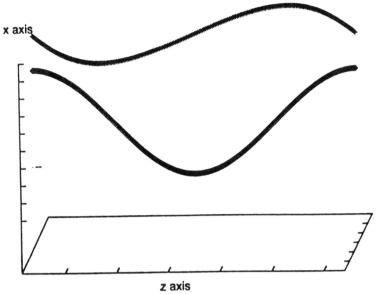

11B)

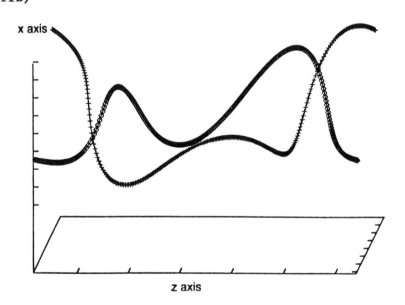

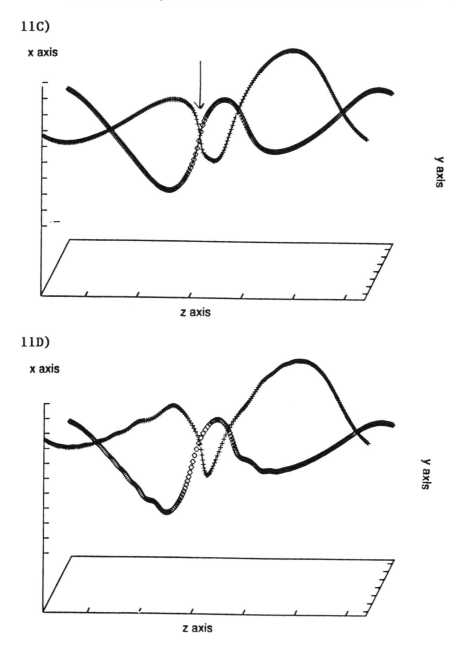

FIGURE 11. Snapshots of the solution with initial data for the filaments involving two plane curves in orthogonal planes at the times A) $t = 0$; B) $t = .300$; C) $t = .908$, the time with smallest separation distance; D) $t = .940$.

Majda have recently made some progress on this topic but much more work remains to be done.

5.2. Mathematical Issues Regarding Properties of the Asymptotic Equations.

#1: As regards the vortex filament pair discussed in 4.2, it would be very interesting to develop mathematical theorems which necessarily prove that finite time collapse must occur for negative circulation ratios for appropriate initial data and cannot occur for positive circulation ratios for arbitrary initial data. The conserved quantities from 4.1 should be useful in this regard. The novel nonlinear Schrödinger equation described in 4.2 for $\Gamma = 1$ is an obvious starting point for showing collapse does not occur.

#2: Are the filament pair equations in (123) a completely integrable Hamiltonian system for at least some special values of the circulation ratio such as $\Gamma = 1$ or $\Gamma = -1$?

#3: It is well known (see Aref) that the motion of 3 point vortices in the plane is integrable but can exhibit finite time collapse while the motion of 4 point vortices can be chaotic. How do genuine three dimensional filament perturbations of these exactly parallel solutions modify this behavior?

5.3. Mathematical Issues Regarding Justification of the Asymptotic Approximation.
The simplest situations to begin justification of the asymptotic approximations occur in sections 3 and 1.

#1: Ting and Klein present a detailed formal asymptotic derivation of the point vortex equations in section 3 from solutions of the Navier-Stokes equations in Chapter 2 of their book. Can this formal work be combined with estimates for the 2-D Navier-Stokes equations to rigorously justify this approximation?

#2: The same issues and references for the self-induction equations in section 1.

Notes. Crow (1970) pioneered the cut-off approximation to compute the finite part of a Biot-Savart integral and this device has been developed further by Moore and Saffman, Widnall, and others. Callegari and Ting (1978) present a succinct critique of this cut-off approximation. One can apply our asymptotic techniques in section 2 involving the Hasimoto transform as well as our techniques in section 4 in conjunction with the cut-off approximation to obtain the same asymptotic equations developed here. We have chosen to emphasize the methods of Callegari and Ting here because they are less "ad hoc" and apply to Navier-Stokes solutions.

On heuristic ground, Zakharov wrote down an equation describing the special case of symmetric perturbations of the anti-parallel pair. Under these very special circumstances, the theory from section 4 recovers a slightly simplified version of Zakharov's equation in a quantitative fashion (see Klein, Majda, Damodaran).

References

Ablowitz, M. & Segur, H. 1981 Solitons and the inverse scattering transform. *SIAM Stud. Appl. Maths* **4**.

Anderson, C. & Greengard, C. 1989 The vortex ring merger problem at infinite Reynolds number. *Commun. Pure Appl. Maths* **42**, 1123.

Aref, H. 1983 Integrable, chaotic and turbulent vortex motion in two-dimensional flows. *Ann. Rev. Fluid Mech.* **15**, 345–389.

Arms, R.J. & Hama, F.R. 1965 Localized-induction concept on a curved vortex and motion of an elliptic vortex ring. *Phys. Fluids* **8**, 553–559.

Arnold, V.I. 1989 *Mathematical Methods of Classical Mechanics.* 2nd edn. Springer.

Ashurst, W., Kerstein, A., Kerr R. & Gibson, C. 1987 Alignment of vorticity and scalar gradient with strain rate in simulated Navier-Stokes turbulence. *Phys. Fluids* **30**, 2343.

Batchelor, G.K. 1967 *An Introduction to Fluid Dynamics.* Cambridge University Press.

Callegari, A.J. & Ting, L. 1978 Motion of a curved vortex filament with decaying vortical core and axial velocity. *SIAM J. Appl. Maths* **35**, 148–175.

Chorin, A.J. 1982 Evolution of a turbulent vortex. *Commun. Math. Phys.* **83**, 517.

Chorin, A.J. 1988 Spectrum, dimension and polymer analogies in fluid turbulence. *Phys. Rev. Lett.* **60**, 1947–1949.

Chorin, A.J. 1994 *Vorticity and Turbulence.* Springer.

Chorin, A.J. & Akao, J. 1991 Vortex equilibria in turbulence theory and quantum analogues. *Physica* D **52**, 403–414.

Chorin, A.J. & Marsden, J.E. 1990 *A Mathematical Introduction to Fluid Mechanics.* Springer.

Corcos, G. & Lin, S. 1984 The mixing layer: deterministic models of a turbulent flow. Part 2. The origin of three-dimensional motion. *J. Fluid Mech.* **139**, 67–95.

Crow, S. 1970 Stability theory for a pair of trailing vortices. *AIAA J.* **8**, 2172–2179.

Hasimoto, H. 1972 A soliton on a vortex filament. *J. Fluid Mech.* **51**, 477–485.

Jimenez, J. 1975 Stability of a pair of co-rotating vortices. *Phys. Fluids* **18**, 1580.

Kerr, R.M. & Hussain, A.K.M.F. 1989 Simulation of vortex reconnection. *Physica* D **37**, 474.

Kida, S. & Takaoka, M. 1991 Breakdown of frozen motion fields and vorticity reconnection. *J. Phys. Soc. Japan* **60**, 2184.

Klein, R. & Knio, O.M. 1995 Asymptotic vorticity structure and numerical simulation of slender vortex filaments. *J. Fluid Mech.* **284**, 275–321.

Klein, R. & Majda, A. 1991*a* Self-stretching of a perturbed vortex filament I. The asymptotic equations for deviations from a straight line. *Physica* D **49**, 323–352.

Klein, R. & Majda, A. 1991 *b* Self-stretching of perturbed vortex filaments II. Structure of solutions. *Physica* D **53**, 267–294.

Klein, R. & Majda, A. 1993 An asymptotic theory for the nonlinear instability of anti-parallel pairs of vortex filaments. *Phys. Fluids* A **5**, 369–387.

Klein, R., Majda, A. & Damodaran, K. 1995 Simplified equations for the interaction of nearly parallel vortex filaments. *J. Fluid Mech.* **288**, 201–248.

Klein, R., Majda, A.J. & McLaughlin, R.M. 1992 Asymptotic equations for the stretching of vortex filaments in a background flow field. *Phys. Fluids* A **4**, 2271–2281.

Lamb, G.L. 1980 *Elements of Soliton Theory*, Wiley-Interscience.

Lamb, H. 1932 *Hydrodynamics*. 6th edn. Cambridge University Press.

Lions, P.L. & Majda, A. (in preparation).

Majda, A. 1991 Vorticity, turbulence and acoustics in fluid flow, *SIAM Rev.* **33**, 349.

Meiron, D., Shelley, M., Ashurst, W. & Orszag, S. 1989 Numerical studies of vortex reconnection. In *Mathematical Aspects of Vortex Dynamics* (ed. R. Caflisch), pp. 183–194. SIAM, Philadelphia, PA.

Melander, M.V. & Zabusky, N. 1987 Interaction and Reconnection of Vortex Tubes via Direct Numerical Simulations. *Proceedings IUTAM Symposium on Fundamental Aspects of Vortex Motion*, Tokyo.

Moore, D.W. & Saffman, P.G. 1972 The motion of a vortex filament with axial flow. *Phil. Trans. R. Soc. Lond.* **272**, 403.

She, Z. Jackson, E. & Orszag, S. 1990 Intermittent vortex structures in homogeneous isotropic turbulence. *Nature* **344**, 226.

Ting, L. & Klein, R. 1991 *Viscous Vortical Flows*. Springer.

Van Dyke, M. 1982 *An Album of Fluid Motion*, Parabolic Press, Stanford, CA.

Widnall, S. 1975 The structure and dynamics of vortex filaments. *Ann. Rev. Fluid Mech.* **7**, 141–165.

Zakharov, V.E. 1988 Wave collapse. *Usp. Fiz. Nauk* **155**, 529–533.

COURANT INSTITUTE OF MATHEMATICAL SCIENCES, NEW YORK UNIVERSITY, NEW YORK, NEW YORK 10012

Proceedings of Symposia in Applied Mathematics
Volume **54**, 1998

Homoclinic Orbits for Pde's

David W. McLaughlin and Jalal Shatah

ABSTRACT. In this lecture, the construction of whiskered tori and homoclinic
orbits is summarized for the sine-Gordon equation as an example of such constructions for completely integrable nonlinear wave equations. The construction uses Bäcklund tranformations which are realized through Lax Pairs. The
persistence of these homoclinic orbits under small perturbations of the wave
equations, of both dissipative and conservative type, is then established. An
analytic perturbation method based on time dependent scattering theory, and
Fredholm theory, is used to establish persistence. The estimates are given in
space-time function spaces, with a certain time decay, which is required for
the existence of a homoclinic orbit.

CONTENTS

1. Introduction

Nonlinear dispersive waves appear throughout a wide class of applications –
including man-made events such as laser beams propagating through nonlinear media, as well as naturally occuring phenomena such as waves in the atmosphere
and oceans. In such applications chaotic, random, and turbulent waves are both
common and important. Even in the absence of randomness and turbulence, deterministic waves can behave chaotically in time, a fact which has the important
consequence of introducing intrinsic and fundamental limitations to predicability
for these systems. Such chaotic systems share mathematical properties such as
sensitivity to initial data, the presence of many instabilities, the existence and persistence of homoclinic orbits, symbol dynamics and strange attractors – the collection of which is described by the modern theory of applied dynamical systems [**22**],

1991 *Mathematics Subject Classification.* Primary ; Secondary .
Funded in part by AFOSR-90-0161 and by NSF DMS 8922717 A01.
Funded in part by by NSF DMS 9401558.

[6], [28]. Usually the mathematical setting for these studies is finite dimensional; however, waves satisfy partial differential equations which are infinite dimensional dynamical systems [27].

There is a class of nonlinear wave equations, known as "soliton equations", which provides a convenient arena in which to study examples of chaotic behavior in infinite dimensional dynamical systems. Examples of such soliton equations include the sine-Gordon equation, the cubically nonlinear Schrödinger equation, and the modified Korteweg-deVries equation. Under periodic boundary conditions, and in the absence of perturbations, solutions of these soliton equations for generic initial data are waves which are very regular (almost periodic) in time [24], [16], a fact which is understood mathematically through the realization that these equations are completely integrable Hamiltonian systems in infinite dimensions. However, some of the soliton equations (including those listed above) admit solutions for very special initial data which are *homoclinic* in time [3], [19]. In these situations, perturbations of the equations (such as small damping and driving terms) produce waves which, as observed numerically, possess very irregular temporal behavior. (See, for example, the review [20].) Thus, near-integrable soliton equations provide natural candidates for a mathematical study of chaotic behavior in nonlinear dispersive waves; that is, for a study of one class of chaotic behavior in infinite dimensional dynamical systems.

We believe that chaotic behavior for this class of near-integrable wave equations can be described and characterized with mathematical precision. In this lecture we describe some initial steps in this direction – namely (i) how integrable theory can be used to classify all instabilities and to provide representations of unstable manifolds and homoclinic orbits, and (ii) how Lyapanov Schmidt methods, together with scattering theory, can be used to establish persistence of homoclinic orbits. Thus, two steps which traditionally initiate studies of chaotic behavior in finite dimensional dynamical systems have been extended to the infinite dimensional setting of near integrable waves.

We focus most of the lecture on the sine-Gordon equation, and perturbations thereof:

$$(1.1) \qquad u_{tt} - c^2 u_{xx} - \sin u \; = \; \epsilon \left(-2\alpha u_t + g(t;u) \right)$$

under Neumann boundary conditions; ie, 2π periodic and odd about $x = 0$. Here ϵ denotes a small perturbation parameter $0 \le \epsilon << 1$. The perturbation $g(t;u)$ is a smooth nonlinear function such that $g(t,0) = g_u(t,0) = 0$.

When $\epsilon = 0$, the sine-Gordon equation is completely integrable. Usually in partial differential equations one can only study the stability of a few very simple solutions; however, for these integrable wave equations, a representation due to Peter Lax [13], known as the "Lax pair", enables one to characterize *all* instabilities for *any* solution. Moreover, a Bäcklund transformation formulated in terms of the Lax pair generates a representation of the unstable and stable manifolds for the unstable solitary waves, as well as representations of their homoclinic orbits [3], [19], [20]. Such explicit representations are very rare, and only available here because of complete integrability. This integrable material is summarized for the sine-Gordon example in Section 2.

What is the fate of these homoclinic orbits as the system is perturbed (for example, as the perturbation parameter ϵ is deformed from 0)? Under which perturbations do the orbits persist, and what mathematical techniques can be used to establish their persistence? These are the topics of Section 3. First, in Section 3.1, we establish persistence of homoclinic orbits for a damped and driven perturbation of equation (1.1). Then in Section 3.2 we consider a simple conservative problem which possesses a homoclinic orbit. We prove that this orbit persist for a large class of non autonomous perturbation. This example illustrates the difficulties which one encounters with conservative perturbations, and how these difficulties can be resolved. In several remarks, we compare these analytical approaches to persistence with the more geometric "shooting methods" of Melnikov, and we also discuss the connection of persistence of homoclinic orbits in the conservative case with the classic problem of the persistence of the sine-Gordon breather.

In our research program, we have been developing mathematical methods to prove the persistence of homoclinic orbits for pde's. In finite dimensional dynamical systems, we distinguish two methods to prove existence of homoclinic orbits: (i) variational principles, and (ii) perturbation methods. For pde's, present variational methods can not be applied due to the presence of infinitely many elliptic directions.

To date, we have adapted from two distinct perturbation methods – (i) Melnikov measurements in [14], [17], [29], and (ii) the analytic methods described in Section 3. Briefly, the analytical methods are formulated in space-time function spaces consisting of functions which decay as $|t| \to \infty$. The procedure is an interesting combination of evolutionary methods initialized at $t = -\infty$, scattering theory to control this initialization, and Lyapanov-Schmidt methods.

Earlier, we studied persistence questions for pde with Melnikov methods [29], [17]. As suggested to us by Jack Hale [7], it would be instructive to compare these Melnikov procedures with the integral equation methods developed by Lin, et. al. Some discussion of this comparison is given in Section 3. For finite dimensional dynamical systems, some references for Melnikov methods include [21], [28], and [6]; while the analytical approach considered by Lin is described in [15], and references therein.

The analytic perturbation method which we use to establish persistence is set in space-time function spaces, which build into the construction that decay in time which is required of a homoclinic orbit. The procedure begins by splitting the perturbed nonlinear wave equation into hyperbolic and elliptic directions. The system which results consists in a nonlinear wave equation coupled to an elliptic equation. The wave equation is then inverted with evolutionary methods, specifically as an initial value problem with data at $t = -\infty$. At this step, one takes what the explicit inverse gives – namely, slowly decaying functions at $t = +\infty$ in the dissipative case, or functions which do not decay and are only bounded in the conservative case. The elliptic equation is then analysed with Lyapanov-Schmidt methods. All appropriate inverses are either given explicitly, or controlled with time dependent scattering theory, depending on the specific case. Finally, a contraction mapping argument in parameter space produces the persistent homoclinic orbit. In the dissipative case, only a finite number of parameters are needed, whereas in the conservative case, a countable sequence of parameters must be adjusted by the contraction argument. This adjustment can be expressed in terms of the vanishing of a certain function of the parameters. The fact that this vanishing is a necessary condition is rather immediate [2] [23]. In this manuscript we establish the sufficiency under nondegeneracy conditions.

2. Integrable Homoclinic Orbits and Lax Pairs

In this section, we consider the sine-Gordon equation,

$$(2.1) \qquad u_{tt} - c^2 u_{xx} - \sin u = 0.$$

This equation is a completely integrable Hamiltonian system [11], [26], [1], a fact which is established from the realization that it is equivalent to the over determined linear system

$$(2.2a) \qquad \left[-ci\sigma_2 \frac{d}{dx} + \frac{iw}{4}\sigma_1 - \frac{1}{16\lambda} e^{i\sigma_3 u} - \lambda I \right] \vec{\psi} = \vec{0},$$

$$(2.2b) \qquad \left[-i\sigma_2 \frac{d}{dt} + \frac{iw}{4}\sigma_1 + \frac{1}{16\lambda} e^{i\sigma_3 u} - \lambda I \right] \vec{\psi} = \vec{0},$$

where $w := cu_x + u_t$, I denotes the identity matrix, and σ_k denote the Pauli matrices. This system is the *Lax pair representation* of the sine-Gordon equation and, as two equations for one unknown $\vec{\psi}$, is overdetermined and has a solution if and only if $\partial_x \partial_t \vec{\psi} = \partial_t \partial_x \vec{\psi}$; that is, if and only if the coefficient $u(x, t)$ satisfies the sine-Gordon equation (2.1). The complete integrability of the sine-Gordon equation, with all of it's miracles, mysteries, and remarkable properties, follow from this observation. A detailed description of this integrable material may be found in [26], [4], [5], [16]. Here we just briefly summarize the use of Lax pairs to classify all instabilities, and to represent their unstable manifolds and homoclinic orbits.

First, we impose periodic boundary conditions,

$$u(x + 2\pi, t) = u(x, t),$$

and consider the first member of the Lax pair (2.2a) as an eigenvalue problem, with the complex parameter λ serving the role as the eigenvalue.

Since its coefficients are periodic in x, this Sturm-Louisville problem is a Floquet spectral problem. As $u(x, t)$ satisfies the sine-Gordon equation, this Floquet spectrum in invariant in t, an invariance which provides a sufficient (countable) number of constants of the motion to make the equation completely integrable. Thus, one expects that the generic level sets of these invariants will be (infinite dimensional) tori,

$$u \in \cdots \times S \times S \times S \times S \times \cdots,$$

and that the solutions $u(x, t)$ will be almost periodic in time [16]. Nongeneric level sets will contain singular tori which arise when one or more of the circles S are "pinched away" [3], [19]. Such singular tori can be unstable with respect to the sine-Gordon flow.

The Floquet spectral theory of (2.2a) characterizes these tori. For generic u, the periodic and antiperiodic eigenvalues of (2.2a) are simple; however, for special u, these eigenvalues can be multiple. To each multiple eigenvalue, a circle S has been "pinched away"; and this correspondence is one-to-one. Thus, multiple eigenvalues of (2.2a) characterize singular degenerate level sets.

Lax pairs can be used to characterize unstable singular tori. In fact, they could be used for a complete stability study; however, most work to date has focused upon the identification of some instabilities. Fix $u(x, t)$, a solution of the sine-Gordon equation on a degenerate torus. The (linear) stability of $u(x,t)$ will be determined from the behavior of the linearization

(2.3) $r_{tt} - c^2 r_{xx} - \cos u(x,t)\, r \; = \; 0.$

The connection of Lax pairs with this linear equation (2.3) is as follows [3]:

THEOREM 2.1. *Fix a solution $u(x,t)$ of the sine-Gordon equation (2.1) and $\lambda \in \mathbb{C}$, and consider any solution $\vec{\psi}(x,t;u,\lambda)$ of the Lax Pair (2.2b) at (u,λ). Then*

$$r(x,t;\lambda) := -i\psi_1\,\psi_2$$

satisfies the linear equation (2.3).

Thus, the Lax pair provides a large collection (by adjusting $\lambda \in \mathbb{C}$) of solutions of the linearization, with which one can characterize instabilities of $u(x,t)$. From these rather explicit linearized solutions, one learns that all exponential instabilities of $u(x,t)$ are associated with complex valued multiple points in the spectrum of (2.2a). Thus, to be unstable, $u(x,t)$ must lie on a critical level set [ie, it must have multiple points in the spectrum of (2.2a)] at u, and at least one of these multiple points must be complex. In this situation, one must verify that the linearized solution labeled by this complex multiple point indeed has exponential growth in time, and this verification has been carried out on many examples.

In this manner, the Lax pair representation can be used to identify hyperbolic tori; that is, critical tori which are unstable. Moreover, as we will now sumarize, Lax pairs generate representations of all orbits homoclinic to these hyperbolic tori, and hence representations of their stable and unstable manifolds. These representations for the sine-Gordon equation are summarized in the following [3]

THEOREM 2.2. *Fix a solution $u(x,t)$ of the sine-Gordon equation which lies on a torus T, with some hyperbolic directions, and an instability associated with a complex double point ν. Let $\vec{\phi}(x,t;u,\nu)$ denote a solution of the Lax pair (2.2b) at (u,ν). Then*

(2.4) $$U(x,t) := u(x,t)\; +\; 2i\ln\left(\frac{i\phi_2}{\phi_1}\right)$$

is also a solution of the sine-Gordon equation, which is 2π periodic in x and which does not lie on the torus T; rather, it is homoclinic to that torus in that it approaches T asymptotically as $|t| \to \infty$, at an exponential rate. In addition, there exists an explicit linear transformation which maps solutions $\vec{\psi}$ of the Lax pair at (u,λ) to solutions $\vec{\Psi}$ at (U,λ).

The proof of this theorem is one of direct verification [3]. (If the double point ν lies in the open first quadrant, one must iterate the transformation, first at ν and then at $-\nu^*$, in order to ensure that a real valued homoclinic orbit results.)

With these formulas, one generates orbits which are homoclinic to degenerate target tori with rather complicated spatial and temporal structure, in which case the homoclinic orbits themselves possess still more complicated spatial and temporal structure. These tori are called "whiskered tori", with the whiskers denoting the stable and unstable manifolds. When the target tori are very simple, the homoclinic orbits also simplify. For this lecture, only the simplist examples will be needed.

The saddle fixed point $u(x,t) = 0$ is the simplist degenerate torus, and indeed the most degenerate. For $c^2 > 1$, under 2π periodic boundary conditions which are *even* about $x = 0$, its unstable manifold is one dimensional with the instability associated with the multiple point at $\lambda_d = i\pi/2$. In this case the formula (2.4) generates the pendulum separatrix:

$$(2.5) \qquad\qquad U_H(x,t) = \pm 4\arctan(\exp t).$$

For $1/4 < c^2 < 1$, under 2π periodic boundary conditions which are *odd* about $x = 0$, the unstable manifold of $u = 0$ is also one dimensional. In this case iterated transformations at $\lambda_d = \exp i\nu/4$ and $-\lambda_d^*$ produce the homoclinic orbit

$$(2.6) \qquad\qquad U_H(x,t) = \pm 4\arctan\left(\tan\nu\frac{\cos(\cos\nu)x}{\cosh(\sin\nu)t}\right).$$

The latter homoclinic orbit is related to a famous soliton solution known as the "breather". Usually, the sine-Gordon equation is studied in the form

$$u_{TT} - c^2 u_{XX} + \sin u = 0,$$

in which case the inverted pendulum position $u = \pi$ is unstable. In this setting, the "breather" is a well known solution which is exponentially localized in X and periodic in T, and can be interpreted as the bound state of a "kink" soliton and an "antikink" soliton. It describes a "particle-like" localized wave, with an internal (periodic) degree of freedom. Upon interchanging X and T,

$$x = cT; \quad t = \frac{X}{c},$$

the usual form of the equation reduces to (2.1), and it's breather solution becomes the homoclinic orbit (2.6). Thus, the problem of persistence of this homoclinic orbit is equivalent that of persistence of the breather.

3. Persistence of Homoclinic Orbits

In the integrable setting, Lax pairs have been used to classify all instabilities, and to construct representations of their homoclinic orbits. The next question is the fate of these objects under perturbation. Do they persist in the presence of perturbations, either damped-driven or conservative? This is the topic of this section, first in the dissipative case and then for a conservative example.

3.1. The Dissipative Case. We study the persistence of homoclinic orbits for the nonlinear wave equation

$$(3.1) \qquad\qquad u_{tt} - c^2 u_{xx} - \sin u = \epsilon\left(-2u_t + g(t; u)\right),$$

under Neumann boundary conditions; ie, 2π periodic and odd about $x = 0$. Here ϵ is a dissipation perturbation parameter $0 \leq \epsilon << 1$, and the constant speed c is restricted to $\frac{1}{4} < c^2 < 1$. The perturbation $g(t; u)$ is a smooth nonlinear function satisfying

$$g(t; u) = O(u^3), \quad u \to 0.$$

($g(\cdot, u)$ can be of order u^2; however, this would have made the subsequent arguments somewhat more involved.)

We denote by H the Sobolev space of functions of x which are 2π periodic, odd, and square integrable with square integrable first derivative. Energy methods [25] show that the Cauchy initial value problem for the wave equation (3.1) is well posed in $C(\mathbb{R}, H) \cap C^1(\mathbb{R}, L_x^2)$.

3.1.1. *Unperturbed Homoclinic Orbits.* As described in Section 2, integrable methods provide homoclinic orbits for the unperturbed ($\epsilon = 0$) case. One of the simplist of these is given by the "breather":

(3.2)
$$b(x,t) = 4 \arctan \frac{\sigma \sin x}{c \cosh \sigma t},$$

where $\sigma := \sqrt{1 - c^2}$. Note that $b(x,t)$ is a solution of the sine-Gordon equation which is 2π periodic and odd in x, and which approaches $u = 0$ exponentially as $t \to \pm\infty$:

(3.3)
$$b(x,t) = O[\exp(-\sigma|t|)], \quad t \to \pm\infty.$$

3.1.2. *Set-up for Persistence.* We seek a solution of equation (3.1) in the form

(3.4)
$$u(x,t) = b(x, t - \alpha) + r(x, t; \epsilon).$$

By translating time, the equation for r takes the form

(3.5) $$r_{tt} - c^2 r_{xx} - r + 2\epsilon r_t + m(x, t; \epsilon, \alpha) r = \epsilon F(x, t; \alpha) + G(x, t, r; \alpha, \epsilon),$$

where

(3.6)
$$
\begin{aligned}
m &:= 1 - \cos b - \epsilon g'(t + \alpha, b) = O[\exp(-2\sigma|t|)] \\
F &:= -2b_t + g(t + \alpha, b) = O[\exp(-\sigma|t|)] \\
G &:= \epsilon[g(t + \alpha, b + r) - g(t + \alpha, b) - g'(t + \alpha, b)r] \\
&\quad + \sin b(\cos r - 1) + \cos b(\sin r - r) \\
&= O[\exp(-\sigma|t|)r^2] + O[r^3]
\end{aligned}
$$

If we can prove the existence of an $r(x, t; \epsilon, \alpha)$, which solves the nonlinear equation (3.4) and decays as $|t| \to \infty$, we will have established the persistence of the homoclinic orbit.

3.1.3. *Properties of the Linearization.* To study the nonlinear boundary value problem (3.5), we start by solving the linear equation

(3.7)
$$\mathcal{L}r := r_{tt} - c^2 r_{xx} - r + 2\epsilon r_t + mr = f(x, t)$$

where $f \in L^1_t(L^2_x)$. In order to invert $\mathcal{L}$ under the given boundary conditions, i.e. odd, 2π periodic in x and $r \to 0$ as $t \to \pm\infty$, we have to distinguish the hyperbolic directions from the elliptic directions. For $\epsilon = 0$, and as $t \to \pm\infty$, equation (3.7) approaches

(3.8)
$$r_{tt} - c^2 r_{xx} - r = 0,$$

which, for $c^2 \in \left(\frac{1}{4}, 1\right)$, has one hyperbolic direction with an exponential growth rate $\sigma = \pm\sqrt{1 - c^2}$, and codimension one center directions with frequency $\omega_k = \sqrt{c^2 k^2 - 1}$ for $k = 2, 3, \ldots$. Therefore to invert $\mathcal{L}$ we introduce the projections

$$
\begin{aligned}
Qr &:= a = (r, \sin x) \sin x \\
Pr &:= v = (I - Q)r,
\end{aligned}
$$

and consider equation (3.7) as a coupled system of equations,

(3.9a)
$$Lv + Ma = f_1 = Pf$$

(3.9b)
$$M^*v + Na = f_2 = Qf$$

where

$$L := P\mathcal{L}P$$
$$M := P\mathcal{L}Q = PmQ$$
$$N := Q\mathcal{L}Q = D + QmQ$$

The operator $D = \partial_t^2 + 2\epsilon\partial_t - \sigma^2$ is a perturbation of a strongly elliptic operator $D_0 = \partial_t^2 - \sigma^2$ which has a well defined inverse under the given boundary conditions, while the operator L is a perturbation of the wave operator $L_0 := \partial_t^2 - c^2\partial_x^2 - 1$ restricted to PH. To define the inverse of L, we solve the Cauchy problem

$$(3.10) \qquad Lv = (\partial_t^2 - c^2\partial_x^2 + 2\epsilon\partial_t - 1 + PmP)v = g,$$

with zero data given at $t = -\infty$.

PROPOSITION 3.1. *Given* $g \in L_t^1(PL_x^2)$ *and* ϵ *sufficiently small, equation (3.10) has a unique finite energy solution* v *such that*

$$|v(t)|_e \le c \int_{-\infty}^t |g(s)|_{L_x^2}\,ds.$$

This solution v *is given by*

$$v(t) = L^{-1}g = \int_{-\infty}^t K(t,s)g(s)\,ds,$$

where K *is the solution operator of the Cauchy problem*

$$Lw = 0, \quad w(s,s) = 0, \quad \partial_t w(s,s) = w_1 \in PL_x^2.$$

Moreover v *satisfies the estimate*

$$|v|_{L_t^3(PH)} \le c\epsilon^{-\frac{1}{3}}|g|_{L_t^1(L_x^2)}$$

PROOF. For ϵ sufficiently small, $|e^{\epsilon t}m(x,t)| \le ce^{-\frac{\sigma}{2}t}$. By standard energy estimates solutions $\tilde{v}$ for the homogenous equation satisfy

$$(3.11) \qquad |\tilde{v}(t)|_e \le ce^{-\epsilon(t-s)}|\tilde{v}(s)|_e,$$

where c in independent of ϵ. A finite energy solution v of equation (3.10) can be represented by

$$v(t) = \tilde{v}(t,t_1) + \int_{t_1}^t K(t,s)g(s)\,ds,$$

where by (3.11)

$$|\tilde{v}(t,t_1)|_e \le ce^{-\epsilon(t-t_1)}|v(t_1)|_e$$
$$K(t,s) : PL_x^2 \to PH, \qquad \|K(t,s)\| \le ce^{-\epsilon(t-s)}.$$

For any solution $v(t) \to 0$ as $t \to -\infty$, we can let $t_1 \to -\infty$ to obtain

$$(3.12) \qquad v(t) = \int_{-\infty}^t K(t,s)g(s)\,ds.$$

This implies uniqueness of solutions that vanish as $t \to -\infty$.

Applying energy estimates to (3.12), we obtain

$$|v(t)|_H \le c \int_{-\infty}^t e^{-\epsilon(t-s)}|g(s)|_{L_x^2}\,ds,$$

and by Hausdorff-Young inequality we obtain

$$|v|_{L_t^3(H)} \le c\epsilon^{-\frac{1}{3}}|g|_{L_t^1(L_x^2)}.$$

□   □

From this proposition it is easy to obtain

COROLLARY 3.1. *Given $e^{a|t|}g \in L^1(L_x^2)$ for any $a > 0$, then v is smooth with respect to ϵ and*

$$|\partial_\epsilon v(t)|_e \le ce^{\frac{a}{2}t}\int_{-\infty}^t e^{-a\tau}|g(\tau)|_{L_x^2}\,d\tau.$$

Using these estimates on L^{-1} we can solve (3.7) by first solving equation (3.9a) for v

$$(3.13) \qquad\qquad v = L^{-1}f_1 - L^{-1}Ma$$

and substituting this expression in (3.9b) to obtain

$$(3.14) \qquad\qquad (D + \tilde{M})a = f_3 := f_2 - M^*L^{-1}f_1.$$

where $\tilde{M} = QmQ - M^*L^{-1}M$. Note that since M decays exponentially, then by corollary 3.1 $\tilde{M}$ is smooth in ϵ.

PROPOSITION 3.2. *The operator $\mathcal{F} = 1 + D^{-1}\tilde{M}$ acting on $L_t^2 \otimes \sin x$ is a Fredholm operator of index 0 and has a one dimensional null space spanned by $\chi = Q\dot{b} + O(\epsilon)$.*

PROOF. Since $\tilde{M}$ is exponentially decaying and D^{-1} maps $L^2(\mathbb{R})$ onto $H^2(\mathbb{R})$, we obtain that $D^{-1}\tilde{M}$ is compact on $L_t^2 \otimes \sin x$, and consequently $\mathcal{F}$ is Fredholm.

The null space of $\mathcal{F}$ can be computed by finding the null space for $\epsilon = 0$ of $\mathcal{F}_0 := \mathcal{F}|_{\epsilon=0}$. By scattering theory, $\dot{b}$ is the only solution of $\mathcal{L}_0 r = 0$ such that $|r(t)|_e \to 0$ as $t \to -\infty$. From proposition 3.2 we conclude that

$$P\dot{b} = -L_0^{-1}M_0Q\dot{b},$$

and that $Q\dot{b}$ is the only solution in the L^2 null space of the unperturbed operator $\mathcal{F}_0$. Since the operator $\mathcal{F}$ is compact and smooth in ϵ we obtain from regular perturbation theory that the null space of $\mathcal{F}$ is one dimensional and is spanned by $\chi = Q\dot{b} + O(\epsilon)$. Finally it is easy to see that the null space of the adjoint is spanned by $\psi = D_0Q\dot{b} + O(\epsilon)$ since

$$(L_0^{-1})^*g(t) = \int_\infty^t K_0(t,s)g(s)\,ds$$

and

$$P\dot{b} = -(L_0^{-1})^*M_0Q\dot{b}.$$

□   □

From propositions 3.1 and 3.2 we deduce that $\mathcal{L}$ has a one dimensional null space on L^2 spanned by

$$\varphi = \chi - L^{-1}M\chi = \dot{b} + O(\epsilon)$$

and that $\mathcal{L}^*$ has a one dimensional null space on L^2 spanned by

$$\phi = \varphi + O(\epsilon) = \dot{b} + O(\epsilon)$$

Moreover since L is invertible and $\mathcal{F}$ is Fredholm, we can state the Fredholm alternative as

THEOREM 3.1. *Given $f \in L_t^1(L_x^2)$, equation (3.7),*

$$\mathcal{L}r = f,$$

has a solution $r \in L_t^\infty(H) \cap L_t^3(H)$ if and only if

$$\langle f, \phi \rangle = \langle f, \dot{b} \rangle + O(\epsilon) = 0,$$

where

$$\langle f, g \rangle := \int_{-\infty}^{+\infty} \int_0^{2\pi} fg \, dx \, dt.$$

Moreover there is a unique solution $r := \mathcal{L}^{-1} f$ that satisfies

$$\langle r, \varphi \rangle = 0$$

$$|r(t)|_e \le c \int_{-\infty}^t |f(s)|_{L_x^2} \, ds$$

$$|r|_{L_t^3(H)} \le c\epsilon^{-\frac{1}{3}} |f|_{L_t^1(L_x^2)}.$$

3.1.4. *The Nonlinear Problem.* With these estimates for the linear equation, we can solve nonlinear problem (3.5) by first solving a modified equation

$$(3.15) \qquad \mathcal{L}r = \epsilon F + G - \frac{\langle \epsilon F + G, \phi \rangle}{\langle \phi, \phi \rangle} \phi,$$

PROPOSITION 3.3. *Given ϵ sufficiently small, equation (3.15) has a unique solution $r^*(\epsilon, \alpha) \in L_t^\infty(H) \cap L_t^3(H)$ such that $\langle r^*, \varphi \rangle = 0$ and*

$$|r^*(t; \epsilon, \alpha)|_e \le c\epsilon$$

$$|r^*(\epsilon, \alpha)|_{L_t^3(H)} \le c\epsilon^{\frac{2}{3}}.$$

Moreover r^* is lipschitz continuous in α, $|r^*(\epsilon, \alpha_1) - r^*(\epsilon, \alpha_2)|_{L_t^3(H)} \le c\epsilon^{\frac{2}{3}} |\alpha_1 - \alpha_2|$.

PROOF. Since the right hand side of equation (3.15) is orthogonal to ϕ we can rewrite the equation as

$$r = \mathcal{K}(r; \epsilon, \alpha) := \mathcal{L}^{-1}[\epsilon F + G - \langle \epsilon F + G, \phi \rangle \phi / \langle \phi, \phi \rangle]$$

From theorem 3.1 we have

- $|\mathcal{K}(r; \epsilon, \alpha)|_{L_t^3(H)} \le c\epsilon^{-\frac{1}{3}} [\epsilon + c|r|_{L_t^3(H)}^2]$
- $\mathcal{K}(.; \epsilon, \alpha)$ maps a ball of radius $|r| \le c\epsilon^{\frac{2}{3}}$ into itself .
- $|\mathcal{K}(r_1; \epsilon, \alpha_1) - \mathcal{K}(r_2; \epsilon, \alpha_2)|_{L_t^3(H)} \le c[\epsilon^{\frac{1}{3}} |r_1 - r_2|_{L_t^3(H)} + \epsilon^{\frac{2}{3}} |\alpha_1 - \alpha_2|]$, for $|r| \le c\epsilon^{\frac{2}{3}}$

Therefore by a contraction mapping argument $\mathcal{K}$ has a unique fixed point in the ball $|r| \le c\epsilon^{\frac{2}{3}}$, and by theorem 3.1 this fixed point satisfies the stated estimates. $\square$ $\qquad\qquad\qquad\qquad\qquad\qquad\qquad\qquad\qquad\qquad\qquad\qquad\qquad \square$

REMARK 3.2. This fixed point is not smooth in ϵ in the space $L_t^\infty(H)$ or $L_t^3(H)$. However from corollary 3.1 it follows that $r(t, \epsilon)$ is differentiable in ϵ and $|\partial_\epsilon r(t, \epsilon)| \le c \exp[\frac{\sigma}{2}t]$.

THEOREM 3.3. *Suppose that*

$$\langle F, \dot{b} \rangle(\alpha) = -2\langle \dot{b}, \dot{b} \rangle + \langle g(\cdot + \alpha, b), \dot{b} \rangle$$

has a simple zero at $\alpha = \alpha_0$. *Then equation (3.5) has a solution* $r^*(\epsilon, \alpha^*) \in L_t^\infty(H) \cap L_t^3(H)$ *for some* $\alpha^* = \alpha_0 + O(\epsilon)$ *such that*

$$|r^*(t; \epsilon, \alpha^*)|_e \le c\epsilon$$

$$|r^*(\epsilon, \alpha^*)|_{L_t^3(H)} \le c\epsilon^{\frac{2}{3}}$$

PROOF. From proposition 3.3 we have a solution to the modified equation (3.15) $r^*(\epsilon, \alpha) \in L_t^\infty(H) \cap L_t^3(H)$, such that $|r^*|_{L_t^\infty(H)} \le c\epsilon$. Moreover since $G = O(e^{-\sigma|t|}r^2 + r^3)$ then $\langle G, \phi \rangle = O(\epsilon^2)$. Therefore by a contraction mappling argument

$$\langle F, \phi \rangle + \frac{1}{\epsilon}\langle G, \phi \rangle = \langle F, \dot{b} \rangle + \mathcal{R}(\epsilon, \alpha) = 0$$

where $\mathcal{R}$ is Lipschitz continuous in α with Lipschitz constant $O(\epsilon)$, has a solution $\alpha^* = \alpha_0 + O(\epsilon)$.   □   □

REMARK 3.4. In this dissipative case, the persistence of the homoclinic orbit under the dissipative perturbation (3.1) could have been obtained by "Melnikov arguments". For our example, these methods [9], [8] are similar to, but easier than, those used in [29] for a damped, driven NLS equation. In this remark, we briefly outline this alternative argument: First, one notes that in the unperturbed case, 0 is an unstable fixed whose unstable manifold $W_{\epsilon=0}^u$ is one dimensional, while its center-stable manifold $W_{\epsilon=0}^{cs}$ is codimension 1 with a single exponentially stable direction. For $\epsilon > 0$, zero persists as a solution, and its stable and unstable manifolds persist under the perturbation, and become W_ϵ^u and W_ϵ^{cs}. In particular, the persistent center-stable manifold W_ϵ^{cs} is of size $O(\sqrt{\epsilon})$ in the many "slow directions" and $O(1)$ in the single "fast direction". The persistence and size of these manifolds are obtained by an implicit function theorem argument, with the $O(\sqrt{\epsilon})$ estimate in the slow direction following from the cubic nonlinearity. For finite T, the distance between the perturbed orbit $b^\epsilon(\cdot) = W_\epsilon^u$ and the unperturbed breather $b(x, t)$ is $O(\epsilon)$:

$$\text{dist}[b^\epsilon(t), b(t)] = O(\epsilon), \quad -\infty < t \le T.$$

Because of the estimate of the size of W_ϵ^{cs}, a "shooting argument" shows

$$\text{dist}[b^\epsilon, W_\epsilon^{cs}] = \epsilon\langle F, \dot{b} \rangle(\alpha) + o(\epsilon),$$

where in this setup $\langle F, \dot{b} \rangle$ is referred to as the Melnikov function.

Finally, if the Melnikov function has a simple zero, an implicit function theorem argument, this time in parameter space, shows that one can solve

$$\text{dist}[b^\epsilon, W_\epsilon^{cs}] = \epsilon\langle F, \dot{b} \rangle(\alpha) + o(\epsilon) = 0,$$

for α; thus, a persistent homoclinic orbit exists for this value of the parameter α.

Two points should be emphasized:

1. The vanishing of the Melnikov distance is the same as the Fredholm alternative condition of Theorem 3.3.

2. This Melnikov argument requires two types of implicit function theorems, one for the persistence (and size) of the manifolds, and a second in parameter space. The analytical methods used in this manuscript also require two distinct types of implicit function theorems.

Finally, since the center-stable manifold W_ϵ^{cs} is codimension 1, a single measurement (of a single parameter α) suffices to determine if the perturbed orbit $b^\epsilon(t)$ intersects W_ϵ^{cs}. However, for conservative perturbations in the absence of dissipation, the "target" is the stable manifold of zero, which is only one dimensional. An *infinite number* of measurements will be required to determine if the unstable manifold W_ϵ^u intersects this stable manifold W_ϵ^u. The geometry required for these Melnikov measurements will be considerably more difficult to develop than in the dissipative case. However, explicit formulas for these measurements are rather natural from the viewpoint of the analytical setup of this manuscript – as we will demonstrate in the next section for a simple conservative example.

3.2. A Conservative Example. In the conservative case, the absence of dissipation presents significant difficulties of a geometric, as well as a technical, nature. In our opinion, the advantages of a this analytical approach only become really apparent when overcoming these difficulties in the conservative case. Here, we restrict to a very simple conservative example which illustrates the essence of the method, without many of the technical complications. For a more realistic problems, we refer the reader to [**18**].

Consider

$$(3.16) \qquad u_{tt} - u_{xx} - m^2 u + 2u^3 = \epsilon[f(x,t) + g(x,t;u;\lambda)]$$

where $m^2 < 1$, f and g are smooth 2π periodic even functions in both x and t, and g a smooth function of $\lambda \in \ell_1^2$ ($|\lambda|_{\ell_1^2}^2 = \sum_k k^2 \lambda_k^2$). We restrict our attention to solutions $u(x,t)$ which are 2π *periodic* in x, and *even* in both x and t, and all function spaces in this section are restricted to even functions.

For $\epsilon = 0$ the zero solution is an unstable fixed point and the restriction $m^2 < 1$ ensures that zero has one dimensional stable and unstable manifolds (consisting of x-independent functions) and a codimension 2 center manifold. Moreover, the stable and unstable manifolds coincide, and produce an orbit homoclinic to the origin

$$(3.17) \qquad\qquad b = m\,\mathrm{sech}\,mt.$$

For $\epsilon > 0$ we will show that the zero solution will perturb to a periodic orbit $\epsilon p(x,t;\epsilon)$ and b will perturb to an orbit homoclinic to $\epsilon p(x,t;\epsilon)$.

3.2.1. *The Linearized equation: Conservative Case.* Linearizing about the homoclinic orbit b yields

$$(3.18) \qquad\qquad r_{tt} - r_{xx} - m^2 r + 6m^2 \mathrm{sech}^2 mt\, r = 0.$$

The function $\dot{b}(t)$ provides the only solution of equation (3.18) which decays exponentially as $t \to \pm\infty$, but is not relevant since it is odd in t. By Fourier expansion we (see [**12**]) obtain that for $k \geq 1$, all solutions can be written as linear combinations of $\{R_{1k}, R_{2k}, k = 1, 2, \cdots\}$:

$$(3.19) \qquad\qquad R_{ik}(x,t) = \hat{R}_{ik}(t)\,\cos kx,$$

where

$$\hat{R}_{1k}(t) = \frac{\omega_k^2 + m^2 - 3m^2\,\tanh^2 t}{\omega^2 + m^2}\cos\omega_k t - \frac{3\omega_k m\,\tanh t}{\omega^2 + m^2}\sin\omega_k t,$$

and where $\omega_k^2 = k^2 - m^2$. There is a similar expression for the odd function $\hat{R}_{2k}(t)$. We refer to $R_{ik}(x,t)$ as *radiation states* because they are only bounded, with no decay as $t \to \pm\infty$. Clearly they are associated with "center directions" for the tangent space of the fixed point at the origin.

As in the dissipative case, we study the linear problem:

(3.20) $$\mathcal{L}r := r_{tt} - r_{xx} - m^2 r + 6b^2(t)r = f(x,t),$$

where $f \in L_t^1(L_x^2)$. To invert $\mathcal{L}$ we introduce projections

$$Qr := a \;=\; \frac{1}{2\pi}\int_0^{2\pi} r(x,t)dx$$
$$Pr := v \;=\; (I - Q)r,$$

and consider equation (3.20) as a pair of equations,

(3.21a) $$Lv = f_1 = Pf$$
(3.21b) $$Da = f_2 = Qf$$

where

$$L := P\mathcal{L}P = \partial_t^2 - \partial_x^2 - m^2 + 6b^2(t),$$
$$D := Q\mathcal{L}Q = \partial_t^2 - m^2 + 6b^2(t).$$

The operator D is a perturbation of a strongly elliptic operator $D_0 = \partial_t^2 - m^2$ which has a well defined inverse under the given boundary conditions. The operator L is a perturbation of the wave operator $L_0 := \partial_t^2 - \partial_x^2 - m^2$ restricted to PH^1. The inverse of L can be computed explicitly using the radiation states $R_{ik}(x,t)$. From the explicit representation of L^{-1} it is easy to prove the following:

PROPOSITION 3.4. *Any finite energy solution of*

(3.22) $$Lv = 0,$$

$$v(s) \in PH^1$$
$$v_t(s) \in PL_x^2,$$

is uniformly bounded in the energy norm for $t \in \mathbb{R}$. Here the energy norm is given by

$$|v(t)|_e^2 := \int [v_t^2 + v_x^2 - m^2 v^2]\,dx.$$

PROPOSITION 3.5. *Any finite energy solution of*

(3.23) $$Lv = f_1$$

where $f_1 \in L_t^1(PL_x^2)$, is uniformly bounded for $t \in \mathbb{R}$, and is uniquely determined by data v_h at $t = -\infty$. That is,

$$v(t) = v_h(t) + \int_{-\infty}^t K(t,s)f(s)ds,$$

where $v_h(t)$ is a solution of the homogeneous equation $Lv_h = 0$ and the Fourier coefficients of K are

$$K_\ell(t,s) = \hat{R}_{1\ell}(s)\hat{R}_{2\ell}(t) - \hat{R}_{2\ell}(s)\hat{R}_{1\ell}(t).$$

This proposition immediately leads to the following corollary about solutions with vanishing data at $t = -\infty$:

COROLLARY 3.2. *Given* $f_1 \in L^1_t(PL^2_x)$, *equation (3.23) has a unique finite energy solution* v *such that*

$$|v(t)|_e \le c \int_{-\infty}^{t} |f_1|_{L^2_x} ds.$$

This solution v *is given by*

$$v(t) = L^{-1}f_1 = \int_{-\infty}^{t} K(t,s)f_1(s)ds.$$

Moreover, $v(t)$ *vanishes as* $t \to +\infty$ *if and only if*

(3.24)
$$\int_{-\infty}^{+\infty} K(t,s)f_1(s)ds = 0.$$

We assume that the force f_1 decays exponentially at a rate $\alpha < m$

(3.25)
$$\exp(\alpha|t|)f_1 \in L^\infty_t(L^2_x),$$

and introduce the spaces of decaying functions,

$$L^\infty_\alpha(L^2) = \left\{ u; \ |u|_{L^\infty_\alpha(L^2_x)} = \sup_t e^{\alpha|t|} |u(t)|_{L^2_x} < \infty \right\}$$

$$L^\infty_\alpha(H^1) = \left\{ u; \ |u|_{L^\infty_\alpha(H^1)} = \sup_t e^{\alpha|t|} |u(t)|_{H^1} < \infty \right\}.$$

PROPOSITION 3.6. *Consider the linear equation*

$$Lv = f_1$$

where $f_1 \in L^\infty_\alpha(L^2_x)$ *for some* $0 < \alpha < m$. *Then* $v(t) = L^{-1}f_1$ *satisfies*

$$\exp(\alpha|t|)v \in L^\infty_t(H^1),$$

if and only if

$$\int_{-\infty}^{+\infty} K(t,s)f_1(s)ds = 0.$$

Finally, to find a solution in the space of even functions, of

$$Da = f_2,$$

where $D = \partial_{tt} - m^2 + 6b^2(t)$ and $\exp(\alpha|t|)f_2 \in L^\infty_t$, we note that the only possible obstruction to the inversion of D is the function Qb_t, which is odd in t. Therefore for $0 < \alpha < m$, we obtain that $D^{-1}f_2 \in L^\infty_\alpha$.

We can summarize these results on the linearized equation with the following Fredholm Alternative theorem:

THEOREM 3.5. *Consider the linear equation (3.20)*

$$\mathcal{L}r := r_{tt} - r_{xx} - m^2r + 6b^2(t)r = f(x,t),$$

where the force f *is an even function and satisfies the decay condition*

$$\exp(\alpha|t|)f \in L^\infty_t(L^2_x),$$

for some $0 < \alpha < m$. This equation has a unique solution $r(t)$ which vanishes as $t \to -\infty$:

$$r(t) = \mathcal{L}^{-1} f.$$

The solution satisfies

$$\exp\left(\alpha|t|\right) r \in L_t^\infty(H^1)$$

if and only if the force $f_1 = Pf$ satisfies

(3.26)
$$\int_{-\infty}^{+\infty} K(t,s) f_1(s) ds = 0,$$

which is equivalent to

(3.27)
$$\langle f, R_{1k} \rangle = 0, \quad \forall k = 1, 2, \cdots$$

3.2.2. *Existence of a Homoclinic Orbit.* Under the perturbation given in equation (3.16) the zero solution will perturb to a periodic solution of size ϵ. This can be seen by noting that on the space of even 2π periodic functions $H^1(\mathbb{S}^1 \times \mathbb{S}^1)$ the linear wave equation $W = \partial_t^2 - \partial_x^2 - m^2$ has a symbol $\hat{W}(j,k) = -j^2 + k^2 - m^2$, where $j, k \in \mathbb{Z}$, and

$$|\hat{W}(j,k)| \geq \left||j^2 - k^2| - m^2\right| \geq \min\left(1 - m^2, m^2\right).$$

This implies that W has a bounded inverse on $H^1(\mathbb{S}^1 \times \mathbb{S}^1)$ and

$$u = W^{-1}[\epsilon f + g - 2u^3].$$

By the implicit function theorem we obtain a unique solution

$$u = \epsilon p(x, t; \lambda, \epsilon),$$

where p is smooth in ϵ and

$$\begin{aligned} p(x,t;\lambda,\epsilon) &= p_0(x,t) + O(\epsilon) \\ \left(\partial_t^2 - \partial_x^2 - m^2\right) p_0 &= f(x,t) \\ |p(\cdot,\cdot;\lambda,\epsilon)|_{H^1} &\leq C \end{aligned}$$

where C depends on $|f|_{H^1}, |\lambda|_{\ell_1^2}$, and ϵ.

Linearize equation (3.16) about $b + \epsilon p$

$$u = b + \epsilon p + \epsilon r$$

to obtain

(3.28) $\quad r_{tt} - r_{xx} - m^2 r + 6m^2 \text{sech}^2 mt\, r = -6p_0 m^2 \text{sech}^2 mt + g(x,t;b;\lambda) + \epsilon G$

where G is a smooth function of $r \in H^1, \lambda \in \ell_1^2$, and ϵ:

$$|G(\cdot, t, r; \lambda, \epsilon)|_{H^1} \leq C \left(|r|_{H^1} + (1 + |\lambda|_{\ell_1^2}) e^{-\sigma|t|}\right)$$

where C depends on $|r|_{H^1}, |\lambda|_{\ell_1^2}$, and ϵ.

To show the existence of an orbit homoclinic to ϵp is equivalent to finding a solution r to equation (3.28) which vanishes as $t \to \pm\infty$. This is done by first choosing $\lambda \in \ell_1^2$ such that the hypothesis of theorem 3.5 are satisfied, and then applying a contraction mapping argument to solve equation (3.28).

PROPOSITION 3.7. *Given $r \in L_\alpha^\infty(H^1)$ and $g \in H^1(\mathbb{S}^1 \times \mathbb{S}^1)$ such that*

$$(3.29) \qquad B(\lambda) := \int_{-\infty}^{\infty} \int_0^{2\pi} R(x,t)[-6p_0 m^2 \mathrm{sech}^2 mt + g]\, dx\, dt$$

has a simple zero at λ_0, i.e. the operator $M = \{M_{ik}\}$ is invertible on ℓ_1^2, where

$$M_{ik} := \int_{-\infty}^{\infty} \int_0^{2\pi} R_i(x,s) \frac{\delta g}{\delta \lambda_k}(x,s;b;\lambda_0)(s)\, ds,$$

and where $R = \{R_k\}$ $k = 1, 2, \ldots$. Then for ϵ sufficiently small $\exists \lambda^*(r) \in \ell_1^2$ lipschitz in r, such that

$$(3.30) \qquad \int_{-\infty}^{\infty} K(t,s)[-6p_0 m^2 \mathrm{sech}^2 mt + g + \epsilon G]\, ds = 0.$$

PROOF. From theorem 3.5 the above equation is equivalent to

$$\int_{-\infty}^{\infty} \int_0^{2\pi} R_{1k}(x,s)\left[-6p_0 m^2 \mathrm{sech}^2 mt + g + \epsilon G\right] dx\, ds = 0$$

for $k = 1, 2, \ldots$. The ϵ independent terms in the above equation are

$$B_k(\lambda) = \int_{-\infty}^{\infty} \int_0^{2\pi} R_{1k}[-6p_0 m^2 \mathrm{sech}^2 mt + g]\, dx\, dt$$

which can be written as

$$B(\lambda) = M(\lambda - \lambda_0) + o(\lambda - \lambda_0)$$

where M acts on ℓ_1^2

$$M_{kl} = \int_{-\infty}^{\infty} \int_0^{2\pi} R_{1k}(x,t)\partial_{\lambda_\ell} g(t)\, dx\, dt$$

Since M has a bounded inverse on ℓ_1^2, equation (3.30) can be written as

$$\lambda = \mathcal{F}(r,\lambda,\epsilon) = \lambda_0 + M^{-1}\epsilon \int_{-\infty}^{\infty} \int_0^{2\pi} RG\, dx\, dt + o(\lambda - \lambda_0).$$

Moreover for any function $F \in L_\alpha^\infty(H^1)$ we have

$$\left|\int_{-\infty}^{\infty} \int_0^{2\pi} RF\right|_{\ell_1^2} \le C|F|_{L_\alpha^\infty(H^1)}$$

which implies that for $|r|_{L_\alpha^\infty(H^1)} \le A$

$$|\mathcal{F}|_{\ell_1^2} \le C(A, |\lambda|_{\ell_1^2}, \epsilon)$$
$$|\mathcal{F}(r_1, \lambda_1, \epsilon) - \mathcal{F}(r_2, \lambda_2, \epsilon)|_{\ell_1^2} \le C(A, \lambda_1, \lambda_2)\epsilon[|\lambda_1 - \lambda_2|_{\ell_1^2} + |r_1 - r_2|_{L_\alpha^\infty(H^1)}].$$

By a contraction mapping argument we conclude that $\exists \epsilon_0(A)$ such that for $|\epsilon| < \epsilon_0(A)$ equation (3.30) has a unique solution

$$\lambda_* = \lambda_0 + 0(\epsilon)$$

which is lipschitz in r.   □   □

THEOREM 3.6. *Given ϵ sufficiently small and a g as in proposition 3.3 $\exists \lambda \in \ell_1^2$ such that equation (3.28) has a solution $r \in L_\alpha^\infty(H^1)$ for $0 < \alpha < m$.*

PROOF. By choosing λ as in proposition 3.7 equation (3.28) can be written

$$r = \mathcal{L}^{-1}\mathcal{G}(r, \epsilon)$$

(3.31)
$$\int_{-\infty}^{\infty} \int_0^{2\pi} R_{1k}\mathcal{G}(r, \epsilon) = 0.$$

From theorem 3.5 we obtain for $|r|_{L_\alpha^\infty(H^1)} \leq A$

$$|\mathcal{L}^{-1}\mathcal{G}(r, \epsilon)|_{L_\alpha^\infty(H^1)} \leq C\left(|p_0|_{H^1(\mathbb{S}^1 \times \mathbb{S}^1)} + |\lambda_*|_{\ell_1^2}\right) + \epsilon C(A)$$

$$|\mathcal{L}^{-1}[\mathcal{G}(r_1, \epsilon) - \mathcal{G}(r_2, \epsilon)]|_{L_{\alpha(H^1)}^\infty} \leq C(A, \epsilon)\,|r_1 - r_2|_{L_\alpha^\infty(H^1)}.$$

By choosing $A = 2(|p_0| + |\lambda_*|)$ and $\epsilon C(A, \epsilon) \leq \frac{1}{2}$ we obtain, from a contraction mapping argument, that equation (3.31) has a unique solution inside a ball in $|r|_{L_{\alpha(H^1)}^\infty} \leq A$. $\qquad\square$ $\qquad\square$

REMARK 3.7. This theorem establishes the existence of a class of perturbation for which the homoclinic orbit persist. This class is not empty as the following example shows. Let $g = \left(\sum_\ell \lambda_\ell \cos \ell x \cos \ell t\right) u^2$; in this case the operator M is given as a diagonal matrix

$$M_{k\ell} = \delta_{k,\ell}\left[1 + 0\left(\frac{1}{\ell}\right)\right]$$

which is invertible on ℓ_1^2.

3.2.3. *Comments on the Persistence of the Sine Gordon Breather.* We have also considered the persistence of the breather $b(x, t)$ for the sine-Gordon equation,

$$u_{tt} - c^2 u_{xx} - \sin u = \epsilon[f(x, t) + g(x, t; \lambda; u)],$$

under 2π periodic, odd boundary conditions in space, Here f and g are 2π periodic in both space and time. The argument proceeds as the one described above, but is somewhat more involved. For example, the linear system is coupled. In addition, the radiation states require the use of complete integrability to write them explicitly and the use of time-dependent scattering theory to prove they are complete. The details of these arguments for the sine-Gordon case will be presented in [**18**].

In conclusion we remark that when considering the persistence of breathers, the most natural conservative perturbations are automnous – for example

(3.32)
$$u_{tt} - c^2 u_{xx} - \sin u = \epsilon \sum_2^\infty \lambda_i u^i.$$

In this setting, the analytical theory for persistence reduces to a consideration of equations for the parameters $\{\lambda_i\}$ of the form

(3.33)
$$M(c)\lambda + O(\epsilon\lambda) = 0,$$

where the $\infty \times \infty$ matrix M is of the form

$$M_{k,l}(c) = \langle R_k, b^l \rangle,$$

and $\{R_k\}$ denotes the radiation states for the linearized sine-Gordon equation, linearized about the unperturbed breather $b(x, t)$. For a breather to persist, M cannot be invertible. Unfortunately, it seems difficult to analyse the invertability of M.

Birnir, McKean and Weinstein [2] analyzed (3.33) as follows: They sought nonlinearities of the form (3.32) for which a one parameter family of breathers persist. In our setting, the parameter may be viewed as the speed c, and the nonlinearity parameters $\{\lambda_k\}$ are to be *independent* of c. This assumption gives (3.33) additional structure, and enables those authors to show that (3.33) has no solutions other that trivial symmetries of the sine-Gordon equation. However, their argument does not show that $M(c)$ is invertible. (See also [10].)

A more natural perturbation than (3.32) would envolve u, u_t, and u_x. For example,

$$(3.34) \qquad u_{tt} - c^2 u_{xx} - \sin u = \epsilon \sum_1^\infty \lambda_i u^i + \epsilon g(u, u_t, u_x).$$

In this case, the equations for the parameters would take the form

$$(3.35) \qquad M(c)\lambda + N(c) + O(\epsilon\lambda) = 0,$$

where the "forcing term" $N(c)$ is of the form

$$N_k = \langle R_k, g(b, b_t) \rangle.$$

The "forcing" $g(u, u_t)$ provides extra freedom, and if the matrix M is invertible, the analytical theory would establish the persistence of the breather.

References

[1] M. Ablowitz, D. Kaup, A. Newell, and H. Segur. Method for solving the sine-Gordon equation. *Phys. Rev. Lett.*, 30:1262, 1973.

[2] B. Birnir, H. McKean, and A. Weinstein. The rigidity of sine-gordon breathers. *Comm. Pure Appl. Math.*, 47:1043–1051, 1994.

[3] N. M. Ercolani, M. G. Forest, and D. W. McLaughlin. Geometry of the modulational instability, III: Homoclinic 0rbits for the periodic sine-Gordon equation. *Physica D*, 43:349–84, 1990.

[4] G. Forest and D. McLaughlin. Spectral theory for the periodic sine gordon equation. *J. Math. Phys.*, 23:1248–1277, 1982.

[5] G. Forest and D. McLaughlin. Modulation of sine gordon and sinh gordon wave trains. *Studies in Appl. Math.*, 68:11–59, 1983.

[6] J. Guckenheimer and P. Holmes. Nonlinear Oscillations, Dynamical Systems, and Bifurcations of Vector Fields. *Appl Math Sci*, 42, 1983.

[7] G. Haller. Orbits homoclinic to resonances: The Hamiltonian pde case. in preparation 1997.

[8] P. Holmes. Space and time periodic perturbations of the sine Gordon equation. *Proc of Warrick Conference on Turbulence and Dynamical Systems in Springer Lecture Notes in Mathematics*, 898:164–191, 1981.

[9] P. Holmes and J. Marsden. A partial differential equation with infinitely many periodic orbits and chaotic oscillations of a forced beam. *Archive for Rational Mechanics and Analysis*, 76:135–165, 1981.

[10] S. Kichenassamy. Breather solutions of the nonlinear wave equation. *Comm. Pure Appl. Math.*, 44:789–818, 1991.

[11] G. Lamb. Coherent optical pulse propagation as an inverse problem. *Phys. Rev. A*, 9:422–430, 1974.

[12] G. Lamb. *Elements of Soliton Theory*. John Wiley and Sons, New York, 1980.

[13] P. D. Lax. Integrals of nonlinear equations of evolution and solitary waves. *Comm. Pure Appl. Math.*, 21:467–, 1968.

[14] Y. Li. *Chaotic Behavior in PDE's*. PhD thesis, Princeton University, 1993.

[15] X. B. Lin. Homoclinic bifurcations with weakly expanding center modes. *Dynamics Reported*, 5:99–189, 1995.

[16] H. McKean. The sine gordon and sinh gordon equations on the circle. *Comm. Pure Appl. Math.*, 34:197–257, 1981.

[17] D. McLaughlin and J. Shatah. Melnikov analysis for pde's. *in preparation*, 1995.

[18] D. McLaughlin and J. Shatah. Persistent homoclinic orbits for sine Gordon equation. *in preparation*, 1997.

[19] D.W. McLaughlin. Whiskered tori and chaotic behavior in nonlinear waves. *Proceeding of the International Congress Zurich*, 1995.

[20] D.W. McLaughlin and E. A. Overman. Whiskered tori for integrable pdes and chaotic behavior in near integrable pdes. *Surveys in Appl Math 1*, 1995.

[21] V. Melnikov. On the stability of the center for time periodic oscillations. *Trans. Moscow Math.*, 12:1–57, 1963.

[22] J. Moser. *Stable and Random Motion in Dynamical Systems*. Princeton University Press, 1987.

[23] G. Cruz Pacheco, D. Levermore, and B. Luce. Melnikov methods for pdes with applications to perturbed NLS equations. *Physica D, submitted*, 1996.

[24] J. Poschel and E. Trubowitz. *Inverse Spectral Theory*. Academic Press, New York, 1987.

[25] W. Strauss. Nonlinear invariant wave equations. In G. Velo and A. S. Wightman, editors, *Invariant Wave Equations*, pages 197–249. Springer-Verlag, Berlin, 1978.

[26] L. Takhtajian and L. Faddev. Essentially nonlinear one-dimensiional model of classical field theory. *Theor. Math. Phys.*, 21:1046, 1974.

[27] R. Teman. *Infinite Dimensional Dynamical Systems in Mechanics and Physics*. Springer-Verlag (New York), 1988.

[28] S. Wiggins. *Global Bifurcations and Chaos: Analytical Methods*. Springer-Verlag (New York), 1988.

[29] D. McLaughlin Y. Li, J. Shatah, and S. Wiggins. Persistent homoclinic orbits for a perturbed NLS equation. *Comm. Pure Appl. Math.*, 49:1175–1255, 1996.

COURANT INSTITUTE OF MATHEMATICAL SCIENCES, 251 MERCER STREET, NEW YORK, NY 10012

E-mail address: dmac@cims.nyu.edu

COURANT INSTITUTE OF MATHEMATICAL SCIENCES, 251 MERCER STREET, NEW YORK, NY 10012

E-mail address: shatah@cims.nyu.edu

Proceedings of Symposia in Applied Mathematics
Volume **54**, 1998

Dedicated to Peter Lax and Louis Nirenberg
on the occasion of their 70th birthday

Lagrangian Metrics on Fractals

Umberto Mosco

ABSTRACT. We describe the dynamics of general self-similar structures by relying on a measure-valued Lagrangian formalism. We define an effective metric of Lagrangian nature, that makes the structure a space of homogeneous type in the sense of abstract harmonic analysis, and we show that a single parameter, the homogeneous dimension, governs the scaling of relevant dynamical quantities. The theory applies both to Euclidean media and to large families of fractals.

1. Introduction

Fractals, both of deterministic and statistic nature, have been widely used in recent years, as useful models of "irregular" media in which general physical effects take place, like electrostatic charge distribution, heat diffusion, wave propagation and so on.

The behavior of most fractal structures, however, has been found — by experiment, theory or numerical simulation — to deviate sharply from that of uniform (Euclidean) media.

The discrepancies come to light when we observe how the relevant quantities behave under the natural symmetries of the structure. The basic concept in this description, on fractals possessing dilational symmetry, is that of *scaling*. Assuming isotropic dilational invariance, the scaling of invariant scalar quantities is expressed by a power law at suitable length scales. The anomalous behavior is then revealed by the scaling exponent, when this is found to be different from its Euclidean counterpart and, in general, non-integer.

The aim of the paper is to show that the dynamics of free fractal structures can be conveniently described by adopting a measure-valued invariant Lagrangian formalism, based on the theory of Dirichlet forms. An "effective" invariant metric of Lagrangian nature on the fractal is seen to have a focussing effect on most relevant physical scaling laws — like spectral asymptotics, subdiffusive behavior, dispersion relations. Indeed, in the intrinsic light of this metric, free fractals exhibit Euclidean-like behavior. A single, possibly non-integer, parameter takes the role of the Euclidean dimension and hence that of a *bona-fide* intrinsic dimension of the structure.

1991 Mathematics Subject Classification. 35P, 58F, 60, 82B.

This intrinsic metric approach to dynamics obeys the prescription requiring *the "geometry" to be the one induced by the dynamics itself.* While certainly a familiar concept, it somehow reverses the prevalent point of view in most fractal theories, where dynamical exponents are derived from fractal geometry.

Preliminaries on self-similarity in Euclidean spaces are given in the following Section 2, mostly taken from the work of Hutchinson [21]. The Lagrangian formalism is described in Section 3. The intrinsic metric, and the related homogeneous structure, is introduced in Section 4.

The main part of the paper is Section 5. Here we describe the notion of *variational fractal*, introduced in [31], and state the basic result, namely, a family of scaled Poincaré inequalities on the intrinsic metric balls. We also discuss the connections with the variational theory developed in [6,7,8,9] and the applications to the most important dynamical power laws on general variational fractals.

We conclude Section 5, and the paper, by sketching a parallel with the theory of sub-elliptic spaces — as for example Heisenberg's group. From the point of view of both fractal and subelliptic theories, the spaces of homogeneous type of abstract harmonic analysis — taken together with self-similar invariant energy forms — appear to be good surrogate "effective" candidates to replace Riemannian (and sub- Riemannian) formalism on structures which are non-differentiable and possibly fractal. On the other hand, we also discuss at the end the limits of the effective approach, in particular in connection with the long range behavior of fractal diffusions.

In each section, we will first explain the main ideas in somehow loose "physical" terms. Then we show how to put them in a more formal mathematical framework.

The Lagrangian formalism on fractal has been described from the physical point of view in [31]. The results of Section 4 and Section 5 have been stated in [32] and proofs will appear in [33].

2. Self-similarity

From the point of view of physics, the fractal structures considered in this section are supposed to be spatially homogeneous and isotropic, with dilational invariance at characteristic (Euclidean) length scales ℓ.

The scale ℓ may be a continuous one, as in the case of uniform structures, or it may be discrete, $\ell = \alpha^{-n}$, $n = 1, 2, \ldots$, for some characteristic length factor $\alpha > 1$ (for simplicity, we only consider scaling down). By $Q(\ell)$ we denote below the quantity Q within an element of the structure of linear size ℓ.

The main physical quantities on which we base our analysis are the *mass* and the *energy*. We suppose that they are known by theory, as in most deterministic self-similar fractals, or are observed by experiment, as it may be the case for statistic fractals and disordered media.

The mass μ will be identified, up to a constant, with the volume, $\mu \propto \text{Vol}$. For simplicity, we denote both measures by μ. The invariance of μ under dilations is expressed by the scaling law

$$\mu(\ell) \approx \ell^{d_{mass}} \mu(1) \, .$$

This defines the (static) exponent d_{mass}.

Our rigorous mathematical formulation will be confined in the present paper to "deterministic" fractals embedded in Euclidean spaces, even if most of our theory could be carried out in more general classes of metric spaces.

Throughout the paper we shall denote the D-dimensional Euclidean space by $\mathbb{R}^D$, $D \geq 1$, the balls $\{y \in \mathbb{R}^D : |x - y| < r\}, x \in \mathbb{R}^D, r > 0$, by $B_e(x, r)$, the Euclidean diameter of $A \subset \mathbb{R}^D$ by $\text{diam}_e A$.

We suppose that $\Psi = \{\psi_1, \ldots, \psi_N\}$ is a given set of *contractive similitudes* ψ_i : $\mathbb{R}^D \to \mathbb{R}^D, i = 1, \ldots, N$, that is,

$$(2.1) \qquad |\psi_i(x) - \psi_i(y)| = a_i^{-1}|x - y|, \quad x, y \in \mathbb{R}^D,$$

and we assume $\alpha_1 \geq \cdots \geq \alpha_N > 1$.

The map

$$K' = \psi(K) := \bigcup_{i=1}^N \psi_i(K)$$

is a contraction in the space of non-empty compact sets of $\mathbb{R}^D$, endowed with the Hausdorff metric. Therefore, there exists a unique (non empty) compact set K in $\mathbb{R}^D$, such that

$$(2.2) \qquad K = \Psi(K) := \bigcup_{i=1}^N \psi_i(K).$$

K is the *invariant* set associated with the family Ψ.

The real number $d_f > 0$, uniquely defined by the relation

$$(2.3) \qquad \sum_{i=1}^N \alpha_i^{-d_f} = 1,$$

is the *similarity (or fractal) dimension* of K.

By using a similar contraction argument in the spaces of measures with bounded support and unit mass — endowed with a metric that induces the weak (vague) convergence of measures — Hutchinson, *loc. cit.*, has proved that there exists also a unique Borel regular measure μ with unit mass, which is *invariant* for Ψ, that is,

$$(2.4) \qquad \int_K \varphi d\mu = \sum_{i=1}^N \alpha_i^{-d_f} \int_K \varphi \circ \psi_i d\mu$$

for every integrable $\varphi : K \to \mathbb{R}$, and such μ is supported on K.

Following Hutchinson, we shall also assume that the family Ψ satisfies in addition the following *open set condition*:

$$(2.5) \qquad \bigcup_{i=1}^N \psi_i(U) \subset U, \quad \psi_i(U) \cap \psi_j(U) = \emptyset \text{ if } i \neq j,$$

where U is a given non-empty bounded open subset of $\mathbb{R}^D$. By c_1 we shall denote the radius of a Euclidean ball contained in U and by c_2 the radius of a ball containing U.

This condition prevents distinct "copies" $\psi_i(K)$ to have overlapping interiors. Then, more specific metric information on K and μ become available. In particular, under the assumption (2.5), the similarity dimension d_f is found to be equal to the Hausdorff dimension of K. Moreover

$$0 < H^{d_f}(K) < \infty;$$

where H^{d_f} is the Hausdorff d_f-dimensional measure in $\mathbb{R}^D$, and

$$(2.6) \qquad \mu \equiv \left(H^{d_f}(K)\right)^{-1} H^{d_f} \mid K \, .$$

This gives a very useful metric characterization of the invariant measure μ.
We shall use the notation

$$\psi_{i_1 \ldots i_n} := \psi_{i_1} \circ \psi_{i_2} \circ \cdots \circ \psi_{i_n} \, ,$$
$$A_{i_1 \ldots i_n} := \psi_{i_1 \ldots i_n}(A) \, ,$$

for arbitrary n-ples of indeces $i_1, \ldots, i_n \in \{1, \ldots, N\}$, $n \geq 1$, and arbitrary $A \subset K$.
The set

$$K_{i_1 \ldots i_n} = \psi_{i_1 \ldots i_n}(K)$$

where $n \geq 1$, $i_1, \ldots, i_n \in \{1, \ldots, N\}$, will be called, following Lindstrøm [30], n-complex. We say that two complexes $K_{i_1 \ldots i_m}$, $K_{j_1 \ldots j_n}$, $m \geq 1$, $n \geq 1$, $i_1, \ldots, i_m \in \{1, \ldots, N\}$, $j_1, \ldots, j_n \in \{1, \ldots, N\}$, are *distinct* if $(i_1, \ldots, i_m) \neq (j_1, \ldots, j_n)$.

For every $n \geq 1$, K is the union of all n-complexes $K_{i_1 \ldots i_n}$ and these mutually intersect only on lower dimensional sets:

$$(2.7) \qquad K = \bigcup_{i_1 \ldots i_n = 1}^{N} K_{i_1 \ldots i_n} \, ,$$

with

$$(2.8) \qquad \mu \left(K_{i_1 \ldots i_n} \cap K_{j_1 \ldots j_n} \right) = 0$$

if $(i_1, \ldots, i_n) \neq (j_1, \ldots, j_n)$, hence

$$(2.9) \qquad \mu(K) = \sum_{i_1 \ldots i_n = 1}^{N} \mu(K_{i_1 \ldots i_n}) \, .$$

Therefore, we see that K can be decomposed in the union of "small copies" $K_{i_1 \ldots i_n}$, of diameter

$$(2.10) \qquad \mathrm{diam}_e K_{i_1 \ldots i_n} = \alpha_{i_1}^{-1} \ldots a_n^{-1} \, \mathrm{diam}_e K \, ,$$

and mass

$$(2.11) \qquad \mu(K_{i_1 \ldots i_n}) = \alpha_{i_1}^{-d_f} \ldots \alpha_{i_n}^{-d_f} \, ,$$

We now show how to get the power law of the mass, previously derived by physical considerations.

Let $0 < \ell < \mathrm{diam}_e K$ be a given (Euclidean) length scale. For every $n \geq 1$ and for every sequence $i_1, \ldots i_n, \cdots \in \{1, \ldots, N\}$, there exists a *least* integer $m \geq 1$, such that

$$\mathrm{diam}_e K_{i_1 \ldots i_m} \equiv \alpha_{i_1}^{-1} \ldots \alpha_{i_m}^{-1} \, \mathrm{diam}_e K \leq \ell \, .$$

Let us denote by I_ℓ the set of all finite sequences $(i_1, \ldots i_m)$ obtained in this way. Then, it is easy to see that for every $(i_i, \ldots i_m) \in I_\ell$ we have:

$$(2.12) \qquad \alpha_1^{-1} \ell < \mathrm{diam}_e K_{i_1 \ldots i_m} \leq \ell \, .$$

If (2.12) holds, we say that $K_{i_1...i_m}$ is of (Euclidean) size ℓ.

Let us now consider, as at the beginning of this section, the special case of an *isotropic fractal*, for which

$$(2.13) \qquad\qquad\qquad \alpha_i \equiv \alpha$$

for every $i \in \{1, \dots, N\}$, and

$$(2.14) \qquad\qquad\qquad d_f = \log N / \log \alpha .$$

Then, every $K_{i_1...i_m}$ of size ℓ has $\text{diam}_e K_{i_1...i_m} \approx \ell$ and mass

$$(2.15) \qquad\qquad \mu(K_{i_1...i_m}) = N^{-m} \mu(K) \approx \ell^{d_f} \mu(K) .$$

This gives back the power law for the scaling of the mass mentioned at the beginning and identifies the mass scaling exponent with the fractal dimension

$$(2.16) \qquad\qquad\qquad d_{\text{mass}} \equiv d_f .$$

We now describe a few important model examples of self-similar structures.

EXAMPLE 1. The Euclidean cube $Q = [0, 1]^D$ of $\mathbb{R}^D$. We take α to be any integer $\alpha > 1$ and $\Psi = \{\psi_1, \dots, \psi_N\}$ to be the similitudes that carry Q into each one of the $N = \alpha^D$ small cubes of size α^{-1} that decompose Q coordinatewise. The fractal dimension, as defined above, coincides with the Euclidean dimension, $d_f = D$, and the invariant measure μ with the normalized D-dimensional Lebesgue measure, $d\mu = dx$, restricted to Q. The power law

$$\mu(\ell) \approx \ell^D \mu(1)$$

holds at all scales $\ell = \alpha^{-n}, n = 1, 2, \dots$, therefore, $d_{\text{mass}} = d_f = D$.

EXAMPLE 2. The Sierpinski gasket K in $\mathbb{R}^D$. Now $\Psi = \{\psi_1, \dots, \psi_N\}$ are the similitudes

$$\psi_i(x) = a_i + \alpha^{-1}(x - a_i) ,$$

where $\{a_1, \dots, a_N\}, N = D+1$, are the vertices of a regular D-simplex and $\alpha = 2$. We have $d_f = \ln N / \ln 2$ and μ is the normalized restriction to K of the d_f-dimensional Hausdorff measure in $\mathbb{R}^D$. The power law

$$\mu(\ell) \approx \ell^{d_f} \mu(1)$$

holds at the length scales $\ell = 2^{-n}, n = 1, 2 \dots$. Now $d_{\text{mass}} = d_f = \ln N / \ln 2 < D$.

This is the most typical example of fractal belonging to the family of so-called *nested fractals*, introduced by Lindstrøm, *loc. cit.* Other well known members of the family, for example, are the *curve of von Koch*, the *snowflake*, the *ice curve* and so on. All these are special self-similar structures, which are intended to be the mathematical models of the so-called *finitely ramified fractals* of physics. These are fractals that can be disconnected by just removing a *finite* number of points. More formally, for the self-similar fractals described before, this property can be expressed by the condition

$$K_{i_1...i_n} \cap K_{j_1...j_n} = F_{i_1...i_n} \cap F_{i_1...i_n}$$

if $(i_1, \dots, i_n) \neq (j_1, \dots, j_n)$, where F is the finite set of all so-called *essential fixed-points* of the maps $\psi_1, \dots, \psi_N$. These are those fixed-points $\xi \neq \eta$ of the maps $\psi_1, \dots, \psi_N$, such that $\psi_i(\xi) \equiv \psi_j(\eta)$ for two distinct $i \neq j$. For the gasket, clearly $F = \{a_1, \dots, a_N\}$.

EXAMPLE 3. The Sierpinski carpet in R^2. This is the invariant set K of the family of $\alpha^3 - 1 = 8$ similitudes, $\alpha = 3$, obtained from the family of $\alpha^3 = 9$ similitudes of Example 1, by taking off the similitude that carries the square $Q = [0,1]^2$ into its interior. K has fractal dimension $d_f = \ln 8 / \ln 3$. Since two distinct n-complexes are two small squares that may intersect along a segment, the carpet is *not* a nested fractal.

Before concluding this section on self-similarity, let us point out that we have considered up to now — and we shall consider also in the following — only *homogeneous* fractals. This restriction can be traced back to our special choice of the coefficients $r_i = \alpha^{-d_f}, i = 1, \ldots, N$, in the identity

$$\int_K \varphi d\mu = \sum_{i=1}^{N} r_i \int_K \varphi \circ \psi_i d\mu ,$$

used in (2.4) to define the invariant measure μ. Indeed, we might have chosen as well any set of "probabilities" $r_i > 0, i = 1, \ldots, N$, that is of positive constants satisfying

$$\sum_{i=1}^{N} r_i = 1 .$$

Then, as also shown by Hutchinson, *loc. cit.*, we would have found a measure μ, invariant in the sense of the identity above. Any such μ describes a *non-homogeneous* distribution of mass and can be taken as a mathematical formulation of the so-called *multifractals* of physics. This class of general invariant measures has been studied recently by many authors, mainly by Fourier analysis methods, see Strichartz [45], Olsen [39] and the references therein. The *dynamical* properties of multifractals, however, seem to remain so far largely unexplored.

3. Lagrangian formalism

For the description of the dynamical properties of self-similar structures we adopt a measure-valued Lagrangian formalism, that we first describe in loose physical terms.

We assume that the *total energy*,

$$W = \int d\mathsf{L} ,$$

is obtained by integrating a measure-valued *Lagrangian* L on the structure, K. Since we have in mind linear field equations, we assume that L, as a measure, is proportional to the *square of the field* $\mathsf{E}^2 \equiv |\operatorname{grad} u|^2$, which in turn depends quadratically on the *potential u*,

$$d\mathsf{L} \propto d\mathsf{E}^2 \equiv |\operatorname{grad} u|^2 .$$

The energy W, as well as the Lagrangian $\mathsf{L}[u]$ and the square of the field $\mathsf{E}^2 \equiv |\operatorname{grad} u|^2$ are expected to be invariant under the dilations of the structure under consideration. This means that the integrated energy W scales according to a power law

$$W(\ell) \approx \ell^{d_{\text{energy}}} W(1) ,$$

where, as before, $W(\ell)$ denotes the energy within a volume element $d\mu$ of (Euclidean) linear size ℓ. This power law defines the (dynamical) exponent d_{energy}.

The local measure $E^2 \equiv |\operatorname{grad} u|^2$ also scales according to

$$E^2(\ell) \approx \ell^{d_{energy}} E^2(1) .$$

On *finite* volume elements $d\mu$, we can then define the *ratio*

$$E^2(d_\mu) \equiv E^2(d\mu)/d\mu$$

to be the *average square field intensity* E^2 within the element $d\mu$. The scaling law of E^2,

$$E^2(\ell) \approx \ell^{-d_{field}} E^2(1) ,$$

now defines the *field exponent*, d_{field}. From our definition of $E^2(d\mu)$ and from the scaling laws for μ and E^2, we obtain the basic relation

$$d_{energy} = d_{mass} - d_{field} .$$

This allows us to derive the field exponent d_{field}, which may not be directly observable, in terms of the two exponent, d_{mass} and d_{energy}, which are, at least in principle, both observable on real physical structures.

Let us identify the scaling exponents d_{energy} and d_{field} in the Euclidean D- dimensional case of Example 1 and check there the validity of the relation above. The energy functional is the classic Dirichlet integral, which under homotheties of ratio ℓ scales like ℓ^{D-2}, therefore $d_{energy} = D - 2$. The field intensity now has pointwise values $E^2 = |Du|^2$ and scales with ℓ as ℓ^{-2}, therefore, $d_{field} = 2$. As seen before, $d_{mass} \equiv D$. Thus, $d_{energy} = d_{mass} - d_{field}$.

Our previous scaling arguments are, of course, rather dubious from a mathematical point of view. However, they go to the hearth of the matter and indeed, as we shall see in a moment, they can be put on more solid ground.

In mathematical terms, with notation from the previous section, we assume that the energy is given by a *strongly local, regular, Dirichlet form* $W \equiv W(u, v)$, in the Hilbert space $H = L^2(K, \mu)$ of square-summable functions on K with respect to the invariant measure μ.

We recall that a *Dirichlet form* in the space H is a densely defined, closed, non-negative, symmetric bilinear form W in H, with domain

$$D_W = \{u \in H : W[u] \equiv W(u, u) < \infty\} ,$$

which enjoies the so-called Markovianity property, namely, $T \circ u \in D_W$ and $W[T \circ u] \leq W[u]$ whenever $u \in D_W$ and $T : \mathbb{R} \to \mathbb{R}$ is such that $T(0) = 0, |T(x) - T(y)| \leq |x - y|$ for all $x, y \in \mathbb{R}$. Moreover, W is *regular* in H if $D_W \cap C(K)$ is dense both in $C(K)$-the space of continuous functions on K with the uniform norm — and in $D(W)$ with the energy norm $\|u\| = \left(W[u] + \|u\|^2_{L^2(K,\mu)}\right)^{1/2}$. W is *strongly local* if $W(u, v) = 0$ for every u constant on the support of v. See [17].

If all these properties are satisfied, we say that W is a form of *diffusion type*. By a well known representation theory of Beurling-Deny [4,5] and LeJan [29], any form W of diffusion type admits the following integral expression

$$(3.1) \qquad\qquad W[u] = \int d\mathsf{L}[u]$$

for every $u \in D_W$, where the measure $L[u]$ is defined by the identity

(3.2) $\int \varphi \, dL[u] = W(\varphi u, u) - (1/2)W(\varphi, u^2)$,   for every $\varphi \in C_0(K)$.

This well defined measure $L[u]$ takes the role, in our mathematical theory, of the measure-valued local Lagrangian $L[u] \propto E^2 \equiv |\operatorname{grad} u|^2$ introduced before.

The identity (3.2) defining the measure $L[u]$ can be easily explained, by noticing that the usual Euclidean square-gradient, taken as a measure $|Du|^2 dx$, can be obtained from the identity

$$\int D(\varphi u) Du \, dx = \int \varphi |Du|^2 dx + (1/2) \int D\varphi \, D(u^2) dx ,$$

for every $\varphi \in C_0(K)$.

An important warning must be given here. On most fractals, like for example the Sierpinski gasket, the measure $L[u]$ does not possess a density with respect to the underlying volume measure μ and the square of the field $E^2 \equiv |\operatorname{grad} u|^2$ cannot be defined pointwise, see *e.g.* [24]. This is one of the main technical difficulties for a rigorous dynamical theory of fractals.

We assume, in addition, that the form W is *invariant* for the family Ψ, in the following sense. There exist real parameters $\rho_i > 0$, $i = 1, \ldots, N$, such that for every $u \in D_W$, we have

(3.3) $W[u] = \sum_{i=1}^{N} \rho_i W[u \circ \psi_i] ,$

moreover,

(3.4) $\rho_i = \mu(K_i)^\sigma ,$ $i = 1, \ldots, N ,$

for some given real constant $\sigma < 1$, independent of i.

The invariance of W is inherited by the local Lagrangian measure L. The proof follows a classic Fourier type argument and consists in replacing rapidly oscillating functions $\sqrt{\varphi} \exp(itu)$, $t > 0$, in the identity (3.3), see [34]. By applying the analogue of the usual Leibniz rule and chain rule — which hold for local Dirichlet forms — and by taking the limit as $t \to \infty$, we obtain the identity

(3.5) $\int_K \varphi dL[u] = \sum_{i=1}^{N} \rho_i \int_K \varphi \circ \psi_i \, dL[u \circ \psi_i]$

for every $\varphi \in C_0(K)$.

This identity can be iterated at all scales and we get

(3.6) $\int_K \varphi \, dL[u] = \sum_{i_1, \ldots, i_n = 1}^{N} \rho_{i_1} \cdots \rho_{i_n} \int_K \varphi \circ \psi_{i_1 \ldots i_n} \, dL[u \circ \psi_{i_1 \ldots i_n}] ,$

for every $n \geq 1$.

The last relation can be transformed into a *change of variable formula* for the Lagrangian measure $L[u] \propto E^2 \equiv |\operatorname{grad} u|^2$, under the transformation $y = \psi_{i_1 \ldots i_n}(x)$ of K into itself.

In order to state this formula precisely, we need to introduce the (intrinsic) *boundary* Γ of K. We define Γ to be the set

$$(3.7) \qquad \Gamma = \bigcup_{i \neq j=1}^{N} \psi_i^{-1}(K_i \cap K_j) .$$

This set is easily seen to be a compact subset of $K \cap \partial U$, and $\mu(\Gamma) = 0$. We assume that the relation

$$K_i \cap K_j = \Gamma_i \cap \Gamma_j \quad \text{for every } i \neq j ,$$

which is equivalent to (3.7), scales down by self-similarity to every pair of distinct n-complexes, that is, we assume that

$$(3.8) \qquad K_{i_1 \ldots i_n} \cap K_{j_1 \ldots j_n} = \Gamma_{i_1 \ldots i_n} \cap \Gamma_{i_1 \ldots i_n}$$

for every $n \geq 1$ and $(i_1, \ldots, i_n) \neq (j_1, \ldots, j_n)$.

Then, from (3.6) we easily obtain the relation

$$(3.9) \qquad \int_{K_{i_1 \ldots i_n} - \Gamma_{i_1 \ldots i_n}} \varphi(y) \left(\mathsf{L}[u] \mid (K_{i_1 \ldots i_n} - \Gamma_{i_1 \ldots i_n}) \right) (dy) =$$

$$= \rho_{i_1} \ldots \rho_{i_n} \int_{K-\Gamma} \varphi \circ \psi_{i_1 \ldots i_n}(x) \left(\mathsf{L}\left[u \circ \psi_{i_1 \ldots i_n} \right] \mid (K - \Gamma) \right) (dx)$$

for every $\varphi \in C(K)$ with support contained in $K_{i_1 \ldots i_n} - \Gamma_{i_1 \ldots i_n}$.

This formula can be also written more loosely as

$$(3.10) \qquad \mathsf{L}[u](dy) = \rho_{i_1} \ldots \rho_{i_n} \mathsf{L}\left[u \circ \psi_{i_1 \ldots i_n} \right] (dx)$$

and gives the transformation rule of the measure $\mathsf{L}[u]$, under the change of variable $y = \psi_{i_1 \ldots i_n}(x)$ that carries $K - \Gamma$ onto $K_{i_1 \ldots i_n} - \Gamma_{i_1 \ldots i_n}$.

In the isotropic case $\alpha_1 \equiv \alpha$, it is natural to assume also that (3.3) holds with $\rho_i \equiv \rho$ for every $i \in \{1, \ldots, N\}$. In this case, (3.10) reduces to

$$(3.11) \qquad \mathsf{L}[u](dy) = \rho^n \mathsf{L}\left[u \circ \psi_{i_1 \ldots i_n} \right] (dx) .$$

For $K_{i_1 \ldots i_n}$ of size $\ell = \alpha^{-n}$, we obtain, in our informal notation,

$$\mathsf{L}(\ell) = \rho^n \mathsf{L}(1) .$$

By comparing this power law with the previous one defining the exponent d_{energy}, we get $\rho^n = \ell^{d_{\text{energy}}}$. This gives d_{energy} as a function of ρ and α,

$$(3.12) \qquad d_{\text{energy}} = -\frac{\ln \rho}{\ln \alpha} .$$

As already noticed, we cannot consider the density $E^2 \equiv d\mathsf{E}^2/d\mu$, therefore we cannot define the exponent d_{field} by means of the power law $E^2(\ell) \approx \ell^{-d_{\text{field}}} E^2(1)$, as we did informally before. Instead, we now use the heuristic relation

$$(3.13) \qquad d_{\text{field}} = d_{\text{mass}} - d_{\text{energy}} ,$$

obtained before, to *define* the exponent d_{field}. In the next section we shall see a metric characterization of d_{field}, that confirms its role as the scaling exponent of the square field intensity $E^2(\ell)$. Always in the isotropic case, $\alpha_i \equiv \alpha$, $\rho_1 \equiv \rho$, from (3.13) we obtain

$$(3.14) \qquad\qquad d_{\text{field}} = \frac{\ln(N\rho)}{\ln \alpha} .$$

Before proceeding, let us check (3.12) and (3.14) in the Euclidean case, when, we recall, $d_{\text{mass}} \equiv d_f = D$ and W is the usual Dirichlet integral. Under homotheties of factor $\ell = \alpha^{-1}$, W scales with $\rho_i \equiv \rho = \ell^{D-2} = \alpha^{2-D}$ and (3.12) gives again the value $d_{\text{energy}} = D - 2$. Moreover, since $N = \alpha^D$, (3.14) gives $d_{\text{field}} = 2$. This is in agreement with the interpretation of d_{field} as the scaling exponent of the square field intensity $E^2(\ell)$ and with the presence of ℓ^2 in the denominator of $E^2 \equiv |\operatorname{grad} u|^2$. We also notice that in this case ρ is actually of the form (3.4), with $\sigma = (D-2)/D$.

The main feature of the measure-valued Lagrangian formalism described so far consists in the fact that — together with classic Euclidean examples — it also covers large families of fractals, for example, nested fractals and their basic model, the Sierpinski gasket.

The so-called "standard" diffusion, or "Brownian motion", on the Sierpinski gasket, was first constructed mathematically by Goldstein [19], Kusuoka [24] and Barlow-Perkins [3], who used probability. The "standard" Laplacean was first described analitically by Kigami [22] and later identified with the generator of the process by Fukushima-Shima [18] and Kusuoka [25], who wrote the associated Dirichlet form. The "standard" diffusions on nested fractals have been first constructed by Lindstrøm, *loc. cit.*, and their analytic description in terms of Dirichlet forms has been given by Kusuoka [25,26] and Fukushima [16].

The probabilistic approach to nested fractals is based on the study of asymptotic random walks on the discrete approximations of the fractal, the so-called *pre-fractals*, while the analytic approach relies on suitable *finite difference schemes* on the same pre-fractals. We would like to point out that both these constructions are reminescent of the old fundamental work of Courant-Friedrichs-Lewy [14], who described classic finite difference equations for the first time not only analitically, in connection with partial differential equations, but also probabilistically, in relation to random walks and Brownian motion.

The main difficulty, in setting up finite difference schemes on the pre-fractals, consists in finding out the correct coefficients and renormalization factors, which are expected to lead in the limit to a non-trivial energy form on the continuous asymptotic fractal. Here we pay the price to the lack of any reasonable differential structure, that prevents us from writing down explicitly some kind of differential operator on the fractal itself, so that — contrary to the classic model — we do not have at hand a differential expression to discretize.

On the Sierpinski gasket, the energy form is obtained as the limit

$$W[u] = \frac{1}{2(N-1)} \lim_{n \to \infty} \left(\frac{N+2}{N}\right)^n \sum_{i_1 \ldots i_n = 1}^{N} \sum_{\xi \neq \eta \in F} \left(u\left(\psi_{i_1 \ldots i_n}(\xi)\right) - u\left(\psi_{i_1 \ldots i_n}(\eta)\right)\right)^2 .$$

Here $F = \{a_1, \ldots, a_N\}$ are the vertices of the gasket, which are also the fixed-points of its self-similitudes.

On more general nested fractals, W is given by a similar asymptotic finite-difference expression,

$$W[u] = \frac{1}{2} \lim_{n \to \infty} \rho^n \sum_{i_1 \ldots i_n = 1}^{N} \sum_{\xi \neq \eta \in F} p(\xi, \eta) \left(u\left(\psi_{i_1 \ldots i_n}(\xi)\right) - u\left(\psi_{i_1 \ldots i_n}(\eta)\right)\right)^2 .$$

Now F is the set of the essential fixed points of the family Ψ, thus $card\,F \leq N$. We recall that, by definition, on a nested fractal our relation (3.6) defining the boundary Γ of K holds with $\Gamma \equiv F$.

In the previous expression the coefficients $p(\xi, \eta)$ and the renormalization factor $\rho > 0$ are, of course, to be chosen conveniently. This choice is not trivial. It was first nicely described in probabilistic terms by Lindstrøm, *loc. cit.*, who found the good matrices $p(\xi, \eta)$, $\xi, \eta \in F$, and the good factors $\rho > 1$. The procedure followed by Lindstrøm is the probabilistic version of the so-called *decimation* procedure of the physical literature, see *e.g.* Rammal-Toulouse, [42]. For a short description of the analytic approach, see also [2] and [36].

All "standard" energy forms that have been constructed on nested fractals — where $\alpha_i \equiv \alpha$ — are Dirichlet forms of diffusion type in the space $L^2(K, \mu)$, which are invariant in the sense of (3.3), with scaling factors $\rho_i \equiv \rho$ for all $i = 1, \ldots, N$. In particular, $\mu(K_i) \equiv 1/N$, therefore our assumption (3.4) is satisfied with

$$\sigma = -\frac{\ln \rho}{\ln N}.$$

Moreover, as already mentioned, $\rho > 1$, therefore $\sigma < 0$. This should be compared with the Euclidean case, where $\sigma = (D - 2)/D$, hence $\sigma < 0$, namely $\sigma = -1$, only if $D = 1$, whereas $0 \leq \sigma < 1$ in all dimensions $D \geq 2$.

4. Lagrangian metrics

We now come to the main point of the paper, the definition of an effective invariant metric of Lagrangian nature. As before, we first proceed heuristically, by relying on physical intuition and scaling considerations.

The intrinsic effective metric r we are looking for must be invariant under the dilations of the structure, so we can assume that it obeys a scaling law of the type

$$r(\ell) \approx \ell^\delta r(1)$$

with respect to the Eucliean length ℓ, for some scaling exponent $\delta > 0$ to be determined.

The idea now is that r must also reflect the intrinsic Lagrangian dynamics of the structure. *We then require r^{-2} to scale like the square of the field,*

$$r^{-2}(\ell) \approx (E^2(\ell)/E^2(1))\,r^{-2}(1).$$

This is motivated by the fact that the (average) square field intensity E^2, as defined in Section 3, has the nature of the square of a gradient, hence is proportional to the reciprocal of the square of a length.

Now, we know that E^2 obeys the power law $E^2(\ell) \approx \ell^{-d_{field}} E^2(1)$. Therefore, the two scaling relations for r^{-2} and E^2 together determine the value of the exponent δ, as

$$\delta = \frac{d_{field}}{2}.$$

The intrinsic effective metric is given by

$$r = \ell^\delta = \ell^{d_{field}/2}.$$

We can now rewrite all scaling laws, previously taken in the Euclidean scale ℓ, in the new metric r. Accordingly, by $Q(r)$ we denote below the quantity Q within a volume element of *intrinsic* size r.

The most important intrinsic scaling exponent is the one that occurs in the power law of the mass with respect to the new metric r, namely, the exponent $\nu > 0$ in the power law

$$\mu(r) \approx r^\nu \mu(1) \,.$$

The importance of this exponent consists, so to say, in its double character: static, because it regards the mass (or volume); dynamic, because it indirectly involves the Lagrangian, *via* the metric r. We call ν the *intrinsic dimension* of the fractal. As we shall see later, this is the fundamental dimensional parameter of the structure.

Before proceeding, let us see how ν is related to its (Euclidean) counterpart d_{mass}, the scaling exponent of μ with respect to ℓ. From the power law $\mu(\ell) \approx \ell^{d_{\text{mass}}}\mu(1)$, and the expression $r = \ell^\delta = \ell^{d_{\text{field}}/2}$ of r, we get

$$\nu = \frac{d_{\text{mass}}}{\delta} = 2\,\frac{d_{\text{mass}}}{d_{\text{field}}} \,.$$

Let us now see how the Lagrangian $\mathsf{L}[u] \propto \mathsf{E}^2 \equiv |\operatorname{grad} u|^2$ scales intrinsically. From the power law of E^2 in the Euclidean scale ℓ, that is, $\mathsf{E}^2(\ell) \approx \ell^{d_{\text{energy}}}\mathsf{E}^2(1)$, we obtain, in the intrinsic notation introduced above, $\mathsf{L}(r) \approx r^{d_{\text{energy}}/\delta}$. Since $d_{\text{energy}}/d = d_{\text{mass}}/\delta - d_{\text{field}}/\delta = \nu - 2$, this gives the intrinsic power law of the Lagrangian

$$\mathsf{L}(r) \approx r^{\nu-2}\mathsf{L}(1) \,.$$

Finally, let us find the intrinsic scaling of the field E^2. From the scaling of E^2 with respect to ℓ, namely, $E^2(\ell) \approx \ell^{-d_{\text{field}}}E^2(1)$, we obtain, always in intrinsic notation, $E^2(r) \approx r^{-d_{\text{field}}/d}E^2(1)$. Therefore, E^2 is governed by the dimensionless power law

$$E^2(r) \approx r^{-2}E^2(1) \,,$$

as in the case of uniform Euclidean media.

We have reached a remarkable conclusion. By introducing the effective metric r in our structure, we see that the basic physical quantities — mass, energy and field — behave under the dilation symmetries of the structure very much as their counterparts in uniform media, with a single parameter, the intrinsic dimension ν, in the role of the Euclidean dimension D.

Once again, let us now turn to mathematics. In defining the effective metric rigorously on general self-similar fractals, we cannot rely — as before — on the notion of intensity E^2 of the square of the field. In fact, we know that on very important fractals the measure $\mathsf{L}[u] \propto \mathsf{E}^2 \equiv |\operatorname{grad} u|^2$ is singular with respect to μ, therefore its density E^2 cannot be defined.

We then proceed as follows. We introduce in K the quasi-distance

$$(4.1) \qquad\qquad d(x, y) = |x - y|^\delta \,, \qquad x, y \in K$$

where $\delta > 0$ is a parameter that we will choose in a moment and $|x - y|$ is, as always, the Euclidean distance of x, y in $\mathbb{R}^D$. We recall that a quasi-distance satisfies the assumptions of a distance, with possibly a multiplicative constant, $c_T \geq 1$, at the right hand side of the triangle inequality.

We want the quasi-distance (4.1) to be related to the Lagrangian, in particular, we want $d(x, y)$ to share with the Lagrangian its scaling properties under the action of the family Ψ. Arguing heuristically, we asked above the reciprocal of the square of metric $r \equiv d$ to scale like the square intensity E^2 of the field. Now, *we simply require*

d^2 *to scale according to the law obeyed by* $L[u]$. Namely, we require the quasi-distance (4.1) to satisfy, in addition, the following identity

$$(4.2) \qquad d^2(x, y) = \sum_{i=1}^{N} \rho_i d^2(\psi_i(x), \psi_i(y))$$

for every $x, y \in K$, where $\rho_1, \ldots, \rho_N$ are the same constants occurring in (3.3) (and (3.5)) of the preceding section.

It is easily seen that (4.1) and (4.2) together uniquely determine $\delta > 0$, as the solution of the identity

$$(4.3) \qquad \sum_{i=1}^{N} \rho_i \alpha_i^{-2\delta} = 1 .$$

We take the quasi-distance (4.1) — where $\delta > 0$ is the value given by (4.3) — to be the (effective) *intrinsic metric* in $K \equiv (K, W)$.

The (quasi-) balls for the intrinsic metric of K will be denoted by $B(x, R)$,

$$B(x, R) = \{y \in K : d(x, y) < R\} ,$$

and the diameter of $A \subset K$ by diam A. Clearly, with notation from Section 3,

$$B(x, R) = B_e(x, R^{1/\delta}) \cap K$$

and diam $A = (\text{diam}_e A)^\delta$.

Let us now estimate the volume (mass) μ of an intrinsic ball $B(x, R)$. We know, from (2.11), the volume $\mu(K_{i_1 \ldots i_n})$ of an n-complex $K_{i_1 \ldots i_n}$. The difficulty in estimating $\mu(B(x, R))$ consists in the fact that we do not know exactly how many n-complexes intersect $B(x, R)$.

It can be proved, see [31], [33], that any self-similar fractal — satisfying (2.1), (2.2), (2.5) of Section 2 — enjoies the following *finite overlapping property* (for a related density result see also [21], *loc.cit.*):

There exists a constant M — that depends only on D, α_1, diam$_e K$ and on the constants c_1, c_2 of Section 1 — such that, for every $x \in K$ and every $\ell > 0$, $K \cap B_e(x, \ell)$ has non empty intersection with at most M n-complexes of (Euclidean) size ℓ.

By relying on this property, $\mu(B(x, R))$ can then be estimated as follows:

For every $x \in K$ and every $0 < r \leq R \leq$ diam$K$, we have

$$(4.4) \qquad M^{-1}\alpha_1^{d_f}\mu(B(x, R))\left(\frac{r}{R}\right)^\nu \leq \mu(B(x, r)) \leq M\alpha_1^{d_f}\mu(B(x, R))\left(\frac{r}{R}\right)^\nu ,$$

where

$$(4.5) \qquad \nu = \frac{d_f}{\delta} .$$

We now can check easily our heuristic derivation of the scaling law of the mass in the intrinsic metric $r \equiv d$. In fact, in the special case $\alpha_i \equiv \alpha$, $\rho_i \equiv \rho$, the identity (4.3) gives

$$(4.6) \qquad \delta = (1/2)\ln(N\rho)/\ln\alpha .$$

We now recall that, by (3.14), $d_{\text{field}} = \ln(N\rho)/\ln\alpha$. Therefore,

$$(4.7) \qquad\qquad\qquad\qquad \delta = \frac{d_{\text{field}}}{2} \, .$$

We point out that this relation between δ and d_{field} is the same that we obtained heuristically earlier. Moreover, since $d_f = d_{\text{mass}}$, (4.5) can be also written as $\nu = 2d_{\text{mass}}/d_{\text{field}}$, confirming the heuristic value of ν found before.

The lower estimate in the inequality (4.4) has a further important consequence, namely, the following *homogeneity property*:

For every $R > 0$ and for every $0 < \varepsilon < 1$, the ball $B(x, R)$, $x \in K$, contains at most $c\varepsilon^{-\nu}$ points whose mutual distance is greater or equal to εR, where $c > 0$ is a constant depending only on D, N, ρ, α and $\text{diam}_e K$.

A pair $K \equiv (K, d)$, where K is a topological space and d is a quasi-distance defined on K, such that: (i) the balls of d form a basis of open neighborhoods at every point of K, (ii) there exist two constants $\nu > 0$ and $c > 0$ such that the homogeneity property above holds, is called a *space of homogeneous type*, briefly, a *homogeneous space*.

A special case in which the homogeneity property is satisfied is when a measure μ exists on K, that has the following *duplication property* with respect to the balls $B(R) = B(x, R)$ of d:

$$\mu(B(2R)) \le c_0\mu(B(R)) \quad \text{for every } R > 0 \, ,$$

where $c_0 > 0$ is some constant, independent of x and R.

Therefore, the fractal $K \equiv (K, W)$ considered so far, endowed with the Lagrangian metric, is a space of homogeneous type. In view of (4.4), its intrinsic dimension ν, given by (4.5), can be naturally interpreted as the *homogeneous dimension* of K.

The notion of space of homogeneous type generalizes that of homogeneous space under the action of a group and plays an important role in the theory of singular integral operators and abstract harmonic analysis, [13], [45]. It is also important in the theory of homogeneous Lie groups and subelliptic operators, [45], see also [7] ad [35]. Thus, the metric Lagrangian approach brings the dynamical theory of fractals in close connection with well established classical theories, rich of methods and results.

Another important power law, derived heuristically above, is the one expressing the scaling of the Lagrangian in the intrinsic metric $r \equiv d$, namely $\mathsf{L}(r) \approx r^{\nu-2}\mathsf{L}(1)$. Let us now check that this law is also confirmed by the mathematical theory.

We let below the α_i's and the ρ_i's to be possibly distinct and we find first the relation between σ and ν. We rewrite the identity (4.3), that defines δ, by taking into account the identity (3.4), that defines σ, and we find

$$\sum_{i=1}^{N} \alpha_i^{-(d_f\sigma + 2\delta)} = 1 \, .$$

By comparing this with the identity (2.3), that uniquely determines d_f, we obtain $d_f = d_f\sigma + 2\delta$. Therefore, by (4.5),

$$(4.8) \qquad\qquad\qquad\qquad \sigma = \frac{\nu - 2}{\nu} \, .$$

We note that (4.8) has the same form as the relation that relates σ to D in the Euclidean case, with ν replacing D.

We now come to the scaling of the energy. We have seen that, under the change of variable $y = \psi_{i_1 \dots i_n}(x)$, the Lagrangian measure $L[u]$ scales according to (3.9), written also as (3.10). On the other hand, we have

$$\rho_{i_1} \dots \rho_{i_n} = \mu(K_{i_1})^\sigma \dots \mu(K_{i_n})^\sigma = \alpha_{i_1}^{-d_f \sigma} \dots \alpha_{i_n}^{-d_f \sigma} .$$

Therefore, by (2.10) and (4.5),

$$\rho_{i_1} \dots \rho_{i_n} = \left(\frac{\mathrm{diam}_e K_{i_1 \dots i_n}}{\mathrm{diam}_e K} \right)^{d_f \sigma} = \left(\frac{\mathrm{diam}\, K_{i_1 \dots i_n}}{\mathrm{diam}\, K} \right)^{v\sigma} .$$

By (4.8), we finally get

(4.9)
$$\rho_{i_1} \dots \rho_{i_n} = \left(\frac{\mathrm{diam}\, K_{i_1 \dots i_n}}{\mathrm{diam}\, K} \right)^{v-2} .$$

The scaling law (3.10) of $L[u]$ can then be written as

(4.10)
$$L[u](dy) = \left(\frac{\mathrm{diam}\, K_{i_1 \dots i_n}}{\mathrm{diam}\, K} \right)^{v-2} L[u \circ \psi_{i_1 \dots i_n}](dx) .$$

By integrating over $K_{i_1 \dots i_n}$ of intrinsic size r, we obtain

(4.11)
$$L[u](K_{i_1 \dots i_n} - \Gamma_{i_1 \dots i_n}) \approx \left(\frac{r}{\mathrm{diam}\, K} \right)^{v-2} L[u](K - \Gamma) .$$

in agreement with the heuristic scaling of L.

Let us now consider, as before, the average square field intensity

$$E^2 \left(K_{i_1 \dots i_n} - \Gamma_{i_1 \dots i_n} \right) = \frac{L[u] \left(K_{i_1 \dots i_n} - \Gamma_{i_1 \dots i_n} \right)}{\mu \left(K_{i_1 \dots i_n} - \Gamma_{i_1 \dots i_n} \right)} ,$$

on a $K_{i_1 \dots i_n}$ of intrinsic size r. By (2.11), (2.12),

$$\mu \left(K_{i_1 \dots i_n} - \Gamma_{i_1 \dots i_n} \right) \approx \left(\frac{r}{\mathrm{diam}\, K} \right)^v \mu(K - \Gamma) .$$

Therefore, from (4.11) we get

(4.12)
$$E^2 \left(K_{i_1 \dots i_n} - \Gamma_{i_1 \dots i_n} \right) \approx r^{-2} E^2(K - \Gamma) ,$$

again in agreement with the heuristics.

The relation (4.12), in particular, is consistent with the meaning of d_{field} — defined by (3.13) — as the exponent that governs the scaling of E^2 at Euclidean scales ℓ and shows at the same time that E^2, in the intrinsic metric r, exhibits the same dimensionless scaling r^{-2} as in Euclidean media.

In the following section we shall consider energy estimates on the whole of an intrinsic ball $B(x, R)$.

We point out once again that we have defined the effective metric d — by means of (4.1) and (4.2) — on arbitrary self-similar structures possessing a measure-valued self-similar Lagrangian, without requiring the Lagrangian measure to be absolutely continuos with respect to μ. Indeed, as already mentioned, on many fractals the local measures $L[u]$ defined by (3.2) are singular with respect to μ.

There are, however, many important cases, covered by the theory developed above, in which the measure $L[u]$ is absolutely continuous with respect to μ, hence the density

$$\frac{dL[u]}{d\mu}$$

exists, allowing us to define the square of the field intensity

$$E^2(x) \equiv |\operatorname{grad} u|^2(x) = \frac{dL[u]}{d\mu}$$

μ-a.e. on K.

For general Dirichlet forms of this kind, a metric theory has been recently developed in [6-9]. In this theory, a distance d_E is defined on K by setting:

$$d_E(x, y) = \sup\{\phi(x) - \phi(y) : |\operatorname{grad} \phi|^2 \le 1 \ \mu\text{-a.e.}, \ \phi \in D_W \cap C_0(K)\},$$

[6]. In other words, $d_E(x, \cdot)$ is taken to be the maximal sub-solution of the *"eiconal equation"* of geometric optics,

$$|\operatorname{grad} \phi|^2 = 1,$$

which keeps sense μ-a.e. on K, with the condition $\phi(x, x) = 0$.

On Riemannian manifolds—where μ is the Riemannian volume and W the Dirichlet form for the Laplace-Beltrami operator — d_E reduces to the usual Riemannian metric. For subelliptic vector fields of Hörmander's type, d_E coincides with the subelliptic metric considered by Nagel-Stein-Wainger [38], Fefferman-Phong [15], see also Jerison-Sánchez Calle [28].

In the present case — K a general self-similar set and W a self-similar form on K satisfying the scaling relation (3.3) — the distance d_E defined above inherits the scaling property of W.

Let us sketch this scaling argument, for simplicity, in the case $\alpha_i \equiv \alpha$, $\rho_i \equiv \rho$. Then, by the scaling relation (3.11), the condition $|\operatorname{grad} \phi|^2(dy) \le \mu(dy)$ on K_i is equivalent to the condition $\rho|\operatorname{grad} \phi \circ \psi_i|^2(dx) \le \alpha^{-d_f}\mu(dx)$ on K. Since $\alpha^{-d_f} = 1/N$, if $\phi' = (N\rho)^{1/2}\phi \circ \psi_i$, then $|\operatorname{grad} \phi'|^2(dx) \le \mu(dx)$ on K. Therefore, $d_E(\psi_i(x), \psi_i(y)) = (N\rho)^{-1/2}d_E(x, y)$, hence

$$d_E^2(x, y) = \sum_{i=1}^{N} \rho d_E^2(\psi_i(x), \psi_i(y)),$$

which is the scaling required in (4.2).

Let us also observe that in the Euclidean case the eiconal equation has the obvious solution $d_E(x, y) \equiv |x - y|$. In this case — Example 1 of Section 2 — the scaling law derived above can be easily checked as follows. Since $|\psi_i(x) - \psi_i(x)| = \alpha^{-1}|x - y|$ for every $i = 1, \ldots, N$, and $N = \alpha^D$, we have $|x - y|^2 = \sum_{i=1}^{N} \alpha^{2-D}|\psi_i(x) - \psi_i(x)|^2$. On the other hand, as already observed, $\alpha^{2-D} = \rho$. This shows that $d_E(x, y)$ satisfies (4.2).

We point out that, due to its effective character, the Lagrangian metric d can be only expected in general to be *equivalent* in the metric sense to the "geodesic" metric d_E, when this is also available. We shall come back on this point in the final remarks of next section, (j).

5. Variational fractals

The Lagrangian formalism described in the preceding sections leads to the notion of *variational fractal*, as a general framework for the study of the dynamics of self-similar structures.

A variational fractal, according to [31], is a pair $K \equiv (K, W)$ where:

(i) K is the invariant set of a family $\Psi = \{\psi_1, \ldots \psi_N\}$ of mappings $\psi_i : \mathbb{R}^D \to \mathbb{R}^D$, $i = 1, \ldots, N$, that satisfy (2.1) with some constants $\alpha_1 \geq \cdots \geq \alpha_N > 1$, (2.5) and (3.8);

(ii) W is a strongly local, regular Dirichlet form in the space $L^2(K, \mu)$, that satisfies conditions (3.3), (3.4) for some constant $\sigma < 1$. Here μ, given by (2.6), is the invariant measure (2.4).

The fractal $K \equiv (K, W)$ is endowed with the metric structure provided by the quasi-distance $d(x, y)$, defined by (4.1), (4.2). As seen in Section 4, $K \equiv (K, d)$ is a space of homogeneous type and the volume $\mu(B(x., R))$ of the (quasi-)balls $B(x, R)$ is estimated by the inequalities (4.4). The constant $\nu > 0$ occurring in (4.4) is the homogeneous (or, intrinsic) dimension of $K \equiv (K, d)$. This is the basic parameter of a variational fractal.

We recall that $\nu = d_f / \delta$, where d_f is the similarity (fractal) dimension — given by (2.3) and equal to the Hausdorff dimension of K — and δ is the constant given by (4.3).

Interesting variational fractals are of course only those that possess a non-trivial energy form W. Therefore, we assume in addition that the form W is *irreducible*, that is, that the only functions u (locally) in D_W with $W[u] = 0$ are the constants.

We actually require that this property holds in the stronger form of the following *Poincaré inequality*

(iii)
$$\int_K |u - u_K|^2 d\mu \leq c_P \int_{K-\Gamma} d|\operatorname{grad} u|^2$$

for every $u \in D_W$, where $|\operatorname{grad} u|^2 \equiv \mathsf{L}[u]$ is the local energy measure of the form W, defined by (3.2), and $c_P > 0$ is a suitable constant.

The example of the Sierpinski gasket and, more generally, of nested fractals shows that distinct complexes $K_i, K_j, i \neq j$, may intersect across a null subset for the measure μ. Therefore, from a measure point of view, they are disconnected. Nevertheless, as the same examples show, the standard diffusions, and the associated Dirichlet forms W, constructed on K are non-trivial. What happens is that the non-empty intersections $K_i \cap K_j \equiv F_i \cap F_j, i \neq j$, while being of μ measure zero, do have *positive capacity* with respect to the form W, or, what is the same, are non-negligible sets for the "Brownian" motion on K.

In our general framework, the finite set F of the essential fixed-points of nested fractals is replaced by the boundary Γ of K, defined in (3.7) and satisfying (3.8).

We shall then assume that $K \equiv (K, W)$ is *connected in the capacity sense*, according to the following condition:

(iv) For every $\ell > 0$, the (finite) family of n-complexes $K_{i_1 \ldots i_n}$ of size $\ell > 0$ that intersect non-trivially $K \cap B_\epsilon(x, \ell), x \in K$, can be suitably ordered, such that each pair of two contiguous $K_{i_1 \ldots i_n}$ satisfies the condition

$$K_{i_1^s \ldots i_n^s} \cap K_{i_1^{s+1} \ldots i_n^{s+1}} = \psi_{i_1^s \ldots i_n^s}(T) \cap \psi_{i_1^{s+1} \ldots_n^{s+1}}(T'),$$

where $T, T' \subset \Gamma$ have positive capacities (relative to the form W) uniformly bounded away from zero (cap $T \geq k > 0$), with the maps $\psi_{i_1^{s+1} \ldots i_n^{s+1}}^{-1} \circ \psi_{i_1^s \ldots i_n^s}$:

$T \to T'$ capacity preserving up to a constant ($k \operatorname{cap} S \leq \operatorname{cap} S' \leq (1/k) \operatorname{cap} S$, S' $= \psi_{i_{t+1}^{-1}...i_s^{+1}}^{-1} \circ \psi_{i_t...i_s}(S), S \subset T$).

We notice that the family of the $K_{i_1...i_n}$'s occurring in the preceding condition is finite, as a consequence of the finite overlapping property of Section 4.

This connectivity condition looks rather involved, however it can be easily verified both in the Euclidean case and on nested fractals. In fact, in the Example 1 of Section 2, $\Gamma = \partial Q$ and the sets T, T' can be always chosen to be just two opposite $(D - 1)$-dimensional faces of the cube Q. These have indeed positive capacity in the Sobolev space $H^1(Q)$. On nested fractals, as already remarked, $\Gamma = F$, which consists of at most N points, each of positive capacity for W.

Under the assumptions (i) to (iv), any variational fractal possesses the following important property:

THEOREM. *There exist two constants $c > 0$ and $q \geq 1$, such that the family of scaled Poincaré inequalities*

$$(5.2) \qquad \int_{B(x,r)} |u - u_{B(x,r)}|^2 d\mu \leq c \left(\frac{r}{\operatorname{diam}K}\right)^2 \int_{B(x,qr)} d|\operatorname{grad}u|^2$$

$0 < r \leq \operatorname{diam}K$, *holds for every $u \in D_W$. The constant c depends only on the structural constants of K and $q = 2^\delta$.*

By structural constants of K we mean the constants $D, N\alpha_1 \dots \alpha_N, \sigma$, the constant M in the finite overlapping property, the constant k in (iv).

The general theory for Dirichlet forms on homogeneous spaces developed in [6-9] generalizes the classic theory of De Giorgi, Nash, Moser for uniformly elliptic operators in divergence form and measurable coefficients. The basic tool in this theory is the distance d_E defined before, together with the family of Poincaré inequalities on the metric balls of d_E, stated in the preceding theorem. The theory applies in all its parts — which include Harnack inequalities, estimates of Green functions and energy's decay — to variational fractals for which the two metrics d and d_E are metrically equivalent (this is of course the case for uniformly elliptic operators in Euclidean spaces).

However, an important part of the theory, in particular, the imbeddings results stated below, can be immediately applied to general variational fractals, satisfying properties (i) to (iv). These results, in fact, [8,9], make only use — together with the homogeneous structure of the underlying (quasi-)metric space K — of the family of scaled Poincaré inequalities given by the theorem above, and hold without assuming the measures $L[u] \equiv |\operatorname{grad}u|^2$ absolutely continuous with respect to μ.

By c we denote below possibly different constants, depending only on the structural constants of K. The main inequalities proved in [8,9] are the following ones:

(a) *Nash inequality*

$$\|u\|_{L^2(K,\mu)}^{2+\frac{4}{\nu}} \leq c \left(W[u] + \|u\|_{L^2(K,\mu)}^2\right) \|u\|_{L^1(K,\mu)}^{\frac{4}{\nu}}$$

in all dimensions ν, for every $u \in D_W$.

(b) When $\nu > 2$, scaled *Sobolev inequalities* on the intrinsic balls. The Sobolev exponent is $2^* = 2\nu/(\nu - 2)$.

(c) If $\nu = 2$, the energy W, in view of (4.10), has "conformal" invariance under the action of Ψ. In this case, *John-Nirenberg inequalities* hold on the balls $B(x, R)$, and functions of finite energy have exponential integrability.

(d) If $v < 2$ — as in the case of nested fractals — Morrey inequalities are satisfied on the balls $B(x, R)$, in particular

$$|u(x) - u(y)| \leq cW[u]^{1/2}d(x, y)^{1-v/2},$$

$$|u(x)| \leq c\left\{W[u] + \|u\|^2_{L^2(K,\mu)}\right\}^{1/2},$$

for every $x, y \in K$.

Thus, if K has intrinsic dimension $v < 2$, functions of finite energy on K are Hölder continuous with exponent $\beta = 1 - (v/2)$, with respect to the intrinsic distance. This means also that they are Hölder continuous with respect to the Euclidean distance, with exponent $\beta_e = \delta\beta = (1/2)(d_{\text{field}} - d_f)$. The exponent β_e was first obtained for the Sierpinski gasket by Barlow-Perkins, *loc cit.*, and Kozlov [23], by direct calculations. On the Sierpinski gasket, $\beta_e = \ln(N+2/N)/\ln 4$, in particular, if $D = 2$, $\beta_e = \ln(5/3)/\ln 4$. However, this Hölder exponent was not related previously to a Morrey imbedding. This interpretation of β_e was first given in [8] and [38].

(e) A further consequence of the scaled Poincaré inequalities, as shown by Biroli-Tchou [10], is the compact imbedding in $L^2(B(x, R), \mu)$ — in all dimensions $v > 0$ — of the space of functions of D_W vanishing on the boundary of $B(x, R)$, that is, belonging to the closure of $C_0(B(x, R)) \cap D_W$ in D_W. If $v < 2$, this property is of course a direct consequence of the Morrey imbedding (d).

(f) As a consequence of (e), the non-negative definite, self-adjoint operator L in the space $L^2(K, \mu)$, uniquely associated with the form W, has a compact resolvent, hence real eigenvalues $\lambda_1 < \lambda_2 \leq \cdots \leq \lambda_n \to \infty$, repeated here with their multiplicity. If $N(x)$, $x > 0$, is the so-called *integrated density of states, i.e.*, $N(x)$ is the number of eigenvalues $\lambda_i \leq x$, then

$$N(x) \approx x^{v/2} \quad \text{as } x \to \infty,$$

see Posta [41]. The classic reference for the spectral asymptotic distribution of the eigenvalues is Weyl's result for the Laplace operator. The analogue problem on fractals has attracted great attention in the physical literature, Alexander-Orbach [1], Rammal-Toulouse, *loc. cit.*, and later also in the mathematical papers on the Sierpinski gasket and nested fractals already mentioned. The Weyl exponent is usually denoted by d_s and is called the *spectral dimension* of K. Its value has been explicitly computed only on the Sierpinski gasket and general nested fractals, where $d_s = 2\ln N/\ln(N\rho)$, see e.g. [16]. Posta's result allows us to identify d_s with the homogeneous dimension v of K in great generality, providing a "geometric" interpretation of d_s. See also [8], [41], [37] for earlier remarks in this sense.

(g) From Nash inequality — as shown by Carlen-Kusuoka-Stroock [12] by semigroup techniques - we can derive the following on-diagonal estimate

$$p_t(x, x) \leq ct^{-v/2}, \qquad x \in K, \quad t > 0,$$

for the transition function $p_t(x, y)$ of the diffusion process X_t on K, if the semigroup $T_t = \exp(-tL)$, $t > 0$ is of Feller type. By (d), this is always the case if $v < 2$.

(h) The standard "Brownian motion" on nested fractals is *sub-diffusive*. In fact, its (Euclidean) mean-square deviation obeys the law

$$\langle \ell^2(t) \rangle \equiv E_x|X_t - x|^2 \approx t^{2/d_w},$$

where d_w — the *diffusion exponent* (or, *walk dimension*) — has the value $d_w = \ln(N\rho)/\ln\alpha > 2$. The space-time scaling of this fractal law clearly differs sharply from the analogous dimensionless scaling $E_x|X_t - x|^2 \approx t$ of the classic Brownian motion. However, if we take the mean-square deviation of X_t in the intrinsic metric $r \equiv d$, this fractal "anomaly" disappears. In fact, we find again the usual dimensionless quadratic expression

$$\langle r^2(t) \rangle \equiv E_x d(X_t, x)^2 \approx t .$$

This follows, for example, from the moment expression $E_x|X_t - x|^{2s} \approx t^{2s/d_w}, s > 0$ in [3], *loc. cit.*, see also [32]. This shows, in particular, that we can identify the diffusion exponent d_w with our exponent d_{field}, obtaining a further link between fractal diffusions and Lagrangian formalism.

(i) Estimates of Green functions in the intrinsic metric can also be obtained, whenever $\nu < 2$, or $\nu \geq 2$ and $d \approx d_E$ as in [6,7]. In these estimates, the homogeneous dimension ν plays again — as in the imbeddings above — the role of the Euclidean dimension in the classic theories.

(j) As observed by M. Biroli, a further consequence of the scaled Poincaré inequalities of the previous theorem — together with well known results by Macias-Segovia concerning Lipschtiz functions on homogeneous spaces — is the estimate $d_E \leq cd$, whenever d_E is available on K.

(k) The dispersion relation of the physics literature is $\omega \propto \Lambda(\omega)^{-d_f/d_w}$, where $\Lambda(\omega)$ is the characteristic length of the exitations in the Euclidean metric, [1], [11].

By measuring the characteristic length in the intrinsic metric, we get the dispersion law

$$\omega \propto \lambda(\omega)^{-1} ,$$

in agreement with the interpretation of λ as wavelength.

We conclude with two general remarks. The first one, (ℓ), suggests connections and possible extensions, the second one, (m), points out limits to the role of ν in the effective geometric approach.

(ℓ) The subelliptic distance d_E, associated with Hörmander's fields of degree m in $\mathbb{R}^D$, satisfies the condition $d_E(x, y) \leq c|x - y|^\delta$, where $\delta = 1/m \leq 1$, as shown by Rothschild-Stein [43]. Endowed with the quasi-distance $d(x, y) = |x - y|^\delta$, $\mathbb{R}^D$ becomes a space of homogeneous type. Moreover, scaled Poincaré inequalities on the homogeneous metric balls are also available, Jerison [27], Jerison-Sánchez Calle, *loc. cit.* In view of these results, the variational theory of [6-9] has been applied to uniformly subelliptic operators defined by Hörmander fields, see also [35]. Therefore, both fractal and subelliptic homogeneous spaces can be studied in a common variational framework. A fundamental parameter distinguishing the two cases is the homogeneous dimension ν, typically smaller than D on fractals (and indeed $\nu < 2 \leq D$ on nested fractals) and instead larger than D on subelliptic spaces, [37].

These connections deserve to be further investigated. Recently, self-similar sets under the action of Hörmander's fields have been studied by Strichartz [46] and Olsen [40], see also the references therein. Dynamical properties should also be considered. In this regard, we remark that the Lagrangian approach can be also developed on fractal structures embedded in more general metric spaces than the Euclidean ones considered here, of the kind of Heisenberg's group.

(m) All properties considered in this paper have a variational character. They express the physical response of *volume* elements of the structure to external forces.

This efficient volume behavior of the structure is adequately described, in our effective geometrical approach, by the intrinsic metric $r \equiv d$ and the intrinsic fractal dimension ν.

As far as the diffusion is concerned, this explains, for example, both the estimate $p_t(x, x) \approx t^{-\nu/2}$, that governs the probability of the so-called "return to the origin" of the diffusion, and the quadratic law $\langle r^2(t) \rangle \approx t$. In fact, the probability of returning to the origin is intuitively proportional to the reciprocal of the volume $V(t)$ available on the fractal within the intrinsic diffusion distance $r(t)$, and this in turn scales like $V(t) \propto \langle r^2(t) \rangle^{\nu/2}$, by the very definition of the exponent ν. This leads to the relations $t^{-\nu/2} \approx p_t(x, x) \propto V^{-1}(t) \propto \langle r^2(t) \rangle^{-\nu/2}$, that show the equivalence of the two laws and their common character as examples of volume behavior of the diffusion.

However, this is not enough to fully understand how fractal diffusions behave. In particular, the effective scaling approach based only on the exponent ν is not adequate to describe the long range behavior of the diffusion, classically given by the Gaussian term in the expression of $p_t(x, y)$ for $x \neq y$. It is known from the physical literature that scaling arguments only predict the functional dependence of p_t on the ratio r^2/t, without deciding the Gaussian form of the dependency. Indeed, pure Gaussian behavior is not expected to generalize to fractals. This has to do with the fact that efficient long range diffusion requires in some sense avoiding to revisit previous sites and therefore reflect efficient *path* behavior, which cannot be described by the volume exponent ν alone, see also, in this regard, the remarks by Stanley [44] and Bunde-Havlin [11].

The full description of the probability density p_t demands a further parameter, related to the so-called *chemical distance* of the physical literature, which, in our framework, is the *intrinsic* fractal dimension of the minimum path connecting two sites on the fractal. Details will be elaborate elsewhere, see also [32]. For a survey of physical studies on fractal diffusions, we refer to Havlin-Ben Havrahm [20] and Bunde-Havlin [11].

6. References

1. Alexander S., Orbach R., *Density of states on fractals: "fractons"*, J. Physique Lett. **43** (1982), L-625.
2. Barlow M.T., *Harmonic analysis on fractal spaces*, Sem. Bourbaki, Vol. 1991/92, Astérisque **206** (1992), 345-368.
3. Barlow M.T, Perkins E.A., *Brownian motion on the Sierpinski gasket*, Prob. Theo. Rel. Fields **79** (1988), 543-624.
4. Beurling A., Deny J., *Espaces de Dirichlet, I. Le cas élémentaire*, Acta Math. **99** (1958), 203-224.
5. Beurling A., Deny J., *Dirichlet spaces*, Proc. Nat. Acad. Sci. U.S.A. **45** (1959), 208-215.
6. Biroli M., Mosco U., *Formes de Dirichlet et estimations structurelles dans les milieux discontinus*, C. R. Acad. Sci. Paris Série I, t. 313 (1991), 593-598.
7. Biroli M., Mosco U., *A Saint-Venant type principle for Dirichlet forms on discontinuous media*, Ann. Mat. Pura Appl. (IV) CLXX (1995), 125-181.
8. Biroli M., Mosco U., *Sobolev and isoperimetric inequalities for Dirichlet forms on homogeneous spaces*, Rend. Mat. Acc. Lincei s. 9 **6** (1995), 37-44.
9. Biroli M., Mosco U., *Sobolev inequalities on homogeneous spaces*, Pot. Anal. 4, **4** (1995), 311-324.
10. Biroli M., Tchou N., *Asymptotic behaviour of relaxed Diriclet problems involving a Dirichlet Poincaré form*, Zeit. für Anal. und ihre Anvend, to appear.

11. Bunde A., Havlin S., *Fractals and Disordered Systems*, Springer-Verlag, Berlin Heidelberg, 1991.

12. Carlen E.A., Kusuoka S., Stroock D.W., *Upper bounds for symmetric Markov transition functions*, Ann. Ins. H. Poincaré, n. 2 (1987), 245-287.

13. Coifman R.R., Weiss G., *Analyse harmonique sur certaines éspaces homogenes*, Lect. Notes in Math. 242, Springer V., Berlin-Heidelberg-New York, 1971.

14. Courant R., Friedrichs K.O., Lewy H., *Über partielle Differenzengleichungen der mathematischen Physik*, Math. Ann. 100 (1928), 32-74.

15. Fefferman C.L., Phong D.H, *Sub-elliptic eigenvalue problems*, Conf. on "Harmonic Analysis" in honour of A. Zygmund, **2**, 590-606, Wadsworth Math. Series, 1983.

16. M. Fukushima, *Dirichlet forms, diffusion processes and spectral dimension for nested fractals*, in "Ideas and Methods in Mathematical Analysis, Stochastics and Applications", S. Albeverio et al. eds., Cambridge Univ. Press, 1992, 151-161.

17. Fukushima M., Oshima Y, Takeda M., *Dirichlet forms and Symmetric Markov Processes*, Walter De Gruyter Co., 1995.

18. Fukushima M., Shima T., *On a spectral analysis for the Sierpinski gasket*, Potential Anal. 1:1-35 (1992), 135.

19. Goldstein S., *Random walks and diffusions on fractals*, in "Percolation theory and ergodic theory of infinite particle systems", Minneapolis, Minn. 1984-85, 121-129; IMA Vol. Math. Appl. 8, Springer, New York- Berlin, 1987.

20. Havlin S., Ben-Avraham D., *Diffusion in disordered media*, Adv. in Phys. **36** (1987), 695-799.

21. Hutchinson J.E., *Fractals and selfsimilarity*, Indiana Univ. Math. J. **30** (1981), 713-747.

22. Kigami J., *A harmonic calculus on the Sierpinski spaces*, Japan J. Appl. Math. **6** (1989), 259-290.

23. Kozlov S.M., *Harmonization and homogenization on fractals*, Commun. Math. Phys. **153** (1993), 159-339.

24. Kusuoka S., *A diffusion process on a fractal*, in "Probabilistic methods in Mathematical Physics", Proc. of Taniguchi Int. Symp., Katata and Kyoto, 1985, K. Ito and N. Ikeda eds., Kinokuniya, Tokio, 1987, 251-274.

25. Kusuoka S., *Dirichlet forms on fractals and product of random matrices*, Publ. RIMS Kyoto Univ. **25** (1989), 659-680.

26. Kusuoka S., *Diffusion processes in nested fractals*, Lect. Notes in Math. 1567, Springer V., 1993.

27. Jerison D., *The Poincaré inequality for vector fields satisfying Hörmander's condition*, Duke Math. J. **53** (1986), 503-523.

28. Jerison D., Sánchez Calle A., *Subelliptic, second order differential operators*, Springer V., Lecture Notes in Math. **1277** (1987), 46-77.

29. LeJan Y., *Mesures associées à une forme de Dirichlet. Applications*, Bull. Soc. Math. France **106** (1978), 61-112.

30. Lindstrøm, T., *Brownian motion on nested fractals*, Memoirs AMS n.420, **83** (1990).

31. Mosco U., *Variational metrics on self-similar fractals*, C. R. Acd. Sci. Paris, t. 321 Série I (1995), 715-720.

32. Mosco U., *Invariant field metrics and dynamical scalings on fractals*, to appear.

33. Mosco U., *Variational fractals*, to appear.

34. Mosco U., *Composite media and asymptotic Dirichlet forms*, J. Funct. Anal. **123** n. 2 (1994), 368-421.

35. Mosco U., *Uniformly subelliptic operators with measurable coefficients*, Proc. Conf. on "Nonlinear Analysis – Calculus of Variations", Atti Sem. Mat. Fis. Univ. Modena **43** (1995), 209-224.

36. Mosco U., *Variations and Irregularities*, in "Second Topological Analysis Workshop on Degree, Singularity and Variations: Developments of the Last 25 Years", Univ. "Tor Vergata", Roma June 1995; M. Matzeu and A. Vignoli eds., Birkhäuser, 1997, to appear.

37. Mosco U., *Metric properties of degenerate and fractal media*, Proc. Conf. "Eur-Homogenization", Nice June 1995, D. Cioranescu, A. Damlamian and P. Donato eds., Gakkotosho, to appear.

38. Mosco U., Notarantonio L., *Homogeneous fractal spaces*, Proc. Conf. on "Variational methods for discontinuous structures", Villa Olmo, Como 1994; R. Serapioni and F. Tomarelli eds., Birkäuser, 1996.

39. Nagel E.M., Stein E.M., Wainger S., *Balls and metrics defined by vector fields I: Basic properties*, Acta Math. **155** (1985), 103-147.

40. Olsen L., *A multifractal formalism*, Adv. in Math. **116** (1995), 82-195.

41. Posta G., *Spectral asymptotics for variational fractals*, in preparation.

42. Rammal R., Toulouse G., *Random walks on fractal structures and percolation clusters*, J. Physique Lettres **44** (1983), L13-L22.

43. Rothschild L.P., Stein E.M., *Hypoelliptic differential operators and nilpotent groups*, Acta Math. **137** (1976), 247-320.

44. Stanley H.E., *Fractal and Multifractals: The Interplay of Physics and Geometry*, in "Fractals and Disordered Systems", by A. Bunde and S. Havlin, Springer-Verlag, Berlin Heidelberg, 1991.

45. Stein E.M., *Harmonic analysis*, Princeton Univ. Series, 1994.

46. Strichartz R.S., *Self-similar measures and their Fourier transforms III*, Indiana U. Math. J. **42** 2 (1993), 367-411.

DIPARTIMENTO DI MATEMATICA
UNIVERSITA DI ROMA "LA SAPIENZA"
I-00185 Roma (Italy)
e-mail address: mosco@mat.uniroma1.it

Proceedings of Symposia in Applied Mathematics
Volume **54**, 1998

Approximate Solutions of Nonlinear Conservation Laws and Related Equations

Eitan Tadmor[†]

To Peter Lax and Louis Nirenberg on their 70th birthday

Abstract

During the recent decades there was an enormous amount of activity related to the construction and analysis of modern algorithms for the approximate solution of nonlinear hyperbolic conservation laws and related problems.

To present some aspects of this successful activity, we discuss the analytical tools which are used in the development of convergence theories for these algorithms. These include classical compactness arguments (based on BV a priori estimates), the use of compensated compactness arguments (based on H^{-1}-compact entropy production), measure valued solutions (measured by their negative entropy production), and finally, we highlight the most recent addition to this bag of analytical tools – the use of averaging lemmas which yield new compactness and regularity results for nonlinear conservation laws and related equations.

We demonstrate how these analytical tools are used in the convergence analysis of approximate solutions for hyperbolic conservation laws and related equations. Our discussion includes examples of Total Variation Diminishing (TVD) finite-difference schemes; error estimates derived from the one-sided stability of Godunov-type methods for convex conservation laws (and their multidimensional analogue – viscosity solutions of demi-concave Hamilton-Jacobi equations); we outline, in the one-dimensional case, the convergence proof of finite-element streamline-diffusion and spectral viscosity schemes based on the div-curl lemma; we also address the questions of convergence and error estimates for multidimensional finite-volume schemes on non-rectangular grids; and finally, we indicate the convergence of approximate solutions with underlying kinetic formulation, e.g., finite-volume and relaxation schemes, once their regularizing effect is quantified in terms of the averaging lemma.

AMS(MOS) Classification. Primary 35L65, 35L60. Secondary 65M06, 65M12, 65M15, 65M60, 65M70.

[†]School of Mathematical Sciences, Tel-Aviv University, Tel-Aviv 69978 Israel, and Department of Mathematics, UCLA, Los-Angeles CA 90095; Email: tadmor@math.ucla.edu

Contents

1   Introduction

The construction, analysis and implementation of approximate solutions to nonlinear conservation laws and related equations were the major focus of an enormous amount of activity in recent decades. Modern algorithms were developed for the accurate computation of shock discontinuities, slip lines, and other similar phenomena which could be characterized by spontaneous evolution of change in scales. Such phenomena pose a considerable computational challenge, which is answered, at least partially, by these newly constructed algorithms. New modern algorithms were devised, that achieve one or more of the desirable properties of high-resolution, efficiency, stability — in particular, lack of spurious oscillations, etc. The impact of these new algorithms ranges from the original impetus in the field of Computational Fluid Dynamics (CFD), to the fields oil recovery, moving fronts, image processing,... [74], [137], [131], [1].

We survey a variety of these algorithms for the approximate solution of nonlinear conservation laws. The presentation is neither comprehensive nor complete — the scope is too wide for the present framework[1]. Instead, we focus our attention of the analysis part – more precisely, we discuss the analytical tools which are used to study the stability and convergence of these modern algorithms. We use these analytical issues as our 'touring guide' to provide a readers' digest on the relevant approximate methods, while studying there convergence properties.

Some general references are in order. The theory of hyperbolic conservation laws is covered in [94], [173],[155], [147]. For the theory of their numerical approximation consult [102],[57],[58],[157]. We are concerned with analytical tools which are used in the convergence theories of such numerical approximations. The monograph [49] could be consulted on recent development regarding weak convergence. The reviews of [167], [122, 123] are recommended references for the theory of compensated compactness, and [39, 40],[17] deal with applications to conservation laws and their numerical approximations. Measure-valued solutions in the context of nonlinear conservation laws were introduced in [41]. The articles [61], [52], [44] prove the averaging lemma, and [110],[111],[77] contain applications in the context of kinetic formulation for nonlinear conservation laws and related equations.

Acknowledgments. I thank R. Spigler and S. Venakides for the hospitality during the 1996 Venice conference on "Recent Advances in Partial Differential equations and Applications" honoring Peter Lax and Louis Nirenberg

---

[1]Among the methods omitted from our discussion are Dafermos' polygonal method, [34], the particle method, [171], relaxation algorithms, [173], [19],[82],[124], and Boltzmann schemes, [38], [133],[136].

on their 70th birthday. The initial step of writing this article was taken few years ago, following a lecture delivered at the University di Roma "La Sapienza", and I thank I. Capuzzo Dolcetta and M. Falcone for their invitation.

Research was supported by ONR Contract N00014-91-J-1076 and NSF Grant DMS91-03104.

2 Hyperbolic Conservation Laws

2.1 A very brief overview — m equations in d spatial dimensions

The general set-up consists of m equations in d spatial dimensions

$$\partial_t \rho + \nabla_x \cdot A(\rho) = 0, \quad (t,x) \in \mathbf{R}^+ \times \mathbf{R}^d. \tag{2.1}$$

Here, $A(\rho) := (A_1(\rho), \ldots, A_d(\rho))$ is the d-dimensional flux, and $\rho := (\rho_1(t,x), \ldots, \rho_m(t,x))$ is the unknown m-vector subject to initial conditions $\rho(0,x) = \rho_0(x)$.

The basic facts concerning such nonlinear hyperbolic systems are, consult [94],[112], [35],[155],[57],[147],

- The evolution of *spontaneous* shock discontinuities which requires weak (distributional) solutions of (2.1);

- The existence of possibly infinitely *many* weak solutions of (2.1);

- To *single* out a unique 'physically relevant' weak solution of (2.1), we seek a solution, $\rho = \rho(t,x)$, which can be realized as a viscosity limit solution, $\rho = \lim \rho^\varepsilon$,

$$\partial_t \rho^\varepsilon + \nabla_x \cdot A(\rho^\varepsilon) = \varepsilon \nabla_x \cdot (Q \nabla_x \rho^\varepsilon), \quad \varepsilon Q > 0; \tag{2.2}$$

- *The entropy condition.* The notion of a viscosity limit solution is intimately related to the notion of an *entropy solution*, ρ, which requires that for all convex entropy functions, $\eta(\rho)$, there holds, [93], [88, §5]

$$\partial_t \eta(\rho) + \nabla_x \cdot F(\rho) \leq 0. \tag{2.3}$$

A scalar function, $\eta(\rho)$, is *an entropy function* associated with (2.1), if its Hessian, $\eta''(\rho)$, symmetrizes the spatial Jacobians, $A_j'(\rho)$,

$$\eta''(\rho) A_j'(\rho) = A_j'(\rho)^\top \eta''(\rho), \qquad j = 1, \ldots, d.$$

It follows that in this case there exists an entropy flux, $F(\rho) := (F_1(\rho), \ldots, F_d(\rho))$, which is determined by the compatibility relations,

$$\eta'(\rho)^\top A'_j(\rho) = F'_j(\rho)^\top, \quad j = 1, \ldots, d. \tag{2.4}$$

The basic questions regarding the existence, uniqueness and stability of entropy solutions for general systems are open. Instead, the present trend seems to concentrate on special systems with additional properties which enable to answer the questions of existence, stability, large time behavior, etc. One-dimensional 2×2 systems is a notable example for such systems: their properties can be analyzed in view of the existence of Riemann invariants and a family of entropy functions, [55], [94, §6], [155], [39, 40]. The system of $m \geq 2$ chromatographic equations, [77], is another example for such systems.

The difficulty of analyzing general systems of conservation laws is demonstrated by the following negative result due to Temple, [170], which states that already for systems with $m \geq 2$ equations, there exists no metric, $\mathcal{D}(\cdot; \cdot)$, such that the problem (2.1), (2.3) is contractive, i.e.,

$$\cancel{\exists}\mathcal{D}: \quad \mathcal{D}(\rho^1(t, \cdot); \rho^2(t, \cdot)) \leq \mathcal{D}(\rho^1(0, \cdot); \rho^2(0, \cdot)), \quad 0 \leq t \leq T, \quad (m \geq 2). \tag{2.5}$$

In this context we state the following.

Theorem 2.1 *Assume the system (2.1) is endowed with a one-parameter family of entropy pairs, $(\eta(\rho; c), F(\rho; c))$, $c \in \mathbf{R}^m$, satisfying the symmetry property*

$$\eta(\rho; c) = \eta(c; \rho), \quad F(\rho; c) = F(c; \rho). \tag{2.6}$$

Let ρ^1, ρ^2 be two entropy solutions of (2.1). Then the following a priori estimate holds

$$\int_x \eta(\rho^1(t, x); \rho^2(t, x)) dx \leq \int_x \eta(\rho_0^1(x); \rho_0^2(x)) dx. \tag{2.7}$$

Couple of remarks is is order.

1. Theorem 2.1 seems to circumvent the negative statement of (2.5). This is done by replacing the metric $\mathcal{D}(\cdot; \cdot)$, with the weaker topology induced by a *family* of convex entropies, $\eta(\cdot; \cdot)$. Many physically relevant systems are endowed with at least one convex entropy function (– which in turn, is linked to the hyperbolic character of these systems, [60],[51],[119]). Systems with "rich" families of entropies like those required in Theorem 2.1 are rare, however, consult [146]. The instructive (yet exceptional...) scalar case is dealt in §2.2. If we relax the contractivity requirement, then we find a uniqueness theory for

one-dimensional systems which was recently developed by Bressan and his co-workers, [11]-[14]; Bressan's theory is based on the L^1-stability (rather than contractivity) of the entropy solution operator of one-dimensional systems.

2. Theorem 2.1 is based on the observation that the symmetry property (2.6) is the key ingredient for Kružkov's penetrating ideas in [88], which extends his scalar arguments into the case of general systems. I have not found a written reference of this extension (though it seems to be part of the 'folklore' familiar to some, [36],[148]). For completeness we therefore turn to

Proof of Theorem 2.1(Sketch). $\rho^1(t,x)$ being an entropy solution of (2.1) satisfies the entropy inequality (2.3). We employ the latter with the entropy pair, $(\eta(\rho^1;c), F(\rho^1;c))$ parameterized with $c = \rho^2(\tau, y)$. This tells us that $\rho^1(t,x)$ satisfies

$$\partial_t \eta(\rho^1(t,x); \rho^2(\tau, y)) + \nabla_x \cdot F(\rho^1(t,x); \rho^2(\tau, y)) \leq 0. \qquad (2.8)$$

Let φ_δ denotes a symmetric C_0^∞ unit mass mollifier which converges to Dirac mass in R as $\delta \downarrow 0$; set $\phi_\delta(t - \tau, x - y) := \varphi_\delta(\frac{t-\tau}{2}) \prod_j \varphi_\delta(\frac{x_j - y_j}{2})$ as an approximate Dirac mass in $R^+ \times R^d$. 'Multiplication' of the entropy inequality (2.8) by $\phi_\delta(t - \tau, x - y)$ yields

$$\partial_t(\phi_\delta \eta(\rho^1; \rho^2)) + \nabla_x \cdot (\phi_\delta F(\rho^1; \rho^2)) \leq (\partial_t \phi_\delta)\eta(\rho^1; \rho^2) + (\nabla_x \phi_\delta) \cdot F(\rho^1; \rho^2). \qquad (2.9)$$

A dual manipulation – this time with (τ, y) as the primary integration variables of $\rho^2(\tau, y)$ and (t, x) parameterizing $c = \rho^1(t, x)$, yields

$$\partial_\tau(\phi_\delta \eta(\rho^2; \rho^1)) + \nabla_y \cdot (\phi_\delta F(\rho^2; \rho^1)) \leq (\partial_\tau \phi_\delta)\eta(\rho^2; \rho^1) + (\nabla_y \phi_\delta) \cdot F(\rho^2; \rho^1). \qquad (2.10)$$

We now add the last two inequalities: by the symmetry property (2.1), the sum of the right-hand sides of (2.9) and (2.10) vanishes, whereas by sending δ to zero, the sum of the left-hand sides of (2.9) and (2.10) amounts to

$$\partial_t \eta(\rho^1(t,x); \rho^2(t,x)) + \nabla_x \cdot F(\rho^1(t,x); \rho^2(t,x)) \leq 0.$$

The result follows by spatial integration. ∎

2.2 Scalar conservation laws ($m = 1, d \geq 1$)

The family of admissible entropies in the scalar case consists of *all* convex functions, and the envelope of this family leads to Kružkov's entropy pairs [88]

$$\eta(\rho; c) = |\rho - c|, \quad F(\rho; c) = sgn(\rho - c)(A(\rho) - A(c)), \qquad c \in R. \qquad (2.11)$$

Theorem 2.1 applies in this case and (2.7) now reads

- L^1-*contraction.* If ρ^1, ρ^2 are two entropy solutions of the scalar conservation law (2.1), then

$$\|\rho^2(t, \cdot) - \rho^1(t, \cdot)\|_{L^1(x)} \leq \|\rho_0^2(\cdot) - \rho_0^1(\cdot)\|_{L^1(x)}. \qquad (2.12)$$

Thus, the entropy solution operator associated with scalar conservation laws is L^1-contractive (– or non-expansive to be exact), and hence, by the Crandall-Tartar lemma [32], it is also monotone

$$\rho_0^2(\cdot) \geq \rho_0^1(\cdot) \Longrightarrow \rho^2(t, \cdot) \geq \rho^1(t, \cdot). \qquad (2.13)$$

Early constructions of approximate solutions for scalar conservation laws, most notably — finite-difference approximations, utilized this monotonicity property to construct convergent schemes, [30], [141]. Monotone approximations are limited, however, to first-order accuracy [71]. (We shall say more on the issue of accuracy in §3.1). At this stage we note that the limitation of first-order accuracy for monotone approximations, can be avoided if L^1-contractive solutions are replaced with (the weaker) requirement of bounded variation solutions.

- *TV bound.* The solution operator associated with (2.1) is translation invariant. Comparing the scalar entropy solution, $\rho(t, \cdot)$, with its translate, $\rho(t, \cdot + \Delta x)$, the L^1-contraction statement in (2.12) yields the TV bound, [172],

$$\|\rho(t, \cdot)\|_{BV} \leq \|\rho_0(\cdot)\|_{BV}, \quad \|\rho(t, \cdot)\|_{BV} := \sup_{\Delta x \neq 0} \frac{\|\rho(t, \cdot + \Delta x) - \rho(t, \cdot)\|_{L^1}}{\Delta x}.$$
$$(2.14)$$

Construction of scalar entropy solutions by TV-bounded approximations were used in the pioneering works of Olĕinik [128], Vol'pert [172], Kružkov [88] and Crandall [28]. In the one-dimensional case, the TVD property (2.14) enables to construct convergent difference schemes with high-order (> 1) resolution; Harten initiated the construction of high-resolution TVD schemes in [69], following the earlier works [6], [98]. A whole generation of TVD schemes was then developed during the beginning of the '80s; some aspects of these developments can be found in §3.2-§3.4.

2.3   One dimensional systems ($m \geq 1, d = 1$)

We focus our attention on one-dimensional hyperbolic systems governed by

$$\partial_t \rho + \partial_x A(\rho) = 0, \quad (t, x) \in R^+ \times R, \qquad (2.15)$$

and subject to initial condition, $\rho(0,x) = \rho_0(x)$. The hyperbolicity of the system (2.15) is understood in the sense that its Jacobian, $A'(\rho)$, has a complete *real* eigensystem, $(a_k(\rho), r_k(\rho)), k = 1, \ldots, m$. For example, the existence of a convex entropy function guarantees the symmetry of $A'(\rho)$ (— w.r.t. $\eta''(\rho)$), and hence the complete real eigensystem. For most of our discussion we shall assume the stronger *strict hyperbolicity*, i.e, distinct real eigenvalues, $a_k(\rho) \neq a_j(\rho)$.

A fundamental building block for the construction of approximate solutions in the one-dimensional case is the solution of *Riemann's problem*.

2.3.1   Riemann's problem

Here one seeks a weak solution of (2.15) subject to the piecewise constant initial data

$$\rho(x,0) = \begin{cases} \rho_\ell, & x < 0 \\ \rho_r, & x > 0. \end{cases} \tag{2.16}$$

The solution is composed of m simple waves, each of which is associated with one (right-)eigenpair, $(a_k(\rho), r_k(\rho))$, $1 \le k \le m$. There are three types of such waves: if the k-th field is *genuinely nonlinear* in the sense that $r_k \cdot \nabla_\rho a_k \neq 0$, these are either k-shock or k-rarefaction waves; or, if the k-th field is *linearly degenerate* in the sense that $r_k \cdot \nabla_\rho a_k \equiv 0$, this is a k-th contact wave.

These three simple waves are *centered*, depending on $\xi = \frac{x}{t}$ (which is to be expected from the dilation invariance of (2.15),(2.16)). The structure of these three centered waves is as follows:

- A k-shock discontinuity of the form

$$\rho(\xi) = \begin{cases} \rho_\ell, & \xi < s \\ \rho_r, & \xi > s; \end{cases}$$

  here s denotes the shock speed which is determined by a Rankine-Hugoniot relation so that $a_k(\rho_\ell) > s > a_k(\rho_r)$.

- A k-rarefaction wave, $\rho(\xi)$, which is directed along the corresponding k-th eigenvector, $\dot{\rho}(\xi) = r_k(\rho(\xi))$. Here r_k is the normalized k-eigenvector, $r_k \cdot \nabla a_k \equiv 1$ so that the gap between $a_k(\rho_\ell) < a_k(\rho_r)$ is filled with a fan of the form

$$a_k(\rho(\xi)) = \begin{cases} a_k(\rho_\ell), & \xi < a_k(\rho_\ell) \\ \xi, & a_k(\rho_\ell) < \xi < a_k(\rho_r) \\ a_k(\rho_r), & a_k(\rho_r) < \xi \end{cases}$$

- A k-contact discontinuity of the form

$$\rho(\xi) = \begin{cases} \rho_\ell, & \xi < s \\ \rho_r, & \xi > s \end{cases}$$

where s denotes the shock speed which is determined by a Rankine-Hugoniot relation so that $a_k(\rho_\ell) = s = a_k(\rho_r)$.

We are concerned with *admissible* systems — systems which consist of either genuinely nonlinear or linearly degenerate fields. We refer to [92] for the full story which concludes with the celebrated

Theorem 2.2 (Lax solution of Riemann's problem) *The strictly hyperbolic admissible system (2.15), subject to Riemann initial data (2.16) with $\rho_\ell - \rho_r$ sufficiently small, admits a weak entropy solution, which consists of shock- rarefaction- and contact-waves.*

For a detailed account on the solution of Riemann problem consult [16]. An extension to a generalized Riemann problem subject to piecewise-linear initial data can be found in [5], [99]. In this context we also mention the *approximate* Riemann solvers, which became useful computational alternatives to Lax's construction. Roe introduced in [138] a linearized Riemann solver, which resolves jumps discontinuities solely in terms of shock waves. Roe's solver has the computational advantage of sharp resolution (at least when there is one dominant wave per computational cell); it may lead, however, to unstable shocks. Osher and Solomon in [130] used, instead, an approximate Riemann solver based solely on rarefaction fans; one then achieves stability at the expense of deteriorated resolution of shock discontinuities.

2.3.2   Godunov, Lax-Friedrichs and Glimm schemes

We let $\rho^{\Delta x}(t, x)$ be the entropy solution in the slab $t^n \le t < t + \Delta t$, subject to piecewise constant data $\rho^{\Delta x}(t = t^n, x) = \sum \rho_\nu^n \chi_\nu(x)$. Here χ denotes the usual indicator function, $\chi_\alpha(x) := 1_{\{|x - \alpha\Delta x| \le \Delta x/2\}}$. Observe that in each slab, $\rho^{\Delta x}(t, x)$ consists of successive *noninteracting* Riemann solutions, at least for a sufficiently small time interval Δt, for which the CFL condition, $\Delta t / \Delta x \max |a_k(\rho)| \le \frac{1}{2}$ is met. In order to realize the solution in the next time level, $t^{n+1} = t^n + \Delta t$, it is extended with a jump discontinuity across the line t^{n+1}, by projecting it back into the finite-dimensional space of piecewise constants. Different projections yield different schemes. We recall the basic three.

Godunov Scheme. Godunov scheme [59] sets

$$\rho^{\Delta x}(t^{n+1}, x) = \sum_\nu \bar{\rho}_\nu^{n+1} \chi_\nu(x),$$

where $\bar{\rho}_\nu^{n+1}$ stands for the cell-average,

$$\bar{\rho}_\nu^{n+1} := \frac{1}{\Delta x} \int_x \rho^{\Delta x}(t^{n+1} - 0, x)\chi_\nu(x)dx,$$

which could be explicitly evaluated in terms of the flux of Riemann solution across the cell interfaces at $x_{\nu\pm\frac{1}{2}}$,

$$\bar{\rho}_\nu^{n+1} = \bar{\rho}_\nu^n - \frac{\Delta t}{\Delta x}\Big\{ A(\rho^{\Delta x}(t^{n+\frac{1}{2}}, x_{\nu+\frac{1}{2}})) - A(\rho^{\Delta x}(t^{n+\frac{1}{2}}, x_{\nu-\frac{1}{2}})) \Big\}. \quad (2.17)$$

Godunov scheme had a profound impact on the field of Computational Fluid Dynamics. His scheme became the forerunner for a large class of upwind finite-volume methods which are evolved in terms of (exact or approximate) Riemann solvers. In my view, the most important aspect of what Richtmyer & Morton describe as Godunov's "ingenious method" ([140, p. 338]), lies in its *global* point of view: one does not simply evolve discrete pointvalues $\{\rho_\nu^n\}$, but instead, one evolves a globally defined solution, $\rho^{\Delta x}(t, x)$, which is realized in terms of its discrete averages, $\{\bar{\rho}_\nu^n\}$.

Lax-Friedrichs Scheme. If the piecewise constant projection is carried out over alternating staggered grids, $\bar{\rho}_{\nu+\frac{1}{2}}^{n+1} := \frac{1}{\Delta x}\int_x \rho^{\Delta x}(t^{n+1}-0, x)\chi_{\nu+\frac{1}{2}}(x)dx$, then one effectively integrates 'over the Riemann fan' which is centered at $(x_{\nu+\frac{1}{2}}, t^n)$. This recovers the Lax-Friedrichs (LxF) scheme, [91], with an explicit recursion formula for the evolution of its cell-averages which reads

$$\bar{\rho}_{\nu+\frac{1}{2}}^{n+1} = \frac{\bar{\rho}_\nu^n + \bar{\rho}_{\nu+1}^n}{2} - \frac{\Delta t}{\Delta x}\Big\{ A(\bar{\rho}_{\nu+1}^n) - A(\bar{\rho}_\nu^n) \Big\}. \quad (2.18)$$

The Lax-Friedrichs scheme had a profound impact on the construction and analysis of approximate methods for time-dependent problems, both linear problems [50] and nonlinear systems [91]. The Lax-Friedrichs scheme was and still is the stable, all purpose benchmark for approximate solution of nonlinear systems.

Both Godunov and Lax-Friedrichs schemes realize the exact solution operator in terms of its finite-dimensional cell-averaging projection. This explains the versatility of these schemes, and at the same time, it indicates their limited resolution due to the fact that waves of different families that are averaged together at each computational cell.

Glimm Scheme. Rather than averaging, Glimm's scheme, [54], keeps its sharp resolution by *randomly sampling* the evolving Riemann waves,

$$\rho^{\Delta x}(t^{n+1}, x) = \sum_\nu \rho^{\Delta x}(t^{n+1} - 0, x_{\nu+\frac{1}{2}} + r^n \Delta x)\chi_{\nu+\frac{1}{2}}(x).$$

This defines the Glimm's approximate solution, $\rho^{\Delta x}(t, x)$, depending on the mesh parameters $\Delta x \equiv \lambda \Delta t$, and on the set of random variable $\{r^n\}$, uniformly distributed in $[-\frac{1}{2}, \frac{1}{2}]$. In its deterministic version, Liu [113] employs equidistributed rather than a random sequence of numbers $\{r^n\}$.

Glimm solution, $\rho^{\Delta x}(t, x)$, was then used to construct a solution for one-dimensional admissible systems of conservation laws. Glimm's celebrated theorem, [54], is still serving today as the cornerstone for existence theorems which are concerned with general one-dimensional systems, e.g. [113],[20],[144].

Theorem 2.3 (Existence in the large) . *There exists a weak entropy solution, $\rho(t, \cdot) \in L^\infty[[0, T], BV \cap L^\infty(R_x)]$, of the strictly hyperbolic system (2.15), subject to initial conditions with sufficiently small variation, $\|\rho_0(\cdot)\|_{BV \cap L^\infty(R_x)} \leq \epsilon$.*

Glimm's scheme has the advantage of retaining sharp resolution, since in each computational cell, the local Riemann solution is realized by a randomly chosen 'physical' Riemann wave. Glimm's scheme was turned into a computational tool known as the Random Choice Method (RCM) in [22], and it serves as the building block inside the front tracking method of Glimm and his co-workers, [56], [21].

2.4   Multidimensional systems ($m > 1, d > 1$)

Very little rigor is known on m conservation laws in d spatial dimensions once $(m-1)(d-1)$ becomes positive, i.e., general multidimensional systems. We address few major achievements.

Short time existence. For H^s-initial data ρ_0, with $s > \frac{d}{2}$, an H^s-solution exists for a time interval $[0, T]$, with $T = T(\|\rho_0\|_{H^s})$, consult e.g, [83],[78, §5.3].

Short time existence – piecewise analytic data. An existence result conjectured by Richtmyer was proved by Harabetian in terms of a Cauchy-Kowalewski type existence result [67].

Short time stability – piecewise smooth shock data. Existence   for piecewise smooth initial data where smoothness regions are separated by shock discontinuities was studied in [117],[106].

Riemann problem. Already in the $d = 2$-dimensional case, the collection of simple waves and their composed interaction in the construction of Riemann solution (– subject to piecewise constant initial data), is considerably more complicated than in the one-dimensional setup. We refer to the recent book [33] for a detailed discussion.

Compressible Euler equations. These system of $m = 5$ equations governing the evolution of density, 3-vector of momentum and Energy in $d = 3$-space variables was – and still is, the prime target for further developments in our understanding of general hyperbolic conservation laws. We refer to Majda, [117], for a definitive summary of this aspect.

3   Finite Difference Methods – TVD Schemes

We begin by covering the space and time variables with a discrete grid: it consists of time-steps of size Δt and rectangular spatial cells of size $\Delta x := (\Delta x_1, \ldots, \Delta x_d)$. Let $\mathcal{C}_\nu$ denotes the cell which is centered around the gridpoint $x_\nu = \nu \Delta x := (\nu_1 \Delta x_1, \ldots, \nu_d \Delta x_d)$, and let $\{\rho_\nu^n\}$ denote the gridfunction associated with this cell at time $t^n = n\Delta t$. The gridfunction $\{\rho_\nu^n\}$ may represent approximate gridvalues, $\rho(t^n, x_\nu)$, or approximate cell averages, $\bar{\rho}(t^n, x_\nu)$ (as in the Godunov and LxF schemes), or a combination of higher moments, e.g., [23].

To construct a finite difference approximation of the conservation law (2.1), one introduce a discrete *numerical flux*, $H(\rho^n) := (H_1(\rho^n), \ldots, H_d(\rho^n))$, where $H_j(\rho^n) = H_j(\rho_{\nu-p}^n, \ldots, \rho_{\nu+q}^n)$ is an approximation to the $A_j(\rho^n)$ flux across the interface separating the cell $\mathcal{C}_\nu$ and its neighboring cell on the x_j's direction, $\mathcal{C}_{\nu+e_j}$. Next, exact derivatives in (2.1) are replaced by divided differences: the time-derivative is replaced with forward time difference, and spatial derivatives are replaced by spatial divided differences expressed in terms of $D_{+x_j}\phi_\nu := (\phi_{\nu+e_j} - \phi_\nu)/\Delta x_j$. We arrive at the finite-difference scheme of the form

$$\rho_\nu^{n+1} = \rho_\nu^n - \Delta t \sum_{j=1}^{d} D_{+x_j} H_j(\rho_{\nu-p}^n, \ldots, \rho_{\nu+q}^n). \tag{3.1}$$

The essential feature of the difference schemes (3.1) is their *conservation form*: perfect derivatives in (2.1) are replaced here by 'perfect differences'. It implies that the change in mass over any spatial domain Ω, $\sum_{\{\nu | x_\nu \in \Omega\}} \rho_\nu^{n+1} |\mathcal{C}_\nu| - \sum_{\{\nu | x_\nu \in \Omega\}} \rho_\nu^n |\mathcal{C}_\nu|$, depends solely on the discrete flux across the boundaries of that domain. This is a discrete analogue for the notion of a weak solution of (2.1). In their seminal paper [96], Lax & Wendroff introduced the notion of conservative schemes, and prove that their *strong* limit solutions are indeed weak solutions of (2.1).

Theorem 3.1 (Lax & Wendroff [96]) *Consider the conservative differ-ence scheme (3.1), with consistent numerical flux so that $H_j(\rho, \ldots, \rho) = A_j(\rho)$. Let $\Delta t \downarrow 0$ with fixed grid-ratios $\lambda_j := \frac{\Delta t}{\Delta x_j} \equiv Const_j$, and let $\rho^{\Delta t} = \{\rho_\nu^n\}$ denote the corresponding solution (parameterized w.r.t. the vanishing grid-size). Assume that $\rho^{\Delta t}$ converges strongly, $s\lim \rho^{\Delta t}(t^n, x_\nu) = \rho(t, x)$, then $\rho(x, t)$ is a weak solution of the conservation law (2.1).*

The Lax-Wendroff theorem plays a fundamental role in the development of the so called 'shock capturing' methods. Instead of tracking jump dis-continuities (– by evolving the smooth pieces of the approximate solution on both sides of such discontinuities), conservative schemes capture a dis-cretized version of shock discontinuities. Equipped with the Lax-Wendroff theorem, it remains to prove strong convergence, which leads us to discuss the compactness of $\{\rho_\nu^n\}$.

3.1   Compactness arguments ($m = d = 1$)

We deal with scalar gridfunctions, $\{\rho_\nu^n\}$, defined on the one-dimensional Cartesian grid $x_\nu := \nu \Delta x, t^n := n \Delta t$ with fixed mesh ratio $\lambda := \frac{\Delta t}{\Delta x}$. The total variation of such gridfunction at time-level t^n is given by $\sum_\nu |\Delta \rho_{\nu+\frac{1}{2}}^n|$, where $\Delta \rho_{\nu+\frac{1}{2}}^n := \rho_{\nu+1}^n - \rho_\nu^n$. It is said to be total-variation-diminishing (TVD) if

$$\sum_\nu |\Delta \rho_{\nu+\frac{1}{2}}^n| \leq \sum_\nu |\Delta \rho_{\nu+\frac{1}{2}}^0|. \tag{3.2}$$

Clearly, the TVD condition (3.2) is the discrete analogue of the scalar TV-bound (2.14). Approximate solutions of difference schemes which respect the TVD property (3.2), share the following desirable properties:

- *Convergence* – by Helly's compactness argument, the piecewise-constant approximate solution, $\rho^{\Delta x}(t^n, x) = \sum_\nu \rho_\nu^n \chi_\nu(x)$, converges strongly to a limit function, $\rho(t^n, x)$ as we refine the grid, $\Delta x \downarrow 0$. This together with equicontinuity in time and the Lax-Wendroff theorem, yield a weak solution, $\rho(t, x)$, of the conservation law (2.1).

- *Spurious oscillations* – are excluded by the TVD condition (2.14).

- *Accuracy* – is not restricted to the first-order limitation of monotone schemes. To be more precise, let us use $\rho^{\Delta t}(t, x)$ to denote a global realization (say – piecewise polynomial interpolant) of the approxi-mate solution $\rho_\nu^n \sim \rho^{\Delta t}(t^n, x_\nu)$. The *truncation error* of the difference scheme is the amount by which the approximate solution, $\rho^{\Delta t}(t, x)$,

fails to satisfy the conservation laws (2.1). The difference scheme is
α-order accurate if its truncation error is, namely,

$$\|\partial_t \rho^{\Delta t} + \nabla_x \cdot A(\rho^{\Delta t})\| = \mathcal{O}((\Delta t)^\alpha). \tag{3.3}$$

(Typically, a strong norm $\|\cdot\|$ is used which is appropriate to the prob-
lem; in general, however, accuracy is indeed a norm-dependent quantity). Consider for example, monotone difference schemes. Monotone
schemes are characterized by the fact that ρ_ν^{n+1} is an increasing function of the preceding gridvalues which participate in its stencil (3.1),
$\rho_{\nu-p}^n, \ldots, \rho_{\nu+q}^n$ (— so that the monotonicity property (2.13) holds) .
A classical result of Harten, Hyman & Lax [71] states that monotone
schemes are at most first-order accurate. TVD schemes, however, are
not restricted to this first-order accuracy limitation (at least in the
one-dimensional case[2]). We demonstrate this point in the context of
second-order TVD difference schemes.

3.2   TVD difference schemes

We follow the presentation in [132]. The starting point is the viscosity reg-
ularization (2.2) with vanishing viscosity of order $\varepsilon = \Delta x / 2\lambda$ (recall that λ
denotes the fixed mesh-ratio, $\Delta t / \Delta x$),

$$\partial_t \rho + \partial_x A(\rho) = \frac{\Delta x}{2\lambda} \partial_x (Q \partial_x \rho). \tag{3.4}$$

We discretize (3.4) with the help of

(i) *An approximate flux,* $\tilde{A}_\nu^n = \tilde{A}(\rho_{\nu-p+1}^n, \ldots, \rho_\nu^n, \ldots, \rho_{\nu+p-1}^n) \approx A(\rho_\nu^n)$;

(ii) *A numerical viscosity coefficient,* $Q_{\nu+\frac{1}{2}}^n = Q(\rho_{\nu-p+1}^n, \ldots, \rho_{\nu+p}^n)$.

These discrete quantities are used to replace the temporal and spatial derivatives in (3.4) by appropriate forward and centered divided differences. The
resulting finite difference method reads

$$\rho_\nu^{n+1} = \rho_\nu^n - \frac{\lambda}{2}\left\{\tilde{A}_{\nu+1}^n - \tilde{A}_{\nu-1}^n\right\} + \frac{1}{2}\left\{Q_{\nu+\frac{1}{2}}^n \Delta\rho_{\nu+\frac{1}{2}}^n - Q_{\nu-\frac{1}{2}}^n \Delta\rho_{\nu-\frac{1}{2}}^n\right\}. \tag{3.5}$$

Observe that (3.5) can be put into conservation form (3.1), in terms of the
numerical flux

$$H_{\nu+\frac{1}{2}}(\rho^n) = \frac{1}{2}(\tilde{A}_{\nu+1}^n + \tilde{A}_\nu^n) - \frac{1}{2\lambda}Q_{\nu+\frac{1}{2}}^n \Delta\rho_{\nu+\frac{1}{2}}^n.$$

---

[2]Consult [64], regarding the first-order accuracy limitation for multidimensional $d > 1$
TVD schemes. This limitation is linked to the lack of a 'proper' isotropic definition for
the total-variation of multidimensional gridfunctions.

Also, the stencil on the right of (3.5) occupies the $(2p+1)$ neighboring grid-values, $\rho_{\nu-p}^n, \ldots, \rho_{\nu+p}^n$. Thus, (3.5) is a $(2p+1)$-points conservative difference-scheme. Harten in [69], was the first to identify a useful sufficient criterion for the TVD property for such scalar difference schemes. Harten's criterion, in its reformulation from [161], states that the difference scheme (3.5) is TVD provided it contains 'enough viscosity' in the sense that

$$\lambda \left| \frac{\Delta \tilde{A}_{\nu+\frac{1}{2}}^n}{\Delta \rho_{\nu+\frac{1}{2}}^n} \right| \leq Q_{\nu+\frac{1}{2}}^n \leq 1. \tag{3.6}$$

We distinguish between two types of TVD schemes, depending on the size of their stencils.

3.2.1   Three-point schemes

Three-point schemes $(p = 1)$ are the simplest ones – their stencil on the right of (3.5) occupies the three neighboring gridvalues, $\rho_{\nu-1}^n, \rho_\nu^n, \rho_{\nu+1}^n$. In this case, $\tilde{A}_\nu^n \equiv A(\rho_\nu^n)$, so that three-point schemes take the form

$$\rho_\nu^{n+1} = \rho_\nu^n - \frac{\lambda}{2} \left\{ A(\rho_{\nu+1}^n) - A(\rho_{\nu-1}^n) \right\} + \frac{1}{2} \left\{ Q_{\nu+\frac{1}{2}}^n \Delta \rho_{\nu+\frac{1}{2}}^n - Q_{\nu-\frac{1}{2}}^n \Delta \rho_{\nu-\frac{1}{2}}^n \right\}. \tag{3.7}$$

Thus, three-point schemes are identified solely by their numerical viscosity coefficient, $Q_{\nu+\frac{1}{2}}^n = Q(\rho_\nu^n, \rho_{\nu+1}^n)$, which characterize the TVD condition (corresponding to (3.6))

$$\lambda |a_{\nu+\frac{1}{2}}^n| \leq Q_{\nu+\frac{1}{2}}^n \leq 1, \qquad a_{\nu+\frac{1}{2}}^n := \frac{\Delta A_{\nu+\frac{1}{2}}^n}{\Delta \rho_{\nu+\frac{1}{2}}^n}. \tag{3.8}$$

The schemes of Roe [138], Godunov [59], and Engquist-Osher (EO) [46], are canonical examples of *upwind* schemes, associated with (increasing amounts of) numerical viscosity coefficients, which are given by,

$$\begin{aligned} Q_{\nu+\frac{1}{2}}^{\text{Roe}} &= \lambda |a_{\nu+\frac{1}{2}}^n|, \\ Q_{\nu+\frac{1}{2}}^{\text{Godunov}} &= \lambda \max_{\zeta \in \mathcal{C}_{\nu+\frac{1}{2}}} \left[\frac{A(\rho_{\nu+1}^n) - 2A(\zeta) + A(\rho_\nu^n)}{\Delta \rho_{\nu+\frac{1}{2}}^n} \right], \\ Q_{\nu+\frac{1}{2}}^{\text{EO}} &= \lambda \frac{1}{\Delta \rho_{\nu+\frac{1}{2}}^n} \int_{\rho_\nu^n}^{\rho_{\nu+1}^n} |A'(\zeta)| d\zeta. \end{aligned}$$

The viscosity coefficients of the three upwind schemes are the same, $Q_{\nu+\frac{1}{2}}^n = \lambda |a_{\nu+\frac{1}{2}}^n|$, except for their different treatment of sonic points (where

$a(\rho_\nu^n) \cdot a(\rho_{\nu+1}^n) < 0$). The Lax-Friedrichs (LxF) scheme (2.18) is the canonical *central* scheme. It has a larger numerical viscosity coefficient $Q_{\nu+\frac{1}{2}}^{LxF} \equiv 1$.

All the three-point TVD schemes are limited to first-order accuracy. Indeed, condition (3.8) is in fact *necessary* for the TVD property of three-point schemes, [160], and hence it excludes numerical viscosity associated with the second-order Lax-Wendroff scheme, [96], $Q_{\nu+\frac{1}{2}}^{LW} = \lambda^2(a_{\nu+\frac{1}{2}}^n)^2$. Therefore, scalar TVD schemes with more than first-order accuracy require at least five-point stencils.

3.2.2 Five-point schemes

Following the influential works of Boris & Book [6], van Leer [98], Harten [69], Osher [129], Roe [138] and others, many authors have constructed second order TVD schemes, using five-point (– or wider) stencils. For a more complete account of these works we refer to the recent books by LeVeque, [102], and Godlewski & Raviart, [57]. A large number of these schemes were constructed as second-order upgraded versions of the basic three-point *upwind* schemes. The FCT scheme of Boris & Book, [6], van Leer's MUSCL scheme [98], and the ULTIMATE scheme of Harten, [69], are prototype for this trend.

We quote here a five-point TVD scheme of Nessyahu-Tadmor (NT) [125], which is a second-order upgraded version of the *central* LxF scheme (2.18): we use the same viscosity coefficient, $Q_{\nu+\frac{1}{2}} \equiv 1$, but we augmented it with a modified approximate flux, $\tilde{A}_\nu$; expressed in terms of the cell averages, $\bar{\rho}_\nu^n$, and the midvalues $\rho_\nu^{n+\frac{1}{2}} := \bar{\rho}_\nu^n - \frac{\lambda}{2}(A(\bar{\rho}_\nu^n))'$, this modified flux is given by $\tilde{A}_\nu = A(\rho_\nu^{n+\frac{1}{2}}) + (\rho_\nu^n)'/2\lambda$. Using these quantities in the viscosity form (3.5) we end up with a second-order predictor-corrector scheme, which admits a LxF-like staggered form (2.18)

$$\rho_\nu^{n+\frac{1}{2}} = \bar{\rho}_\nu^n - \frac{\lambda}{2}(A(\bar{\rho}_\nu^n))', \tag{3.9}$$

$$\bar{\rho}_{\nu+\frac{1}{2}}^{n+1} = \frac{\bar{\rho}_\nu^n + \bar{\rho}_{\nu+1}^n}{2} + $$
$$- \frac{(\rho_\nu^n)' - (\rho_{\nu+1}^n)'}{8} - \frac{\Delta t}{\Delta x}\left\{ A(\rho_{\nu+1}^{n+\frac{1}{2}}) - A(\rho_\nu^{n+\frac{1}{2}}) \right\}. \tag{3.10}$$

Here, $\{w_\nu'\}$ denotes the *discrete numerical derivative* of an arbitrary grid-function $\{w_\nu\}$. The choice $w_\nu' \equiv 0$ recovers the original first-order LxF scheme (2.18). Second-order accuracy requires $w_\nu' \sim \Delta x \partial_x w(x_\nu)$; a prototype example is the so called min-mod limiter,

$$w'_\nu = \frac{1}{2}(s_{\nu-\frac{1}{2}} + s_{\nu+\frac{1}{2}}) \cdot \min\{|\Delta w_{\nu-\frac{1}{2}}|, |\Delta w_{\nu+\frac{1}{2}}|\}, \quad s_{\nu+\frac{1}{2}} := sgn(\Delta w_{\nu+\frac{1}{2}}).$$
(3.11)

With this choice of a limiter, the central NT scheme (3.9)-(3.10) satisfies the TVD condition (3.6), and at the same time, it retains formal second order accuracy (at least away from extreme gridvalues, ρ_ν, where $\rho'_\nu = s_{\nu-\frac{1}{2}} + s_{\nu+\frac{1}{2}} = 0$).

We conclude we few additional remarks.

Limiters A variety of discrete TVD limiters like (3.11) was explored during the '80s, e.g, [159] and the references therein. For example, a generalization of (3.11) is provided by the family of min-mod limiters depending on tuning parameters, $0 < \theta_{\nu\pm\frac{1}{2}} < 1$,

$$w'_\nu(\theta) = \frac{1}{2}(s_{\nu-\frac{1}{2}} + s_{\nu+\frac{1}{2}}) \times$$
$$\min\{\theta_{\nu-\frac{1}{2}}|\Delta w_{\nu-\frac{1}{2}}|, \frac{1}{2}|w_{\nu+1} - w_{\nu-1}|, \theta_{\nu+\frac{1}{2}}|\Delta w_{\nu+\frac{1}{2}}|\} \quad (3.12)$$

An essential feature of these limiters is *co-monotonicity*: they are 'tailored' to produce piecewise-linear reconstruction of the form $\sum[w_\nu + \frac{1}{\Delta x}w'_\nu(x - x_\nu)]\chi_\nu(x)$, which is co-monotone with (and hence, share the TVD property of –) the underlying piecewise-constant approximation $\sum w_\nu\chi_\nu(x)$. Another feature is the limiting property at extrema gridvalues (where $\rho'_\nu = 0$), which is necessary in order to satisfy the TVD property (3.2). In particular, these limiters are necessarily *nonlinear* in the sense of their dependence on the discrete gridfunction.

Systems (One-dimensional problems). The question of convergence for approximate solution of hyperbolic systems is tied to the question of existence of an entropy solution – in both cases there are no general theories with $m > 1$ equations[3]. Nevertheless, the ingredients of scalar high-resolution schemes were successfully integrated in the approximate solution of system of conservation laws.

Many of these high-resolution methods for systems, employ the Godunov approach, where one evolves a global approximation which is realized as a piecewise polynomial,

$$\rho(x, t^n) = \sum_j p_j(x)\chi_j(x), \quad \bar{p}_\nu(x_\nu) = \bar{\rho}^n_\nu. \quad (3.13)$$

---

[3]There is a large literature concerning two equations – the 2×2 p-system and related equations are surveyed in [155].

Typically, this piecewise polynomial approximate solution is recon-structed from the previously computed cell averages, $\{\bar{\rho}_\nu^n\}$, and in this context we may distinguish between two main classes of methods: *up-wind* and *central* methods.

Upwind schemes evaluate cell averages at the center of the piecewise polynomial elements; integration of (2.15) over $\mathcal{C}_\nu \times [t^n, t^{n+1}]$ yields

$$\bar{\rho}_\nu^{n+1} = \bar{\rho}_\nu^n - \frac{1}{\Delta x}\left[\int_{\tau=t^n}^{t^{n+1}} f(\rho(\tau, x_{\nu+\frac{1}{2}},))d\tau - \int_{\tau=t^n}^{t^{n+1}} f(\rho(\tau, x_{\nu-\frac{1}{2}},))d\tau\right].$$

This in turn requires the evaluation of fluxes along the discontinuous cell interfaces, $(\tau \times x_{\nu+\frac{1}{2}})$. Consequently, upwind schemes must take into account the characteristic speeds along such interfaces. Special attention is required at those interfaces in which there is a combina-tion of forward- and backward-going waves, where it is necessary to decompose the "Riemann fan" and determine the separate contribu-tion of each component by tracing "the direction of the wind". These characteristic decompositions (– using exact or approximate Riemann solvers) enable to solve with high resolution the corresponding charac-teristic variables. At the same time, It is the need to follow these characteristic variables which greatly complicates the upwind algo-rithms, making them difficult to implement and generalize to complex systems. The original first-order accurate Godunov scheme (2.17) is the forerunner for all other upwind Godunov-type schemes. A vari-ety of second- and higher-order sequels to Godunov upwind scheme were constructed, analyzed and implemented with great success dur-ing the seventies and eighties, starting with van-Leer's MUSCL scheme [98], followed by [138, 69, 129, 26]. These methods were subsequently adapted for a variety of nonlinear related systems, ranging from incom-pressible Euler equations, [4], [45], to reacting flows, semiconductors modeling, ... We shall say more about these methods in §3.4 below. At this point we refer to [58, 102] and the references therein a for a more complete accounts on these developments.

In contrast to upwind schemes, central schemes evaluate staggered cell averages at the breakpoints between the piecewise polynomial el-ements,

$$\bar{\rho}_{\nu+\frac{1}{2}}^{n+1} = \bar{\rho}_{\nu+\frac{1}{2}}^n - \frac{1}{\Delta x}\left[\int_{\tau=t^n}^{t^{n+1}} f(\tau, \rho(x_{\nu+1}))d\tau - \int_{\tau=t^n}^{t^{n+1}} f(\rho(\tau, x_\nu))d\tau\right].$$

Thus, averages are integrated over the entire Riemann fan, so that the corresponding fluxes are now evaluated at the smooth centers of the cells, (τ, x_ν). Consequently, costly Riemann-solvers required in the upwind framework, can be now replaced by straightforward quadrature rules. The first-order Lax-Friedrichs (LxF) scheme (2.18) is the canonical example of such central difference schemes. The LxF scheme (like Godunov's scheme) is based on a piecewise constant approximate solution, $p_\nu(x) = \bar{p}_\nu$. Its Riemann-solver-free recipe, however, is considerably simpler. Unfortunately, the LxF scheme introduces excessive numerical viscosity (already in the scalar case outlined in §3.2.1 we have $Q^{LxF} \equiv 1 > Q^{\mathrm{Godunov}}$), resulting in relatively poor resolution. The central scheme (3.9)-(3.10) is a second-order sequel to LxF scheme, with greatly improved resolution. An attractive feature of the central scheme (3.9)-(3.10) is that it avoids Riemann solvers: instead of characteristic variables, one may use a componentwise extension of the non-oscillatory limiters (3.12).

Multidimensional systems There are basically two approaches.

One approach is to reduce the problem into a series of one-dimensional problems. Alternating Direction (ADI) methods and the closely related dimensional splitting methods, e.g., [140, §8.8-9], are effective, widely used tools to solve multidimensional problems by piecing them from one-dimensional problems – one dimension at a time. Still, in the context of nonlinear conservation laws, dimensional splitting encounters several limitations, [31]. A particular instructive example for the effect of dimensional splitting errors can be found in the approximate solution of the weakly hyperbolic system studied in [48],[81, §4.3].

The other approach is 'genuinely multidimensional'. There is a vast literature in this context. The beginning is with the pioneering multidimensional second-order Lax-Wendroff scheme, [97]. To retain high-resolution of multidimensional schemes without spurious oscillations, requires one or more of several ingredients: a careful treatment of waves propagations ('unwinding'), or alternatively, a correctly tuned numerical dissipation which is free of Riemann-solvers ('central differencing'), or the use of adaptive grids (which are not-necessarily rectangular), ... Waves propagation in the context of multidimensional upwind algorithms were studied in [25, 103, 139, 154] .... Another 'genuinely multidimensional' approach can be found in the positive schemes of [95]. The pointwise formulation of ENO schemes due to Shu & Osher, [151, 152], is another approach which avoids dimensional splitting: here, the reconstruction of cell-averages is bypassed by the reconstruction *pointvalues* of the fluxes in each dimension; the semidiscrete fluxed are then integrated in time using non-oscillatory ODEs solvers (which are briefly mentioned in §3.4.2 below). Multidimensional non-oscillatory *central* scheme was presented in [81], generalizing the one-dimensional (3.9)-(3.10); consult [105],[89] for applications to the multidimensional incompressible Euler equations. Finite volume methods, [85, 86, 24, 29]... , and finite-element methods (the streamline-diffusion and discontinuous Galerkin schemes, [76, 79, 80, 146, 121]...) have the advantage of a 'built-in' recipe for discretization over general triangular grids (we shall say more on these methods in §7 below). Another 'genuinely multidimensional' approach is based on a relaxation approximation was introduced in [82]. It employs a central scheme of the type (3.9)-(3.10) to discretize the relaxation models models, [173], [19], [124],....

3.3   TVD filters

Every discretization method is associated with an appropriate finite-dimensional projection. It is well known that *linear* projections which are monotone (or equivalently, positive), are at most first-order accurate, [59]. The lack of monotonicity for higher order projections is reflected by spurious oscillations in the vicinity of jump discontinuities. These are evident with the second-order (and higher) centered differences, whose dispersive nature is responsible to the formation of binary oscillations [63],[104]. With highly-accurate spectral projections, for example, these $\mathcal{O}(1)$ oscillations reflect the familiar Gibbs phenomena.

TVD schemes avoid spurious oscillations — to this end they use the necessarily *nonlinear* projections (expressed in terms of nonlinear limiters like those in (3.12)). TVD filters, instead, suppress spurious oscillations. At each time-level, one post-process the computed (possibly oscillatory) solution $\{\rho^n\}$. In this context we highlight the following.

 • **Linear filters**. Consider linear convection problems with discontinuous initial data. Approximate solutions of such problems suffer from loss of accuracy due to propagation of singularities and their interference over domain of dependence of the numerical scheme. Instead, one can show, by duality argument, that the numerical scheme retains its original order of accuracy when the truncation in (3.3) is measured w.r.t. sufficiently large negative norm, [120]. Linear filters then enable to accurately recover the exact solution in any smoothness region of the exact solution, bounded away from its singular support. These filters amount to finite-order mollifiers [120], or spectrally accurate mollifiers, [118], [66], which accurately recover pointvalues from high-order moments. (We outline such technique in §4.2).

 • **Artificial compression**. Artificial compression was introduced by

Harten [68] as a method to sharpen the poor resolution of contact discontinuities. (Typically, the resolution of contacts by α-order schemes diffuses over a fan of width $(\Delta t)^{(\alpha)/(\alpha+1)}$). The idea is to enhance the focusing of characteristics by adding an anti-diffusion modification to the numerical fluxes: if we let $H_{\nu+\frac{1}{2}}$ denote the numerical flux of a three-point TVD scheme (3.7), then one replaces it with a modified flux, $H_{\nu+\frac{1}{2}} \longrightarrow H_{\nu+\frac{1}{2}} + \tilde{H}_{\nu+\frac{1}{2}}$, which is expressed in terms of the min-mod limiter (3.11)

$$\tilde{H}_{\nu+\frac{1}{2}} := \frac{1}{\lambda}\{\rho'_\nu + \rho'_{\nu+1} - sgn(\Delta\rho_{\nu+\frac{1}{2}})|\rho'_{\nu+1} - \rho'_\nu|\}. \tag{3.14}$$

Artificial compression can be used as a second-order TVD filter as well. Let $Q_{\nu+\frac{1}{2}}$ be the numerical viscosity of a three-point TVD scheme (3.7). Then, by adding an artificial compression modification (3.14) which is based on the θ-limiters (3.12), $\rho'_\nu = \rho'_\nu(\theta)$ with $\theta_{\nu+\frac{1}{2}} := Q_{\nu+\frac{1}{2}} - \lambda^2 a^2_{\nu+\frac{1}{2}}$, one obtains a second-order TVD scheme, [69], [132]. Thus, in this case the artificial compression (3.14) can be viewed as a second-order anti-diffusive TVD filter of first-order TVD schemes

$$\rho_\nu^{n+1} \longleftarrow \rho_\nu^{n+1} - \{\tilde{H}_{\nu+\frac{1}{2}}(\rho^n) - \tilde{H}_{\nu-\frac{1}{2}}(\rho^n)\}. \tag{3.15}$$

• **TVD filters.** A particularly useful and effective, general-purpose TVD filter was introduced by Engquist et. al. in [47]; it proceeds in three steps. {i} (Isolate extrema). First, isolate extrema cells where $\Delta\rho^n_{\nu-\frac{1}{2}} \cdot \Delta\rho^n_{\nu+\frac{1}{2}} < 0$. {ii} (Measure local oscillation). Second, measure local oscillation, osc_ν, by setting

$$osc_\nu := \min\{m_\nu, \frac{1}{2}M_\nu\}, \qquad \left\{\begin{matrix} m_\nu \\ M_\nu \end{matrix}\right\} = \left\{\begin{matrix} \min \\ \max \end{matrix}\right\}(\Delta\rho^n_{\nu-\frac{1}{2}}, \Delta\rho^n_{\nu+\frac{1}{2}})$$

{iii} (Filtering). Finally, oscillatory minima (respectively – oscillatory maxima) are increased (and respectively, increased) by updating $\rho^n_\nu \rightarrow \rho^n_\nu + sgn(\Delta\rho^n_{\nu+\frac{1}{2}})osc_\nu$, and the corresponding neighboring gridvalue is modified by subtracting the same amount to retain conservation. This post-processing can be repeated, if necessary, and one may use a local maximum principle, $\min_j\rho^n_j \leq \rho^n_\nu \leq \max_j\rho^n_j$ as a stopping criterion. In this case, the above filter becomes TVD once the binary oscillations are removed, [153].

3.4   TVB approximations

3.4.1   Higher resolution schemes (with three letters acronym)

We have already mentioned the essential role played by nonlinear limiters in TVD schemes. The mechanism in these nonlinear limiters is switched on in extrema cells, so that the zero discrete slope $\rho' = 0$ avoids new spurious extrema. This, in turn, leads to deteriorated first-order local accuracy at non-sonic extrema, and global accuracy is therefore limited to second-order[4].

To obtain an improved accuracy, one seeks a more accurate realization of the approximate solution, in terms of higher (than first-order) piecewise polynomials

$$\rho^{\Delta x}(t^n, x) = \sum_\nu p_\nu(x)\chi_\nu(x), \quad p_\nu(x) = \sum_j \rho_\nu^{(j)}(\frac{x - x_\nu}{\Delta x})^j/j!. \qquad (3.16)$$

Here, the exact solution is represented in a cell C_ν in terms of an r-order polynomial p_ν, which is reconstructed from the its neighboring cell averages, $\{\bar{\rho}_{\nu_\mu}\}$. If we let $\rho^{\Delta x}(t \geq t^n, \cdot)$ denote the entropy solution subject to the reconstructed data at $t = t^n$, $P^{\Delta x}\rho(t^n, \cdot)$, then the corresponding Godunov-type scheme governs the evolution of cell averages

$$\bar{\rho}_\nu^{n+1} := \frac{1}{\Delta x} \int_x \rho^{\Delta x}(t^{n+1} - 0, x)\chi_\nu(x)dx. \qquad (3.17)$$

The properties of Godunov-type scheme are determined by the polynomial reconstruction should meet three contradicting requirements:

{i} *Conservation*: $p_\nu(x)$ should be cell conservative in the sense that $\fint_{C_\nu} p_\nu(x) = \fint_{C_\nu} \rho_\nu(x)$. This tells us that $P^{\Delta x}$ is a (possibly nonlinear) projection, which in turn makes (3.17) a conservative scheme in the sense of Lax-Wendroff, (3.1).

{ii} *Accuracy*: $\rho_\nu^{(j)} \sim (\Delta x \partial_x)^j \rho(t^n, x_\nu)$.
At this stage, we have to relax the TVD requirement. This brings us to the third requirement of

{iii} *TVB bound*: we seek a bound on the total variation on the computed solution. Of course, a bounded variation, $\|\rho^{\Delta x}(t^n, \cdot)\|_{BV} \leq$ Const. (and in fact, even the weaker $(\Delta x)^\theta\|\rho^{\Delta x}\|_{BV} \leq$ Const.) will suffice for convergence by L^1-compactness arguments.

The (re-)construction of non-oscillatory polynomials led to new high-resolution schemes. In this context we mention the following methods (which were popularized by their trade-mark of three-letters acronym ...): the Piecewise-Parabolic Method (PPM) [26], the Uniformly Non-Oscillatory (UNO) scheme [73], and the Essentially Non-Oscillatory schemes (ENO) of

---

[4]The implicit assumption is that we seek an approximation to *piecewise-smooth* solutions with finitely many oscillations, [165]. The convergence theories apply to general BV solutions. Yet, general BV solutions cannot be resolved in *actual* computations in terms of 'classical' macroscopic discretizations – finite-difference, finite-element, spectral methods, etc. Such methods can faithfully resolve piecewise smooth solutions.

Harten et. al. [70]. There is large numerical evidence that these highly-accurate methods are TVB (and hence convergent), at least for a large class of piecewise-smooth solutions. We should note, however, that the convergence question of these schemes is open. (It is my opinion that new characterizations of the (piecewise) regularity of solutions to conservation laws, e.g., [37], together with additional tools to analyze their compactness, are necessary in order to address the questions of convergence and stability of these highly-accurate schemes).

There are alternative approach to to construct high-resolution approximations which circumvent the TVD limitations. We conclude by mentioning the following two.

One approach is to evolve more than one-piece of information per cell. This is fundamentally different from standard Godunov-type schemes where only the cell average is evolved (and higher order projections are *reconstructed* from these averages – one per cell). In this context we mention the quasi-monotone TVB schemes introduced in [23]. Here, one use a TVD evolution of cell averages together with additional higher moments. Another instructive example for this approach is found in the third-order TVB scheme, [142]: in fact, Sanders constructed a third-order non-expansive scheme (circumventing the first-order limitation of [71]), by using a 2×2 system which governs the first two moments of the scalar solution. More recently, Bouchut et. al. [8], constructed a second-order MUSCL scheme which respects a discrete version of the entropy inequality (2.3) w.r.t *all* Kružkov's scalar entropy pairs in (2.11); this circumvents the second-order limitation of Osher & Tadmor [132, Theorem 7.3], by evolving *both* – the cell average and the discrete slope in each computational cell.

Another approach to enforce a TVB bound on higher(> 2)-resolution schemes, makes use of gridsize-dependent limiters, $\rho^{(j)} = \rho^{(j)}\{\bar{\rho}^n, \Delta x\}$, such that the following holds, e.g., [149],

$$\|\rho^{\Delta x}(t^{n+1}, \cdot)\|_{BV} \leq \|\rho^{\Delta x}(t^{n+1}, \cdot)\|_{BV} + \text{Const} \cdot \Delta x.$$

Such Δx-dependent limiters fail to satisfy, however, the basic dilation invariance of (2.15)-(2.16), $(t, x) \rightarrow (ct, cx)$.

3.4.2 Time discretizations

One may consider separately the discretization of time and spatial variables. Let P_N denote a (possibly nonlinear) finite-dimensional spatial discretization of (2.1); this yields an N-dimensional approximate solution, $\rho_N(t)$, which is governed by the system of N nonlinear ODEs

$$\frac{d}{dt}\rho_N(t) = P_N(\rho_N(t)). \tag{3.18}$$

System (3.18) is a *semi-discrete* approximation of (2.1). For example, if we let $P_N = P^{\Delta x}$, $N \sim (\Delta x)^{-d}$, to be one of the piecewise-polynomial reconstructions associated with Godunov-type methods in (3.16), then one ends up with a semi-discrete finite-difference method, the so called method of lines. In fact, our discussion on streamline-diffusion and spectral approximations in §5 and §6 below will be primarily concerned with such semi-discrete approximations.

An explicit time discretization of (3.18) proceeds by either a multi-level or a Runge-Kutta method. A CFL condition should be met, unless one accounts for wave interactions, consult [101]. For the construction of non-oscillatory schemes, one seeks time discretizations which retain the non-oscillatory properties of the spatial discretization, P_N. In this context we mention the TVB time-discretizations of Shu & Osher, [150],[151, 152]. Here, one obtains high-order multi-level and Runge-Kutta time discretizations as *convex combinations* of the standard forward time differencing, which amounts to the first-order accurate forward Euler method. Consequently, the time discretizations [151, 152] retain the nonoscillatory properties of the low-order forward Euler time differencing — in particular, TVD/TVB bounds, and at the same time, they enable to match the time accuracy with the high-order spatial accuracy.

4 Godunov Type Methods

4.1 Compactness arguments cont'd — one-sided stability estimates

We prove convergence and derive error bounds using one-sided stability estimates. The one-sided stability estimates restrict our discussion to scalar equations – one-dimensional convex conservation laws in §4.2 and multidimensional convex Hamilton-Jacobi equation in §4.3. (We refer to [100] for a recent contribution concerning the one-sided stability of one-dimensional systems). We begin with the case $d = 1$.

Let $\{\rho^\varepsilon(t,x)\}$ be a family of approximate solutions tagged by their small-scale parameterization, ε. To upper-bound the convergence rate of such approximations, we shall need the usual two ingredients of stability and consistency.

- *Lip^+-stability.* The family $\{\rho^\varepsilon\}$ is Lip^+-stable if

$$\|\rho^\varepsilon(t,\cdot)\|_{Lip+} := \sup_x \partial_x \rho^\varepsilon(t,x) \le Const. \tag{4.1}$$

This notion of Lip^+-stability is motivated by Olĕinik's One-Sided Lipschitz Condition (OSLC), $\rho_x(t,\cdot) \le Const$, which uniquely identifies the entropy solution of *convex* conservation laws, (2.15), with scalar $A'' > 0$. Since the Lip^+-(semi)-norm dominates the total-variation,

$$\|\rho^\varepsilon(t,\cdot)\|_{BV} \le Const.\|\rho^\varepsilon(t,\cdot)\|_{Lip+} + \|\rho_0^\varepsilon(\cdot)\|_{L^1}, \quad Const = 2|\mathrm{supp}_x \rho^\varepsilon(t,\cdot)|,$$

$\{\rho^\varepsilon\}$ have bounded variation and convergence follows. Equipped with Lip^+-stability, we are able to *quantify* this convergence statement. To this end, we measure the local truncation error in terms of

- *Lip'-consistency.* The family $\{\rho^\varepsilon\}$ is Lip'-consistent of order ε if

$$\|\partial_t \rho^\varepsilon + \partial_x A(\rho^\varepsilon)\|_{Lip'(t,x)} \sim \varepsilon. \tag{4.2}$$

It follows that the stability+consistency in the above sense, imply the convergence of $\{\rho^\varepsilon\}$ to the entropy solution, ρ, and that the following error estimates hold [163], [126],

$$\|\rho^\varepsilon(t,\cdot) - \rho(t,\cdot)\|_{W^s(L^p(x))} \sim \varepsilon^{\frac{1-sp}{2p}}, \quad -1 \le s \le 1/p. \tag{4.3}$$

The case $(s,p) = (-1,1)$ corresponds to a sharp Lip'-error estimate of order ε — the Lip'-size of the truncation in (4.2); the case $(s,p) = (0,1)$ yields an L^1-error estimate of order one-half, in agreement with Kuznetsov's general convergence theory, [90]. (We shall return to it in §7.3). Moreover, additional *local* error estimates follow, and we illustrate this in the context of Godunov-type schemes.

4.2   Godunov type methods revisited $(m = d = 1)$

Godunov type schemes form a special class of transport projection methods for the approximate solution of nonlinear conservation laws, [72].

Let $E(t_2 - t_1)$ denote the entropy solution operator associated with the convex conservation law (2.15). A Godunov-type method yields a globally defined approximate solution, $\rho^{\Delta x}(t,x)$, which is governed by iterating the evolution-projection cycle,

$$\rho^{\Delta x}(\cdot,t) = \begin{cases} E(t - t^{n-1})\rho(\cdot, t^{n-1}), & t^{n-1} < t < t^n = n\Delta t, \\ P^{\Delta x}\rho(\cdot, t^n - 0), & t = t^n, \end{cases} \tag{4.4}$$

subject to initialization step, $\rho(t = 0, \cdot) = P^{\Delta x} u_0(\cdot)$.

Here, $P^{\Delta x}$ is an *arbitrary, possibly nonlinear* conservative projection, which

depends on a small spatial scale Δx. For example, the piecewise polyno-
mial projection (3.16), $P^{\Delta x}\rho(x) = \sum p_\nu(x)\chi_\nu(x)$, where the χ_ν's are the
characteristic functions of cells $\mathcal{C}_\nu$ with possibly variable sizes, $\Delta x \leq |\mathcal{C}_\nu| \leq$
Const.Δx.

The question of Lip'-consistency for Godunov-type schemes based on
cell-conservative projections, $P^{\Delta x}$, could be answered in terms of the L^1-
size of $I - P$ over all BV functions [127]. Together with Lip^+-stability we
conclude

$$\|\rho^{\Delta x}(t,\cdot) - \rho(t,\cdot)\|_{W^{s,p}} \leq Const.\|I - P\|_{BV \to L^1}^{\frac{1-sp}{2p}}. \qquad (4.5)$$

The last error bound, (4.5), tells us that the convergence rate of a
Godunov-type scheme depends solely on the properties of $P^{\Delta x}$. First, Lip^+
stability is guaranteed if $P^{\Delta x}$ retains the OSLC of the exact solution operator; the OSLC property of such projections was studied in [128],[10], [126].
Second, the convergence rate depends on measuring $P^{\Delta x}$ as an approxi-
mate identity. Typically, $\|I - P^{\Delta x}\|_{BV \to L^1}$ is of order $\mathcal{O}(\Delta x)$, and (4.5)
yields the familiar L^1 rate of order $\mathcal{O}(\sqrt{\Delta x})$, [114], [30], [141], [145],... (and
[24, 27, 85, 86, 7] in the multidimensional case). Moreover, one can inter-
polate between the weak $W^{-1}(L^1)$-error estimate of order $\mathcal{O}(\Delta x)$, and the
one-sided Lipschitz bounds of ρ and $\rho^{\Delta x}$ to conclude, [163]

$$|\rho^{\Delta x}(t,x) - \rho(t,x)| \leq Const.[1 + \max_{\Omega_x}|\rho_x(t,x)|](\Delta x)^{1/3}.$$

This shows a pointwise convergence which depends solely on the *local* smooth-
ness of the entropy solution in $\Omega_x := \{y|\ |y - x| \leq C(\Delta x)^{1/3}\}$.

4.3   Hamilton-Jacobi equations $(m = 1, d \geq 1)$

We consider the multidimensional Hamilton-Jacobi (HJ) equation

$$\partial_t\rho + H(\nabla_x\rho) = 0, \quad (t,x) \in R^+ \times R^d, \qquad (4.6)$$

with convex Hamiltonian, $H'' > 0$. Its unique viscosity solution is identified
by the one-sided concavity condition, $D_x^2\rho \leq$ Const., consult [87], [108].
Given a family of approximate HJ solutions, $\{\rho^\varepsilon\}$, we make the analogous
one-sided stability requirement of

- *Demi-concave stability.* The family $\{\rho^\varepsilon\}$ is demo-concave stable if

$$D_x^2\rho^\varepsilon \leq Const. \qquad (4.7)$$

We then have the following.

Theorem 4.1 ([107]) *Assume $\{\rho_1^\varepsilon\}$ and $\{\rho_2^\varepsilon\}$ are two demi-concave stable families of approximate solutions. Then*

$$\|\rho_1^\varepsilon(t,\cdot) - \rho_2^\varepsilon(t,\cdot)\|_{L^1(x)} \leq Const.\|\rho_1^\varepsilon(0,\cdot) - \rho_2^\varepsilon(0,\cdot)\|_{L^1(x)} +$$

$$+ Const. \sum_{j=1}^{2} \|\partial_t \rho_j^\varepsilon + H(\nabla_x \rho_j^\varepsilon)\|_{L^1(t,x)}. \quad (4.8)$$

If we let $\rho_1^\varepsilon \equiv \rho^1$, $\rho_2^\varepsilon \equiv \rho^2$ denote two demi-concave viscosity solutions, then (4.8) is an L^1-stability statement (compared with the usual L^∞-stability statements of viscosity solutions, [29]). If we let $\{\rho_1^\varepsilon\} = \{\rho^\varepsilon\}$ denote a given family of demi-concave approximate HJ solutions, and let ρ_2^ε equals the exact viscosity solution ρ, then (4.8) yields the L^1-error estimate

$$\|\rho^\varepsilon(t,\cdot) - \rho(t,\cdot)\|_{L^1(x)} \leq Const.\|\partial_t \rho^\varepsilon + H(\nabla_x \rho^\varepsilon)\|_{L^1(t,x)} \sim O(\varepsilon). \quad (4.9)$$

This corresponds to the Lip'-error estimate of (4.3) with $(s,p) = (-1,1)$. One can then interpolate from (4.9) an L^p-error estimates of order $\mathcal{O}(\varepsilon^{\frac{1+p}{2p}})$. For a general L^∞-convergence theory for approximate solutions to HJ equations we refer to [3] and the references therein.

5   Streamline Diffusion Finite Element Schemes

5.1   Compensated compactness ($m \leq 2, d = 1$)

We deal with a family of approximate solutions, $\{\rho^\varepsilon\}$, such that

(i) *It is uniformly bounded, $\rho^\varepsilon \in L^\infty$, with a weak* limit, $\rho^\varepsilon \rightharpoonup \rho$;*

(ii) *The entropy production, for all convex entropies η, lies in a compact subset of $W_{loc}^{-1}(L^2(t,x))$,*

$$\forall \eta'' > 0: \qquad \partial_t \eta(\rho^\varepsilon) + \partial_x F(\rho^\varepsilon) \hookrightarrow W_{loc}^{-1}(L^2(t,x)). \quad (5.1)$$

The conclusion is that $A(\rho^\varepsilon) \rightharpoonup A(\rho)$, and hence ρ is a weak solution; in fact, there is a strong convergence, $\rho^\varepsilon \to \rho$, on any nonaffine interval of $A(\cdot)$. For a complete account on the theory of compensated compactness we refer to the innovative works of Tartar [167] and Murat [123]. In the present context, compensated compactness argument is based on a clever application of the div-curl lemma. First scalar applications are due to Murat-Tartar, [122],[167], followed by extensions to certain $m = 2$ systems by DiPerna [39] and Chen [17].

The current framework has the advantage of dealing with L^2-type estimates rather than the more intricate BV framework. How does one verify the
$W_{loc}^{-1}(L^2)$-condition (5.1)? we illustrate this point with canonical viscosity
approximation (2.2). Multiplication by η' shows that its entropy production
amounts to $\varepsilon(\eta'Q\rho_x^\varepsilon)_x - \varepsilon\eta''Q(\rho_x^\varepsilon)^2$. By entropy convexity, $\varepsilon\eta''Q > 0$[5], and
space-time integration yields

- *An entropy production bound*

$$\sqrt{\varepsilon}\|\frac{\partial\rho^\varepsilon}{\partial x}\|_{L_{loc}^2(t,x)} \leq Const. \tag{5.2}$$

Though this bound is too weak for strong compactness, it is the key estimate behind the $W_{loc}^{-1}(L^2)$-compactness condition (5.1). We continue with
the specific examples of streamline-diffusion in §5.2 and spectral viscosity
methods in §6.

5.2   The streamline diffusion method

The Streamline Diffusion (SD) finite element scheme, due to Hughes, Johnson, Szepessy and their co-workers [76], [79], [80], was one of the first methods
whose convergence was analyzed by compensated compactness arguments.
(Of course, finite-element methods fit into L^2-type Hilbert-space arguments).
In the SD method, formulated here in several space dimensions, one seeks a
piecewise polynomial, $\{\rho^{\Delta x}\}$, which is uniquely determined by requiring for
all piecewise polynomial test functions $\psi^{\Delta x}$,

$$\langle\partial_t\rho^{\Delta x} + \nabla_x \cdot A(\rho^{\Delta x}),\ \psi^{\Delta x} + |\Delta x| \cdot \boxed{(\psi_t^{\Delta x} + A'(\rho^{\Delta x})\psi_x^{\Delta x})}\rangle = 0. \tag{5.3}$$

Here, Δx denotes the spatial grid size (for simplicity we ignore time discretization). The expression inside the framed box on the left represents a
diffusion term along the streamlines, $\dot{x} = A'(\rho^{\Delta x})$. Setting the test function,
$\psi^{\Delta x} = \rho^{\Delta x}$, (5.3) yields the desired entropy production bound

$$\sqrt{\Delta x}\|\partial_t\rho^{\Delta x} + \nabla_x \cdot A(\rho^{\Delta x})\|_{L_{loc}^2(t,x)} \leq Const. \tag{5.4}$$

Thus, the spatial derivative in (5.2) is replaced here by a streamline-directional
gradient. This together with an L^∞-bound imply $W_{loc}^{-1}(L^2)$-compact entropy
production, (5.1), and convergence follows [79],[80],[158]. We note in passing
that the extension of the SD method for systems of equations is carried out
by projection into entropy variables, [119], which in turn provide the correct
interpretation of (5.4) as an entropy production bound.

---

[5]Observe that the viscosity matrix is therefore required to be positive w.r.t. the Hessian
η''.

5.3   TVD schemes revisited ($d = 1$)

We replace the streamline diffusion term inside the framed box on the left of (5.3) by a *weighted* spatial diffusion expression,

$$\langle \partial_t \rho^{\Delta x} + \partial_x A(\rho^{\Delta x}), \psi^{\Delta x} \rangle + \Delta x \cdot \boxed{\langle \rho_x^{\Delta x}, \psi_x^{\Delta x} \rangle_Q} = 0.$$

This yields a semi-discrete finite-difference scheme in its viscosity form (3.5), and one may carry an alternative convergence analysis based on compensated compactness arguments [169].

6   Spectral Viscosity Approximations

6.1   Compensated compactness cont'd ($m \leq 2, d = 1$)

Let P_N denote an appropriate spatial projection into the space of N-degree polynomials,

$$P_N \rho(t, x) = \sum_{|k| \leq N} \hat{\rho}_k(t) \phi_k(x);$$

here $\{\phi_k\}$ stands for a given family of orthogonal polynomials, either trigonometric or algebraic ones, e.g., $\{e^{ikx}\}$, $\{L_k(x)\}$, $\{T_k(x)\}$, etc. The corresponding N-degree approximate solution, $\rho_N(t, x)$, is governed by the spectral viscosity (SV) approximation

$$\partial_t \rho_N + \partial_x P_N A(\rho_N) = \frac{1}{N} \partial_x (Q * \partial_x \rho_N). \tag{6.1}$$

The left hand side of (6.1) is the standard spectral approximation of the conservation law (2.1). The expression on the right represents the so called spectral viscosity introduced in [162]. It contains a minimal amount of high-modes regularization which retains the underlying spectral accuracy of the overall approximation,

$$\frac{1}{N} \partial_x (Q * \partial_x \rho_N) := \frac{1}{N} \sum_{|k| > N^\theta} \hat{Q}_k \hat{\rho}_k(t) \phi_k''(x).$$

It involves a viscous-free zone for the first N^θ modes, $0 < \theta < \frac{1}{2}$. High modes diffusion is tuned by the viscosity coefficients $\hat{Q}_k$.

Spurious Gibbs oscillations violate the strict TVD condition in this case. Instead, an entropy production estimate, analogous to (5.2) is sought,

$$\frac{1}{\sqrt{N}} \left\| \frac{\partial \rho_N}{\partial x} \right\|_{L^2_{loc}(t,x)} \leq Const.$$

This together with an L^∞-bound carry out the convergence analysis by compensated compactness arguments, [162], [116]. Extensions to certain $m = 2$ systems can be found in [143].

6.2   Hyper-viscosity approximations

The second-order high-modes diffusion on the right of (6.1) is replaced by higher $2s$-order diffusion,

$$\frac{(-1)^{s+1}}{N^{2s-1}}\partial_x^s(Q * \partial_x^s \rho_N) := \frac{(-1)^{s+1}}{N^{2s-1}} \sum_{|k|>N^\theta} \hat{Q}_k \hat{\rho}_k(t)\phi_k^{(2s)}(x). \qquad (6.2)$$

This allows for a larger viscosity-free zone of size N^θ, with $0 < \theta < \frac{2s-1}{2s}$ (with possibly $s = s_N \leq \sqrt{N}$), consult [164]. The underlying hyper-viscosity approximation (for say $s = 2$) reads

$$\partial_t \rho^\varepsilon + \partial_x A(\rho^\varepsilon) + \varepsilon^3 \partial_x^4 \rho^\varepsilon = 0. \qquad (6.3)$$

The solution operator associated with (6.3) is not monotone, hence L^1-contraction and the TVD condition fail in this case. Instead, compensated compactness arguments show, under the assumption of an L^∞- bound[6], the hyper-viscosity approximation (6.3) and its analogous spectral-viscosity approximations, converge to the entropy solution.

7   Finite Volume Schemes $(d \geq 1)$

7.1   Measure-valued solutions $(m = 1, d \geq 1)$

We turn our attention to the multidimensional scalar case, dealing with a families of uniformly bounded approximate solutions, $\{\rho^\varepsilon\}$, with weak* limit, $\rho^\varepsilon \rightharpoonup \rho$. DiPerna's result [41] states that if the entropy production of such a family tends weakly to a negative measure, $m \leq 0$,

$$\forall \eta'' > 0 : \qquad \partial_t \eta(\rho^\varepsilon) + \nabla_x \cdot F(\rho^\varepsilon) \rightharpoonup m \leq 0, \qquad (7.1)$$

then the measure-valued solution ρ coincides with the entropy solution, and convergence follows. This framework was used to prove the convergence of multidimensional finite-difference schemes [27], streamline diffusion method [79],[80], spectral-viscosity approximations [18] and finite-volume schemes [24], [86],[85]. We focus our attention on the latter.

---

[6]The L^∞ boundedness of (6.3) is to the best of my knowledge, an open question, [62].

7.2   Finite-volume schemes

We are concerned with finite-volume schemes based on possibly *unstructured* triangulation grid $\{T_\nu\}$ (for simplicity we restrict attention to the $d = 2$ case). The spatial domain is covered by a triangulation, $\{T|\nu\}$, and we compute approximate averages over these triangles, $\bar{\rho}_\nu^n \sim \frac{1}{|T_\nu|} \int_{T_\nu} \rho(t^n, x)dx$, governed by the finite volume (FV) scheme

$$\bar{\rho}_\nu^{n+1} = \bar{\rho}_\nu^n - \frac{\Delta t}{|T_\nu|} \sum_\mu \tilde{A}_{\nu_\mu}(\rho_\nu^n, \rho_{\nu_\mu}^n). \tag{7.2}$$

Here $\tilde{A}_{\nu_\mu}$ stand for approximate fluxes across the interfaces of T_ν and its neighboring triangles (identified by a secondary index μ).

Typically, the approximate fluxes, $\tilde{A}_{\nu_\mu}$ are derived on the basis of approximate Riemann solvers across these interfaces, which yield a monotone scheme. That is, the right hand side of (7.2) is a monotone function of its arguments $(\rho_\nu^n, \rho_{\nu_\mu}^n)$, and hence the corresponding FV scheme is L^1-contractive. However, at this stage one cannot proceed with the previous compactness arguments which apply to TVD schemes over fixed Cartesian grid: since the grid is unstructured, the discrete solution operator is not translation invariant and L^1-contraction need not imply a TV bound. Instead, an entropy dissipation estimate yields

$$\sum_n \Delta t \sum_{\nu,\mu} |\rho_\nu^n - \rho_{\nu_\mu}^n|(\Delta x)^\theta \leq Const, \quad 0 < \theta < 1. \tag{7.3}$$

Observe that (7.3) is weaker than a TV bound (corresponding to $\theta = 0$), yet it suffices for convergence to a measure-valued solution, consult [24], [85].

7.3   Error estimates — compactness arguments revisited

Kuznetsov [90] was the first to provide error estimates for scalar approximate solutions, $\{\rho^\varepsilon\}$, of *multidimensional* scalar conservation laws. Subsequently, many authors have used Kuznetsov's approach to prove convergence + L^1-error estimates; we refer for the detailed treatments of [141], [115], [166],... . A more recent treatment of [24] employs the entropy dissipation estimate (7.3), which in turn, by Kuznetsov arguments, yields an L^1-convergence rate estimate of order $(\Delta x)^{\frac{1-\theta}{2}}$ (independently of the BV bound).

Kuznetsov's approach employs a *regularized* version of Kružkov's entropy pairs in (2.11), $\eta^\delta(\rho; c) \sim |\rho - c|$, $F^\delta(\rho; c) \sim sgn(\rho - c)(A(\rho) - A(c))$. Here, one measures by how much the entropy dissipation rate of $\{\rho^\varepsilon\}$ fails to satisfy the entropy inequality (2.3), with Kružkov's regularized entropies.

Following the general recent convergence result of [7], we consider a family
of approximate solutions, $\{\rho^\varepsilon\}$, which satisfies

$$\partial_t |\rho^\varepsilon - c| + \nabla_x \cdot \{sgn(\rho^\varepsilon - c)(A(\rho^\varepsilon) - A(c))\} \leq \partial_t R_0(t, x) + \nabla_x \cdot R(t, x), \quad (7.4)$$

with

$$\|R_0(t, x)\|_{\mathcal{M}_{t,x}} + \|R(t, x)\|_{\mathcal{M}_{t,x}} \leq \text{Const} \cdot \varepsilon. \quad (7.5)$$

Then, the convergence rate proof proceeds along the lines of Theorem 2.1:
Using the key property of symmetry of the regularized entropy pairs, ($\eta^\delta :=$
$\phi_\delta \eta, F^\delta := \phi_\delta F$), one finds $\int_x \eta^\delta(\rho^\varepsilon; \rho) dx \leq \text{Const}.\varepsilon/\delta$. In addition, there is
a regularization error, $\|\eta^\delta - \eta\|_{L^1(x)}$, of size $\mathcal{O}(\delta)$, and an L^1 error estimate
of order $\mathcal{O}(\sqrt{\varepsilon})$ follows (under reasonable assumptions on the L^1-initial error
w.r.t. BV data), consult [7]

$$\|\rho^\varepsilon(t, \cdot) - \rho(t, \cdot)\|_{L^1_{loc}(x)} \leq \text{Const}.\sqrt{\varepsilon}.$$

Observe that this error estimate, based on (7.4)-(7.5) is the multidimensional
analogue of the Lip'-consistency requirement we met earlier in (4.2).

8 Kinetic Approximations

8.1 Velocity averaging lemmas ($m \geq 1, d \geq 1$)

We deal with solutions to transport equations

$$a(v) \cdot \nabla_x f(x, v) = \partial_v^s g(x, v). \quad (8.1)$$

The averaging lemmas, [61], [52], [44], state that in the generic non-degenerate
case, averaging over the velocity space, $\bar{f}(x) := \int_v f(x, v) dv$, yields a gain of
spatial regularity. The prototype statement reads

Lemma 8.1 ([44], [110]) . *Let $f \in L^p(x, v)$ be a solution of the transport
equation (8.1) with $g \in L^q(x, v), 1 \leq q < p \leq 2$. Assume the following
non-degeneracy condition holds*

$$meas_v\{v| \ |a(v) \cdot \xi| < \delta\}_{|\xi|=1} \leq Const \cdot \delta^\alpha, \quad \alpha \in (0, 1). \quad (8.2)$$

Then $\bar{f}(x) := \int_v f(x, v) dv$ belongs to Sobolev space $W^\theta(L^r(x))$,

$$\bar{f}(x) \in W^\theta(L^r(x)), \qquad \theta < \frac{\alpha}{\alpha(1 - \frac{p'}{q'}) + (s + 1)p'}, \quad \frac{1}{r} = \frac{\theta}{q} + \frac{1 - \theta}{p}. \quad (8.3)$$

Variants of the averaging lemmas were used by DiPerna and Lions to construct global weak (renormalized) solutions of Boltzmann, Vlasov-Maxwell and related kinetic systems, [42], [43]; in Bardos et. al., [2], averaging lemmas were used to construct solutions of the incompressible Navier-Stokes equations. We turn our attention to their use in the context of nonlinear conservation laws and related equations.

8.1.1   Scalar conservation laws

The following result, adapted from [110], is in the heart of matter.

Theorem 8.1 ([110]) *Consider the scalar conservation law (2.1) whose flux satisfies the non-degeneracy condition (consult (8.2))*

$$\exists \alpha \in (0,1): \ meas_v\{v| \ |\tau + A'(v) \cdot \xi| < \delta\} \le Const \cdot \delta^\alpha, \quad \forall \tau^2 + |\xi|^2 = 1. \tag{8.4}$$

Let $\{\rho^\varepsilon\}$ be a family of approximate solutions satisfying the entropy condition (2.3),

$$\partial_t \eta(\rho^\varepsilon) + \nabla_x \cdot F(\rho^\varepsilon) \le 0, \quad \forall \eta'' > 0. \tag{8.5}$$

Then $\rho^\varepsilon(t,x) \in W_{loc}^{\frac{\alpha}{\alpha+4}}(L^r(t,x)), \quad r = \frac{\alpha+4}{\alpha+2}$.

Proof. To simplify notations, we use the customary 0th index for time direction,

$$x = (t \leftrightarrow x_0, x_1, \dots, x_d), \qquad A(\rho) = (A_0(\rho) \equiv 1, A_1(\rho), \dots, A_d(\rho)).$$

The entropy condition (8.5) with Kružkov entropy pairs (2.1), reads

$$\nabla_x \cdot [sgn(\rho^\varepsilon - v)(A(\rho^\varepsilon) - A(v))] \le 0.$$

This defines a family of non-negative measures, $m^\varepsilon(x,v)$,

$$\nabla_x \cdot [sgn(v)A(v) - sgn(\rho^\varepsilon - v)(A(\rho^\varepsilon) - A(v))] =: m^\varepsilon(x,v). \tag{8.6}$$

Differentiate (8.6) w.r.t. v: one finds that the indicator function, $f(x,v) = \chi_{\rho^\varepsilon}(v)$, where

$$\chi_{\rho^\varepsilon}(v) := \begin{cases} +1 & 0 < v < \rho^\varepsilon \\ -1 & \rho^\varepsilon < v < 0 \\ 0 & |v| > \rho^\varepsilon \end{cases}, \tag{8.7}$$

satisfies the transport equation (8.1) with $g(x,v) = m^\varepsilon(x,v) \in \mathcal{M}_{x,v}$ [7]. We now apply the averaging lemma with $(s = q = 1, p = 2)$, which tells us that $\rho^\varepsilon(t,x) = \int \chi_{\rho^\varepsilon}(v)dv \in W_{loc}^{\frac{\alpha}{\alpha+4}}(L^r(t,x))$ as asserted. ∎

---

[7]Once more, it is the symmetry property (2.6) which has a key role in the derivation of the transport kinetic formulation (8.1).

Couple of remarks is in order.

1. The last theorem quantifies the regularity of entropy satisfying approximate solutions, $\{\rho^\varepsilon\}$, in terms of the non-degeneracy (8.5). In particular $\{\rho^\varepsilon\}$ is compact and strong convergence follows.

    In fact more can be said if the solution operator associated with $\{\rho^\varepsilon\}$ is translation invariant: a bootstrap argument yields an improved regularity, [110],

    $$\rho^\varepsilon(t > 0, \cdot) \in W^{\frac{\alpha}{\alpha+2}}(L^1(x)). \tag{8.8}$$

    This shows that due to nonlinearity, (8.4), the corresponding solution operator has a *regularization* effect beyond the initial layer at $t = 0$.

2. In particular, Theorem 8.1 provides an alternative route to analyze the entropy stable multi-dimensional schemes whose convergence proof was previously accomplished by measure-valued arguments; here we refer to finite-difference, finite-volume, streamline-diffusion and spectral approximations ..., which were studied in [29, 24, 85, 86, 79, 80, 18],..... Indeed, the feature in the convergence proof of all these methods is the $W^{-1}_{loc}(L^2)$-compact entropy production, (8.11). Hence, if the underlying conservation law satisfies the non-linear degeneracy condition,

    $$meas_v\{v|\ \tau + A'(v) \cdot \xi = 0\} = 0,$$

    then the corresponding family of approximate solutions, $\{\rho^\varepsilon(t > 0, \cdot)\}$ becomes compact. Moreover, if the entropy production is bounded measure, then there is actually a *gain* of regularity indicated in Theorem 8.1 (– and in (8.8) for the translation invariant case).

8.1.2   Degenerate parabolic equations

The above results can be extended in several directions, consult [110] (and [111] for certain $m = 2$ systems). As an example one can treat convective equations together with (possibly degenerate) diffusive terms

$$\partial_t \rho^\varepsilon + \nabla_x \cdot A(\rho^\varepsilon) = \nabla_x \cdot (Q\nabla_x \rho^\varepsilon), \quad Q \geq 0. \tag{8.9}$$

Assume the problem is not linearly degenerate, in the sense that

$$meas_v\{v|\ \tau + A'(v) \cdot \xi = 0,\ \langle Q(v)\xi, \xi \rangle = 0\} = 0. \tag{8.10}$$

Let $\{\rho^\varepsilon\}$ be a family of approximate solutions of (8.1) with $W^{-1}_{loc}(L^2)$-compact entropy production,

$$\partial_t \eta(\rho^\varepsilon) + \nabla_x \cdot F(\rho^\varepsilon) \hookrightarrow W^{-1}_{loc}(L^2(t,x)), \quad \forall \eta'' > 0. \tag{8.11}$$

Then $\{\rho^\varepsilon\}$ is compact in $L^2_{loc}(t, x)$, [110].

The case $Q = 0$ corresponds to a multidimensional extension of Tartar's compensated compactness arguments in §5.1, and it quantifies the regularity of DiPerna's measure-valued solutions in §7.1. The case $A = 0$ correspond possibly degenerate parabolic equations (consult [84] and the references therein, for example). According to (8.10), satisfying the ellipticity condition, $\langle Q(v)\xi, \xi \rangle > 0$ on a set of non-zero measure, guarantees regularization, compactness ...

8.2   Kinetic schemes

We restrict our attention to the scalar case (— and refer to [15],[109], [133] for a comprehensive rigorous treatment of Boltzmann equation). Here, we demonstrate an application of Theorem 8.1 in the context of the BGK-like relaxation model introduced in [135] following the earlier works [9],[53],

$$\partial_t f^\varepsilon + A'(v) \cdot \nabla_x f^\varepsilon = \frac{1}{\varepsilon} \left(\chi_{\rho^\varepsilon} - f^\varepsilon \right). \tag{8.12}$$

As before, the indicator function $\chi_{\rho^\varepsilon(t,x)}(v)$ denotes the 'pseudo-Maxwellian' (8.8) associated with $\rho^\varepsilon := \bar{f}^\varepsilon$. The relaxation term on the right of (8.12) belongs to $W^{-1}(\mathcal{M}_{t,x})$, [9], and the averaging lemma 8.1 applies with ($s = q = 1$, $p = 2$). It follows that if the conservation law is linearly nondegenerate in the sense that (8.5) holds, then $\{\rho^\varepsilon\}$ is compact – in fact $\{\rho^\varepsilon(t > 0, \cdot)\}$ gains Sobolev regularity of order $\frac{\alpha}{\alpha+2}$, [110]. The relaxation model (8.12) was analyzed previously by BV-compactness arguments, e.g., [53], [135].

There is more than one way to convert microscopic kinetic formulations of nonlinear equations, into macroscopic algorithms for the approximate solution of such equations. We conclude by mentioning the following three (in the context of conservation laws). Brenier's transport collapse method, [9], is a macroscopic projection method which preceded the BGK-like model (8.12) (see also [53]). Another approach is based on Chapman-Enskog asymptotic expansions, and we refer to [145], for an example of macroscopic approximation other than the usual Navier-Stokes-like viscosity regularization, (3.4). Still another approach is offered by Godunov-type schemes, (4.4), based on projections of the Maxwellians associated with the specific kinetic formulations. These amount to specific Riemann solvers which were studied in [38], [134], [136].

References

[1] L. ALVAREZ AND J.-M. MOREL *Formulation and computational aspects of image analysis*, Acta Numerica (1994), 1–59.

[2] C. BARDOS, F. GOLSE AND D. LEVERMORE, *Fluid dynamic limits of kinetic equations II: convergence proofs of the Boltzmann equations*, Comm. Pure Appl. Math. XLVI (1993), 667–754.

[3] G. BARLES AND P.E. SOUGANIDIS, *Convergence of approximation schemes for fully nonlinear second order equations*, Asympt. Anal. 4 (1991), 271–283.

[4] BELL J.B., COLELLA. P., AND GLAZ. H.M. *A Second-Order Projection Method for the Incompressible Navier-Stokes Equations*, J. Comp. Phys. 85 (1989), 257–283.

[5] M. BEN-ARTZI AND J. FALCOVITZ, *Recent developments of the GRP method*, JSME (Ser.b) 38 (1995), 497–517.

[6] J.P. BORIS AND D. L. BOOK, *Flux corrected transport: I. SHASTA, a fluid transport algorithm that works*, J. Comput. Phys. 11 (1973), 38–69.

[7] F. BOUCHUT AND B. PERTHAME *Kruzkov's estimates for scalar conservation laws revisited*, Universite D'Orleans, preprint, 1996.

[8] F. BOUCHOT, CH. BOURDARIAS AND B. PERTHAME, *A MUSCL method satisfying all the numerical entropy inequalities*, Math. Comp. 65 (1996) 1439–1461.

[9] Y. BRENIER, *Résolution d'équations d'évolution quasilinéaires en dimension N d'espace à l'aide d'équations linéaires en dimension N + 1*, J. Diff. Eq. 50 (1983), 375–390.

[10] Y. BRENIER AND S.J. OSHER, *The discrete one-sided Lipschitz condition for convex scalar conservation laws*, SIAM J. Numer. Anal. 25 (1988), 8–23.

[11] A. BRESSAN, *The semigroup approach to systems of conservation laws*, 4th Workshop on PDEs, Part I (Rio de Janeiro, 1995). Mat. Contemp. 10 (1996), 21–74.

[12] A. BRESSAN, *Decay and structural stability for solutions of nonlinear systems of conservation laws*, 1st Euro-Conference on Hyperbolic Conservation Laws, Lyon, Feb. 1997.

[13] A. BRESSAN AND R. COLOMBO *The semigroup generated by 2 × 2 conservation laws*, Arch. Rational Mech. Anal. 133 (1995), no. 1, 1–75.

[14] A. BRESSAN AND R. COLOMBO *Unique solutions of 2 × 2 conservation laws with large data*, Indiana Univ. Math. J. 44 (1995), no. 3, 677–725.

[15] C. CERCIGNANI, The Boltzmann Equation and its Applications, Appl. Mathematical Sci. 67, Springer, New-York, 1988.

[16] T. CHANG AND L. HSIAO, *The Riemann Problem and Interaction of Waves in Gasdynamics*, Pitman monographs and surveys in pure appl. math, 41, John Wiley, 1989.

[17] G.-Q. CHEN *The theory of compensated compactness and the system of isentropic gas dynamics*, Preprint MCS-P154-0590, Univ. of Chicago, 1990.

[18] G.-Q. CHEN, Q. DU AND E. TADMOR, *Spectral viscosity approximation to multidimensional scalar conservation laws*, Math. of Comp. 57 (1993).

[19] CHEN, D. LEVERMORE AND LIU, *Hyperbolic conservation laws with stiff relaxation terms and entropy* Comm. Pure Appl. Math. 47 (1994) 787–830.

[20] I. L. CHERN, *Stability theorem and truncation error analysis for the Glimm scheme and for a front tracking method for flows with strong discontinuities*, Comm. Pure Appl. Math. XLII (1989), 815–844.

[21] I.L. CHERN, J. GLIMM, O. MCBRYAN, B. PLOHR AND S. YANIV, *Fromt Tracking for gas dynamics* J. Comput. Phys. 62 (1986) 83–110.

[22] A. J. CHORIN, *Random choice solution of hyperbolic systems*, J. Comp. Phys. 22 (1976), 517–533.

[23] B. COCKBURN *Quasimonotone schemes for scalar conservation laws. I. II, III.* SIAM J. Numer. Anal. 26 (1989) 1325–1341, 27 (1990) 247–258, 259–276.

[24] B. COCKBURN, F. COQUEL AND P. LEFLOCH, *Convergence of finite volume methods for multidimensional conservation laws*, SIAM J. Numer. Anal. 32 (1995), 687–705.

[25] P. COLELLA *Multidimensional upwind methods for hyperbolic conservation laws*, J. Comput. Phys. 87 (1990), 87–171.

[26] P. COLELLA AND P. WOODWARD, *The piecewise parabolic method (PPM) for gas-dynamical simulations*, JCP 54 (1984), 174–201.

[27] F. COQUEL AND P. LEFLOCH, *Convergence of finite difference schemes for conservation laws in several space dimensions: a general theory*, SIAM J. Numer. Anal. (1993).

[28] M. G. CRANDALL, *The semigroup approach to first order quasilinear equations in several space dimensions*, Israel J. Math. 12 (1972), 108–132.

[29] M. G. CRANDALL AND P. L. LIONS, *Viscosity solutions of Hamilton-Jacobi equations*, Trans. Amer. Math. Soc. 277 (1983), 1–42.

[30] M. G. CRANDALL AND A. MAJDA, *Monotone difference approximations for scalar conservation laws*, Math. of Comp. 34 (1980), 1–21.

[31] M. G. CRANDALL AND A. MAJDA, *The method of fractional steps for conservation laws*, Numer. Math. 34 (1980), 285–314.

[32] M. G. CRANDALL AND L. TARTAR, *Some relations between non expansive and order preserving mapping*, Proc. Amer. Math. Soc. 78 (1980), 385–390.

[33] T. CHANG AND S. YANG, Two-Dimensional Riemann Problems for Systems of Conservation Laws, Pitman Monographs and Surveys in Pure and Appl. Math., 1995.

[34] C. DAFERMOS, *Polygonal approximations of solutions of initial-value problem for a conservation law*, J. Math. Anal. Appl. 38 (1972) 33–41.

[35] C. DAFERMOS, *Hyperbolic systems of conservation laws*, in "Systems of Nonlinear PDEs", J. M. Ball, ed, NATO ASI Series C, No. 111, Dordrecht, D. Reidel (1983), 25–70.

[36] C. DAFERMOS, private communication.

[37] R. DEVORE AND B. LUCIER, *High order regularity for conservation laws*, Indiana Univ. Math. J. 39 (1990), 413–430.

[38] S. M. DESHPANDE, *A second order accurate, kinetic-theory based, method for inviscid compressible flows*, NASA Langley Tech. paper No. 2613, 1986.

[39] R. DIPERNA *Convergence of approximate solutions to conservation laws*, Arch. Rat. Mech. Anal. 82 (1983), 27–70.

[40] R. DIPERNA, *Convergence of the viscosity method for isentropic gas dynamics*, Comm. Math. Phys. 91 (1983), 1–30.

[41] R. DiPerna, *Measure-valued solutions to conservation laws*, Arch. Rat. Mech. Anal. 88 (1985), 223-270.

[42] R. DiPerna and P. L. Lions, *On the Cauchy problem for Boltzmann equations: Global existence and weak stability*, Ann. Math. 130 (1989), 321–366.

[43] R. DiPerna and P.L. Lions, *Global weak solutions of Vlasov-Maxwell systems*, Comm. Pure Appl. Math. 42 (1989), 729–757.

[44] R. DiPerna, P.L. Lions and Y. Meyer, L^p *regularity of velocity averages*, Ann. I.H.P. Anal. Non Lin. 8(3-4) (1991), 271–287.

[45] W. E. and C.-W.Shu *A numerical resolution study of high order essentially non-oscillatory schemes applied to incompressible flow* J. Comp. Phys. 110, (1993) 39-46.

[46] B. Engquist and S.J. Osher *One-sided difference approximations for nonlinear conservation laws*, Math. Comp. 36 (1981) 321–351.

[47] B. Engquist, P. Lotstedt and B. Sjogreen, *Nonlinear filters for efficient shock computation*, Math. Comp. 52 (1989), 509–537.

[48] B. Engquist and O. Runborg, *Multi-phase computations in geometrical optics*, J. Comp. Appl. Math. 74 (1996) 175–192.

[49] C. Evans, Weak Convergence Methods for Nonlinear Partial Differential equations, AMS Regional Conference Series in Math. 74, Providence R.I. 1990.

[50] K. Friedrichs *Symmetric hyperbolic linear differential equations*, CPAM 7 (1954) 345–.

[51] K. O. Friedrichs and P. D. Lax, *Systems of conservation laws with a convex extension*, Proc. Nat. Acad. Sci. USA 68 (1971), 1686–1688.

[52] P. Gérard, *Microlocal defect measures*, Comm. PDE 16 (1991), 1761–1794.

[53] Y. Giga and T. Miyakawa, *A kinetic construction of global solutions of first-order quasilinear equations*, Duke Math. J. 50 (1983), 505–515.

[54] J. Glimm, *Solutions in the large for nonlinear hyperbolic systems of equations*, Comm. Pure Appl. Math. 18 (1965), 697–715.

[55] J. Glimm and P. D. Lax, Decay of solutions of systems of nonlinear hyperbolic conservation laws, Amer. Math. Soc. Memoir 101, AMS Providence, 1970.

[56] J. Glimm, B. Lindquist and Q. Zhang, *Front tracking, oil reservoirs, engineering scale problems and mass conservation* in *Multidimensional Hyperbolic Problems and Computations*, Proc. IMA workshop (1989) IMA vol. Math Appl 29 (J. Glimm and A. Majda Eds.), Springer-Verlag, New-York (1991), 123-139.

[57] E. Godlewski and P.-A. Raviart, Hyperbolic Systems of Conservation Laws, Ellipses, Paris, 1991.

[58] E. Godlewski and P.-A. Raviart, Numerical Approximation of Hyperbolic Systems of Conservation Laws, Springer, 1996.

[59] S. K. Godunov, *A difference scheme for numerical computation of discontinuous solutions of fluid dynamics*, Mat. Sb. 47 (1959), 271–306.

[60] S. K. Godunov, *An interesting class of quasilinear systems*, Dokl. Akad. Nauk. SSSR 139(1961), 521–523.

[61] F. Golse, P. L. Lions, B. Perthame and R. Sentis, *Regularity of the moments of the solution of a transport equation*, J. of Funct. Anal. 76 (1988), 110–125.

[62] J. Goodman, private communication.

[63] J. GOODMAN AND P. D. LAX, *On dispersive difference schemes. I*, Comm. Pure Appl. Math. 41 (1988), 591–613.

[64] J. GOODMAN AND R. LEVEQUE, *On the accuracy of stable schemes for 2D scalar conservation laws*, Math. of Comp. 45 (1985), 15–21.

[65] J. GOODMAN AND XIN, *Viscous limits for piecewise smooth solutions to systems of conservation laws*, Arch. Rat. Mech. Anal. 121 (1992), 235–265.

[66] D. GOTTLIEB AND E. TADMOR, *Recovering Pointwise Values of Discontinuous Data within Spectral Accuracy*, in "Progress and Supercomputing in Computational Fluid Dynamics", Progress in Scientific Computing, Vol. 6 (E. M. Murman and S. S. Abarbanel, eds.), Birkhauser, Boston, 1985, 357–375.

[67] E. HARABETIAN, *A convergent series expansion for hyperbolic systems of conservation laws*, Trans. Amer. Math. Soc. 294 (1986), no. 2, 383–424.

[68] A. HARTEN, *The artificial compression method for the computation of shocks and contact discontinuities:I. single conservation laws*, CPAM 39 (1977), 611–638.

[69] A. HARTEN, *High resolution schemes for hyperbolic conservation laws*, J. Comput. Phys. 49 (1983), 357–393.

[70] A. HARTEN, B. ENGQUIST, S. OSHER AND S.R. CHAKRAVARTHY, *Uniformly high order accurate essentially non-oscillatory schemes. III*, JCP 71, 1982, 231–303.

[71] A. HARTEN M. HYMAN AND P. LAX, *On finite-difference approximations and entropy conditions for shocks*, Comm. Pure Appl. Math. 29 (1976), 297–322.

[72] A. HARTEN P.D. LAX AND B. VAN LEER, *On upstream differencing and Godunov-type schemes for hyperbolic conservation laws*, SIAM Rev. 25 (1983), 35–61.

[73] A. HARTEN AND S. OSHER, *Uniformly high order accurate non-oscillatory scheme. I*, SIAM J. Numer. Anal. 24 (1982) 229–309.

[74] C. HIRSCH, Numerical Computation of Internal and External Flows, Wiley, 1988.

[75] D. HOFF AND J.S. SMOLLER, *Error bounds for the Glimm scheme for a scalar conservation law*, Trans. Amer. Math. Soc. 289 (1988), 611–642.

[76] T. J. R. HUGHES AND M. MALLET, *A new finite element formulation for the computational fluid dynamics: III. The general streamline operator for multidimensional advective-diffusive systems*, Comput. Methods Appl. Mech. Engrg. 58 (1986), 305–328.

[77] F. JAMES, Y.-J PENG AND B. PERTHAME, *Kinetic formulation for chromatography and some other hyperbolic systems*, J. Math. Pures Appl. 74 (1995), 367–385.

[78] F. JOHN, Partial Differential Equations, 4th ed. Springer, New-York, 1982.

[79] C. JOHNSON AND A. SZEPESSY, *Convergence of a finite element methods for a nonlinear hyperbolic conservation law*, Math. of Comp. 49 (1988), 427–444.

[80] C. JOHNSON, A. SZEPESSY AND P. HANSBO, *On the convergence of shock-capturing streamline diffusion finite element methods for hyperbolic conservation laws*, Math. of Comp. 54 (1990), 107–129.

[81] G.-S. JIANG AND E. TADMOR, *Nonoscillatory Central Schemes for Multidimensional Hyperbolic Conservation Laws*, SIAM J. Sci. Compt., in press.

[82] S. JIN AND Z. XIN, *The relaxing schemes for systems of conservation laws in arbitrary space dimensions*, Comm. Pure Appl. Math. 48 (1995) 235–277.

[83] T. KATO, *The Cauchy problem for quasi-linear symmetric hyperbolic systems*, Arch. Rat. Mech. Anal. 58 (1975), 181–205.

[84] Y. KOBAYASHI, *An operator theoretic method for solving $u_t = \Delta\psi(u)$*, Hiroshma Math. J. 17 (1987) 79–89.

[85] D. KRÖNER, S. NOELLE AND M. ROKYTA, *Convergence of higher order upwind finite volume schemes on unstructured grids for scalar conservation laws in several space dimensions*, Numer. Math. 71 (1995) 527–560.

[86] D. KRÖNER AND M. ROKYTA, *Convergence of Upwind Finite Volume Schemes for Scalar Conservation Laws in two space dimensions*, SINUM 31 (1994) 324–343.

[87] S.N. KRUŽKOV, *The method of finite difference for a first order non-linear equation with many independent variables*, USSR comput Math. and Math. Phys. 6 (1966), 136–151. (English Trans.)

[88] S.N. KRUŽKOV, *First order quasilinear equations in several independent variables*, Math. USSR Sbornik 10 (1970), 217–243.

[89] R. KUPFERMAN AND E. TADMOR, *A fast high-resolution second-order central scheme for incompressible flows*, Proc. Nat. Acad. Sci.

[90] N.N. KUZNETSOV, *Accuracy of some approximate methods for computing the weak solutions of a first-order quasi-linear equation*, USSR Comp. Math. and Math. Phys. 16 (1976), 105–119.

[91] P.D. LAX, *Weak solutions of non-linear hyperbolic equations and their numerical computations*, Comm. Pure Appl. Math. 7 (1954), 159–193.

[92] P. D. LAX, *Hyperbolic systems of conservation laws II*, Comm. Pur Appl. Math. 10 (1957), 537–566.

[93] P.D. LAX, *Shock waves and entropy*, in *Contributions to nonlinear functional analysis*, E.A. Zarantonello Ed., Academic Press, New-York (1971), 603–634.

[94] P.D. LAX, Hyperbolic Systems of Conservation Laws and the Mathematical Theory of Shock Waves (SIAM, Philadelphia, 1973).

[95] P.D. LAX AND X.-D. LIU *Positive schemes for solving multi-dimensional hyperbolic systems of conservation laws*, Courant Mathematics and Computing Laboratory Report NYU, 95-003 (1995), Comm. Pure Appl. Math.

[96] P. LAX AND B. WENDROFF *Systems of conservation laws*, Comm. Pure Appl. Math. 13 (1960), 217–237.

[97] P. LAX AND B. WENDROFF *Difference schemes for hyperbolic equations with high order of accuracy*, Comm. Pure Appl. Math. 17 (1964), 381–.

[98] B. VAN LEER, *Towards the ultimate conservative difference scheme. V. A second-order sequel to Godunov's method*, J. Comput. Phys. 32 (1979), 101–136.

[99] F. LEFLOCH AND P.A. RAVIART, *An asymptotic expansion for the solution of the generalized Riemann problem, Part I: General theory*, Ann. Inst. H. Poincare, Nonlinear Analysis 5 (1988), 179–

[100] P. LEFLOCH AND Z. XIN, *Uniqueness via the adjoint problem for systems of conservation laws*, CIMS Preprint.

[101] R. LEVEQUE *A large time step generalization of Godunov's method for systems of conservation laws*, SIAM J. Numer. Anal. 22(6) (1985), 1051–1073.

[102] R. LEVEQUE, Numerical Methods for Conservation Laws, Lectures in Mathematics, Birkhäuser, Basel 1992.

[103] R. LeVeque, *Wave propagation algorithms for multi-dimensional hyperbolic systems*, Preprint.

[104] D. Levermore and J.-G. Liu, *Oscillations arising in numerical experiments*, NATO ARW seies, Plenum, New-York (1993), To appear.

[105] D. Levy and E. Tadmor, *Non-oscillatory central schemes for the incompressible 2-D Euler equations*, Math. Res. Lett., 4 (1997) 1-20.

[106] D. Li, *Riemann problem for multi-dimensional hyperbolic conservation laws*, Free boundary problems in fluid flow with applications (Montreal, PQ, 1990), 64–69, Pitman Res. Notes Math. Ser., 282.

[107] C.-T. Lin and E. Tadmor, L^1-*stability and error estimates for approximate Hamilton-Jacobi solutions*, in preparation.

[108] P. L. Lions, Generalized Solutions of Hamilton-Jacobi Equations, Pittman, London 1982.

[109] P.L. Lions, *On kinetic equations*, in Proc. Int'l Congress of Math., Kyoto, 1990, Vol. II, Math. Soc. Japan, Springer (1991), 1173-1185.

[110] P. L. Lions, B. Perthame and E. Tadmor, *Kinetic formulation of scalar conservation laws and related equations*, J. Amer. Math. Soc. 7(1) (1994), 169–191

[111] P. L. Lions, B. Perthame and E. Tadmor *Kinetic formulation of the isentropic gas-dynamics equations and p-systems*, Comm. Math. Phys. 163(2) (1994), 415–431.

[112] T.-P. Liu, *The entropy condition and the admissibility of shocks*, J. Math. Anal. Appl. 53 (1976), 78–88.

[113] T. P. Liu, *The deterministic version of the Glimm scheme*, Comm. Math. Phys. 57 (1977), 135–148.

[114] B. Lucier, *Error bounds for the methods of Glimm, Godunov and LeVeque*, SIAM J. Numer. Anal. 22 (1985), 1074–1081.

[115] *B. Lucier, Lecture Notes*, 1993.

[116] Y. Maday, S. M. Ould-Kaber and E. Tadmor *Legendre pseudospectral viscosity method for nonlinear conservation laws*, SIAM J. Numer. Anal. 30 (1993), 321–342.

[117] A. Majda, Compressible Fluid Flow and Systems of Conservation Laws in Several Space Variables, Springer-Verlag New-York, 1984.

[118] A. Majda, J. McDonough and S. Osher, *The Fourier method for nonsmooth initial data*, Math. Comp. 30 (1978), 1041–1081.

[119] M.S. Mock *Systems of conservation laws of mixed type*, J. Diff. Eq. 37 (1980), 70–88.

[120] M. S. Mock and P. D. Lax, The computation of discontinuous solutions of linear hyperbolic equations, Comm. Pure Appl. Math. 31 (1978), 423–430.

[121] K.W. Morton *Lagrange-Galerkin and characteristic-Galerkin methods and their applications*, 3rd Int'l Conf. Hyperbolic Problems (B. Engquist & B. Gustafsson, eds.), Studentlitteratur, (1991), 742–755.

[122] F. Murat, *Compacité par compensation*, Ann. Scuola Norm. Sup. Pisa 5 (1978), 489–507.

[123] F. Murat, *A survey on compensated compactness*, in 'Contributions to Modern calculus of variations' (L. Cesari, ed), Pitman Research Notes in Mathematics Series, John Wiley New-York, 1987, 145–183.

[124] R. NATALINI, *Convergence to equilibrium for the relaxation approximations of conservation laws*, Comm. Pure Appl. Math. 49 (1996), 1–30.

[125] H. NESSYAHU AND E. TADMOR, Non-oscillatory central differencing for hyperbolic conservation laws. J. Comp. Phys. 87 (1990), 408–463.

[126] H. NESSYAHU AND E. TADMOR, *The convergence rate of approximate solutions for nonlinear scalar conservation laws*, SIAM J. Numer. Anal. 29 (1992), 1–15.

[127] H. NESSYAHU, E. TADMOR AND T. TASSA, *The convergence rate of Godunov type schemes*, SIAM J. Numer. Anal. 31 (1994), 1–16.

[128] O. A. OLĔINIK *Discontinuous solutions of nonlinear differential equations*, Amer. Math. Soc. Transl. (2), 26 (1963), 95–172.

[129] S. OSHER, *Riemann solvers, the entropy condition, and difference approximations*, SIAM J. Numer. Anal. 21 (1984), 217-235.

[130] S. OSHER AND F. SOLOMON, *Upwind difference schemes for hyperbolic systems of conservation laws*, Math. Comp. 38 (1982), 339–374.

[131] S. OSHER AND J. SETHIAN, *Fronts propagating with curvature dependent speed: Algorithms based on Hamilton-Jacobi formulations*, J. Comp. Phys. 79 (1988), 12–49.

[132] S. OSHER AND E. TADMOR, *On the convergence of difference approximations to scalar conservation laws*, Math. of Comp. 50 (1988), 19–51.

[133] B. PERTHAME, *Global existence of solutions to the BGK model of Boltzmann equations*, J. Diff. Eq. 81 (1989), 191-205.

[134] B. PERTHAME, *Second-order Boltzmann schemes for compressible Euler equations*, SIAM J. Num. Anal. 29, (1992), 1–29.

[135] B. PERTHAME AND E. TADMOR, *A kinetic equation with kinetic entropy functions for scalar conservation laws*, Comm. Math. Phys.136 (1991), 501–517.

[136] K. H. PRENDERGAST AND K. XU, *Numerical hydrodynamics from gas-kinetic theory*, J. Comput. Phys. 109(1) (1993), 53–66.

[137] A. RIZZI AND B. ENGQUIST, Selected topics in the theory and practice of computational fluid dynamics, J. Comp. Phys. 72 (1987), 1–69.

[138] P. ROE, *Approximate Riemann solvers,parameter vectors, and difference schemes*, J. Comput. Phys. 43 (1981), 357–372.

[139] P. ROE, *Discrete models for the numerical analysis of time-dependent multidimensional gas dynamics*, J. Comput. Phys. 63 (1986), 458–476.

[140] R. RICHTMYER AND K.W. MORTON, Difference Methods for Initial-Value Problems, Interscience, 2nd ed., 1967.

[141] R. SANDERS, *On Convergence of monotone finite difference schemes with variable spatial differencing*, Math. of Comp. 40 (1983), 91–106.

[142] R. SANDERS, *A third-order accurate variation nonexpansive difference scheme for single conservation laws*, Math. Comp. 51 (1988), 535–558.

[143] S. SCHOCHET, *The rate of convergence of spectral viscosity methods for periodic scalar conservation laws*, SIAM J. Numer. Anal. 27 (1990), 1142–1159.

[144] S. SCHOCHET, *Glimm's scheme for systems with almost-planar interactions*, Comm. Partial Differential Equations 16(8-9) (1991), 1423–1440.

[145] S. SCHOCHET AND E. TADMOR, *Regularized Chapman-Enskog expansion for scalar conservation laws*, Archive Rat. Mech. Anal. 119 (1992), 95–107.

[146] D. SERRE, *Richness and the classification of quasilinear hyperbolic systems*, in "Multidimensional Hyperbolic Problems and Computations", Minneapolis MN 1989, IMA Vol. Math. Appl. 29, Springer NY (1991), 315–333.

[147] D. SERRE, Systemés de Lois de Conservation, Diderot, Paris 1996.

[148] D. SERRE, private communication.

[149] C.W. SHU, *TVB uniformly high-order schemes for conservation laws*, Math. Comp. 49 (1987) 105–121.

[150] C. W. SHU, *Total-variation-diminishing time discretizations*, SIAM J. Sci. Comput. 6 (1988), 1073–1084.

[151] C. W. SHU AND S. OSHER, *Efficient implementation of essentially non-oscillatory shock-capturing schemes*, J. Comp. Phys. 77 (1988), 439–471.

[152] C. W. SHU AND S. OSHER, *Efficient implementation of essentially non-oscillatory shock-capturing schemes. II*, J. Comp. Phys. 83 (1989), 32–78.

[153] W. SHYY, M.-H CHEN, R. MITTAL AND H.S. UDAYKUMAR, *On the suppression on numerical oscillations using a non-linear filter*, J. Comput. Phys. 102 (1992), 49–62.

[154] D. SIDILKOVER, *Multidimensional upwinding: unfolding the mystery*, Barriers and Challenges in CFD, ICASE workshop, ICASE, Aug, 1996.

[155] J. SMOLLER, Shock Waves and Reaction-Diffusion Equations, Springer-Verlag, New York, 1983.

[156] G. SOD, *A survey of several finite difference methods for systems of nonlinear hyperbolic conservation laws*, JCP 22 (1978) 1–31.

[157] G. SOD, Numerical Methods for Fluid Dynamics, Cambridge University Press, 1985.

[158] A. SZEPESSY, *Convergence of a shock-capturing streamline diffusion finite element method for scalar conservation laws in two space dimensions*, Math. of Comp. (1989), 527–545.

[159] P. R. SWEBY, *High resolution schemes using flux limiters for hyperbolic conservation laws*, SIAM J. Num. Anal. 21 (1984), 995–1011.

[160] E. TADMOR *The large-time behavior of the scalar, genuinely nonlinear Lax-Friedrichs scheme* Math. Comp. 43 (1984), no. 168, 353–368.

[161] E. TADMOR *Numerical viscosity and the entropy condition for conservative difference schemes* Math. Comp. 43 (1984), no. 168, 369–381.

[162] E. TADMOR, *Convergence of Spectral Methods for Nonlinear Conservation Laws*, SIAM J. Numer. Anal. 26 (1989), 30–44.

[163] E. TADMOR, *Local error estimates for discontinuous solutions of nonlinear hyperbolic equations*, SIAM J. Numer. Anal. 28 (1991), 891–906.

[164] E. TADMOR, *Super viscosity and spectral approximations of nonlinear conservation laws*, in "Numerical Methods for Fluid Dynamics", Proceedings of the 1992 Conference on Numerical Methods for Fluid Dynamics (M. J. Baines and K. W. Morton, eds.), Clarendon Press, Oxford, 1993, 69–82.

[165] E. TADMOR AND T. TASSA, *On the piecewise regularity of entropy solutions to scalar conservation laws*, Com. PDEs 18 91993), 1631-1652.

[166] T. TANG AND Z. H. TENG, *Error bounds for fractional step methods for conservation laws with source terms*, Simon Fraser University Preprint.

[167] L. TARTAR, *Compensated compactness and applications to partial differential equations*, in *Research Notes in Mathematics 39*, Nonlinear Analysis and Mechanics, Heriott-Watt Symposium, Vol. 4 (R.J. Knopps, ed.) Pittman Press, (1975), 136–211.

[168] L. TARTAR, *Discontinuities and oscillations*, in Directions in PDEs, Math Res. Ctr Symposium (M.G. Crandall, P.H. Rabinowitz and R.E. Turner eds.) Academic Press (1987), 211-233.

[169] T. TASSA, *Applications of compensated compactness to scalar conservation laws*, M.Sc. thesis (1987), Tel-Aviv University (in Hebrew).

[170] B. TEMPLE, *No L^1 contractive metrics for systems of conservation laws*, Trans. AMS 288(2) (1985), 471–480.

[171] Z.-H. TENG *Particle method and its convergence for scalar conservation laws*, SIAM J. Num. Anal. 29 (1992) 1020–1042.

[172] A. I. VOL'PERT, *The spaces BV and quasilinear equations*, Math. USSR-Sb. 2 (1967), 225–267.

[173] G.B. WHITHAM, Linear and Nonlinear Waves, Wiley-Interscience, 1974.

Proceedings of Symposia in Applied Mathematics
Volume **54**, 1998

THE SMALL DISPERSION KDV EQUATION
WITH DECAYING INITIAL DATA.

S. Venakides, Duke University

Dedicated with affection to Peter Lax and
Louis Nirenberg on their 70th birthday

INTRODUCTION

The initial value problem for the Korteweg-de Vries (KdV) equation

$$(0.1) \qquad u_t - 6uu_x + \epsilon^2 u_{xxx} = 0, \quad u(x,0) = u_0(x), \quad \epsilon > 0,$$

was solved by Gardner, Greene, Kruskal and Miura [GGKM] for initial data $u_0(x)$ that decay sufficiently rapidly as $|x| \to \infty$. Their critical observation was that the KdV is the solvability condition for the linear system

$$(0.2) \qquad Lf = \lambda f, \qquad \partial_t f = Bf,$$

in which λ is an arbitrary complex parameter, and where the operators L and B are given by

$$(0.3a) \qquad L \equiv -\epsilon^2 \partial_{xx} + u,$$
$$(0.3b) \qquad B \equiv -4\epsilon^2 \partial_{xxx} + 3(u\partial_x + \partial_x u).$$

Lax [L] placed this discovery under the general principle of the Lax pair of operators that led to the discovery and solution of many more integrable systems. The pair consists of two operators $L = L(t)$ and $B = B(t)$, such that for each complex parameter λ, the linear system (0.2) has a nontrivial solution $f = f(\lambda, x, t)$. The compatibility of the two equations (0.2) reduces easily to the operator equation

$$(0.4) \qquad \partial_t L = [B, L] \equiv BL - LB,$$

When the Lax pair is given by (0.3a) and (0.3b), equation (0.4) is equivalent to the KdV equation. It reduces to other integrable systems for different and appropriate choices of the Lax pair.

The solution to (0.1) now proceeds as follows. The second equation (0.2) governs the time evolution of the eigenfunctions f of the operator L for each fixed value of λ. This evolution equation becomes asymptotically linear and can be solved explicitly in the limit of u tending to zero. Thus, for potentials u that decay as $|x| \to 0$ and for fixed λ in the spectrum of L, the large x asymptotic behavior of f, referred to as the scattering data of u, can be explicitly calculated from the scattering data at $t = 0$. The function $u(x, t)$, which is the potential of the Schrödinger operator L, is reconstructed from the scattering data at time t by the inverse scattering procedure.

We study the behavior of the solution of (0.1) in the limit $\epsilon \downarrow 0$, the so-called zero dispersion problem for KdV. For $\epsilon > 0$, no matter how small, solutions of (0.1) with smooth, decaying initial data, exist and remain smooth for all time. As $\epsilon \downarrow 0$, the KdV equation (0.1) reduces (formally) to the inviscid Burger's equation, $u_t - 6uu_x = 0$, the solutions of which may develop shocks. It is a central question in the theory to understand how the dispersive term $\epsilon^2 u_{xxx}$, prevents the formation of shocks. In the neighborhood of the point of shock formation for the inviscid Burger's equation, it turns out that as $\epsilon \downarrow 0$ the solution $u = u(x, t, \epsilon)$ develops rapid oscillations which transport energy away from the shock in the form of highly oscillating, modulated wave trains (see formula(2.34) below). This situation should be contrasted with the well-known dissipative regularization of Burger's equation, $u_t - 6uu_x - \epsilon^2 u_{xx} = 0$, $\epsilon > 0$; here the dissipative term $\epsilon^2 u_{xx}$ "burns" away the shock energy and results in the generation of heat.

The small dispersion problem for KdV was first analyzed by Lax and Levermore [LL1,LL2] for the case of a well, $u_0(x) < 0$. The case of a "bump", $u_0(x) > 0$, was later analyzed by Venakides[V1]. In both cases, it was possible to describe the small dispersion limit using variational methods that average out the rapid oscillations. In further developments, Venakides derived the form (2.34) of the rapid oscillations, first by taking the continuum limit of Abel sums [V2,V3] and later by introducing a "quantum condition" for the variational problem [V4]. We outline some of these results in Part 1 of this article. The methods outlined in this article have been used to obtain new results for a variety of singular limits of integrable systems (see [LLV] for a description of the history of these problems and a survey of results).

In Part 2 of the article, we present recent results that we have obtained in collaboration with Deift and Zhou [DVZ2]. Our approach is based on the Riemann-Hilbert (RH) formulation of IST, and our analysis is motivated by the steepest descent method for oscillatory RH problems introduced by Deift and Zhou in [DZ1], and developed further, in particular, in [DZ2] and [DVZ1]. The analysis in [DVZ2],

(a) Extends the RH steepest descent method by introducing a systematic procedure for determining the contour of main contribution to the solution of the RH problem, when this contour is not obvious. This procedure leads to an associated scalar RH problem for a certain phase function that now plays the role of the solution of the variational problem in Lax-Levermore theory.

(b) Reduces the initial value problem of the modulation equations to solving sets of algebraic equations constrained by algebraic inequalities. These equations and inequalities are obtained in terms of the initial data u_0, directly from the RH problem. Thus, the modulation equations are completely bypassed (see Section 2.2 for further discussion).

(c) Derives the Lax-Levermore-Venakides theory, in a direct and natural way using the RH problem, and in addition obtains precise and rigorous phase-shift information on the rapid oscillations.

The above approach has recently been used in [DKMVZ](see also [DKM]), to make significant progress in the theory of orthogonal polynomials and prove a form of the universality conjecture for random matrices.

PART 1

1.1 Radiationless Initial Data: Lax-Levermore Theory.

The initial data $u_0(x)$ is a single negative bump with a minimum value of $-\eta_{max}^2$. The scattering data for $u_0(x)$ are computed asymptotically for small ϵ by the semiclassical (WKB) method. The discrete eigenvalues are packed in the range of the potential $-\eta_{max}^2 < \lambda < 0$. In terms of the transformed spectral variable $\eta = (-\lambda)^{1/2}$, as was first proved by Weyl, the asymptotic density of eigenvalues $\eta_1, \eta_2, \eta_3, ...$ is given by the formula

$$\text{density of eigenvalues } \sim \frac{1}{\pi\epsilon}\phi(\eta)$$

(1.1) $$\phi(\eta) = \int_{x_-(\eta)}^{x_+(\eta)} \frac{\eta}{\sqrt{v(x) - \eta^2}}\, dx\,, \qquad 0 < \eta < \eta_{max}\,,$$

where $x_-(\eta) < x_+(\eta)$ are defined by $u_0(x_\pm) = \eta^2$. The total number of eigenvalues N_ϵ is given asymptotically by

(1.2) $$N_\epsilon \sim \frac{1}{\pi\epsilon} \int_0^{\eta_{max}} \phi(\eta)\, d\eta\,.$$

The norming exponent χ_j (definition: the L^2 normalized eigenvector f_j of L, corresponding to η_j, satisfies $f_j \sim e^{-(\eta_j x - \chi_j)/\epsilon}$ as $x \to +\infty$) obtained from the WKB analysis at $t = 0$ is given by

(1.3) $$\chi_j \sim \chi(\eta_j)\,, \qquad \chi(\eta) = \eta x_+(\eta) + \int_{x_+(\eta)}^{\infty} \left(\eta - \sqrt{\eta^2 - v(x)}\right)\, dx\,.$$

The reflection coefficient that constitutes the scattering data for energies $\lambda \geq 0$ is calculated to be zero to all orders of the WKB expansion and is neglected.

The KdV solution corresponding to this reflectionless initial data is given by

(1.4) $$u(x, t, \epsilon) = -2\epsilon^2 \partial_{xx} \ln \tau_\epsilon(x, t)\,.$$

where $\tau_\epsilon(x, t)$ is obtained from the formula

(1.5) $\tau_\epsilon(x, t)$

$$= \sum_{\Lambda_\epsilon} \frac{\pi^{-|\Lambda_\epsilon|}}{|\Lambda_\epsilon|!} \exp\left((2/\epsilon) \sum_{\eta_j \in \Lambda_\epsilon} (-\eta_j x + 4\eta_j^3 t + \chi_j) + \sum_{\eta_i \in \Lambda_\epsilon} \sum_{\eta_j \in \Lambda_\epsilon}' \ln \left|\frac{\eta_i - \eta_j}{\eta_i + \eta_j}\right|\right)\,.$$

Here, Λ_ϵ ranges over all the subsets of the set of eigenvalues $\{\eta_1, \eta_2, \eta_3, ...\}$ and $|\Lambda_\epsilon|$ denotes the number of elements of the subset Λ_ϵ. We observe from the formula the well known evolution of the norming exponents $\chi_j(t) = \chi_j(0) + 4\eta_j^3 t$.

Lax and Levermore show, using a steepest descent type argument, that,

$$(1.6) \qquad \lim_{\epsilon \downarrow 0} \epsilon^2 \ln \tau_\epsilon(x, t) = \lim_{\epsilon \downarrow 0} \epsilon^2 \max_{\Lambda_\epsilon} \{\text{parenthesis in (1.5)}\} = O(1).$$

as $\epsilon \downarrow 0$.

In an insightful move, they represent the maximizing subset $\Lambda_\epsilon = \Lambda^*(\epsilon)$ by the distribution

$$(1.7) \qquad \psi_\epsilon^*(\eta) = \pi \epsilon \sum_{\eta_j \in \Lambda^*} \delta(\eta - \eta_j),$$

and write the sums in the parenthesis in (1.5) as integrals using the measure $\psi_\epsilon^*(\eta) d\eta$. Taking the limit $\epsilon \downarrow 0$ they show that

$$(1.8) \qquad \begin{aligned} \lim_{\epsilon \downarrow 0} \epsilon^2 \ln \tau_\epsilon(x, t) &= \frac{1}{\pi} \max \left\{ 2(a, \psi) + (L\psi, \psi) : \psi \in A \right\} = Q(x, t), \\ A &\equiv \left\{ \psi \in L^1([0, \eta_{max}]) : 0 \le \psi \le \phi \right\}, \end{aligned}$$

the maximum being attained within the admissible set. Here,

$$(1.9a) \qquad a(\eta, x, t) = -\eta x + 4\eta^3 t + \chi(\eta),$$

$$(1.9b) \qquad L\psi(\eta) \equiv \frac{1}{\pi} \int_0^{\eta_{max}} \log\left| \frac{\eta - \mu}{\eta + \mu} \right| \psi(\mu)\, d\mu,$$

arise from the expressions in the exponent of (1.5) and $(\ ,\ )$ is the standard inner product. The maximization is posed over nonnegative L^1 functions. The maximizer $\psi^*(\eta)$ of (1.8) is the distribution limit of (1.7) as $\epsilon \downarrow 0$.

The variational problem is strictly convex and therefore has a unique solution that depends continuously on (x, t) in the weak topology of measures. This in turn implies that $Q(x, t)$ is continuously differentiable with

$$(1.10) \qquad \partial_x Q(x, t) = -\frac{2}{\pi}(\eta, \psi^*(x, t)), \qquad \partial_t Q(x, t) = \frac{8}{\pi}(\eta^3, \psi^*(x, t)).$$

The convexity of $\log \tau_\epsilon$ as a function of x and t then yields the limits

$$(1.11) \qquad \partial_x Q(x, t) = \lim_{\epsilon \to 0} \epsilon^2 \partial_x \log \tau_\epsilon(x, t), \qquad \partial_t Q(x, t) = \lim_{\epsilon \to 0} \epsilon^2 \partial_t \log \tau_\epsilon(x, t),$$

uniformly over compact subsets of (x, t).

Taking the derivative of (1.11) with respect to x and utilizing (1.4), Lax and Levermore obtain

$$(1.12) \qquad \underset{\epsilon \downarrow 0}{\text{weak}\lim}\, u(x, t, \epsilon) = -2\partial_{xx} Q(x, t) = \frac{4}{\pi} \partial_x (\eta, \psi^*).$$

The limit of the solution $u(x, t, \epsilon)$ as ϵ tends to zero is weak, since the derivation of (1.12) involves the interchange of the order of differentiation and the taking of the limit.

The maximization problem (1.8) is attacked analytically by solving its variational conditions given by

$$
\begin{array}{lll}
\text{(1.13)} & L\psi(\eta) + a(\eta, x, t) \leq 0, & \text{when } \psi(\eta, x, t) = 0, \\
& L\psi(\eta) + a(\eta, x, t) = 0, & \text{when } 0 < \psi(\eta, x, t) < \phi(\eta), \\
& L\psi(\eta) + a(\eta, x, t) \geq 0, & \text{when } \psi(\eta, x, t) = \phi(\eta).
\end{array}
$$

Lax and Levermore make the generic Ansatz that the set $S(x, t) = \{\eta \in \mathbf{R} : 0 < \psi(\eta) < \phi(\eta)\}$ is a finite union of $N + 1$ intervals ($N = 0, 1, 2, ...$). As we will see below, the $2N + 2$ end-points of the intervals $\beta_1 > \beta_2 > ... > \beta_{2N+1} \geq 0$ enter the analysis in an important way. Differentiating the equality in (1.13) with respect to x and t they obtain the much simpler set of equalities

$$
\begin{array}{llll}
\text{(1.14)} & L\psi_x(\eta) = \eta, & L\psi_t(\eta) = -12\eta^2, & \text{when } \eta \in S, \\
& \psi_x(\eta) = 0, & \psi_t(\eta) = 0, & \text{when } \eta \notin S.
\end{array}
$$

They then extend $\psi(\eta)$ to be an odd function over $\mathbf{R}$ that vanishes when $|\eta| \geq \eta_{max}$, and observe that the η-derivative of L is the Hilbert transform,

$$
\text{(1.15)} \qquad \frac{d}{d\eta} L\psi(\eta) = \frac{1}{\pi} \int_{-\infty}^{\infty} \frac{\psi(\mu)}{\eta - \mu} \, d\mu \equiv H\psi(\eta) \, .
$$

This observation allows them to solve (1.14) explicitly for ψ_x and ψ_t, essentially by Riemann-Hilbert problem technique, in terms of the hyperelliptic curve

$$
\text{(1.16)} \qquad R(\eta, x, t) = \left(\prod_{i=1}^{2N+1} (\beta_i^2 - \eta^2) \right)^{1/2} \, .
$$

The compatibility condition $\partial_t \psi_x = \partial_x \psi_t$ provides a system of $2N + 1$ nonlinear hyperbolic partial differential equations in Riemann invariant form for the $2N + 1$ unknown $\beta_j's$,

$$
\text{(1.17)} \qquad \partial_t \beta_i + s_i(\beta_1, \cdots, \beta_{2N+1}) \, \partial_x \beta_i = 0, \qquad \text{for } i = 1, \cdots, 2N + 1.
$$

The emergence of a hyperelliptic function in the calculation marks an important connection. KdV has multiphase quasiperiodic wave solutions that are described with the aid of hyperelliptic functions. Indeed, each of these waves is completely characterized up to phase-shifts by the branch points of the corresponding hyperelliptic function. An N-phase wave has $2N + 1$ branch points from which the mean wave value plus a wavenumber and a frequency for each phase can be calculated. Here, $N = 1, 2, 3, ...$, while one can associate $N = 0$ with the constant solution. When equations (1.17) were derived, it was immediately noticed that they are identical to the modulation equations, first derived by Whitham [W] (for $N = 1$) and then, at about the same time as (1.17), by Flaschka, Forest, McLaughlin [FFM] (for $N > 1$), that describe the slow space and time evolution of the branch points of modulating multi-phase KdV waves. It was immediately realized that the solution of the small dispersion problem develops fast space-time oscillations, and that the hyperelliptic curve that describes these oscillations is exactly (1.16). The fact that this is so, was first shown with the help of a "quantum condition" imposed as an Ansatz by Venakides [V4]. It follows rigorously from our calculation in part 2 of this paper.

The formal integrability of (1.17) was proved by Tsarev[Ts], with further developments by Krichever[K], and Wright[W]; Tian[T] and Gurevich, Krylov and El'[GKE] constructed the global solution of the modulation equations for given initial data and $t - t_{\text{crit}}$ small. For such t, Tian proves that $N = N(x, t) \leq 1$ for all x. In Part 2 of this article, we derive $2N + 1$ local, algebraic conditions for the $2N + 1$ unknown β_i's, (following the notation of [DVZ2], we label them α_i, and β_i) from which the latter can be calculated without the need of solving a PDE. The derivation of $N = N(x, t)$ is also discussed.

Remark. In recent results by Ercolani, Levermore and Zhang [ELZ1][ELZ2], the Lax Levermore maximizer is characterized as the limiting density of half-line Dirichlet spectra of the associated Schroedinger operator. These results are presented in the article by Dave Levermore in these proceedings.

1.2 Resolving the Oscillations: Leading Correction to Lax-Levermore Theory.

As explained above, the Lax-Levermore calculation of the weak limit of the solution reveals the nature of the small-scale oscillations only indirectly, through the fact that the β_i's satisfy the same modulation equations as the branch points of the multi-phase KdV waves. To derive the small scale oscillations directly, Venakides [V4] proposed an Ansatz that has the effect of sharpening the Lax-Levermore calculation in the following way. According to Lax-Levermore theory, equation (1.4) produces the weak limit of the solution if the function τ_ϵ given by (1.5) is replaced by the single term

$$(1.18) \qquad \exp\left(\frac{2(a, \psi^*) + (L\psi^*, \psi^*)}{\pi \epsilon^2} \right),$$

where ψ^* is the the maximizer of (1.18). Recall that ψ^* is the distribution limit of the discrete maximizer ψ^*_ϵ of the parenthesis in (1.5), and observe that, according to (1.7), the integral of ψ^*_ϵ must be an integral multiple of $\pi\epsilon$. This observation plays a crucial role in the derivation of the leading correction to the theory.

Ansatz. To obtain the asymptotic behavior of the solution as ϵ tends to zero, we replace the function τ_ϵ with

$$(1.19) \qquad \tau_\epsilon \sim K(\epsilon) \sum_\psi \exp\left(\frac{2(a, \psi) + (L\psi, \psi)}{\pi \epsilon^2} \right)$$

summation taking place over all functions ψ that solve the maximization problem (1.8) subject to the additional *quantum condition*,

$$(1.20) \qquad \frac{1}{\pi \epsilon} \int_c \psi(\eta) \, d\eta = \text{integer},$$

where C is any topological component of the support of ψ.

Observing that an $O(\epsilon)$ perturbation of ψ^* will satisfy the quantum condition, we set $\psi(\eta) = \psi^*(\eta) + \epsilon\,\bar\psi(\eta) + \cdots$ and perform the maximization with respect to $\bar\psi$ subject to the additional constraint (1.20). We obtain, after a calculation, the variational conditions

(1.21)
$$\begin{cases} L\bar\psi(\eta) = \sum_{i=0}^{g} \frac{1}{\pi} c_i \mathbf{1}_i(\eta) & \text{when } \eta \in S(x,t), \\[2mm] \bar\psi(\eta) = 0 & \text{when } \eta \notin S(x,t), \\[2mm] \frac{1}{\pi} \int_{\beta_{2i}}^{\beta_{2i-1}} \bar\psi(\eta)\, d\eta = m_i - \frac{1}{\pi\epsilon} \int_{\beta_{2i}}^{\beta_{2i-1}} \psi^*(\eta)\, d\eta & \text{for } i = 1, \cdots, g, \end{cases}$$

where $\mathbf{1}_i(\eta)$ is the characteristic function of the interval $(\beta_{2i}, \beta_{2i-1})$, the constant c_i is a Lagrange multiplier, and $S(x,t)$ is the set defined after (1.13) for $\psi = \psi^*$.

Condition (1.20) allows a free *integer* parameter for each connected component of S. The maximizing ψ 's are then parametrized by the multi-integer $(m_1, m_2, ..., m_N)$. The summation in (1.19) is effectively over this multi-integer.

The solutions of (1.21) are obtained with the aid of the hyperelliptic curve (1.16). They form a vector space over the multi-integers $(m_1, \cdots, m_N)$ and are easily identified with the holomorphic differentials corresponding to the hyperelliptic curve (1.16) in the variable $-\eta^2$. Upon inserting them into the right hand side of (1.19), we obtain [V4] a sum over all multi-integers which turns out to be the well known theta function expression (2.34) for multi-phase KdV waves. More accurately, the series on the right side of (1.19) turns out to be the product of the theta function with a factor that is quadratic in x. When the second x-derivative of the logarithm is taken in (1.4), this factor produces the constant in the formula (2.34).

1.3 The Case of Pure Radiation Initial Data. When the initial data $u_0(x)$ of (0.1) is a positive bump (we normalize it to have unit height), the solution contains no solitons and consists of pure radiation. This is exactly the problem that we analyze in Part 2 by the method of steepest descent [DVZ2]. In this section, we outline our procedure [V1] of renormalizing the problem to make the calculation of the weak limit of the solution amenable to the Lax-Levermore approach. We obtain the following results.

(1.22)
$$\text{weak-}\lim_{\epsilon\downarrow 0} u(x, t, \epsilon) = 1 - 2\partial_{xx} W(x, t).$$

$W(x,t)$ is derived from the maximization problem,

(1.23)
$$W(x,t) = \frac{1}{\pi} \max \left\{ 2(a, \psi) - 2(\gamma, \psi_+) + (L\psi, \psi) : \psi \in A \right\},$$
$$A \equiv \left\{ \psi \in L^1([0,1]) \right\},$$

where

(1.24)
$$a(\eta, x, t) = \eta(x_m - x - 6t) + 4\eta^3 t + \tilde\chi(\eta),$$

(1.25) $$\tilde{\chi}(\eta) = \int_{x_m}^{x_+} (\eta^2 - 1 + u_0(x))^{1/2} dx,$$

(1.26) $$\gamma(\eta) = \int_{x_-}^{x_+} (\eta^2 - 1 + u_0(x))^{1/2} dx,$$

the point x_m being the maximizer of the initial data $(u_0(x_m) = 1)$, while $x_- \leq x_+$ are the two solutions to the equation $u_0(x) = 1 - \eta^2$. Finally, the operator L is as defined in (1.9b) with $\eta_{max} = 1$.

As in the Lax-Levermore calculation, the maximization problem (1.23) is attacked analytically by solving its variational conditions that are now given by

(1.27)
$$\begin{aligned}
L\psi(\eta) + a(\eta, x, t) = 0, \qquad & \text{when } \psi(\eta, x, t) < 0, \\
0 \leq L\psi(\eta) + a(\eta, x, t) \leq \gamma(\eta), \qquad & \text{when } \psi(\eta, x, t) = 0, \\
L\psi(\eta) + a(\eta, x, t) = \gamma(\eta), \qquad & \text{when } \psi(\eta, x, t) > 0.
\end{aligned}$$

The maximizer ψ^* is calculated and is an admissible function. Inserting the value of W, expressed as a functional of ψ, into (1.22) we abtain after a short calculation,

(1.28) $$\text{weak-}\lim_{\epsilon \downarrow 0} u(x, t, \epsilon) = 1 + \frac{4}{\pi} \partial_x (\eta, \psi^*)$$

We will now outline the derivation of these results. We first make the change of the dependent variable

(1.29a) $$u(x, t) = 1 + w(\xi, t), \quad \text{where } \xi = x + 6t$$

and we observe that w satisfies the KdV initial value problem,

(1.29b) $$w_t - 6ww_\xi + \epsilon^2 w_{\xi\xi\xi} = 0, \quad w(\xi, 0) = u_0(\xi) - 1.$$

In order to have a scattering problem that is amenable to the Lax-Levermore approach, we modify the initial data $w(\xi, 0)$ by adding to it a Heaviside function that equals unity when $x \geq x_0$ and is zero otherwise. We call this the "barrier". The time evolution of pure barrier initial data was first derived through modulation theory by Gurevich and Pitaevski[GP] and later, rigorously by Lax and Levermore[LL]. To avoid non-negligible interaction of the barrier with our original initial data within the time frame of our calculation, we take x_0 to be as large as we please. We find that x_0 drops out of our calculation eventually. We now have initial data that tend to different limits as $x \to \pm\infty$. The scattering theory for such potentials has been derived by Buslaev and Fomin [BF]. Following Lax and Levermore, we neglect reflection corresponding to energies that are greater than the potential peak; the WKB approximation to the corresponding reflection coefficient is zero to all orders. The scattering data for energies $-\eta^2$ in the range of the initial potential $(u_0(\xi) - 1 + \text{Barrier})$, i.e. between its ground state $-\eta^2 \sim \inf w_0(\xi) = -1, (\eta = 1)$ and its peak $-\eta^2 \sim \max w_0(\xi) = 0, (\eta = 0)$ are the norming exponents $\chi(\eta)$. The latter are defined as for pure soliton data. The normalized eigenfunctions that are now not in L^2, decay as $e^{-(\eta\xi - \chi(\eta))/\epsilon}$ in the limit $x \to +\infty$ and are sinusoidal of unit amplitude as $x \to -\infty$.

For the inverse scattering step, we have developed a continuum analogue of the formula (1.5) for the Kay-Moses determinant ; formula (1.4) that gives the potential w in terms of a function τ_ϵ is still valid, while formula (1.5) is replaced by its continuum analogue,

$$(1.30) \qquad \tau_\epsilon(x,t) = 1 + \sum_{m=1}^{\infty} \frac{d_m}{m!}$$

$$(1.31) \quad d_m = A_m \int_0^1 \cdots \int_0^1 \exp\left((2/\epsilon) \sum_{j=1}^{m} (-\eta_j \xi + 4\eta_j^3 t + \chi(\eta_j)) \right.$$
$$\left. + \sum_{j=1}^{m} \sum_{j=1}^{m} \ln \left| \frac{\eta_i - \eta_j}{\eta_i + \eta_j} \right| \right) d\eta_1 ... d\eta_m.$$

The coefficients A_m play no role in our calculation, and the norming exponent $\chi(\eta)$ has been defined above.

We now calculate the scattering data $\chi(\eta)$ at $t = 0$. For large negative ξ, the eigenfunction is a pure sinusoid of unit amplitude, a standing wave that is the sum of an incident right traveling wave and its total reflection. For most values of η, the wave is attenuated by a factor of $O(1)e^{\gamma(\eta)/\epsilon}$ as it penetrates the bump of our initial data. It is a standing wave of smaller amplitude $O(1)e^{-\gamma(\eta)/\epsilon}$ in the potential well that is formed between the bump and the barrier; finally, it decays as $O(1)e^{-\eta(\xi-x_0)/\epsilon-\gamma(\eta)/\epsilon}$ when $\xi \geq x_0$. Thus, for most values of η we have $\chi(\eta) \sim x_0\eta - \gamma(\eta)$. There are exceptional η's, however, the resonances of the well between the bump and the barrier. These are packed in the energy range of the potential well. In terms of the transformed spectral variable $\eta = (1 - \lambda)^{1/2}$ (λ is the eigenvalue parameter of the operator L of the original problem (2.1); $\lambda - 1$ is the eigenvalue parameter for the reduced problem (1.29)) the asymptotic density of resonances $\bar{\eta}_1, \bar{\eta}_2, \bar{\eta}_3, ...$ is given by the WKB formula

$$\text{density of resonances} \sim \frac{1}{\pi\epsilon}\varphi(\eta),$$
$$(1.32) \qquad \varphi(\eta) = \int_{x_+(\eta)}^{x_0} \frac{\eta}{(1 - u_0(\xi) - \eta^2)^{1/2}} d\xi, \qquad 0 < \eta < 1.$$

At the resonant values of η, the eigenfunction is "almost" a bound state of the well, decaying as it tunnels from the well through the bump into the region on the left where it has $O(1)$ amplitude. Thus, at the resonances, the decay of the eigenfunction on the right ($x \to +\infty$) is $O(1)e^{-\eta(\xi-x_0)/\epsilon+\gamma(\eta)/\epsilon}$. An approximation of the norming exponent that is sufficient for our calculation shows that for values of η that extend from the resonance $\bar{\eta}_j$ half-way to any of the two neighboring resonances

$$(1.33) \qquad e^{2\chi(\eta)/\epsilon} = \frac{e^{(2\eta x_0 - 2\gamma(\eta))/\epsilon}}{(\eta - \bar{\eta}_j)^2 + e^{-4\gamma(\eta)/\epsilon}},$$

where the index j runs over the resonances. Observe that the right hand side equals $e^{2\eta x_0/\epsilon}$ multiplied by a sharp spike that is an approximation to the Dirac delta function. We make three further observations.

(a) If η is not exponentially close to a resonance, then as noted earlier $\chi(\eta) \sim x_0(\eta) - \gamma(\eta)$.

(b) The integral with respect to η of the left hand side of (1.33) over a small interval $<< \epsilon$ centered at the resonance $\bar{\eta}_i$ is essentially the integral of a delta function centered at $\eta = \bar{\eta}_i$. Its singular contribution to χ equals $\bar{\eta}_i x_0$. Thus, resonant values of η_j make singular contributions to the integrals (1.31).

(c) The contribution of more than one of the integration variables, say η_j and η_k to the integral (1.31), when both are in the immediate vicinity of the same resonance is negligible, because of the term $\ln|\eta_j - \eta_k|$ in (1.31).

A careful analysis [V1] of (1.31) and (1.33) reveals that as $\epsilon \downarrow 0$, the following formula holds.

(1.34) $\epsilon^2 \log \tau_\epsilon(x,t) \sim \epsilon^2 \max\{\text{parenthesis in (1.31)}\} = O(1),$

with the understanding that $\{\eta_1, \eta_2, ...\}$ is any finite subset of the interval $(0,1)$, that $\chi(\eta)$ is given by,

$$\chi(\eta) = \eta x_0 , \qquad \text{when } \eta \text{ is resonant,}$$

(1.35) $\chi(\eta) = \eta x_0 - \gamma(\eta_j) , \qquad \text{when } \eta \text{ is nonresonant.}$

and that if $k \neq j$, η_j and η_k cannot be chosen to equal the same resonance (the parenthesis in (1.31) would then be $-\infty$ because of the logarithmic term).

We are now in a position to outline the derivation of (1.22). Let $\pi \epsilon \sum_j \delta(\eta - \eta_j^*)$ (see (1.7)) be the distribution that maximizes the parenthesis in (1.31). Clearly, we want as many of the $\eta - j^*$ to be resonances because resonances contribute more. We are, however, restricted by the fact that there are only finitely many resonances available. As $\epsilon \downarrow 0$, let the continuum limit of the distribution $\pi \epsilon \sum_j \delta(\eta - \eta_j^*)$ equal the density of resonances φ plus an unknown density ψ. The continuum limit of $\pi \epsilon^2$ times the parenthesis will then be

(1.36) $2(-\xi\eta + 4\eta^3 t, \varphi + \psi) + 2(\chi(\eta), \varphi + \psi) + (L(\varphi + \psi), \varphi + \psi).$

Now, we can place the η_j's on resonances when $\psi \leq 0$; we can do this only for density φ of them when $\psi > 0$. Thus we have

$$\chi(\eta) = \eta x_0 , \qquad\qquad\qquad \text{when } \psi(\eta) \leq 0 ,$$

(1.37)

$$(\chi(\eta), \varphi + \psi) = (\eta x_0, \varphi + \psi) - (\gamma(\eta_j), \psi) , \qquad \text{when } \psi(\eta) > 0.$$

The two equations in (1.37) can be combined into one,

(1.38) $(\chi(\eta), \varphi + \psi) = (\eta x_0, \varphi + \psi) - (\gamma(\eta_j), \psi_+),$

where ψ_+ is the positive part of ψ, i.e., equal to ψ when ψ is positive and equal to zero when ψ is negative. Inserting into (1.36) we obtain the expression

(1.39) $2(-\xi\eta + 4\eta^3 t, \varphi + \psi) + 2(\eta x_0, \varphi + \psi) - 2(\gamma(\eta_j), \psi_+) + (L(\varphi + \psi), \varphi + \psi).$

We rearrange this, throwing away terms that are either independent of or linear in ξ. They would be anihilated by taking the second logarithmic derivative in ξ and would have no effect in the weak limit of w. We obtain

(1.40) $-2(\gamma, \psi_+) + 2(-\xi\eta + 4\eta^3 t + x_0\eta + L\varphi, \psi) + (L\psi, \psi),$

Finally, recalling that $\xi = x + 6t$ and utilizing the calculus equality $L\varphi(\eta) = \eta x_m + \tilde{\chi}(\eta)$ (see 1.25), we identify this expression with the quantity to be maximized in (1.23).

PART 2

2.0 Inverse Scattering Formulated as a Riemann-Hilbert Problem.

We study the initial value problem

(2.1) $$u_t - 6uu_x + \epsilon^2 u_{xxx} = 0, \quad u(x,0) = u_0(x) \geq 0.$$

We begin our exposition with a brief outline of the results of the inverse scattering transformation, reformulated as a Riemann-Hilbert (RH) problem by Shabat [S] (see also [BDT]). After changing from the $\sqrt{\lambda}$–plane to the λ–plane, taking the right half $\sqrt{\lambda}$–plane onto $\mathbf{C} \setminus (-\infty, 0]$, the procedure is as follows.

Let $r(\lambda; \epsilon)$, $\lambda > 0$ be the reflection coefficient for the Schrödinger equation $-\epsilon^2 f_{xx} + u_0(x)f = \lambda f$ associated with the initial data $u_0(x)$, which we assume to be positive, and hence free of solitons for all $\epsilon > 0$. Define the matrix

(2.2) $$v^{(0)}(\lambda; \epsilon) = \begin{cases} \sigma_1, & \lambda < 0, \\ \begin{pmatrix} 1 - |r|^2 & -\bar{r}e^{-2i\alpha/\epsilon} \\ re^{2i\alpha/\epsilon} & 1 \end{pmatrix}, & \lambda > 0, \end{cases}$$

where $\sigma_1 = \begin{pmatrix} 0 & 1 \\ 1 & 0 \end{pmatrix}$ and $\alpha = 4t\lambda^{3/2} + x\lambda^{1/2}$. Here $\lambda^{1/2}$ denotes the branch cut from $-\infty$ to zero, which is positive for λ positive. The goal is to find a row vector-valued function $m(\lambda) = m(\lambda; x, t, \epsilon) = (m_1, m_2)$ analytic for complex λ off the real axis, satisfying the jump and asymptotic conditions

(2.3) $$m_+(\lambda; x, t, \epsilon) = m_-(\lambda; x, t, \epsilon)v^{(0)}(\lambda; x, t, \epsilon),$$

(2.4) $$m(\lambda; x, t, \epsilon) \to (1, 1) \quad \text{as } \lambda \to \infty$$

where $m_\pm(\lambda; x, t, \epsilon) \equiv \lim_{\delta \downarrow 0} m(\lambda \pm i\delta; x, t, \epsilon)$. The appearance of σ_1 in the jump matrix reflects the $\sqrt{\lambda}$ to $-\sqrt{\lambda}$ symmetry of the RH problem in the $\sqrt{\lambda}$–plane (see [S], [BDT]).

The Riemann-Hilbert problem (RH) of finding $m(\lambda)$ given the jump matrix $v^{(0)}(\lambda)$ has a unique solution m in the space $(1,1) + L^2(|d\lambda^{1/2}|)$. The space and time variables x and t are simply parameters in the solution process. The solution to the initial value problem (1) is given by

(2.5) $$u(x, t, \epsilon) = -2i\epsilon \partial_x m_{11}(x, t, \epsilon)$$

where

(2.6) $$m_1(\lambda; x, t, \epsilon) = 1 + m_{11}\lambda^{-1/2} + O(\lambda^{-1}), \quad \lambda \to \infty,$$

(see again [S] or [BDT]).

For convenience, we consider initial data $u_0(x)$ that consists of a single positive "bump" as in Figure 3a below, and we normalize the bump's height to equal unity (see, however, the Remark at the end of the paper). For technical reasons we assume the initial data $u_0(x)$ to be real analytic, but this is not an essential restriction. We use the WKB approximation with two turning points (see for example [BO]) to calculate the reflection coefficient given by

$$(2.7) \qquad r(\lambda; \epsilon) \sim -ie^{-2i\rho(\lambda)/\epsilon}\chi_{[0,1]}(\lambda),$$

$$(2.8) \qquad \rho(\lambda) = x_+\lambda^{1/2} + \int_{x_+}^{\infty}[\lambda^{1/2} - (\lambda - u_0(x))^{\frac{1}{2}}]dx$$

and the squared modulus of the transmission coefficient $(0 < \lambda < 1)$,

$$(2.9\text{-}2.10) \qquad 1 - |r|^2 \sim e^{-2\tau(\lambda)/\epsilon}, \quad \tau(\lambda) = \int_{x_-}^{x_+}(u_0(x) - \lambda)^{1/2}dx$$

the quantities $x_-(\lambda) < x_+(\lambda)$ being defined by the relation $u_0(x_\pm(\lambda)) = \lambda$. As usual, $\chi_{[0,1]}$ denotes the characteristic function of the interval [0,1].

In the following, we identify r with its WKB approximation. As a result, the jump matrix reduces to the identity for $\lambda > 1$ and our problem reduces to a RH problem on the interval $(-\infty, 1]$. We will make a series of reductions on the original RH problem (2.2). The result of the nth reduction is a RH problem $m_+^{(n)} = m_-^{(n)}v^{(n)}$ with the jump matrix $v^{(n)}$ supported on an oriented contour $\Sigma^{(n)}$. Here $m^{(n)}$ is analytic outside the contour $\Sigma^{(n)}$, and $m_+^{(n)}$ and $m_-^{(n)}$ denote, respectively, the left and right boundary values of $m^{(n)}$ on the contour. For each n, we have $(-\infty, 0) \subset \Sigma^{(n)}$ and $v^{(n)}(\lambda) = \sigma_1$ for $\lambda \in (-\infty, 0)$, while m will be analytic on $(1, \infty)$. Thus we only give the expression for $v^{(n)}(\lambda)$ for $0 < \lambda < 1$.

Our method proceeds in several steps.

2.1 Introduction of the complex phase function $g(\lambda)$.

We introduce a change of the dependent variable m

$$(2.11) \qquad m^{(1)}(\lambda) = m(\lambda)e^{ig(\lambda)\sigma_3/\epsilon}, \quad \sigma_3 = \begin{pmatrix} 1 & 0 \\ 0 & -1 \end{pmatrix},$$

where the scalar function $g(\lambda) = g(\lambda; x, t)$, to be determined below, is analytic in λ off the set $(-\infty, 1]$ and satisfies $g_+(\lambda) + g_-(\lambda) = 0$ for $\lambda \in (-\infty, 0)$, in order to maintain the jump σ_1 for $m^{(1)}$ across $(-\infty, 0)$, and $g(\lambda) \to 0$ as $\lambda \to \infty$. From (2.5) and (2.11),

$$(2.12) \qquad u = -2i\epsilon\partial_x m_{11}^{(1)} - 2\partial_x g_1.$$

where $g(\lambda) = g_1/\lambda^{1/2} + O(1/\lambda)$ and $m_1^{(1)}(\lambda) = 1 + m_{11}^{(1)}/\lambda^{1/2} + O(1/\lambda)$. The RH problem in the new variable becomes $m_+^{(1)} = m_-^{(1)} v^{(1)}$. By a simple calculation,

$$(2.13) \qquad v^{(1)} = \begin{pmatrix} e^{(ig_+ - ig_- - 2\tau)/\epsilon} & -ie^{-ih/\epsilon} \\ -ie^{ih/\epsilon} & e^{-i(g_+ - g_-)/\epsilon} \end{pmatrix},$$

where $h = g_+ + g_- - 2\rho + 2\alpha$.

Our strategy is the following. For each value of $0 < \lambda < 1$, we reduce the jump matrix $v^{(1)}$ to one of the following three forms with exponentially small errors as $\epsilon \searrow 0$,

$$(2.14) \quad \begin{pmatrix} 1 & -ie^{-ih(\lambda)/\epsilon} \\ -ie^{ih(\lambda)/\epsilon} & 0 \end{pmatrix}, \quad \begin{pmatrix} 0 & -ie^{-ih(\lambda)/\epsilon} \\ -ie^{ih(\lambda)/\epsilon} & 0 \end{pmatrix},$$

$$\begin{pmatrix} 0 & -ie^{-ih(\lambda)/\epsilon} \\ -ie^{ih(\lambda)/\epsilon} & 1 \end{pmatrix},$$

by imposing the additional requirements listed below:

The interval $0 < \lambda < 1$ is partitioned into finitely many intervals so that on each interval one of the following three conditions is satisfied, leading directly in (2.13) to one of the three matrix forms (2.14). The requirement on h' below is used for a further reduction of the RH problem as explained in the next paragraph.

(2.15a) $-\tau = (g_+ - g_-)/2i$ and $h' < 0$.

(2.15b) $-\tau < (g_+ - g_-)/2i < 0$ and $h' = 0$; thus $g_+ + g_- - 2\rho + 2\alpha = -\Omega_j$, where Ω_j is some constant of integration.

(2.15c) $g_+ - g_- = 0$ and $h' > 0$.

Recall that we have also required that

(2.15d) $g_+ + g_- = 0$ when $\lambda < 0$ and $g_+ - g_- = 0$ when $\lambda > 1$.

We now show that when $\epsilon \searrow 0$, we may replace the first and the third forms by the identity matrix. As a result, the RH problem for $m^{(1)}$ reduces to one with a piecewise constant jump matrix $(h' = 0)$ of the second form. Indeed, the first form of the jump matrix $v^{(1)}$ above admits the lower–upper triangular factorization,

$$(2.16) \qquad v^{(1)} = v_2 v_1 \equiv \begin{pmatrix} 1 & 0 \\ -ie^{ih(\lambda)/\epsilon} & 1 \end{pmatrix} \begin{pmatrix} 1 & -ie^{-ih(\lambda)/\epsilon} \\ 0 & 1 \end{pmatrix}$$

Let (a, b) be an interval in which condition (2.15a) holds. Consider the new RH problem $(m^{(2)}, v^{(2)}, \Sigma^{(2)})$ described in the following figure

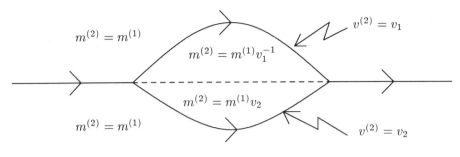

$$m^{(2)} = m^{(1)}$$
$$v^{(2)} = v_1$$
$$m^{(2)} = m^{(1)}v_1^{-1}$$
$$m^{(2)} = m^{(1)}v_2$$
$$m^{(2)} = m^{(1)}$$
$$v^{(2)} = v_2$$

Figure 1

The lense–shaped region in the figure is drawn within the domain of analyticity of h, v_1 and v_2. (The analyticity of h follows from the analyticity of our initial data). The jump matrices v_1 and v_2 shown on the arcs in the figure are the analytic continuation of the corresponding matrices on the interval (a, b). One easily checks that the definitions of $m^{(2)}$ and $v^{(2)}$ in the figure are consistent. For instance, on (a, b), $m_+^{(2)} = m_+^{(1)}v_1^{-1}$ and $m_-^{(2)} = m_-^{(1)}v_2$. It follows immediately from (2.16) that $m_+^{(2)} = m_-^{(2)}$ on (a, b), and hence the interval can be removed from $\Sigma^{(2)}$. By the Cauchy–Riemann equations, and the fact that $h(\lambda)$ is real on (a, b), the condition $h'(\lambda) < 0$ implies that $\pm \operatorname{Im} h(\lambda) < 0$ for $\pm \operatorname{Im} \lambda > 0$. Thus, the jump matrix $v^{(2)}(\lambda) \to I$ exponentially as $\epsilon \to 0$ for λ on the two arcs. This allows us to also remove the arcs from the contour. We have effectively removed the interval (a, b) from the original contour $\Sigma^{(1)} = \Sigma^{(0)} = (-\infty, 1)$.

Similarly, using an upper–lower factorization for the third form of the jump matrix, we remove from the contour any intervals on which the jump matrix takes the third form.

We consider the case in which the remaining RH problem $(m^{(3)}, v^{(3)}, \Sigma^{(3)})$ has contour $\Sigma^{(3)}$ consisting of $(-\infty, 0]$ together with a finite set of disjoint intervals $I_j = (\alpha_j, \beta_j)$, $j = 0, \ldots, N$, ordered from left to right and lying in the open interval $(0, 1)$, with $\alpha_0 = 0$. For $\lambda \in I_j$, the jump matrix $v^{(3)}(\lambda) = -i\sigma_1 e^{-i\Omega_j/\epsilon\sigma_3}$, where $\Omega_j = -h(\lambda)$ is constant (independent of λ). Without loss of generality we choose $\Omega_0 = 0$.

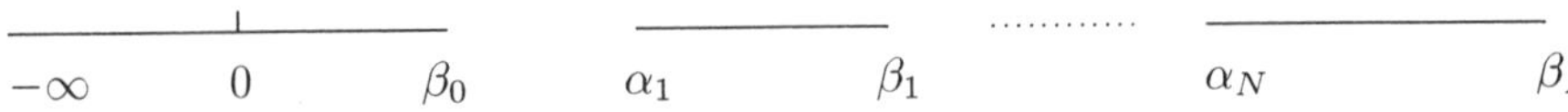

$$\overline{}_{\substack{| \\ }} \qquad \overline{} \qquad \cdots\cdots\cdots \qquad \overline{}$$

$-\infty \qquad 0 \qquad \beta_0 \qquad\qquad \alpha_1 \qquad\qquad \beta_1 \qquad\qquad\qquad \alpha_N \qquad\qquad \beta_N$

Figure 2

To complete the solution of the initial value problem (1), two tasks remain. Firstly, we must determine the phase function g from conditions (2.15a,b,c); this includes determining the value of N. the intervals I_j, the values Ω_j, and dg_1/dx, which is needed in (2.12). Secondly, we must solve the RH problem for $m^{(3)}$.

2.2 Determination of $g'(\lambda)$ and of the intervals I_j.

We observe that for each fixed x and t, g' satisfies a scalar RH problem on the real axis. Indeed, $g'_+ + g'_- = 0$ when $\lambda < 0$, and $g'_+ - g'_- = 0$ when $\lambda > 1$. When $0 < \lambda < 1$, the equalities in conditions (2.15a), (2.15b), and (2.15c) specify $g'_+ + g'_- - 2\rho' + 2\alpha' = 0$ on each interval I_j, while $g'_+ - g'_-$ either equals 0 or equals $-2i\tau'$ outside these intervals. For reasons that will become clear below, we only consider $g'_+ - g'_- = -2i\tau'$ when λ lies between any two of the I_j's while on the remaining interval $(0,1) \setminus (0, \beta_N)$, we examine both possibilities i.e. $g'_+ - g'_- = -2i\tau'$ (case A) or $g'_+ - g'_- = 0$ (case B).

We now impose the additional requirement that

(2.17) The functions $(\sqrt{\lambda} g'(\lambda))_\pm$ are continuous for real λ.

Necessarily $g'(\lambda)$ has the following form,

(2.18)
$$g'(\lambda) = \sqrt{p(\lambda)} \left(\int_{\cup I_j} \frac{2\rho'(\mu) - 2\alpha'(\mu)}{\sqrt{p(\mu)}_+ (\mu - \lambda)} \frac{d\mu}{2\pi i} + \int_{(0,E) \setminus \cup I_j} \frac{-2i\tau'(\mu)}{\sqrt{p(\mu)}_+ (\mu - \lambda)} \frac{d\mu}{2\pi i}, \right),$$

where

$$p(\lambda) = (\lambda - \beta_0) \prod_{j=1}^{N} (\lambda - \alpha_j)(\lambda - \beta_j),$$

and $E = 1$ in case A and $E = \beta_N$ in case B. Here, $\sqrt{p(\lambda)}$ is analytic in the complement of $\Sigma^{(3)}$ and positive for $\lambda > \beta_N$. Also, $\sqrt{p(\lambda)}_+$ denotes the boundary value from above.

A sufficient number of conditions to determine the endpoints of the intervals I_j can now be written down. Indeed, the condition $g(\lambda) = O(\lambda^{-1/2})$ for large λ implies $g'(\lambda) = O(\lambda^{-3/2})$, which leads to the following moment conditions,

(2.19) $\displaystyle \int_{\cup I_j} \frac{\rho'(\lambda) - \alpha'(\lambda)}{\sqrt{p(\lambda)}_+} \lambda^k d\lambda + \int_{(0,E) \setminus \cup I_j} \frac{-i\tau'(\lambda)}{\sqrt{p(\lambda)}_+} \lambda^k d\lambda = 0, \quad k = 0, \dots, N.$

A second set of conditions is obtained by integrating g' around each I_j and using (15a): In case A we obtain

(2.20) $\displaystyle \int_{I_j} (g'_+(\lambda) - g'_-(\lambda)) d\lambda = -2i(\tau(\beta_j) - \tau(\alpha_j)), \quad j = 1, \dots N.$

In case B, $\tau(\beta_j)$ must be replaced by zero.

Conditions (2.19) and (2.20) consist of a system of $(N+1) + N = 2N + 1$ independent equations for the $2N+1$ unknowns β_0, α_j, β_j, $j = 1, \dots N$. Conversely suppose that for given x, t and some N, the quantities β_0, α_j, β_j, $j = 1, \dots N$ satisfy conditions (2.19) and (2.20), giving rise to an explicit expression (2.18) for g', and hence for g by integration. Suppose further that the function g so constructed, also satisfies the inequalities in (2.15a,b,c). Then g is the desired solution of the scalar RH problem. For given x and t, we expect that, generically, the foregoing procedure will be successful for some finite $N = N(x,t)$.

For any given t, we now outline a systematic procedure for obtaining $N = N(x,t)$ for all x. The result of this procedure is to construct for each t the *separatrix* $F(t) = \{(x,\lambda) : \lambda = \lambda(x,t), -\infty < x < \infty, \lambda = \beta_0(x,t), \alpha_1(x,t), \ldots, \beta_N(x,t)\}$ (see figures 3a and 3b).

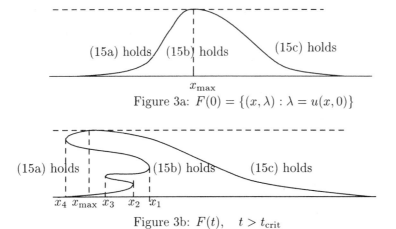

Figure 3a: $F(0) = \{(x,\lambda) : \lambda = u(x,0)\}$

Figure 3b: $F(t), \quad t > t_{\text{crit}}$

For times t less than some critical value t_{crit} one takes $N = N(x,t) = 0$ and solves (2.19) for β_0; the time t_{crit} corresponds precisely to the time at which (2.1) with $\epsilon = 0$ breaks down. It turns out that the associated function g constructed as above indeed satisfies the auxilliary inequalities (2.15a,b,c) and hence g is the desired solution. Thus, for $t < t_{\text{crit}}$, we obtain a separatrix of the shape of figure 3a; moreover, we find that $F(t) = (x,\lambda) : \lambda = u(x,t)$, where $u(x,t)$ is the solution of (2.1) with $\epsilon = 0$. For $x > x_{\text{max}}(t)$, we are in case B and for $x < x_{\text{max}}(t)$, we are in case A. When $t > t_{\text{crit}}$, one again sets $N = N(x,t) = 0$ for $x >> 1$, computes β_0 from (2.19) and verifies once again that the side conditions are satisfied in case B. However, as we move x to the left, we find that at least one of the inequalities (2.15a,b,c) breaks down. In figure 3b below the first inequality in (2.15b), $-\tau < (g_+ - g_-)/2i$, breaks down at $x = x_1$ and for $x < x_1$ the interval I_0 breaks up into two intervals I_0 and I_1. For such x one solves the system (2.19) and (2.20) for β_0, α_1, and β_1, and verifies that indeed the associated g satisfies the inequalities (2.15a,b,c). In the scenario of figure 3b, as we move x further to the left, the same inequality for τ again breaks down at some $x = x_2(t)$, for some $\lambda \in I_0(x_2(t),t)$. Again I_0 splits up into two intervals yielding a total of three intervals I_0, I_1 and I_2, etc. As we continue to move towards x_3, the interval I_1 shrinks to a point and eventually disappears. For x between x_4 and x_3 we are again in the two interval case and for $x < x_4$ we return to the one interval case $N = 0$. For $x > x_{\text{max}}$, we are always in case B and for $x < x_{\text{max}}$ we are always in case A. The scenario of figure 3b is representative of the generic case in which any finite number of intervals may be open at some x. For more complicated, indeed pathological initial data infinitely many folds may appear.

We present a nontrivial example of the above procedure in the next section, in which explicit calculation is possible.

In the context of our theory, the modulation equations are equivalent to the compatibility condition $\partial_x \partial_t g' = \partial_t \partial_x g'$, exactly as in the Lax, Levermore theory. Equations (2.19) and (2.20) should be viewed as providing , for any fixed x and t, an implicit formula for the solution of the modulation equations, with the initial conditions for (2.1) built in. In our procedure, we do not need to write the modulation equations at all. Furthermore, in [M], K. McLaughlin was able to reproduce Tian's result directly from (2.19) and (2.20).

2.3 Example: Tent Problem. Consider the tent shaped initial data $u_0(x) = 1 - |x|$ for $x \in (-1, 1)$, and zero elsewhere. As $x_+(\lambda) = 1 - \lambda$, $\rho = \lambda^{1/2} - \frac{2}{3}\lambda^{3/2}$, and $\tau = \frac{4}{3}(1 - \lambda)^{3/2}$. We have the following results.

(i) For $t < 1/6$, $N = 0$ and β_0 has a triangle shape for fixed t,

$$(2.21a) \qquad \beta_0 = \begin{cases} 0, & |x| > 1, \\ (1 + x)/(1 - 6t), & -1 < x < -6t, \\ (1 - x)/(1 + 6t), & -6t < x < 1. \end{cases}$$

(ii) For $t > 1/6$, denote $x_0 = 1 - (4(1 + 6t))^{1/3}$ and $x_1 = 1 - 12t < x_0$. We consider three regions, I$= (-\infty, x_1) \cup (1, \infty)$, II$= (x_1, x_0)$, III$= (x_0, 1)$. In region I, $N = 0$, $\beta_0 = 0$. In region III, $N = 0$, $\beta_0 = (1 - x)/(1 + 6t)$. In region II, $N = 1$, $\beta_0 = 0$, α_1 and β_1 are uniquely determined from (19) and (20) in the case B, which reduce respectively to

$$(2.21b) \qquad -1 + x + (1 + 6t)(\alpha_1 + \beta_1) - \frac{4}{\pi}\int_0^{\alpha_1} \sqrt{\frac{s(1 - s)}{(\alpha_1 - s)(\beta_1 - s)}}\,ds = 0,$$

and

$$(2.21c) \qquad \frac{2}{3}(1 - \alpha_1)^{3/2} - \frac{(1 + 6t)}{2}\int_{\alpha_1}^{\beta_1} \sqrt{\frac{(s - \alpha_1)(\beta_1 - s)}{s}}\,ds$$
$$= \frac{1}{\pi}\int_{\alpha_1}^{\beta_1} \sqrt{\frac{(s - \alpha_1)(\beta_1 - s)}{s}}\left(\int_0^{\alpha_1} \sqrt{\frac{u(1 - u)}{(\alpha_1 - u)(\beta_1 - u)}}\frac{du}{u - s}\right)ds.$$

We have proved that this system of equations can be inverted to give α_1 and β_1, and furthermore, that the inequalities in (15a,b,c) are satisfied, but we omit the proof.

2.4 Determination of the Ω_j's, of $g(\lambda)$ and of $\partial_x g_1$.

We begin by introducing some quantities that appear in the calculation below. Let X be the Riemann surface for $\sqrt{p(\lambda)}$, and let a_j, b_j, $1 \leq j \leq N$, denote the standard homology basis drawn in figure 4 below.

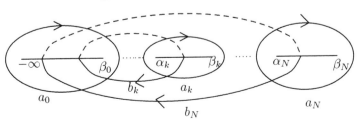

Figure 4

A basis of the holomorphic differentials for X is given by $\omega_i = \frac{p_i(\lambda)}{\sqrt{p(\lambda)}}d\lambda$, $i = 1, \cdots, N$, where $p_i(\lambda)$ are polynomials of degree $\leq N - 1$ that are uniquely determined from the relations $\int_{a_j} \omega_i = \delta_{ij}$. We introduce further the meromorphic differential

$$(2.22) \qquad (q/\sqrt{p})d\lambda = \frac{\lambda^N + q_{N-1}\lambda^{N-1} + \cdots + e_0}{\sqrt{p(\lambda)}}d\lambda,$$

where the N coefficients $q_0, \ldots q_{N-1}$ are uniquely determined by the N relations

$$(2.23) \qquad \int_{a_j} (q/\sqrt{p})d\lambda = 0, \quad j = 1, \ldots, N.$$

We now observe from the relations (15a,b,c), that $g(\lambda)$ also satisfies a RH problem. Solving this RH problem in exactly the same way as the problem for g' in section 2.2, we obtain similarly to (2.18)

$$(2.24) \quad g(\lambda) = \sqrt{p(\lambda)} \sum_{j=0,\cdots,N} \left(\int_{I_j} \frac{2\rho(\mu) - 2\alpha(\mu) - \Omega_j}{\sqrt{p(\mu)}_+ (\mu - \lambda)} \frac{d\mu}{2\pi i} \right.$$
$$\left. + \int_{(0,E)\backslash \cup I_j} \frac{-2i\tau(\mu)}{\sqrt{p(\mu)}_+ (\mu - \lambda)} \frac{d\mu}{2\pi i} \right).$$

The determination of the Ω_j's is straightforward. Indeed, since $g(\lambda) = O(\lambda^{-1/2})$ for large λ, the following moment conditions are derived anologous to (2.19),

$$(2.25) \quad \sum_{j=0,\cdots,N} \int_{I_j} (2\rho(\lambda) - 2\alpha(\lambda) - \Omega_j)\omega_k - \int_{(0,E)\backslash \cup I_j} 2i\tau(\lambda)\omega_k = 0, \quad k = 1, \ldots, N.$$

Hence, recalling that $\Omega_0 = 0$, we obtain for $k = 1, \ldots, N$,

$$(2.26) \quad \Omega_k = 4\int_{\cup I_j} (\rho(\lambda) - \alpha(\lambda))\omega_k - 4\int_{(0,1)\backslash \cup I_j} i\tau(\lambda)\omega_k$$
$$= -2x\int_{a_\infty} \lambda^{1/2}\omega_k - 8t\int_{a_\infty} \lambda^{3/2}\omega_k + 4\int_{\cup I_j} \rho\omega_k - 4\int_{(0,1)\backslash \cup I_j} i\tau\omega_k$$
$$\equiv x\Omega_{k1} + t\Omega_{k2} + \Omega_{k3},$$

where a_∞ is a clockwise oriented circle surrounding the interval $(0,1)$. We find

$$(2.27) \qquad \Omega_{k1} = -\operatorname{Res}_\infty \lambda^{1/2}\omega_k, \quad \Omega_{k2} = -4\operatorname{Res}_\infty \lambda^{3/2}\omega_k.$$

Repeating the argument above for $\partial_x g$ and $\partial_t g$, we see that

$$(2.28) \qquad \partial_x \Omega_k = \Omega_{k1}, \quad \partial_t \Omega_k = \Omega_{k2}.$$

Observe that these relations are not obvious at all from (2.26), since Ω_{k1}, Ω_{k2}, and Ω_{k3} all depend on x and t through ω_k.

Finally, to determine $\partial_x g_1$, we consider the RH problem for $\partial_x g$. We obtain in analogy to (2.25),

$$\sum_{j=0,\cdots,N} \int_{I_j} (-2\alpha_x - \Omega_{jx}) \frac{\lambda^k}{\sqrt{p}_+} d\lambda = 0, \quad k = 0, \ldots, N-1,$$

and

$$(2.29) \qquad g_{1x} = \frac{1}{2\pi i} \sum_{j=0,\cdots,N} \int_{I_j} \frac{(2\alpha_x + \Omega_{jx})qd\lambda}{\sqrt{p}_+} = \frac{1}{2\pi i} \int_{\cup I_j} \frac{2\alpha_x qd\lambda}{\sqrt{p}_+}$$

$$= -2\operatorname{Res}_\infty(\lambda^{1/2} q/\sqrt{p}) = -\sum_{j=0}^{N} (\alpha_j + \beta_j)/2 - q_{N-1}.$$

2.5 Solution of the RH problem for $m^{(3)}$.

Introduce the period matrix $\phi_{ij} = \int_{b_j} \omega_i = \phi_{ji} \in i\mathbf{R}$ and the theta function $\theta(y) = \theta(y, \phi) = \sum_{m \in \mathbf{Z}^N} e^{2\pi i(m,y) + \pi i(m, \phi m)}$. The theta function satisfies $\theta(s) = \theta(-s)$, $\theta(s + e_j) = \theta(s)$, $\theta(s \pm \tau_j) = e^{\pm 2\pi i s_j - \pi i \tau_{jj}} \theta(s)$,

where e_j is the j–th column of I and $\tau_j = \tau e_j$. The period lattice for the theta function is given by $\Lambda = \mathbf{Z}^N + \tau \mathbf{Z}^N$.

Let $w(\lambda) = \int_\infty^\lambda \omega$ and
(2.30)
$$m^{(4)} = (m_1^{(4)}, m_2^{(4)}) = \left(\frac{\theta(w(\lambda) - \Omega/(2\pi\epsilon) + d)}{\theta(w(\lambda) + d)}, \frac{\theta(-w(\lambda) - \Omega/(2\pi\epsilon) + d)}{\theta(-w(\lambda) + d)} \right),$$

where $\Omega = (\Omega_1, \ldots, \Omega_N)^t$ and d is to be determined below. Using the properties of theta functions, one checks that $m^{(4)}$ is well defined (single valued) and meromorphic off the slits and satisfies $m_+^{(4)} = m_-^{(4)} v^{(4)}$ with $v^{(4)}(\lambda) = iv^{(3)}(\lambda)$ for $\lambda \in \cup_{j=0}^N I_j$, and $v^{(4)}(\lambda) = v^{(3)}(\lambda) = \sigma_1$ for $\lambda < 0$.

To solve the RH problem for $v^{(3)}$, let $\gamma = (\prod_0^N \frac{\lambda - \alpha_j}{\lambda - \beta_j})^{1/4} \to 1$ as $\lambda \to \infty$. Clearly, $\gamma_+ = -i\gamma_-$ on $\cup_0^N I_j$ and therefore $\gamma m^{(4)}$ has the jump matrix $v^{(3)}$.

We now show that $m^{(3)} = c\gamma m^{(4)}$ for some λ–independent c and some properly chosen d. Observing that $\gamma(\lambda)$ has zeros at α_j, $j = 1, \ldots, N$, of order $1/4$, we want to choose a d such that $\theta(w(\lambda) \pm d)$ also have these points as zeros. By the theory of θ functions [FK], we can choose $d = 0$, and under such a choice, $\theta(w(\lambda)) = \theta(w(\lambda) \pm d)$ has zeros of order $1/2$ at these points. Hence $m^{(3)}$ is locally L^2 near these points. It follows that

$$m_\pm^{(3)} - (1,1) \in L^2(|d\sqrt{\lambda}|).$$

Now choose $c = \theta(0)/\theta(\Omega/(2\pi\epsilon))$ so that $m^{(3)} \to (1,1)$ as $\lambda \to \infty$. An elementary calculation using (2.27) and (2.28), shows that
(2.31)
$$m_1^{(3)} = 1 - \frac{1}{2\pi i} \left(\nabla \log \theta(\Omega/(2\pi\epsilon)) - \nabla \log \theta(0) \right) \cdot \Omega_x \lambda^{-1/2} + O(\lambda^{-1}), \quad \text{as } \lambda \to \infty.$$

where $\Omega_x = \partial_x \Omega$. Using (2.5) and (2.29), we obtain as $\epsilon \to 0$,

$$(2.32) \quad u(x,t,\epsilon) \sim \sum_{j=0}^{N}(\alpha_j + \beta_j) + 2q_{N-1} + \frac{\epsilon}{\pi}\frac{\partial}{\partial x}\left(\nabla \log \theta(\Omega/(2\pi\epsilon)) - \nabla\theta(0)\right)\cdot\Omega_x,$$

so that

$$(2.33) \qquad u(x,t,\epsilon) \sim \sum_{j=0}^{N}(\alpha_j + \beta_j) + 2q_{N-1} - 2\epsilon^2\frac{\partial^2}{\partial x^2}\log\theta(\Omega/(2\pi\epsilon)).$$

Alternatively, in a space–time window of scale ϵ about (x,t), we have

$$(2.34) \quad u(x + \epsilon x_1, t + \epsilon t_1, \epsilon) \sim \sum_{j=0}^{N}(\alpha_j + \beta_j) + 2q_{N-1}$$
$$- 2\frac{\partial^2}{\partial x_1^2}\log(\theta(\Omega/(2\pi\epsilon) + (2\pi)^{-1}x_1\Omega_x + (2\pi)^{-1}t_1\Omega_t).$$

Using (2.27) and (2.28), we see that (2.34) is precisely the form of solutions of themodulated wave equations of [FFM]. Formula (2.34) was first derived for the small dispersion limit of KdV in [V4] as described in the section on the resolution of the oscillations above, but without precise information on the phase shift. Our calculations above determine the phase shift to order ϵ.

2.6 A Variational Principle for $g'(\lambda)$. We make the change of variable $\lambda = 1 - \eta^2$, taking the upper complex half-plane onto $\mathbf{C}\setminus(-\infty, 1]$. The function g transforms to $G(\eta) = g(\lambda)$. We extend the definition of G onto the lower complex η half-plane, by the relation $G(-\eta) = -G(\eta)$ and we observe that G is analytic off the real η axis. We transform the RH problem for g into a problem for G, observing that $G_+ - G_- = g_+ + g_-$ and $G_+ + G_- = -(g_+ - g_-)\mathrm{sgn}\eta$. On the real axis, we define the function $\psi(\eta)$ to equal zero outside the interval $[-1, 1]$ while on $(-1, 1)$

$$(2.35) \qquad \psi(\eta) = \partial_\eta[\alpha(\lambda(\eta)) - \rho(\lambda(\eta)) + \frac{1}{2}(G_+ - G_-)],$$

and one verifies that

$$(2.36) \quad \psi(\eta) = \partial_\eta \,\mathrm{Re}[(1-\eta^2)^{1/2}(x - x_{\max}) + 4(1-\eta^2)^{3/2}t +$$
$$\int_{x_{\max}}^{\infty}((1-\eta^2)^{1/2} - (1 - \eta^2 - u_0(x)^{1/2})dx] + \frac{1}{2}\partial_\eta(G_+ - G_-).$$

After a nontrivial calculation conditions (2.15a,b,c) transform to the variational conditions (1.13). This allows us to use a variational formulation for the derivation of ψ. Indeed, the function $\psi(\eta)$, restricted to the interval $[0,1]$ is the unique maximizer of the functional $Q(\psi) = (2a, \psi) - (2\gamma, \psi_+) + (L\psi, \psi)$ where

$$L\psi(\eta) = \int_0^1 \ln\frac{|\eta - \mu|}{|\eta + \mu|}\psi(\mu)d\mu,$$

$$a(\eta, x, t) = 4t\eta^3 + (x_{\max} - x - 6t)\eta + \int_{x_{\max}}^{x_+}(\eta^2 - 1 + u_0(x))^{1/2}dx,$$

$$\gamma(\eta) = \int_{x_-}^{x_+}(\eta^2 - 1 + u_0(x))^{1/2}dx,$$

$\psi_+(\eta)$ is the positive part of the function ψ. The benefit of the variational formulation is that, due to the convexity of the maximization problem, the uniqueness of ψ, and hence of G and of g, are guaranteed.

Remark:. The R/H steepest descent method can also be used to analyze potential wells $u_0(x) < 0$: here the critical behavior of the RH problem takes place on the interval $[\min_x u_0(x), 0]$, rather than the interval $[0, 1 = \max_x u_0(x)]$ as in the case of a "bump". If u_0 consists of a "bump" separated from a well, then the critical region for the RH problem is the interval $[\min_x u_0(x), \max_x u_0(x)]$.

Acknowledgements I wish to thank my collaborators Percy Deift and Xin Zhou. The entire part 2 of this article consists of our joint work [DVZ2]. I also wish to thank Peter Lax, Dave Levermore, Dave McLaughlin, Thomas Kriecherbauer and Ken McLaughlin for many very useful conversations. I gratefully acknowledge support by NSF Grant DMS-9500623 and ARO Grant DAAH04-96-1-0157.

References.

[AKNS] M.J. Ablowitz, D.J. Kaup, A.C. Newell, and H. Segur, Method for solving the sine–Gordon equation, stud. Appl. Math. 74, 249–315 (1974).

[BC] B. Beals and R. Coifman, Scattering and inverse scattering for first order systems. Comm. Pure Appl. Math. 37, 39-90, (1984).

[BDT] R. Beals, P. Deift and C. Tomei. Direct and Inverse Scattering on the Line, AMS, Math. Surveys and Monographs No. 28 (1988).

[BF] V.S. Buslaev and V.N. Fomin, An Inverse Scattering Problem for the One-Dimensional Schroedinger Equation on the Entire Axis, Vestnik Leningrad Univ. **17**, 56–64 (in Russian), 1962.

[BO] C.M. Bender and S.A. Orszag, Advanced Mathematical methods for scientists and engineers, McGraw–Hill, (1978).

[CK] A. Cohen and T. Kappeler, Scattering and Inverse Scattering for Steplike Potentials in the Schroedinger Equation, Indiana U. Math. J. **34(1)**, 127–180, 1985.

[DM] P. Deift and K. T-R McLaughlin, A continuum limit of the Toda lattice; to appear in Mem. Amer. Math. Soc.

[DKM] P. Deift, T. Kriecherbauer and K. T-R McLaughlin, New Results for the Asymptotics of Orthogonal Polynomials and Related Problems via the Inverse Spectral Method, preprint, 1996.

[DKMVZ] P. Deift, T. Kriecherbauer, K. T-R McLaughlin, S. Venakides, X. Zhou, Asymptotics for Polynomials Orthogonal with Respect to Varying Exponential Weights, to appear in IMRN, 1997.

[DMN] B.A. Dubrovin, V.B. Matveev, and S.P. Novikov, Nonlinear Equations of Korteweg-de Vries Type, Finite Zoned Linear Operators, and Abelian Varieties, Uspekhi Mat. Nauk **31**, 55–136, 1976.

[DM] P. Deift and K. T-R McLaughlin, A continuum limit of the Toda lattice, Memoirs of the AMS, to appear.

[DVZ1] P. Deift, S. Venakides and X. Zhou, The collisionless shock region for the

long–time behavior of solutions of the KdV equation. Comm. Pure Appl. Math.
47, 199–206 (1994).

[DVZ2] P. Deift, S. Venakides and X. Zhou, New results in small dispersion KdV by
an extension of the steepest descent method for Riemann-Hilbert problems. IMRN,
1997, No 6, 285-299.

[DZ1] P. Deift and X. Zhou, A steepest descent method for oscillatory Riemann–
Hilbert problems: Asymptotics for the MKdV equation, Ann. of Math. 137, 295-368
(1993).

[DZ2] P. Deift and X. Zhou, Asymptotics for the Painleve II equation, Comm. Pure
Appl. Math. 48, 277–337 (1995).

[EGLS] N. Ercolani, I. Gabitov, D. Levermore and D. Serre eds., Singular Limits
of Dispersive Waves, Proceedings of the NATO Advanced Research Workshop at
Lyon during 8-12 July 1991, Plenum, New York, 1994 .

[EGK] Breaking Problems in Dispersive Hydrodynamics, in [EGLS], 1994.

[ELZ1] N. Ercolani, C.D. Levermore, T. Zhang, The Behavior of the Weyl Function
in the Zero-Dispersion KdV limit, Comm. Math. Phys. 183, 119-143, 1997.

[ELZ1] N. Ercolani, C.D. Levermore, T. Zhang, Weyl Functions via Dirichlet Spectrum and the KdV Zero-Dispersion Limit, preprint, 1997.

[FK] H. Farkas and I. Kra, Riemann Surfaces, 2nd ed. Springer, New York, 1992.

[FM] M. Gregory Forest and K. T-R McLaughlin, Onset of oscillations in nonsoliton
pulses in nonlinear dispersive fibers, preprint, 1996.

[GGKM] C.S. Gardner, J.M. Greene, M.D. Kruskel and R.M. Miura, Method for
Solving the Korteweg de Vries Equation, Phys. Rev. Lett., 19, 1967, 1095-1097.

[GK]A.V. Gurevich and A.L. Krylov, Dissipationless Shock Waves in Media with
Positive Dispersion, Sov. Phys. JETP 65, 944, 1987.

[GKE] A.V. Gurevich, A.L. Krylov and G.A. El', *Riemann Wave Breaking in Dispersive Hydrodynamics*, JETP Lett. 54, 102 (1991).

[GP] A.V. Gurevich and L.P. Pitaevski, Nonstationary Structure of a Collisionless
Shock Wave, Sov. Phys. JETP 38, 291–297, 1974.

[KM] I. Kay and H.E. Moses, Reflectionless Transmission through Dielectrics and
Scattering Potentials, J. Appl. Phys. 27, 1503–1508, 1956.

[KW] Y. Kodama and S. Wabnitz, Analytic theory of guiding-centre nonreturn-to-
zero and return-to-zero signal transmission in normally dispersive nonlinear optical
fibers, Optical Lett., 20, 2291-2293 (1995).

[L] P.D. Lax, Integrals of Nonlinear Equations of Evolution and Solitary Waves,
Comm. Pure Appl. Math. 21, 467–490, 1968.

[LL1] P.D. Lax and C.D. Levermore, The zero dispersion limit of the Korteweg-deVries equation, Proc. Nat. Acad. Sci USA 76(8), 3602–3606, (1979).

[LL2] P.D. Lax and C.D. Levermore, The small dispersion limit of the Korteweg-deVries equation, I, II, III, comm. Pure Appl. Math. 36, 253–290, 571–593, 809–829 (1983).

[LLV] P.D. Lax, C.D. Levermore and S. Venakides, The Generation and propagation of oscillations in dispersive initial value problems, in "Important Developments in Soliton Theory", eds. A.S. Fokas and V.E. Zakharov, Springer-Verlag, Berlin, Heidelberg, New York, 205-241, 1993.

[Le] C.D. Levermore, The Hyperbolic Nature of the Zero Dispersion KdV Limit, Comm. P.D.E. **13**(4), 495–514, 1988.

[M] K. T-R McLaughlin, A continuum limit of the Toda lattice, Ph.D. Thesis, New York University, 1995.

[MT] H.P. McKean and E. Trubowitz, Hill's Operator and Hyperelliptic Function Theory in the Presence of Infinitely Many Branch Points, Comm. Pure Appl. Math. **29**, 143–226, 1976.

MV] H.P. McKean and P. Van Moerbeke, The Spectrum of Hill's Equation, Invent. Math. **30**, 217–274, 1975.

[MS] D.W. McLaughlin and J. Strain, Calculating the Weak Limit of KdV by Quadratic Programming, in Singular Limits of Dispersive Waves, N. Ercolani, I. Gabitov, D. Levermore and D. Serre eds., NATO ARW series, Plenum, New York, 1994.

[S] A.B. Shabat, One dimensional perturbations of a differential operator and the inverse scattering problem, Problems in Mechanics and Mathematical Physics, Nauka, Moscow, 1976.

[T] F.R. Tian, Oscillations of the Zero Dispersion Limit of the Korteweg-de Vries Equation, Comm. Pure Appl. Math **46**, 1093–1129, 1993.

[Ts] S.P. Tsarev, On Poisson brackets and one-dimensional systems of hydrodynamic type, Soviet Math. Dokl. 31, 488-491, (1985).

[V1] S. Venakides, The zero dispersion limit of the Korteweg-deVries equation for initial potentials with nontrivial reflection coefficient, Comm. Pure Appl. Math. 38, 125–155 (1985).

[V2] S. Venakides, The generation of modulated wavetrains in the solution of the Korteweg-de Vries Equation, Comm. Pure Appl. Math. 38, 883–909, (1985).

[V3] S. Venakides, The zero dispersion limit of the Korteweg-de Vries equation with periodic initial data, AMS Transactions, 301,#1, 189–226, (1987).

[V4] S. Venakides, The Korteweg-deVries equation with small dispersion: higher order Lax–Levermore theory, Comm. Pure Appl. Math. 43, 335–361 (1990).

[W] O. Wright, Korteweg–de Vries zero dispersion limit: through first breaking for cubic–like analytic initial data, Comm. Pure Appl. Math. 46, 421–438, (1993).

[ZV] T. Zhang and S. Venakides, Periodic Limit of Inverse Scattering, CPAM **46**, 819–865, 1993.

Selected Titles in This Series

(Continued from the front of this publication)

ISBN 0-8218-0657-2

9 780821 806579